Agriculture and the Environment

A World Bank Symposium

Agriculture and the Environment
Perspectives on Sustainable Rural Development

Edited by
Ernst Lutz

with the assistance of
Hans P. Binswanger, Peter Hazell, and Alexander McCalla

The World Bank
Washington, D.C.

Manufactured in the United States of America
First printing October 1998

The findings, interpretations, and conclusions expressed in this study are entirely those of the authors and should not be attributed in any manner to the World Bank, to its affiliated organizations, or to members of its Board of Executive Directors or the countries they represent.

The cover was designed by Tomoko Hirata and Alan Kahan, using photographs by Curt Carnemark (center), Ernst Lutz (top right and bottom left, in Togo), and Stefano Pagiola (top left and bottom right, in Bangladesh).

For a copy of the "Update" describing new publications, contact the Office of the Publisher, The World Bank, 1818 H Street, N.W., Washington, D.C. 20433 U.S.A. A catalog and ordering information are also available on the Internet at http://www.worldbank.org.

Library of Congress Cataloging-in-Publication Data

Agriculture and the environment: perspectives on sustainable rural
development / edited by Ernst Lutz; with the assistance of Hans P.
Binswanger, Peter Hazell, and Alexander McCalla.
p. cm. — (A World Bank symposium)
Includes bibliographical references.
ISBN 0-8213-4249-5
1. Agriculture—Environmental aspects—Developong countries.
2. Rural development—Developing countries. 3. Sustainable
development—Developing countries. 4. Agriculture and state—
Developing countries. 5. Rural development—International
cooperation. 6. Sustainable development—International cooperation.
I. Lutz, Ernst, 1949- . II. World Bank. III. Series.
S589.76.D44A46 1998
338. 1—dc21 98-36512
CIP

Contents

Foreword

Agriculture in developing countries has been remarkably successful during the past few decades in enabling food production to keep pace with the growing population. This success has been achieved in part at the cost of placing more stress on natural resources and the environment. As we look to the future, the need to continue increasing food production, while at the same time minimizing environmental damages, conserving the resource base, and reducing poverty, hunger, and malnutrition, poses an enormous challenge.

During the 1970s, in part through the leadership of Robert McNamara, the World Bank placed emphasis on reaching the rural poor through integrated rural development projects. But as such projects produced mixed results and prices of agricultural commodities fell in real terms, international interest in agricultural and rural matters waned during the 1980s. This trend was reflected in a decline in the World Bank's rural activities. Nevertheless, roughly 70 percent of the poorest people in developing countries still live in rural areas, and focusing on the rural sector must again become a top priority of development assistance. Therefore, rather than abandon rural development work because earlier work had limited success, we must learn from past mistakes and proceed with renewed vigor. This means reinventing rural development to fully integrate the environmental and social dimensions that were given insufficient attention earlier.

Over the past two years, the World Bank has intensively revisited its rural development strategy and developed a new one. The vision and specific steps to be taken are outlined in *Rural Development: From Vision to Action. A Sector Strategy* (1997). With a renewed commitment to sound development in the rural sector, the Bank seeks to play its part in a broadly based effort encompassing farmers, their families, communities, and policymakers at the national levels, as well as the donor community.

This volume brings together state-of-the-art applied, practical research related to agriculture and environment in the developing world. It distills current knowledge and summarizes it for development practitioners. Where possible, authors use specific examples to indicate which approaches have worked and which have not, under which conditions, and why.

The general observations and findings in this volume indicate that, in the future, environmentally sound increases in productivity may be harder to achieve than in the past. But a significant potential for achieving sustainable increases still exists. To realize that potential, it is becoming even more important to undertake necessary reforms—even politically sensitive ones like land reforms—so that policy and institutional frameworks fully support sustainable agricultural intensification by creating a conducive environment in which appropriate technological innovation can flourish.

A conducive policy and institutional framework includes the political will for reform and a commitment by those in power to seek the common good rather than let policy be driven by special interests and influenced by corruption. Such reform needs to be supported by broadly based policy analysis that not only considers agriculture narrowly defined but also integrates natural resources and the environment as well as the social dimensions of rural development work.

It is clear that, in contrast to some of the earlier centralized efforts, the new sustainable rural development approach must be much more decentralized and participatory. It will entail more on-the-ground testing with pilot projects before scaling up to larger efforts. Farmers, local communities, nongovernmental organizations, and other parts of civil society will need to be involved in rural development activities to a greater extent than before.

Farmers, although operating with incomplete knowledge and discounting the future in accordance with their level of income and wealth, will be viewed as rational managers who seek to achieve many objectives (such as increased food security and income and decreased risks) given the many constraints they face. They therefore need to be given choices and encouraged to innovate and experiment rather than be pushed to adopt preselected technologies by extension services that often lack cutting-edge knowledge or messages. Many of the sustainable increases in productivity are expected to come from innovative farmers working within a policy and institutional framework that provides the right incentives for environmentally sound rural development, such as more efficient use of inputs. In many cases, this will be good for productivity, input cost, and the government budget, as well as the environment. Where externalities are involved, appropriate government interventions will be needed to achieve more optimal results for the whole society.

Although we emphasize work on the policy and institutional side rather than "technological fixes," we also stress the continued need for top-quality, long-term research such as that undertaken over the past several decades by the various international centers supported by the Consultative Group for International Agricultural Research (CGIAR). Such basic research, now with a more serious effort to integrate environmental, social, and small-farmer concerns, remains essential for the future.

The challenges for the rural sector loom large, and the constraints are serious, but with good governance and a real commitment to environmentally and socially sustainable rural development at all levels, much can be achieved. We at the World Bank and those of us involved with CGIAR will do everything we can to alleviate rural poverty and hunger and to assist with participatory rural development, while safeguarding natural resources and the environment.

Ismail Serageldin
Vice President for Special Programs
and Chairman, CGIAR
The World Bank

Ian Johnson
Vice President
Environmentally and Socially
Sustainable Development
The World Bank

Contributors

KYM ANDERSON
University of Adelaide
Adelaide, Australia

JACQUELINE A. ASHBY
Center for Tropical Agriculture (CIAT)
Cali, Colombia

HANS P. BINSWANGER
World Bank
Washington, D.C.

HARVEY BLACKBURN
Agricultural Research Service
United States Department of Agriculture
Washington, D.C.

DANIEL W. BROMLEY
University of Wisconsin
Madison

LYNN R. BROWN
International Food Policy Research Institute
Washington, D.C.

ROLAND BUNCH
Asociación de Consejeros para una Agricultura
Sostenible, Ecológica y Humana (COSECHA)
Tegucigalpa, Honduras

NEIL BYRON
Center for International Forestry Research (CIFOR)
Bogor, Indonesia

JULIAN CALDECOTT
Consultant
Salisbury, United Kingdom

UJJAYANT CHAKRAVORTY
University of Hawaii and East-West Center
Honolulu

Sanjiva Cooke
World Bank
Washington, D.C.

DEAN CURRENT
Tropical Agricultural Research
and Higher Education Center (CATIE)
Turrialba, Costa Rica

CORNELIS DE HAAN
World Bank
Washington, D.C.

KLAUS DEININGER
World Bank
Washington, D.C.

John Farrington
Overseas Development Institute
London

Douglas A. Forno
World Bank
Washington, D.C.

Roberta V. Gerpacio
University of California
Davis

Lawrence Haddad
International Food Policy Research Institute
Washington, D.C.

Peter Hazell
International Food Policy Research Institute
Washington, D.C.

John Heath
World Bank
Washington, D.C.

Stein T. Holden
Agricultural University of Norway
Ås

Steen Joffe
Consultant
Linton, Kent, United Kingdom

David Kaimowitz
Center for International Forestry Research (CIFOR)
Bogor, Indonesia

Edwin B. Knapp
Center for Tropical Agriculture (CIAT)
Cali, Colombia

Ernst Lutz
World Bank
Washington, D.C.

Alexander McCalla
World Bank
Washington, D.C.

Ruth Meinzen-Dick
International Food Policy Research Institute
Washington, D.C.

Deepa Narayan
World Bank
Washington, D.C.

Stefano Pagiola
World Bank
Washington, D.C.

Prabhu L. Pingali
International Maize and Wheat Improvement Center (CIMMYT)
Mexico City, Mexico

Jules N. Pretty
University of Essex
Colchester, United Kingdom

Agnes R. Quisumbing
International Food Policy Research Institute
Washington, D.C.

Helle Munk Ravnborg
Center for Development Research
Copenhagen, Denmark

Carlos Reiche
Interamerican Institute for Agricultural Cooperation (IICA)
San José, Costa Rica

Sara Scherr
University of Maryland
Washington, D.C.

Tjaart W. Schillhorn van Veen
World Bank
Washington, D.C.

Nigel Smith
University of Florida
Gainesville

Jitendra Srivastava
World Bank
Washington, D.C.

Henning Steinfeld
Food and Agriculture Organization of the United Nations (FAO)
Rome

WILLIAM SUNDERLIN
Center for International Forestry Research (CIFOR)
Bogor, Indonesia

GRAHAM THIELE
Programa de Investigación de la Papa (Potato Research Program; PROIMPA)
Cochamba, Bolivia

DINA UMALI-DEININGER
World Bank
Washington, D.C.

Acknowledgments

We are grateful to the Swiss Trust Fund for Studies for cofinancing the production of this compendium. Special thanks go to Paul Egger, director of the Agricultural Service in the Swiss Agency for Development Cooperation, who supported the idea of a volume on sustainable rural development and gave valuable advice.

We would also like to thank each contributor to the volume. For most of them, this was a volunteer effort based on each author's commitment to rural development.

Editing and proofreading by Elizabeth R. Forsyth, composition by Patricia Hord, and indexing by Patricia Deminna are gratefully acknowledged. Special thanks go to James Wolfensohn, president of the World Bank, for supporting a renewed emphasis on the rural sector.

1

Introduction

Ernst Lutz, Hans P. Binswanger, Peter Hazell, and Alexander McCalla

Even though urbanization is increasing at a rapid pace, most of the world's poor people still live in rural areas. Thus anyone concerned for the poor should have a strong interest in rural development. A second reason for giving appropriate attention to the rural sector is the enormous challenge facing developing countries, which have to satisfy the nutritional needs of their growing populations.

The production increases of recent decades have been significant, as a result in part of expansion in area and in part of increases in productivity. However, negative side effects—deforestation, desertification, salinization—may impair the capacity to produce increasing amounts of food in the future. Clearly, much more attention needs to be paid to increasing productivity without negatively affecting natural resources and the environment. In this book, we have included chapters that summarize applied research on ways to meet the challenge.

We have grouped the chapters into three parts. The first one takes a policy perspective, the second takes an institutional and social perspective, and the third looks at rural development mainly from a technical perspective. In actuality, all three are intertwined. Each of the three parts opens with an introduction and summary.

The chapters in part I discuss policy perspectives and make the case that an appropriate policy framework is essential to achieving sustainable rural development. We start with a chapter by Hazell and Lutz (chapter 2) that deals with definitional issues surrounding the concept of sustainability, surveys resource problems and indicative cost estimates, and broadly discusses ways in which these problems can be addressed. In chapter 3, Heath and Binswanger show how natural resource degradation can often be policy-induced. Therefore, as a matter of priority, policy distortions need to be removed to achieve better incentives for higher growth and, in many cases, for environmental reasons. Pretty, in chapter 4, discusses in more detail what a policy framework looks like that is conducive to sustainable agriculture. As is argued, given the diversity of situations, one does not need a fixed model; rather the social process of innovation itself must be made sustainable. In chapter 5, Holden and Binswanger extensively survey recent research findings on decisionmaking by small farmers, given the market imperfections and risks they face, and examine the implications for sustainable rural development policies. Trade liberalization is one part of improving the overall incentive framework. Anderson, in chapter 6, discusses the economic and environmental impacts of this. In chapter 7, Bromley distills key lessons from research and project experience dealing with property regimes and development and assesses the policy implications. His chapter would fit equally well into part II. A short chapter by Hazell (chapter 8) concludes part I. It is almost too short to be a free-standing chapter, but it serves to summarize part I by discussing the key ingredients of sustainable rural development policies in terms of growth, equity, poverty, and environmental concerns.

The chapters in part II approach rural development mainly from an institutional and social perspective. They make the case that sustainable rural development must be participatory and decentralized and that it must address gender issues (for productivity and other reasons) and not shy away from sensitive issues like land reform. Narayan (chapter 9) outlines broadly what participatory rural development looks like and what the project experience has been. Her principles are illustrated and applied to watershed management by Ashby and others in chapter 10. Similarly, Farrington and Thiele (chapter 11) examine what local, participatory approaches to technology generation and dissemination look like and how they could be scaled up. Bunch (chapter 12) applies participatory principles to innovative extension. Deininger (chapter 13) explains the rationale for market-based or "negotiated" land reform and provides actual experiences from Colombia, Brazil, and South Africa. Caldecott and Lutz (chapter 14) discuss decentralization trends in developing countries and their implications for biodiversity conservation and, in particular, the conditions under which a certain degree of decentralization might be beneficial to conservation efforts. Part II concludes with a survey of applied research by Quisumbing and others (chapter 15) that shows the importance of properly addressing gender issues for productivity growth and well-being.

Part III of the volume addresses a number of topics mainly from a technical perspective, while demonstrating the importance of appropriate policy and institutional frameworks as well. It starts with a chapter by Lutz and others (chapter 16) that looks at soil conservation from a farmer's perspective. It argues for more empirical work that assesses the on-farm profitability of soil conservation measures and for less rhetoric. Srivastava and others (chapter 17) assess the importance of biodiversity conservation for agricultural development and propose ways to mainstream agrobiodiversity into agricultural development. Overuse and misuse of pesticides have been a major reason for negative environmental impacts connected with productivity increases. The ways in which pest management can be improved with integrated pest management and minimal use of pesticides are addressed in chapter 18 by Schillhorn and others and in chapter 19 by Pingali and Gerpacio. Overuse of irrigation water and insufficient drainage have led to waterlogging and salinity problems in many parts of the world. Chakravorty, in chapter 20, makes the case that more market-based and basinwide approaches would use the increasingly scarce water more optimally. A comprehensive treatment of livestock-environment interactions is presented by Steinfeld and others in chapter 21. How to distinguish between appropriate and inappropriate deforestation and how to reduce inappropriate deforestation are discussed by Kaimowitz and others in chapter 22. Lessons from agroforestry projects in Central America and the Caribbean are presented by Current and others in chapter 23. We draw conclusions from the whole volume in part IV.

Part I

Policy Perspectives

This part of the volume includes chapters that primarily take a policy perspective, even though institutional ones inevitably arise because policy and institutional issues are often intertwined. Here we describe the contents of the chapters of part I (as a guide to the reader) and summarize key points made by the various authors.

Choosing among the many possible definitions of sustainable development, Hazell and Lutz, in chapter 2, define sustainable agricultural development as agricultural income traditionally measured in national income accounts corrected for changes in the value of the stock of capital, which includes natural, manmade, and human capital. According to their definition of sustainability, capital can be changed from one category into another, while keeping the total constant. Yet they also recognize that there may be critical levels of each type of capital (such as critical ecosystems) beyond which capital should not be substituted. The chapter discusses externality issues, environmental valuation techniques, and the design of appropriate incentive schemes that internalize external costs.

Although there is great diversity in agroecological conditions around the developing world, the authors contrast two important types of systems because the environmental problems related to them are quite different. In *irrigated agriculture*, productivity has grown rapidly as a result of increased use of modern inputs including irrigation water, high-yielding varieties, fertilizers, pesticides, and machinery. Problems are particularly acute when inappropriate incentive systems encourage the overuse of such inputs. In contrast, in many *rainfed agricultural areas*, the main problem is resource degradation.

In the absence of comprehensive estimates of the costs of resource degradation in agriculture in developing countries, the authors present indicative estimates of the costs of deforestation, water depletion, waterlogging and salinization, fish stock depletion, soil degradation, health effects, and biodiversity losses. They then discuss policy, institutional, and technical options that would reduce those costs. In particular, they consider factors that form the overall incentive system for natural resource management: prices and subsidies, nonprice interventions, externalities, information, poverty reduction through investments in resource-poor areas, property rights, costs of collective action, and technology design and research. They conclude that environmental problems are not the inevitable outcome of agricultural growth. Instead, they are usually associated with inappropriate incentive systems, insufficient investment in resource-poor areas, inadequate attention to poverty and social concerns, and political economy systems where the rich and powerful extract rents.

In chapter 3, Heath and Binswanger state that reducing population growth and assisting poor farmers with resource management problems are urgently needed in many countries. But they argue that the removal of adverse policies should be given even higher priority, because those policies may, as their *joint consequence*, reduce economic efficiency, increase poverty, and cause natural resource degradation. These policies also tend to favor the modernization of large-scale farming at the expense of more efficient and employment-intensive family farms. Their removal thus would be a *win-win-win* situation in which the same policy changes could lead to more growth, less poverty, and more sustainable natural resource management.

Heath and Binswanger begin by discussing the Boserup hypothesis, which suggests that, under the right conditions, higher population and market access can lead to improvements in natural resources rather than their deterioration. They note that the Boserup effects are far from automatic. Rather they are an outcome of investment decisions made by farmers, who require a positive incentive regime and access to soil and water resources and markets. If adverse policy regimes deny these conditions, impoverished peasants are forced to mine the land resources rather than augment them. This is especially damaging if they only have access to marginal land or to humid tropical forest frontiers. Although the authors discuss the importance of agricultural policies, they also stress the importance of proper infrastructure (related to Boserup's market access condition), education (in particular the education of girls), health, and macroeconomic stability.

Using recent studies of the Machakos district in Kenya and soil degradation in Ethiopia, the authors illustrate the dependence of the Boserup effects on the policy environment that governs the investment incentives of farmers. They also examine growth, the use of land and labor, and the degradation of marginal natural resources in Colombian agriculture.

Heath and Binswanger admit that eliminating the adverse policies and privileges of rural elites, which reduce efficiency and increase poverty and resource degradation, is an extremely difficult task. These policies did not come about by accident, but as the consequence of a historical evolution that involved bargaining among politically strong groups looking out for their own interests at each point in time. Analyzing the joint consequences of the policies may support the reform effort but is unlikely to be sufficient. Peasants have a low collective action potential, and steps must be taken to provide them with a greater input into policymaking. Unless some political actors with enlightened hearts and minds see it as in their interest to bring poor peasants more strongly into the policy process, policy and land reforms will continue to be slow.

In chapter 4, Jules Pretty makes a case for sustainable intensification of agriculture. He describes five main groups of people with different views of future agricultural development: business-as-usual optimists, environmental pessimists, industrial-world-to-the-rescue group, new modernists, and the group arguing for sustainable intensification of agriculture. As part of the last group, Pretty shows that regenerative and low-input (but not necessarily zero-input) agriculture can be highly productive if farmers participate fully in all stages of technology development and extension. He also suggests that the productivity of agricultural and pastoral lands is as much a function of human capacity and ingenuity as it is of biological and physical processes. Such agriculture seeks to integrate the use of a wide range of pest, nutrient, soil, and water management technologies. It also aims to create diverse enterprises within farms so that by-products or wastes from one component or enterprise become inputs to another, thereby reducing the negative impacts on the environment.

Pretty stresses the importance of not prescribing a concretely defined set of technologies, practices, or policies, because, as conditions and knowledge change, so must farmers and communities be encouraged and allowed to change. In Pretty's view, therefore, sustainable agriculture is not simply a model to be imposed; *it is more a process for learning*, an idea that is also supported by Bunch (chapter 12). What needs to be made sustainable is the social process of innovation itself, which implies an enhanced capacity to adapt to unexpected changes and emerging uncertainties. For such innovation to happen, it is important to have an enabling policy environment. In this context, Pretty recognizes that in practice, policy is the net result of the actions of different interest groups pulling in complementary and opposing directions, often making the outcome uncertain.

In chapter 5, Holden and Binswanger analyze small-farmer decisionmaking in the context of imperfect markets and its implications for efficiency and for sustainability of natural resource management. They argue that new policies to stimulate sustainable rural development are much needed and that, to succeed, these policies have to build on an understanding of the decisionmaking environment and behavioral responses of small farmers.

Small farmers represent a major link between the economy and the environment because their livelihoods depend directly on the use of land resources (soil and vegetation). The fact that small farmers represent both production and consumption units complicates matters, particularly when market imperfections cause their production and consumption decisions to be nonseparable. Nonseparability implies that consumption needs and asset distribution may have significant impacts on production decisions and thus the management of natural resources.

Even though small farmers tend to be only partly integrated into markets, they are affected by market imperfections. These include missing markets, partly missing markets (rationing, seasonality), thin markets (imperfect competition), and interlinked markets. Covariate risk, asymmetric information, and transaction costs are the causes of pervasive market imperfections. These problems are even more acute in areas with low population density where the population derives a living from marginal land resources.

Small farmers largely are rational and respond to changes in the set of constraints and opportunities they face. But inefficiencies may still accrue due to economies of scale, poverty and subsistence constraints, unequal access to credit and markets, and various policies disfavoring small farmers. It is thus possible that small farmers are too poor to be efficient from society's perspective. These inefficiencies may also lead to underinvestments in conservation.

Market imperfections are more likely to hurt small and poor farmers than large and rich farmers because it is more probable that poor farmers will be rationed out of credit markets, for example, and will be less able to solve their problems through consumption smoothing (coping strategies). This explains why the most vulnerable households are most likely to diversify their production. There may thus be a case for intervention on the basis of efficiency as well as equity. The possibility of a reinforcing relationship between poverty and environmental degradation strengthens the case for appropriate intervention.

According to Holden and Binswanger, population growth may stimulate sustainable intensification in certain conditions, while in others it may lead to land degradation. They believe that incentive structures and how these are affected by policies have a strong impact on whether small farmers are able to choose a sustainable and welfare-improving development path or are forced onto a nonsustainable and welfare-reducing one.

Particularly in Africa where policy regimes have been very adverse and in other areas that are poorly integrated into markets, small farmers appear to degrade their environment because population growth fragments scarce land resources and leads to unsustainable intensification given stagnant technology. These rural pockets appear to be in a poverty-environment trap that resembles a neo-Malthusian scenario. They represent one of the most significant challenges to the developing world. Induced technological and institutional innovation appears to be the best means, together with stable macroeconomic conditions, of switching from a Malthusian to a Boserupian development path. Agricultural research and extension, infrastructure development, basic education, and family planning are also important elements, together with land tenure policies and other measures to provide secure livelihoods.

Given the complexity of these issues, Holden and Binswanger suggest a strategy with pilot projects and flexible project design accompanied by systematic monitoring and evaluation. This approach would allow for maximum learning, adjustment, and scaling up as confidence is gained.

In chapter 6, Anderson notes that the world, during the past decade, at last began reducing distortions to agricultural incentives in significant ways. The long-term growth in agricultural protection in high-income countries slowed, and the heavy direct and indirect taxation of agriculture was reduced in numerous lower-income countries.

Anderson addresses the concern that agricultural trade liberalization in industrial countries may have negative effects on rural and natural environments, on agriculture's resource base, and on food safety. He is confident that the removal of trade distortions will in almost all situations boost economic welfare. In the case of agricultural policy reform, it would also improve the environment on balance in many instances. Where it would not, the reason is commonly that governments do not have appropriate environmental policies in place. But Anderson also shows that many policy changes can have negative environmental effects even when the net welfare effect might be positive. Hence, more attempts will need to be made to quantify the environmental changes likely to be associated with trade and other economic reforms.

Another concern is expressed by the aid community, which is worried that the reduction in government attention to agriculture in high-income countries is contributing to a loss of interest in funding agricultural development assistance programs, including international agricultural research. To date, comparative static economic modelers have paid little attention to these environmental or research and development effects of reform.

Chapter 6 then examines the consequences for incentives to invest in agricultural research of a unilateral removal of policies that keep domestic food prices artificially low. At higher food prices, the environmental cost—degradation of the farm sector's natural resource base—of not reducing distortions in the markets for farm chemicals and irrigation water is higher than it would be if product prices and hence input intensities were lower. Another implication is that the returns to developing countries from investing in agricultural research will be higher in the future as a result of those higher prices. The international agricultural research centers too will contribute more from their research. Insofar as developing countries' reforms involve raising prices not only for farm products but also for chemical inputs and water, the rewards will be especially high from generating environmentally friendlier precision technologies as distinct from those that use heavy doses of chemicals and water.

Anderson expects that environmental groups will pressure the World Trade Organization to place the issue of trade and environment linkages on the agenda of its next round of talks. It is too early to tell what form that might take, but it is unlikely to result in agreement to do anything as extreme as allow countries to use trade measures against countries with less-stringent environmental policies. Even calls to have an environmental impact assessment of what is negotiated in the way of reducing trade restrictions are impractical, because, if such an assessment were to produce amendments, the whole agreement could unravel. Such calls will be less likely, however, as more empirical research demonstrates the (expected) absence of significant adverse environmental consequences of trade liberalization.

In chapter 7, Bromley addresses property relations in the context of developmental efforts. His thesis is that property relations are central in development because they connect people to one another with respect to land and related natural resources. Property relations are simply socially constructed contractual arrangements among a group of people for mediating individual and collective behaviors regarding objects and circumstances of value to members of the community. Development projects are sometimes less successful than they would otherwise be precisely because the existing property relations have been ignored or misunderstood.

Bromley offers nine main lessons for development practitioners. The first one is quite obvious but merits being recorded: *property regimes are part of the larger institutional structure*. Property regimes are structural attributes of an economy that provide agents with domains of choice within which they may act; property regimes must be understood as social constructs whose nature and existence are antecedent to what we ordinarily regard as "economic" behavior. They must be seen as occurring logically and socially prior to the economizing behavior we attempt to analyze and explain.

All property regimes require external legitimacy. To have a property right is to know that the authority system will defend you against the predatory behavior of others. Degradation of natural resources often occurs as a result of a breakdown of authority over the resource, caused by internal conflict or by intrusion into the area by outsiders.

Ambiguous rights regimes have ambiguous efficiency and distributional consequences. Ambiguity of some general meager resources can work to the advantage of the poor. However, when development projects increase the income flow from natural resources over which rights were previously ambiguous, then powerful individuals at the local level will usually figure out a way to expropriate at least a part—if not the majority—of these new income streams.

Property relations must be specified prior to the implementation of development projects. As a general rule, projects are placed in poor areas that warrant outside assistance precisely on account of the incoherence and contentious nature of local institutional arrangements—of which property relations are of paramount importance. Therefore, every development intervention must be preceded by a concerted effort to ensure that the institutional arrangements have been modified so as to ensure that the benefits go to the intended beneficiaries.

Resource degradation is the result of problems that precede property regimes. The most serious resource degradation problems occur in situations of open access in which no property rights exist. But such situations are themselves manifestations of larger institutional problems, only one of which may relate to the particular property regime in place.

Ecological variability demands flexible institutions and actions. Ecological settings that exhibit great variability require property regimes that allow quick and low-cost adaptations to new circumstances. Livestock, because of their mobility, provide this flexibility in a way that land, which is immobile, cannot possibly equal.

Resource degradation is contextual. Once we recognize that the social and economic meaning of various assets differs across ecological and social circumstance, then it follows automatically that the social and economic meaning of "resource" also differs.

Security of tenure is necessary for efficient investments, but a title, as such, is not necessary. Security of tenure is a necessary condition for investment in, or the "wise" management of, land and related natural resources. The contrary proposition—investment in land is a necessary condition to secure "title" in that land—is equally probable.

Mobilizing local interests can improve the chances of a program's success. In natural resource projects, there is growing recognition that "community-based conservation" is an important innovation in creating the conditions for sustaining both natural resources and local commitment to that sustainability. But although creating a sense of local ownership in particular resources can solve much of the incentive problem, such arrangements may still require an oversight role for some higher authority if there is a national or international interest in the particular resource or if externalities are present.

Bromley, in chapter 7, essentially argues that the outcomes of development activities are fundamentally dependent on institutional arrangements in general and property relations in particular.

Hazell, in chapter 8, states that agricultural development should simultaneously contribute to four principal goals: growth, poverty reduction, food security, and sustainable natural resource management. Experience suggests that there can be a high degree of complementarity between these four goals if agricultural development (a) is broadly based and involves small- and medium-size farms, (b) is market driven, (c) is participatory and decentralized, and (d) is driven by productivity-enhancing technological change that does not degrade the resource base.

Back in the 1950s and 1960s, agricultural development experts tended to be interested in agricultural growth, and the lessons that emerged from their experience can be summarized as the five "I"s for agricultural growth: innovation, infrastructure, inputs, institutions, and incentives. It later became evident that governments also need to promote the role of the private sector in agricultural production, marketing, and service provision.

In the 1970s and 1980s, attention focused on using agricultural development to reduce poverty and food insecurity as well as to spur growth. The lessons that emerged from that era can be summarized as follows: (a) broadly based agricultural development should be promoted, (b) participation of affected stakeholder groups is essential, (c) land reforms may be necessary, (d) investments in human capital are very important, (e) women's concerns must be addressed, and (f) the rural nonfarm economy should be actively encouraged.

The new priority for environmental sustainability that has emerged in the 1990s does not negate the need for agriculture to continue contributing to growth, poverty alleviation, and increased food security. It is just that agriculture is now required to do this in ways that do not degrade the environment. Hazell argues that two fundamentally different types of environmental problems are associated with agriculture. The first type arises in intensive farming systems and is associated with the misuse of modern inputs (irrigation water, fertilizers, and pesticides). The second type arises in extensive farming systems and is associated with rapid population growth, poverty, and growth in agricultural productivity insufficient to meet the increasing need for food and livelihood (elaborated in chapter 2).

A general finding of the part on policy perspectives is that significant strides can be made toward sustainability. But clearly, appropriate policies and institutions are a precondition. These are important for growth, equity, and environmental reasons; they are also needed to find appropriate technical solutions. A key question that remains is whether the political process can produce the needed policies and institutions.

2

Integrating Environmental and Sustainability Concerns into Rural Development Policies

Peter Hazell and Ernst Lutz

Many developing countries have achieved impressive growth rates in agriculture in recent decades. India, for example, which was threatened by hunger and mass starvation in the 1960s, is now self-sufficient in staple foods even though her population has more than doubled. Yet in spite of this success, serious concerns remain for the future. First, hunger and malnutrition persist in many countries, often because past patterns of agricultural growth failed to benefit the poor adequately. Second, agricultural demand will grow along with population growth and rising per capita incomes, and this will require continuing increases in agricultural productivity. Yet growth in yields appears to be slowing while the prospects for expanding cropped and irrigated areas are limited. Third, if not checked, environmental problems associated with agriculture could threaten future levels of agricultural productivity as well as the health and well-being of rural people.

Two basic types of environmental problems are associated with agriculture. Most of the successful breakthroughs in productivity have occurred in more-favored agroecological zones and have been based on intensive use of irrigation water, fertilizers, pesticides, and other modern inputs. Agriculture based on intensive use of these inputs is prone to mismanagement that leads to environmental degradation (particularly when the system of incentives is inappropriate). However, where governments have neglected to intensify agricultural production through the use of modern technology, population growth has worsened poverty and hunger, driving rural people to expand cultivation into less-favored, often environmentally fragile areas, such as forests, hillsides, and wetlands, and to reduce fallow periods to the point of depressing soil fertility.

Continued agricultural growth is a necessity, not an option, for most developing countries. Further, this growth must be achieved on a sustainable basis so as not to jeopardize the underlying base of natural resources, and it must be equitable if it is to contribute to the alleviation of poverty and food insecurity. Government policies, institutional development, agricultural research, and projects at local, regional, and national levels need to be designed and implemented with these objectives in mind.

What Is Sustainable Agricultural Development?

There are many alternative definitions of sustainability and of sustainable agricultural development (Pezzey 1992). Conceptually, those most relevant for our purposes are measures of agricultural income that are corrected for changes in the value of the capital stock, especially the stock of natural resources used in agricultural production. As Sir John Hicks has argued, sustainable income is the maximum value that a person or society can consume during a specific period of time and still expect to be as well off at the end of the period as at the beginning (Hicks 1946: 172). Applying this concept, an agricultural system can be said to be sustainable if the amount of income extracted for consumption each year can be sustained over time. This requires that the value of the capital stock not be depleted over time; hence sufficient income must be set aside or otherwise forgone to replenish any capital depreciation or losses incurred in the production process. For this purpose, capital is defined to include all natural resource, human, and man-made capital assets.

A further distinction is made between weak and strong sustainability, where the difference depends on whether the definition permits substitution between different types of capital assets. With weak sustainability, substitution of different forms of capital is allowed, but the total value of capital must remain intact or increase. In contrast, with strong sustainability, substitution of different forms of capital is not allowed, and an effort is made to keep constant the stock of each form of capital, including the value of natural resources (Pearce, Markandya, and Barbier 1989). Serageldin (1996) has suggested a compromise definition called sensible sustainability. This allows for capital transformations—that is, of natural capital into human capital (through education) or into man-made capital—but also recognizes that critical levels of each type of capital may exist beyond which concerns about substitutability can arise.[1] Because the exact boundaries of the critical limits for each type of capital are not known, "it behooves the sensible person to err on the side of caution in depleting resources (especially natural capital)" (Serageldin 1996: 8).

"Weak" and "sensible" definitions of sustainability have the advantage that they do not suggest freezing or preserving all natural resources at existing levels. Rather, they permit tradeoffs between growth and environmental objectives: resources can be degraded or depleted to increase production, *if* compensatory investments are made in other forms of capital to sustain the stream of consumable income over time. Decisions about the degradation or depletion of natural resources should, therefore, depend on whether the social benefits accruing to the users are greater or smaller than the social costs (Turner 1993: figs. 1.2 and 1.3). For example, because some resources are renewable (trees, soil nutrients) or have adequate substitutes (farm trees can replace woodlands for fuelwood), they need not all be preserved at current levels or at all points in time. In cases of critical habitats, however, the costs of degradation are likely to be so high that conversion to agriculture is not a socially optimal decision.

To achieve agricultural sustainability defined in this way raises three sets of issues. First, at what level of society are environmental costs and benefits to be measured? Because resource degradation often has spillover or externality costs for nonusers (as when the degradation of watershed protection areas leads to the silting of irrigation systems downstream or when water pollution causes health problems throughout river basin systems), the more aggregated the concept of society used, the greater the total environmental costs. Thus levels of environmental degradation that may be considered acceptable to farmers or rural communities may be unacceptable at the national or international levels if the externality costs are significant. The reverse can also be true: resources that generate environmental benefits may be undervalued at the farmer or community levels because their benefits are captured at broader national or international levels. For example, farmers may convert forests in excess because they do not capture the forest's full value in modulating climate or in preserving biodiversity. Clearly, assessing sustainability at the farm or community levels is not sufficient in the presence of important externalities, and regional and national assessments are also needed to guide policy decisions. International or global assessments are also required where important international externalities are involved (such as the impacts of agriculture on global climate change), even though these may be perceived as less relevant for decisionmakers in individual countries.

The second issue concerns methods of valuing environmental costs and benefits. Environmental valuation techniques have been increasingly used in empirical studies during the past decade or two. Values can be grouped into direct use values, indirect use values, option values, and existence values. Direct use values are associated with food production and consumption, biomass, recreation, and health. Indirect use values are associated, for example, with ecological functions, flood control, and storm protection. Option values are associated with the preservation of resources that have probable but uncertain value in the future, for example, biodiversity conservation. Finally, existence values include the value that people place on the mere existence of resources from which they may never directly or indirectly benefit (from the existence of parks or species, which they may never actually visit or see). As a rule, the more tangible the costs and benefits, the more reliable the estimates that can be produced. Various methods have been developed for estimating environmental costs and benefits for use in project and policy analyses or environmental accounting (see, for example, Dixon and others 1992, Lutz 1993, Lutz and Munasinghe 1994, and Hamilton and Lutz 1996).

Empirical valuation work in developing countries at the conjunction of agriculture and the environment is still quite limited. Continued development of databases and practical methods of valuing environmental costs and benefits at the farm, community, and national levels will be important for conducting a rigorous analysis of policy and agricultural research priorities for promoting sustainable agricultural development.

The third issue concerns the design of incentive schemes that induce farmers and communities to factor important externality costs or benefits into their own resource management decisions. As a general finding, undistorted macroeconomic and sectoral policies tend to provide better incentives from an environmental perspective than highly distorted policies (Warford, Munasinghe, and Cruz 1997 and Lutz and Young 1992). Subsidies for agricultural inputs tend to be particularly costly in terms of economic efficiency, government budgets, and the environment (World Bank 1997: chap. 4).

Environmental Problems in Agriculture

Despite considerable diversity in agroecological conditions around the developing world, it is useful to distinguish between the environmental problems related to intensive, irrigated farming areas and those related to more extensive, rainfed farming areas.

Irrigated Agriculture

Productivity has grown fastest in irrigated agriculture because of the increased use of modern inputs: irrigation, fertilizers, pesticides, high-yielding varieties, and machinery. However, this intensification has also increased the potential for the inappropriate use of modern inputs, particularly when inappropriate incentives prevail.

The following major environmental problems are associated with intensification in irrigated areas (Pingali and Rosegrant 1994):

- Intensive use of irrigation water in areas with poor drainage can lead to waterlogged soils and a rise in the water table. In arid and semiarid areas, this in turn causes salt to build up in the soil. Salinization reduces yields and can eventually lead to abandonment of land.
- Perennial flooding of rice paddies and continuous rice culture lead to a buildup of micronutrient deficiencies and soil toxicities, formation of hard pans in the soil, and a reduction in the nitrogen-carrying capacity of the soil. Work at the International Rice Research Institute shows that farmers are having to increase the amount of fertilizer they use over time simply to maintain existing yields in intensive paddy fields (Pingali 1992).
- Excessive and inappropriate use of pesticides deteriorates the quality of water, poses health hazards for humans, and leads to resistance of pests to pesticides. Farmers can become trapped into using more and more frequent sprays to control pest damage.
- An increasing reliance on a few carefully bred crop varieties contributes to a loss of genetic diversity and to a common vulnerability to the same pest- and weather-related risks. In some cases, millions of hectares of land have been planted to the same wheat or rice varieties, and widespread losses have occurred because of the outbreak of a single pest or disease. The loss of traditional varieties also reduces the pool of genes available for breeding plants capable of resisting evolving plant pests.

Rainfed Agriculture

Population growth and poverty in many rainfed lands are causing serious resource degradation. Until recently, natural resources were generally abundant in these areas, and, once used, farmers could allow resources time to recover through rotations and shifting cultivation. Moreover, many of the more fragile lands were not farmed at all. Today, rainfed lands frequently must support moderate to high population densities, providing not only increasing amounts of food but basic essentials such as fuelwood, water, and housing. In the absence of adequate increases in agricultural productivity to secure their livelihoods, farmers reduce needed fallows and expand into new areas, many of which are environmentally fragile and easily degraded. Environmental problems associated with rainfed farming include (Scherr and Hazell 1994):

- Conversion of primary forest to agriculture, with loss of biodiversity, climate change, and exposure of fragile soils
- Expansion into steep hillsides causing soil erosion and lowland flooding
- Degradation of watershed protection areas, with downstream siltation of dams and irrigation systems and increased flooding

- Shortened fallows with loss of soil nutrients and organic matter, resulting in declining yields
- Increased pressure on common property resources (woodlands and grazing areas), with breakdown of indigenous institutions that regulate and manage these resources, leading to open-access regimes and resource degradation
- Declining resilience in ecosystems, with reduced ability to rebound from stresses such as droughts.

Overuse of fertilizers and pesticides is much less common in rainfed agriculture because the current levels of use are still low. Exceptions include a few high-value cash crops such as tobacco and vegetables. A bigger problem is insufficient use of fertilizers to replace soil nutrients lost through increased cropping and reduced fallows.

Costs of Resource Degradation

The full extent of environmental damage from agriculture is difficult to assess. Here, we attempt to summarize indicative estimates from various studies. As will be shown throughout this volume, many of these costs are associated with market, institutional, and policy failures.

Deforestation and Forest Degradation

According to the Food and Agriculture Organization of the United Nations, deforestation is the sum of all transitions from natural forest classes (both continuous and fragmented) to all other classes. Based on this definition, deforestation during 1980–90 amounted to 8.2 percent in Asia, 6.1 percent in Latin America, and 4.8 percent in Africa (see chapter 22 in this volume). Not all of this can be attributed to agricultural conversion.

Forest degradation represents a decrease of density or increase of disturbance in forest classes. Even when there may be little deforestation as such, degradation can imply major losses of forest products and ecological services, such as reductions in watershed and climate services, biodiversity, and carbon storage. Because private agents generally do not consider these costs in their decisions about logging and forest conversion, rates of depletion can be excessive from a societal view.

Water Depletion, Water Degradation, and Waterlogging and Salinization

Water is a vital economic asset that is already becoming scarce in many countries. Moreover, water shortages are often compounded by wasteful and inefficient use and by pollution from the dumping of human and animal wastes and industrial pollutants. Water quality is inadequate for an estimated 1 billion people, and related health costs and fatalities are high. Agriculture is the major user of water in many developing countries and is a significant contributor to water quality problems.

Water problems are in large part policy-induced: cost recovery rates in irrigation amount to between 20 and 25 percent of true costs, and subsidies cost governments some $20 billion to $25 billion each year (de Moor and Calamai 1997). Electricity for pumping water is also subsidized in many countries. These subsidies encourage excessive and inefficient use of water, and this not only aggravates water scarcity problems but also leads to waterlogging and salinization of irrigated land. In China, for example, salinization already affects 23 percent of the total irrigated area (Umali 1993: table 3.1). In Pakistan, which has the world's largest contiguous surface distribution system, an estimated 3.5 million hectares (or 25 percent of the irrigated area) is affected by waterlogging and salinity, and 8 percent of these lands are seriously degraded (Pinstrup-Andersen and Pandya-Lorch 1994). The detrimental effects on crop productivity can be serious. Yields of major crops in the affected areas of Egypt and Pakistan have declined significantly (Barghouti and Le Moigne 1991). Further, there are also negative social impacts because rich farmers generally benefit from water-related subsidies more than the poor.

Fish Stock Depletion

Fish protein is an important part of the diet in many countries. Fish production from aquaculture is increasing, but output from the marine fisheries sector appears to have peaked at around 85 million to 90 million tons. Fish stocks are being depleted due to ineffective management of common property resources. An important problem is that in many countries, the government subsidizes the maintenance of excessively large fishing fleets, and this encourages overfishing. Estimates of the total subsidies involved range from $11 billion to $54 billion (Milazzo 1997). Uncontrolled and wasteful fishing (including the throwing away of unwanted catches) have depleted marine stocks, driving once-common species like cod and halibut to commercial extinction and threatening the livelihood of millions of people. Water pollution also affects fisheries.

Soil Degradation

Good soil is a crucial factor of production in agriculture, and maintaining the depth and quality of soil is important for preserving agricultural productivity in the future. The costs of soil degradation can be measured by lower yields or, where farmers can compensate through more intensive use of manure and fertilizers, by higher production costs. What is not so easily measured is the reduction in value of the stock of this natural capital. Inasmuch as a reduction in value has a direct impact on future agricultural production, the loss in future yields can be measured. But it is not just the amount of soil eroded that is the key to future yield losses; the amount (depth) and quality of the soil that remains, including the soil that is deposited in other fields used for agriculture, are also important. Other important factors that are generally not recorded also affect agricultural productivity (such as soil compaction, pH, and organic matter content). Field studies showing the relationships between soil loss, degradation, and deposition and agricultural productivity in developing countries are rare and, even where such information exists, cannot easily be extrapolated because soil degradation differs from site to site because of differences in the type of soil, slope, vegetative cover, and cultivation practices. Not surprisingly, therefore, estimates of the costs of soil erosion and degradation, particularly at national levels, are few and vary widely.

In work on Indonesia, Repetto and others (1989) have estimated that capitalized losses for future productivity are approximately 40 percent of the annual value of upland farm production in Java, or between $340 million to $406 million a year. Of this amount, $315 million are estimated to be on-farm losses of productivity and the remaining $25 million to $80 million to be downstream damages from siltation. For Mali, Bishop and Allen (1989), assuming a 10-year time horizon and a 10 percent discount rate, have estimated that the present value of current and future net farm income forgone nationwide due to one year of average soil loss amounts to between 4 and 16 percent of agricultural gross domestic product. The World Resources Institute (1991) has estimated that soil depreciation amounts to almost 10 percent of Costa Rica's annual agricultural production. But in our view, all of these estimates appear to be high; in any case, more empirical work is needed.

In addition to on-site effects of soil erosion and degradation, one must consider off-site effects. Some of these costs are relatively easy to measure, such as the costs of dredging water channels or the costs of reduced fish catches as a result of water with high turbidity. In other cases, costs are more hidden (like sedimentation of reservoirs used for hydro power or irrigation) or not easily accounted for (like the damage to coral reefs from silt deposits).

In assessing the depreciation of a capital stock—in this case the stock of soil—by discounting the future income losses, the discount rate assumed is highly influential on the actual results. For nations as a whole and for national income and accounting purposes, one may assume discount rates of 5 to 10 percent. Higher rates are relevant for individual farmers, particularly for those with small farms (Cuesta, Carlson, and Lutz 1997).

Research on soil erosion and degradation is complicated by the influence of stochastic variables, such as rainfall intensity. Much more work is needed, particularly work that extends the focus from plot-level experiments to the watershed and from purely technical aspects to economic and social aspects as well.

Health Effects

Injudicious use of pesticides and a lack of safe spray equipment and protective clothing suitable for tropical conditions are causing significant short-term as well as long-term chronic health problems. Studies in Indonesia, Malaysia, and Sri Lanka in the early 1980s suggest that 12–15 percent of farmers who use pesticides have been poisoned at least once in their careers (Jeyaratnam, Lun, and Phoon 1987). But long-term health effects can also be significant (Pingali and Roger 1995). In addition to exposure during spraying, the unsafe storage, handling, and disposal of pesticides subject farmers and farm families to high levels of health risk and can contaminate the ecosystem.

Biodiversity Losses

There has been serious loss of traditional crop varieties, while large areas are planted to a few modem varieties (for example, IR36 rice has been planted on more than 10 million hectares in Asia). This varietal specialization has occasionally led to widespread crop losses due to outbreaks of diseases and pests (brown planthopper in rice) and requires devoting more shares of research resources to "maintenance" research and germ plasm conservation programs (Anderson, Hazell, and Evans 1987).

Policy, Institutional, and Technical Options to Reducing Resource Degradation

In thinking about possible solutions, it should be recognized that resources cannot be protected or conserved without finding acceptable means of livelihood for the people who use them. This is particularly true in fragile, rainfed areas where poverty is a major force driving the degradation of many resources. Sustainable and poverty-reducing agricultural intensification in problem areas is often key to solving resource degradation problems, particularly because solutions involving interregional migration and economic diversification into nonfarm activities are limited.

It also needs to be recognized that future food needs will be difficult to meet if countries revert to low-input, low-yield agricultural technologies. Continued increases in yield will be critical, with limited possibilities for reducing dependence on water and fertilizers. Moreover, short of any major biotechnology breakthroughs, the scope for meeting all additional food needs from irrigated lands is limited, and rainfed lands will need to become increasingly important sources of agricultural output. This will require that increasing shares of the resources available for agricultural research be devoted to rainfed areas and that these areas receive higher priority in public investment programs.

Policy interventions that seek to overcome environmental problems in agriculture need to be based on a proper understanding of why farmers degrade natural resources. Why, for example, do farmers often seem to overgraze rangeland, deplete soil nutrients and organic matter, and overuse irrigation water, pesticides, and nitrogen, when these actions cause health problems and reduce future incomes for themselves, their children, and the communities in which they live?

The answer lies with incentives including discount rates. Farmers are not irrational. To the contrary, they maximize income and minimize risks in a dynamic context and often under harsh conditions and serious constraints. So, for example, they degrade resources when there are good economic and social reasons for doing so (when the benefits they obtain exceed the perceived costs that they, as individuals, must bear). If the management of natural resources is to be improved, these economic and social incentives will need to be changed in appropriate ways.

Policy Issues and Options

Several factors impinge on the incentives for managing natural resources: technology design, poverty, property rights, externalities, costs of collective action, prices, government nonprice interventions, and access to information about the condition of resources. Focusing on these factors provides a useful way of discussing the kinds of changes needed to move toward environmentally sustainable agricultural development. In the following paragraphs, we discuss these factors in the same policy, institutional, and technology groupings used in this book.

Prices and Subsidies

Inappropriate prices for inputs and outputs can encourage farmers to degrade resources by making unsustainable practices more profitable. Inappropriate prices can arise from externality problems that distort market prices from their socially correct values. For example, rural market prices for fuelwood and charcoal generally undervalue the true cost of the wood, because the market price does not capture the environmental benefits of trees for soil conservation, modulation of local climate, and so forth. But governments also distort market prices. Inputs (especially water, fertilizers, and pesticides) are often subsidized, encouraging excessive use. De Moor and Calamai (1997) offer the following estimates for developing countries: water subsidies, \$20 billion to \$25 billion; fertilizer subsidies, about \$6 billion; pesticide subsidies, about \$4 billion. Many governments have also kept agricultural output prices too low through export taxes and overvalued exchange rates, reducing farmers' profits and their returns to investing in the conservation and improvement of natural resources. (Fortunately, the average degree of taxation of agriculture in developing countries has been reduced during the past decade, see chapter 6.)

In many cases, market liberalization, removal of subsidies, and realignment of domestic and border prices will improve incentives for sustainable farming practices. A good example is the favorable impact in Indonesia of removing subsidies on pesticides. Not only has pesticide use dropped dramatically, but the adoption of integrated pest management practices has spread widely and yields have increased (Ruchijat and Sukmaraganda 1992). Yet market solutions alone are not always adequate. Externality problems, as for example with tree products, may require adjustments in domestic prices from border prices to reflect environmental costs and benefits to the country. But this is not always practical, especially for export products, unless competing countries also make similar price adjustments or importing countries are willing to pay a price premium through green labeling schemes. This requires more effective international collaboration on environmental problems than has been achieved so far.

Nonprice Government Interventions

Government nonprice interventions can have significant effects on the incentives and opportunities available to farmers in making choices about technology and resource use. Public investments in, for example, rural roads, schools, clean water, health centers, family planning, and soil erosion control can create new opportunities in farm and nonfarm activities for rural people and reinforce positive incentives for sustainable resource management. Education and improved health can also help increase opportunities for migration, reducing the population pressure on resources and providing capital flows through remittances for investments in agriculture. Drought-relief interventions, such as food-for-work programs, can be particularly helpful in relieving the pressure on resources when they are most vulnerable.

But inappropriate government interventions can be environmentally destructive. Construction of new roads or settlement schemes in environmentally fragile areas can be destructive (for example, some of Indonesia's transmigration projects). Rigid regulation of land and forest use can also inhibit development of sustainable but more profitable patterns of resource use (farmers in the Philippines and Thailand were initially prevented from investing in new types of productive trees in protected hillside areas). Drought-relief interventions can also backfire if they are heavily subsidized. Subsidized drought insurance, for example, increases the profitability of more risky farming practices, some of which may be environmentally unsuitable for drought-prone areas.

A key issue for policy is determining the relative weight to be given to resource-poor areas in allocating public investments among rural areas. Resource-poor areas have been sadly neglected in the past, yet the prevalence of poverty and environmental degradation in those areas implies that increasing the social and environmental benefits of public investments may help to offset their lower efficiency returns.

Externalities

Externalities arise when the costs associated with resource degradation are not fully borne by the individuals causing the problem. For example, removing trees that protect watersheds may be privately beneficial to the individual farmers who do it but can lead to soil erosion and downstream flooding and siltation of irrigation works that are costly to society. Externalities are prevalent in agriculture and are a particularly important factor in explaining water pollution, overuse of agrochemicals, deforestation, and loss of biodiversity.

In countries with well-developed institutions, carefully crafted taxes, subsidies, and government regulations can be effective in overcoming externality problems. These instruments are typically less effective in developing countries, particularly in remote rural areas. Some externalities can be reduced by pricing marketed goods appropriately. But in most cases, solutions must be sought through local governments and organizations. Central governments can help by providing environmental guidelines, contributing to resource monitoring systems (providing aerial photographs, remote-sensing data, downstream water testing), and protecting designated conservation areas (parks and other environmentally valued sites). But effective action requires the joint involvement of people who misuse resources and of people who are affected most immediately by that misuse. Empowerment of local action groups is important, and this can sometimes be reinforced by appropriate changes in property rights. For example, ownership rights over watershed protection areas and waterways can be bestowed on communities or local organizations rather than on individuals or the public sector.

Information on the Condition of Natural Resources

We believe that farmers respond rationally to economic and social incentives but that sometimes they may be poorly informed about the consequences of their actions. For example, they may not be fully aware of the longer-term consequences of soil degradation of particular farming practices, the effect of removing trees on soil erosion on neighboring farms, or the effect of siltation problems downstream. Integrated pest management programs have shown that many farmers cannot distinguish between harmful and beneficial insect species and are often inclined to spray both.

Better information needs to be provided; programs to increase public awareness that use schools, nongovernmental organizations (NGOs), and the media can be effective. Simple resource-monitoring systems also need to be established to help communities track important changes. These might include systems that track biodiversity, beneficial species, and soil erosion. Such monitoring systems need to be maintained by the community itself, perhaps by a farm cooperative or the village school.

Institutional Issues and Options

In this section, we discuss select institutional issues that are addressed much more extensively in later chapters.

Poverty Reduction through Investments in Resource-Poor Areas

Poor people are more desperate and more likely to trade off tomorrow's production in order to eat today than people who are not poor. Chronic poverty, exacerbated by population growth and occasional droughts, underlies much of the resource degradation observed in rainfed areas. It is much less of a factor in intensive agricultural areas, although it can still be a problem where there is a high incidence of landlessness or a highly inequitable distribution of land.

Alleviating poverty is fundamental to redressing the environmental problems of most resource-poor areas. Appropriate strategies need to include targeted assistance programs for the poor, economic diversification strategies for rural areas, assistance programs for small farms, and public investments in rural infrastructure, health facilities, schools, and the like.

Agricultural growth also needs to play a major role in the development of many resource-poor areas. Despite outmigration, the absolute number of people living in these areas will continue to grow in the next decades. Rural nonfarm economies are too small and too constrained by local demand to create significant employment opportunities in the absence of agricultural growth.

There was some hesitancy in the past to invest in resource-poor areas, primarily because the returns to investment were perceived to be much higher in resource-rich areas. But these higher returns are less assured today. Productivity growth in intensive agricultural systems is becoming more difficult and costly to achieve, particularly when the environmental costs are considered. At the same time, the returns from investments in resource-poor areas are becoming more attractive. Recent developments in research on farming systems, soil management, and agroforestry suggest that there may be potential for greater productivity in many resource-poor areas than previously thought (Scott and Scherr 1995), and this is supported by recent research in India (Fan and Hazell 1996). Moreover, the worsening poverty and resource degradation that are occurring in many resource-poor areas as populations reach critical levels suggest that the social and environmental benefits from investing in these areas are now potentially attractive.

Unfortunately, investments in resource-poor areas are complicated and risky. The diversity of agroclimatic, social, and economic conditions that exist demands considerable site specificity in approach. Moreover, genetic improvement of individual crops is not likely to be sufficient, and productivity growth must typically be achieved through improved management of landscapes and farming systems with limited reliance on purchased inputs. Successful approaches, as exemplified by the efforts of several NGOs, are based on participatory approaches with farmers and local organizations, not on the more traditional top-down approach that characterized many integrated rural development projects of the past. Success is also difficult to measure in the short run, not only because of a multiplicity of economic, social, and environmental goals, but also because resource-poor areas are typically subject to considerable climatic risk from year to year.

Property Rights

The property rights that farmers have over natural resources can be important in determining whether they take a short- or long-term perspective in managing resources. For example, farmers who feel that their tenure is insecure, with or without formal rights, are less likely to be interested in conserving resources or in making investments that improve the long-term productivity of resources.

Property rights are often problematic during the transition from extensive to intensive agricultural systems, when they typically must evolve from indigenous, community-based tenure systems to registered and legally recognized, private property arrangements. Property rights are also becoming more problematical in the management of common property resources, such as open rangeland or forest, because the local institutions that traditionally control and regulate the use of such resources are breaking down.

Research has shown that secure property rights over land are an important factor in determining land-improving investments, such as tree planting, continuous manuring, and terracing and contouring. However, this does not necessarily imply that governments should immediately invest in ambitious land registration programs where this has not already been done. Many of the indigenous, community-based land tenure systems that exist in long-settled areas can be surprisingly effective, and these systems are spontaneously evolving, with the growth of populations and commercialization, toward systems of privatized land rights (Migot-Adholla and others 1991). These traditional systems can also be effective in recognizing multiple-user rights over the same land and are often more equitable than fully privatized systems. The appropriate role for governments in these cases is not to replace the indigenous systems abruptly, but to seek ways of strengthening them and facilitating their adaptation to changing circumstances. Legal registration of blocks of community or village-held land and simple voluntary systems for recording land transactions may sometimes increase security by reducing land disputes between and within communities.

In contrast, land registration may be economically worthwhile in areas of high population density or commercialized agriculture, particularly when formal lending institutions are also well developed and land is already effectively privatized. Feder and others (1988) have found this to be true for Thailand.

Traditional management systems for many common property resources face increasing challenges as rural populations grow. Not only must more local users be accommodated, but common property resources are increasingly threatened by encroachment and unregulated use by outsiders. Full privatization of common property resources is sometimes the relevant solution, but often there are good economic and social reasons for keeping some resources as common property. A good example is open rangeland in drought-prone areas. The ability to move animals over wide areas is an integral part of risk management, especially in drought years. The costs of fencing and the existence of limited water holes also constrain privatization. Some resources are better managed as common property because of economies of scale or because this is an effective way of managing multiple-user rights.

Where resources are to remain common property, effective organizations are needed to manage them. Often, governments have undermined indigenous institutions by nationalizing important common property resources, such as forests and rangelands, and then failed to manage them effectively. As a result, these resources have degenerated into open-access areas. The most successful institutions for managing common properties are local organizations dominated by the resource users themselves. It helps if local organizations can be built on the vestiges of indigenous rules and institutions and if the group as a whole has secure rights to own the resource. Successful examples include social forestry groups in India and rangeland management groups in parts of West Africa.

Property rights over irrigation water are in most need of reform in many countries. The ownership and allocation of water continue to be dominated by the public sector, and a pernicious combination of bureaucratic inertia, low water charges (including limited cost recovery), and subsidized electricity for pumping is responsible for considerable inefficiency in the use of water for agriculture as well as inappropriate water management practices that lead to polluted water and waterlogged and salinized land.

Evidence is growing that more appropriate pricing of water and electricity can significantly reduce water and environmental problems in irrigated agriculture as well as provide new sources of revenue for maintaining and improving irrigation infrastructure. Unfortunately, progress has been slow because of farmer resistance and vested political interests. In most countries, cost recovery does not cover expenditures for operations and maintenance.

More promising approaches lie in the devolution of many water allocation decisions to water user groups and in the creation of tradable water rights for farmers. Water rights are particularly attractive because they lead to the full economic pricing of water; water is valued at its full opportunity cost, not just a charge, based on cost recovery, set by an irrigation authority. Moreover, creation of property rights in water bestows a new asset on farmers and is therefore more likely to be accepted than a new water charge or tax.

Cost of Collective Action

Conserving or improving natural resources often requires collective action by groups of users. Examples include the management of common property resources or the organization of adjacent farmers to invest labor in land terracing or bunding. Organizing farmers into effective and stable groups for collective action is difficult, and success is conditioned on a range of physical, social, and institutional factors (Uphoff 1986, Ostrom 1994, and Rasmussen and Meinzen-Dick 1995).

A resource that is to be managed or improved collectively should be reasonably small, naturally bounded, and accessible to group members to facilitate the control and exclusion of outsiders. It helps too if use by one member has limited effect on the availability of the resource to other members (low "subtractability"). Greater social cohesion within the group is facilitated if the number of users is small, the members are homogeneous in their values and economic dependence on the resource, and the net benefits from group membership are substantial and distributed equitably.

Institutional design is also important. Ostrom (1994) has identified five principles for designing effective local organizations: (a) the members and the boundaries of the resource to be managed or improved must be clearly defined, (b) a clear set of rules and obligations should be established that are adapted to local conditions, (c) members should collectively be able to modify those rules to changing circumstances, (d) an adequate monitoring system should be in place, with enforceable sanctions, preferably graduated to match the seriousness and context of the offense, and effective mechanisms for resolving conflicts, and (e) the organization, if not empowered or recognized by government authorities, should at least not be challenged or undermined by them.

An important line of research on the determinants of successful collective action highlights the changing dynamics of the net benefits. In areas with low population densities, resource shortages and degradation are rare, and there is little incentive to organize for investments in improving and conserving resources. But in areas with growing populations, resources become scarcer and, in the absence of investments, are subject to worsening degradation. Eventually, the economic and social benefits of organizing for investment exceed the costs, and user groups can be expected to emerge spontaneously to make the necessary investments. Within the context of this "induced innovation" theory, it is futile to try to organize farmers prematurely. Moreover, the purpose of outside interventions should be to create enabling conditions for the emergence of spontaneous group action, not to supplant it with externally planned organizations.

Technological Issues and Options

When poorly designed, or inappropriately used, technologies can lead farmers to increase production in ways that degrade natural resources. New technologies have often been developed with a narrow focus on short-term profitability to farmers and without due consideration to their longer-term sustainability. For example, the development of powerful pesticides and herbicides reduced costs and improved yields but often had negative effects on the environment and on long-term yields. A related problem is the spread of new technologies from the agroclimatic zones or farming systems for which they were developed to other, less-suitable areas where they may degrade resources.

Many national agricultural research and extension systems have yet to integrate environmental concerns successfully into their agenda. Too little consideration is given to the sustainability features of recommended technologies, to broader aspects of natural resource management, and to the technology and management problems of more fragile rainfed areas where resource degradation is considerable.

Public expenditure on agricultural research has declined in recent years, and there are serious questions about whether the national, regional, and international research systems that serve developing countries have the capacity to rise to the productivity and environmental challenges that lie ahead. During recent years, these systems have been reduced in size and effectiveness through declining financial support, they have limited capacity to capitalize on modern biotechnology research, and many need major institutional reform if they are to respond to the concerns of natural resource management and the needs of rainfed farming.

This decline has not been, and will not be, adequately offset by private sector investment. The private sector makes money from focusing on the problems of commercial farmers and higher-value crops. It has little or no incentive to work on the problems of small-scale farmers growing basic food crops, largely for home consumption, or to address many of the environmental problems in agriculture that have off-site rather than on-site costs. Publicly funded national agricultural research systems must be reformed to address natural resource management problems more effectively within the context of agricultural intensification.

Research at the International Food Policy

Research Institute and elsewhere shows that the economic payoff from public investments in agricultural research is high. On economic grounds alone, a good case can be made for investing more in agricultural research. And this does not consider the social benefits that properly directed agricultural research can generate by helping to reduce poverty and environmental degradation.

World Bank grant support for the Consultative Group for International Agricultural Research grew gradually from $12 million in fiscal 1980 to $45 million in fiscal 1997, with additional grant support provided during the fiscal 1994–95 period. Bank funding of national research systems during the same period fluctuated within a range of between $130 million and $390 million, with no discernible trend.

Conclusions

Past patterns of agricultural growth have sometimes been associated with negative environmental effects and with inequities that have pushed small farmers and landless persons to the margin of society. But this is not an inevitable outcome of agricultural growth. Rather, we believe, it reflects inappropriate economic incentives for managing modern inputs in intensive farming systems, insufficient investment in many resource-poor areas and in social and poverty concerns, and political economy systems that allow the rich and powerful to extract rents by legal or illegal means. The chapters in this volume address these policy issues, institutional approaches, and technical issues. The challenges are great, and the constraints are significant, but much can be done with good governance and a real commitment to environmentally and socially sustainable rural development.

Note

1. The main problem is not with transformations as such. It is that assets are sometimes drawn down or liquidated and the revenues consumed rather than invested in other productive assets.

References

The word "processed" describes informally reproduced works that may not be commonly available through libraries.

Anderson, Jock R., Peter B. R. Hazell, and L. T. Evans. 1987. "Variability of Cereal Yields: Sources of Change and Implications for Agricultural Research and Policy." *Food Policy* 12(3, August):199–212.

Barghouti, Shawki, and Guy Le Moigne. 1991. "Irrigation and the Environmental Challenge." *Finance and Development* 28(June):32–33.

Bishop, Joshua, and Jennifer Allen. 1989. "The On-Site Costs of Soil Erosion in Mali." Environment Working Paper 21. Environment Department, World Bank, Washington, D.C. Processed.

Cuesta, Mauricio, Gerald Carlson, and Ernst Lutz. 1997. "An Empirical Assessment of Farmers' Discount Rates in Costa Rica." Work in Progress Paper. Environment Department, World Bank, Washington, D.C. Processed.

de Moor, A., and P. Calamai. 1997. "Subsidizing Unsustainable Development: Undermining the Earth with Public Funds." Earth Council, San José, Costa Rica. Processed.

Dixon, John, Richard A. Carpenter, Louise Fallon, Paul B. Sherman, and S. Manipomoke. 1992. *Economic Analysis of the Environmental Impacts of Development Projects*. London: Earthscan.

Fan, Shenggen, and Peter B. R. Hazell. 1996. "Should the Indian Government Invest More in Less-Favored Areas?" Environment and Production Technology Division Discussion Paper 33. International Food Policy Research Institute, Washington, D.C. Processed.

Feder, Gershon, T. Unchain, Y. Chalamwong, and C. Hongladaron. 1988. *Land Policies and Farm Productivity in Thailand*. Baltimore, Md.: Johns Hopkins University Press.

Hamilton, Kirk, and Ernst Lutz. 1996. "Green National Accounts: Policy Uses and Empirical Experience." Environment Department Paper 39. Environment Department, World Bank, Washington, D.C. July. Processed.

Hicks, John. 1946. *Value and Capital*, 2d ed. Oxford: Oxford University Press.

Jeyaratnam, J., K. C. Lun, and W. O. Phoon. 1987. "Survey of Acute Pesticide Poisoning among Agricultural Workers in Four Asian Countries." *Bulletin of the World Health Organization* 65(4):521–27.

Lutz, Ernst, ed. 1993. *Toward Improved Accounting for the Environment*. An UNSTAT–World Bank Symposium. Washington, D.C.: World Bank.

Lutz, Ernst, and Mohan Munasinghe. 1994. "Integration of Environmental Concerns into

Economic Analyses of Projects and Policies in an Operational Context." *Ecological Economics* 10:37–46.

Lutz, Ernst, and M. Young. 1992. "Integration of Environmental Concerns into Agricultural Policies of Industrial and Developing Countries." *World Development* 20(2):241–53.

Migot-Adholla, Frank, E. Shem, Peter B. R. Hazell, Benoit Blarel, and Frank Place. 1991. "Indigenous Land Rights Systems in Sub-Saharan Africa: A Constraint on Productivity?" *The World Bank Economic Review* 5(1):155–75.

Milazzo, M. J. 1997. "Reexamining Subsidies in World Fisheries." Office of Sustainable Fisheries, U.S. Department of Commerce, Washington, D.C. Processed.

Ostrom, Elinor. 1994. *Neither Market nor State: Governance of Common-Pool Resources in the Twenty-First Century.* Lecture Series 2. Washington, D.C.: International Food Policy Research Institute.

Pearce, David, Anil Markandya, and Edward Barbier. 1989. *Blueprint for a Green Economy.* London: Earthscan Publications.

Pezzey, John. 1992. "Sustainable Development Concepts: An Economic Analysis." Environment Paper 2. Environment Department, World Bank, Washington, D.C. Processed.

Pingali, Prabhu L. 1992. "Diversifying Asian Rice-Farming Systems: A Deterministic Paradigm." In Shawki Barghouti, L. Garbux, and Dina Umali, eds., *Trends in Agricultural Diversification: Regional Perspectives*, pp. 107–26. World Bank Technical Paper 180. Washington, D.C.: World Bank.

Pingali, Prabhu L., and P. A. Roger, eds. 1995. *Impact of Pesticides on Farmer Health and the Rice Environment.* Norwell, Mass.: Kluwer Academic Publishers; Los Baños, Philippines: International Rice Research Institute.

Pingali, Prabhu L., and M. W. Rosegrant. 1994. *Confronting the Environmental Consequences of the Green Revolution in Asia.* Environment and Production Technology Division Discussion Paper 2. Washington, D.C.: International Food Policy Research Institute.

Pinstrup-Andersen, Per, and R. Pandya-Lorch. 1994. *Alleviating Poverty, Intensifying Agriculture, and Effectively Managing Natural Resources.* Food, Agriculture, and the Environment Discussion Paper 1. Washington, D.C.: International Food Policy Research Institute.

Rasmussen, L. N., and Ruth Meinzen-Dick. 1995. *Local Organizations for Natural Resource Management: Lessons from Theoretical and Empirical Literature.* Environment and Production Technology Division Discussion Paper 11. Washington, D.C.: International Food Policy Research Institute.

Repetto, Robert, William Magrath, Michael Wells, Christine Beer, and Fabrizio Rossini. 1989. *Working Assets: Natural Resources in the National Income Accounts.* Washington, D.C.: World Resources Institute.

Ruchijat, E., and T. Sukmaraganda. 1992. "Rational Integrated Pest Management in Indonesia: Its Successes and Challenges." In P. Ool, G. Lim, T. Ho, P. Manalo, and J. Waage, eds., *Integrated Pest Management in the Asia-Pacific Region.* Oxford: C.A.B International.

Scherr, Sara, and Peter B. R. Hazell. 1994. *Sustainable Agricultural Development Strategies in Fragile Lands.* Environment and Production Technology Division Discussion Paper 1. Washington, D.C.: International Food Policy Research Institute.

Serageldin, Ismail. 1996. *Sustainability and the Wealth of Nations: First Steps in an Ongoing Journey.* Environmentally Sustainable Development Studies and Monographs Series 5. Washington, D.C.: World Bank.

Templeton, Scott R., and Sara Scherr. 1995. *Population Pressure and the Microeconomy of Land Management in Hills and Mountains of Developing Countries.* Environment and Production Technology Division Discussion Paper 26. Washington, D.C.: International Food Policy Research Institute.

Turner, Kerry. 1993. *Sustainable Environmental Economics and Management: Principles and Practice.* London: Belhaven Press.

Umali, Dina L. 1993. *Irrigation-Induced Salinity: A Growing Problem for Development and the Environment.* World Bank Technical Paper 215. Washington, D.C.: World Bank.

Uphoff, Norman. 1986. *Local Institutional Development: An Analytical Sourcebook with Cases.* West Hartford, Conn.: Kumarian Press.

Warford, Jeremy, Mohan Munasinghe, and Wilfrido

Cruz. 1997. *The Greening of Economic Policy Reform.* 2 vols. Washington, D.C.: World Bank.

World Bank. 1997. *Expanding the Measure of Wealth: Indicators of Environmentally Sustainable Development.* Washington, D.C.

World Resources Institute. 1991. *Accounts Overdue: Natural Resource Depreciation in Costa Rica.* Washington, D.C.

3

Policy-Induced Effects of Natural Resource Degradation: The Case of Colombia

John Heath and Hans P. Binswanger

Observers are often struck by the joint occurrence of growing rural populations, rural poverty, and degradation of the natural resource base used by the poor. As a result, population growth and poverty are often seen as the causes of natural resource degradation. Reducing the rates of population growth and migration to urban areas is seen as a possible solution, along with assisting farmers with soil conservation.

Of course, reducing population growth and assisting poor farmers with resource management problems are urgently needed in many countries. In this chapter, however, we argue that many other policy options exist. They consist of the removal of adverse policies that have as their *joint consequence* a reduction in economic efficiency, an increase in poverty, and the degradation of natural resources. Their removal thus would be a win-win-win situation in which the same policy changes could lead to more growth, less poverty, and more sustainable natural resource management.

We start with a discussion of the Boserup hypothesis, which suggests that higher population and market access lead to improvements in—not deterioration of—natural resources. Using recent studies of the Machakos district in Kenya and soil degradation in Ethiopia, we illustrate the dependence of the Boserup effects on the policy environment that governs the investment incentives of farmers.

With this background, we examine growth, the use of land and labor, and the degradation of marginal natural resources in Colombian agriculture. We demonstrate with data that the use of land and labor in Colombia has been driven in highly inefficient directions by a variety of agricultural, land, and rural finance policies and programs. These have prematurely and dramatically reduced employment opportunities in the sector and, as a result, have concentrated poverty in rural areas and increased resource degradation on hillsides and on the Amazon frontier. Binswanger (1989) has described similar phenomena in the Brazilian Amazon. We show that labor policies have not contributed in major ways to these adverse trends but instead that the misallocation of land and labor and an exceptionally high female unemployment rate in rural Colombia are consequences of the same policy factors. We then turn to various policy options for correcting the misallocation of resources, reducing poverty, and relieving the pressure of unsustainable farming on hills and in tropical forest areas with marginal land resources.

The Boserup Effects

Judgments about the sustainability of farming often give too much weight to assumptions about the "carrying capacity" of the resource base (see Cleaver and Schreiber 1994). In fact, there is no clear evidence about what constitutes a threshold level of population pressure. Nevertheless, it is indisputable that population growth influences farming techniques, investment in land, and land use on the toposequence.

As a statement about the nature of the population dynamic in agricultural systems, Boserup's hypothesis is compelling: as land becomes more scarce in relation to labor, and access to markets improves, agriculture is intensified, with the net result being higher agricultural production per unit of area (Boserup 1965). Rather than deteriorating, the base of land resources improves in the process.

The Boserup effects of population growth and improved market access lead to:

- The intensification of land use
- A shift from hand hoes to plows
- The increasing use of organic and inorganic fertilizer
- The shift to integrated crop-livestock systems
- Investment in land and irrigation facilities
- An increase in the use of agricultural labor
- Higher agricultural production per unit of area.

This hypothesis is consistent with much research on the development of farming systems and has received empirical support from an Africa-wide study, a study of erosion in the rangelands of Botswana and Tanzania and of smallholder settlement areas of Zimbabwe, and an in-depth longitudinal study of the Machakos district of Kenya (Ruthenberg 1980, Pingali, Bigot, and Binswanger 1987, and Tiffen, Mortimore, and Gichuki 1994).

Much of the literature on the Boserup effects treats these beneficial relationships as if they were mechanical, brought about by compelling physical and biochemical relationships and by individual utility-maximizing behavior within the constraints set by the natural world. We argue that the Boserup effects are far from automatic. We view them as the outcome of investment decisions made by farmers. In order to come about, the investments require a positive incentive regime, access to soil and water resources, and access to markets. If adverse policy regimes deny these conditions, impoverished peasants are forced to mine, rather than augment, the land resources. This is especially damaging if they only have access to marginal land or to humid tropical forest frontiers.

Various authors have analyzed soil conservation as an investment strategy, with consideration of the economic and policy factors underlying erosion. Anderson and Thampapillai (1990) and Barbier and Bishop (1995) review these factors. In this chapter, the discussion of policy effects is consistent with this tradition.

Before turning to Colombia, we use an African comparison of the workings of the Boserup effects. The Machakos study of Tiffen, Mortimore, and Gichuki (1994) is the most recent of a wealth of studies documenting the Boserup sequences. Machakos is a semiarid district in Kenya with poor to middling agroclimatic conditions. In Kenya, the best land was reserved for white settlers. Colonial policy thereby obliged the native population to derive their food supply from a greatly reduced land base. The natives were forbidden to grow the most remunerative cash crops and therefore lacked incentives from the market. In the 1930s, the region was characterized by heavy soil erosion and declining yields. In contrast, today Machakos supports a population almost six times as large as it was in 1932, and agricultural output per unit of area (in constant maize units) has increased almost tenfold. Crop yields have risen. Cash crops and horticultural products have been successfully introduced. There are more trees, more soil conservation works, and greater use of organic manure now than in the 1930s.

The investment incentives needed to bring about these effects are associated with the following conditions, which have been assured since independence:

- An agricultural policy that, compared with that of other African countries, taxes the sector lightly
- Access to international markets for coffee and domestic markets for other cash crops
- Construction of infrastructure associated with rural development projects
- Access to nonfarm and urban employment opportunities in Nairobi
- Ability to finance investments from sales revenues and labor incomes
- Security of tenure, provided initially by the traditional communal tenure and later via land titles
- New food production technology, especially for maize
- Locally adapted soil conservation technology and farmer-led initiatives to implement it.

Recent cross-sectional studies of Ethiopia, in contrast, find that areas where population density significantly exceeds carrying capacity have high indexes of soil degradation (Grepperud 1994).[1] Longitudinal studies similar to those undertaken in Machakos would be required to prove that these areas have not experienced the beneficial sequence of Boserup effects. Nevertheless, a comparison of policies and programs prevailing in the two countries over the past 30 to 40 years is instructive.

Ethiopian farmers were heavily taxed throughout the period via a great diversity of methods. The construction of rural infrastructure was limited. Access to international markets and even to domestic markets was often disrupted. Employment opportunities in the rural nonfarm and urban economies were extremely limited by the lack of agricultural and economywide growth. Periodic famines depleted assets and further undermined the peasant's ability to mobilize investment resources.

There was no security of tenure during the period. Under Haile Selassie, farmers were tenants at will rather than holders of secure ownership or usufruct rights. Under the communist regime, the state owned the land, usufruct rights were never secure, and peasants were dislocated as a result of villagization programs and forced migration and were pressured to join collectives. Improved technology in food grains was scarce. Very few programs were aimed at developing and disseminating improved soil conservation techniques among peasant farmers.

The sharp contrast in policy regimes and natural resource outcomes between Kenya and Ethiopia shows that Boserup effects are not an automatic response to population growth or market access. Instead they require a policy and institutional regime that provides peasants and commercial farmers with favorable incentives as well as opportunities to earn income on and off the farm to enable farmers to invest in production inputs, technology, and conservation activities. With this background, we now turn to Colombia, our main example of the impact of policy on land degradation and poverty.

The Case of Colombia

In Colombia, unsustainable farming of the Andean slopes has long been recognized as a problem.[2] In 1950 Lauchlin Currie led a World Bank mission to Colombia. This mission "noted for the first time on a national scale the extent to which flat, apparently rich, bottomlands were occupied by low-intensity livestock ranching estates, while slopes steep enough to make cultivation a hazard to life were occupied for crop farming" (Blakemore and Smith 1971: 232; see also Currie 1965). The situation has not improved since then. In many areas, the Andean slopes are being denuded of vegetative and soil cover, and the resulting loss of moisture retention is having an adverse effect on the flow of streams and reducing the availability of water for agriculture, both for poor farmers on the slopes and for richer farmers located in the valley bottoms. In the scattered indigenous reserves (what remains from the colonial *resguardos*), there is an acute problem of fragmented holdings: deprived of access to land elsewhere, indigenous farmers are carving up their land into *microfundias*. In the Amazon and Orinoco basins, and on the Pacific coast, pressure is rising to put unstable lands of intrinsically limited fertility under annual crops.

According to Currie (1965), 40 years ago the problem was caused by too many poor, inefficient farmers working on the slopes. The solution—both to rural poverty and to resource degradation—was to encourage migration of the surplus rural population to the towns, leaving the land to be worked by fewer, more technically sophisticated farmers. More than 40 years later, the population of poor people in rural Colombia has more than doubled, despite massive rural-urban migration (World Bank 1995a). Three out of four poor people in Colombia now reside in rural areas. Poor farmers are still working the slopes.

Recently the natural rate of increase of the rural population has slowed substantially, helping to raise wages and reduce the absolute level of pressure on the resource base. Since 1985 the rural population has begun to decline for the first time in absolute terms. But the overall pattern of growth has failed to absorb what Currie perceived as "surplus rural labor." Poor farmers continue to have limited access to good land and therefore continue to exert pressure on the slopes (and on equally fragile land in the Amazon-Orinoco basin).

Are there other solutions? Improving the techniques that hillside farmers use on marginal land may be part of the story. However, although low-cost technologies exist, little progress has been made in diffusing them, largely because the incentives are lacking: on marginal lands small farmers may have insufficient incentives to invest in soil conservation (Ashby 1985). Even if better, more cost-effective techniques for hillside farming are developed, they will be a palliative at best. To devise an appropriate solution, it is necessary to look beyond the hillsides. The challenge is to improve the rural poor's access to less fragile, more fertile lands at lower elevations and to extend the opportunities for off-farm rural employment. We show here that it is necessary to change the agricultural policy framework that influences the operation of land and labor markets.

Currie's prescription of accelerating rural-urban migration and turning farming over to large-scale, modernized commercial farmers has now been pursued in Colombia for more than 40 years. It has not reduced rural poverty significantly but rather has further concentrated the poor in the countryside. It has also led to an extremely low use of both land and labor in this country and to low overall productivity of the agricultural sector compared with its enormous potential. The urban migration approach to solving rural poverty and resource degradation has been a complete failure.

The Low Propensity of Colombian Agriculture to Absorb Labor

Agricultural development in Colombia has involved substantial misallocation of resources: land has been very unevenly exploited, and the growth of farm output has absorbed less labor than might have been expected. Over the past 40 years or so, compared with other countries in the same range of per capita income, the growth of farm employment has been exceptionally low in Colombia (Misión de Estudios del Sector Agropecuario 1990, which extends earlier work by Syrquin and Chenery 1989). Between 1925 and mid-century, by international standards, the decline in Colombian agriculture's share of the gross domestic product (GDP) and its share of the labor force was comparable to that of other countries in the same range of per capita income. But from the early 1950s onward, this picture changed abruptly: the relative importance of farm employment fell off at a much faster rate than projected on the basis of international evidence concerning the correlation between employment and GDP shares (see figure 3.1).

In Colombia, the growth path of agriculture has been extremely capital- and labor-intensive. Between 1950 and 1987, agriculture's annual growth rate averaged an impressive 3.5 percent. Capital inputs to agriculture grew at an average annual rate of 2.8 percent, land area devoted to agriculture and livestock grew 1.4 percent, and employment grew only 0.6 percent (see table 3.1).

The sector's relatively low propensity to absorb labor is reflected in the pattern of land use. Crop farming has captured a relatively small proportion of the natural resource base: 16 percent of Colombia's land area is suitable for crops, but less than 4 percent is cultivated. Livestock rearing is overextended: 13 percent of the territory is deemed appropriate for pasture, but 35 percent is put to this use. Small farmers with limited access to good-quality flatland may end up deforesting marginal land on the Andean slopes. Only about half of Colombia's agricultural and forestry lands is still forested, whereas two-thirds are only suitable for forestry and should be left under tree cover (IGAC 1988).

Agricultural policymaking in Colombia shows substantial elements of "large-farmer bias." First, the patterns of public investment and orientation of the trade regime have combined to favor livestock and grain crops, neither of which uses labor intensively. Second, credit policies have tended to discriminate against small farmers. Only about one-third of small farmers are able to obtain a loan from the formal sector. There is no sign that past practices favoring large-scale farms in credit policies are being corrected.[3]

Figure 3.1. Share of Primary Sector in Total Employment in Colombia, 1925–86

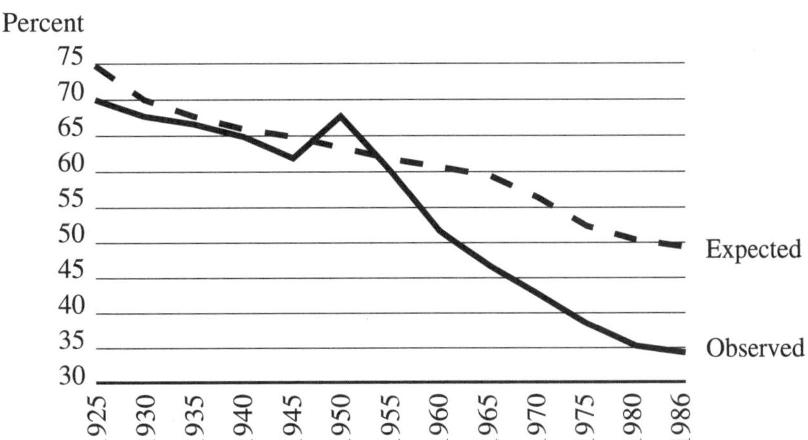

Source: Misión de Estudios del Sector Agropecuario 1990. These findings extend earlier work by Syrquin and Chenery 1989.

Table 3.1. Sources of Farm Sector Growth in Colombia, 1950–87

	Growth rates				*Contributions to growth*				
Period	*Farm GDP*	*Area*	*Capital*	*Employment*	*Area*	*Capital*	*Employment*	*All factors*	*Productivity*
1950–55	3.03	1.15	–0.37	2.09	0.22	–0.13	0.96	1.05	1.98
1955–60	4.08	0.68	1.26	0.21	0.13	0.44	0.10	0.67	3.41
1960–65	2.77	1.31	2.44	0.67	0.22	0.81	0.34	1.36	1.41
1965–70	4.94	2.17	4.68	1.29	0.41	1.87	0.53	2.81	2.13
1970–75	4.33	2.03	6.51	–3.91	0.38	2.60	–1.60	1.39	2.94
1975–80	4.58	1.67	4.96	2.75	0.28	2.18	1.05	3.51	1.07
1980–84	0.89	1.80	0.93	0.66	0.27	0.45	0.24	0.96	–0.07
1984–87	3.53	0.48	1.97	0.76	0.07	0.95	0.28	1.30	2.23
Average, 1950–87	3.52	1.41	2.80	0.57	0.25	1.15	0.24	1.63	1.89

Source: World Bank 1995b.

Third, tax policies have converted agriculture into a tax shelter for both income and capital gains taxation, providing incentives for holding land as a tax shelter rather than for agricultural production.

By subsidizing capital inputs and the livestock sector, the Colombian government has sponsored a strategy of agricultural development that discriminates against small farmers. Tractor subsidies have encouraged shedding of labor: already, by the early 1970s, one-quarter of the cultivated area was mechanized. Public irrigation schemes have tended to favor larger rather than small producers; and, because much of the land in the irrigation districts is devoted to pasture and grain crops—which use less labor than higher-margin crops—employment generation has been much less significant than it might have been.

Cattle rearing spread rapidly in the immediate postwar period. The area under pasture grew from 12.1 million hectares in 1950 to 26.7 million hectares in 1986. (In the same period, the area under crops increased from 2.6 to 4.3 hectares). The Caribbean region accounted for 38 percent of the growth in pastures, while the Andean valley bottoms accounted for a further 30 percent. There was also rapid growth of the pastoral area in the eastern savanna (centered on Meta): by the mid-1980s, no less than half of all cattle pasture was located in the sparsely populated area east of the Andean ranges (Misión de Estudios del Sector Agropecuario 1990).

The livestock sector has been more protected than the crop sector, helping to account for its rapid expansion. Between 1980 and 1992, beef and milk absorbed 82 percent of total support (price plus nonprice interventions) that the government conferred on a group of nine farm commodities (World Bank, LACTAD 1994). Expansion of cattle rearing has been land-extensive, favoring the creation of large estates: the rate of growth of the herd has only slightly exceeded the rate of growth of the area in pasture. The off-take rate remains very low—equal to 50 or 60 percent of that achieved in Argentina and the United States. The extensive livestock sector occupies large amounts of arable land on "lower slopes," which could be developed for intensive agriculture by the construction of road infrastructure. Much of these lands will also need drainage. The large estates favored by policy simply do not have the labor needed to make these investments.

Narrow access to land and employment in rural areas has increased the propensity for violence. Although narrow access has not been the only cause of rural violence, it has contributed significantly to Colombia's extremely violent history over the past half century. The propensity for violence has a negative feedback on investment and employment. Rural insecurity probably reduces the incentive to invest in agriculture and, more important, skews the pattern of investment toward activities that are relatively nonintensive in the use of labor. The supervision costs and incentive losses that are typically associated with hired labor are even greater in an environment where hired workers may initiate, or collude in, violent reprisals against employers. This disincentive may encourage landowners in areas of arable potential to invest in livestock rather than crops; and it may reinforce the tendency for larger irrigated holdings to be placed under pasture or grain crops rather than high-margin crops that require careful supervision and are intensive in the use of hired labor.

The Negative Impact of Past Agricultural and Land Policies

A number of policy interventions have circumscribed the poor's access to land. First, the scope for tenancy has been reduced. Beginning with Law 200 of 1936, and culminating with the Ley de Aparcería of 1975, a series of legal measures have had the effect—intentional or otherwise—of reducing the incentive for large landowners to lease out land to tenants (see table 3.2). The right of landowners to employ sharecroppers, other tenants, and *colonos* was formally outlawed by Law 1a of 1968. This led to a precipitous fall in the use of sharecroppers and *colonos*, although they had not completely disappeared even by 1988. A less dramatic reduction occurred in the number of other tenants. These trends occurred even while agricultural area expanded in the country as a whole. Much of the decline in sharecropping centered on the coffee-growing region (associated with an increase in the importance of small owner-operated farms) and on areas of the Caribbean lowlands, following clearance of the land for cattle rearing.

The steepest fall in rural employment—3.9 percent a year—occurred between 1970 and 1975. This suggests that the 1968 law was the critical event and that, by the time the ban on sharecropping was reaffirmed in 1975, most of the tenants and squatters had already been expelled.

The negative impact on employment that resulted from closing the tenancy option could potentially have been offset by redistributing land toward smaller farmers. But colonization and redistribution by the land reform agency, INCORA, have done little to change the overall shape of the agrarian structure. Since its inception in 1961, INCORA has redistributed relatively little land in areas of established settlement, concentrating its efforts on the colonization and titling of frontier lands. Because land on the frontier has tended not to be handed out in small parcels, colonization initiatives on this agricultural margin have failed to alter the overall picture of concentration. Between 1960 and 1988, the area occupied by holdings under 5 hectares declined from 6 to 5 percent, the area in medium-size farms (5–50 hectares) rose from 24 to 26 percent, and the area in larger farms (more than 50 hectares) fell from 70 to 69 percent.

The lack of tenancy and land reform options encouraged land invasion. During the 1970s, there was a wave of illegal farm occupations, affecting 1,500–2,000 farms and roughly two-thirds of departments. From this point on, INCORA's work tended to center on regularizing the claims of illegal invaders. But even this avenue for land acquisition was closed: Law 30 of 1988 banned INCORA from acquiring illegally occupied land.

The literature on farm size and productivity has established that factor productivity on farms operated primarily with family labor is typically larger than that of larger farms operated primarily with hired labor or tenants. Around 1970, the situation in Colombia was consistent with this generalization (see Berry and Cline 1979). Moreover, for any given level of resource use and output, the small family-based sector generates much more employment than the large-scale sector. When there is technical change, the optimal size of family-based operations tends to increase. But once the size of farms has adjusted to the new technologies, the typical negative relation-

Table 3.2. Changes in Land Tenure in Colombia, 1960–88
(thousands of hectares unless otherwise noted)

Tenure	*1960*	*1988*	*Growth (percent)*
Owners	18,995	29,117	53.3
Sharecroppers	1,100	273	–75.2
Other tenants	1,231	829	–32.7
Colonos[a]	2,889	554	–80.8
Other[b]	526	1,123	113.5

a. Occupants of untitled land (internal and external frontier).
b. Includes squatters.
Source: World Bank 1995b.

ship reappears.[4]

Between 1976 and 1988, the yields of small farmers increased on average by 82 percent, compared with an increase of only 2 percent for medium and larger farmers. Much of this small-farmer yield response was based on the adoption of scale-neutral input packages (primarily involving agrochemicals) successfully promoted by the extension component of the integrated rural development program that was launched in 1976. The yield response is striking given that the volume of public resources channeled to the small-farm sector was small in relation to the volume garnered by larger farms. The level and speed of the response by small farmers suggest that the relationship between farm size and productivity observed for the 1960s and early 1970s continues to hold.

This raises an important question. If small family farms are more efficient than large farms relying on hired labor, why do large farmers not find it more profitable to subdivide their properties and rent or sell these parcels to smaller farmers? The restriction on *leasing* has already been examined. Large farms in Colombia generally are not parceled out and *sold* to small farmers because the market price of agricultural land typically exceeds the capitalized value of farm profits. This occurs because the value of farmland is only partly based on its agricultural potential. In all areas, land serves as a hedge against inflation. Its immobility makes it a preferred form of collateral in credit markets, conferring additional utility, particularly where production risk cannot be insured. In periurban areas, land holds out the prospect of higher returns from real estate development than from farming. Finally, credit subsidies and tax write-offs are likely to be capitalized into land values (Binswanger, Deininger, and Feder 1995). Many of these nonfarming benefits do not apply to small farmers, who rarely have access to subsidized credit and do not benefit from the value of landownership as a tax shelter.

If the market price of land exceeds the capitalized value of farm profits, a poor smallholder or landless worker will not be able to finance the purchase of land out of farm profits, even if the owner, or a mortgage bank, is willing to advance him or her a loan covering the full purchase price of the land. This means that purchase will only be feasible if the productivity differential between small and large farms is huge, if recourse is made to nonfarm income, or if the purchaser is willing to exploit unpaid family labor, devoting the imputed labor earnings to the purchase of land.

This hypothesis is borne out by a study of 15 *municipios* located in distinct agricultural and agrarian systems in Colombia (Suárez and others 1993). The study demonstrates that the market for land is highly segmented, failing to transfer land from large to small farmers. There is a very active land market among large landowners, more active indeed than in the United States or other countries.[5] There is also an active market for the sale of small parcels between smallholders. But there are few transactions between the two groups. However, the evidence presented in this study may not be statistically representative of the different regions covered. Thus, there are good grounds for making a closer, more precise evaluation of this phenomenon of segmentation.

One serious flaw of past approaches to land reform in Colombia and elsewhere was failure to recognize that the market segmentation and inflated prices of land are themselves policy consequences. They are affected by macroeconomic conditions, nonfarm investment options, tax policies, and agricultural policies (for a consideration of these issues in another context, see Binswanger 1989). Correcting the maldistribution of land in Colombia and elsewhere must start by eliminating these distortions. Doing so would enable land markets to work better and would make any land reform program more affordable to governments and beneficiaries. In Colombia, for example, improving the collection of agricultural income taxes and capital gains taxes on farmland would reduce the attractiveness of land as a tax shelter, therefore tending to lower its price.

A further problem is that past agricultural credit policies and recent credit initiatives are not conducive either to holding down the price of land or to bringing more land onto the market. The agricultural investment incentives recently adopted (Law 101 of 1993) will mainly be captured by large farmers, giving them a competitive edge in relation to small farmers. Past instances of blanket rescheduling of credit have helped to keep alive inefficient large farms that could fruitfully be parceled out and sold off to small farmer groups. A better approach would be to provide financial assistance for restructuring the assets and liabilities of potentially viable large farms, coupled with financial incentives for the owners of nonviable farms to leave the agricultural sector, making their land available for purchase by small farmers.

An important factor tending to push up the price of land, one that is specific to Colombia, is the laundering of money from drug trafficking. It is estimated that, in the late 1980s, traffickers were investing between 8 and 23 percent of their cocaine revenues in the purchase of land and had accumulated nearly 1 million hectares (equivalent to 3 percent of the farmed area). Most of this land was located in the departments of Córdoba, northern Antioquia, Caquetá, Meta, Sucre, Atlántico, and Casanare and was primarily used for cattle rearing (Bejerano 1988). One way to tackle this problem would be to place the onus on would-be buyers of large tracts to demonstrate that the cash used to buy land was obtained by licit means; failure to provide satisfactory proof would result in an embargo on the sale of land.

Labor Policies

To what extent have inflexible and distorted labor markets contributed to the misallocation of land and labor in Colombia and therefore aggravated the deterioration of resources on slopes and at the frontiers of the rain forest? Labor markets may be said to be efficient when there are signs of integration between urban and rural sectors and between different rural localities. Integration suggests a trend toward lowering the wage differentials between localities and sectors for workers with comparable skills.

The urban-rural wage differential in Colombia is low: wages in construction are only about 10 percent higher than those in agriculture. The recent social security reform will probably tend to push up the cost of hiring workers in the urban formal sector. This will reduce the take-up from the informal sector, a trend that may induce a greater slackness in the urban labor market, tending to keep wages for unskilled work close to those prevailing in the rural sector.

Interregional rural to rural migration has also been substantial, and there is little evidence of regional segmentation of rural labor markets. In addition, a substantial amount of temporary interregional migration is associated with the contracting of casual labor for harvesting sugarcane, coffee, cotton, and other crops, but this involves men much more than women. As a consequence, male rural unemployment rates are very low, on the order of 2 percent.

In contrast, rural female unemployment rates are very high. According to a 1993 household survey, in the countryside, 12 percent of women looking for work were unable to find it compared with only 2 percent of men (results of the CASEN survey are reported in Reyes and Martínez 1994). Rural unemployment was particularly acute among young women: 25 percent of women under 25 years of age were unable to find work. In towns, the gender disparity was much narrower: 6 percent of men and 9 percent of women were unemployed. The rural male-female unemployment differential is also extremely high by international standards.

There is a long tradition of women working as day laborers (in the coffee harvest). Also, between 1973 and 1985, the share of women in the economically active population of rural areas rose from 14 to 32 percent (Berry 1992). However, females are less able to respond with alacrity to demand for short-term workers, because they often must care for young children, which tends to reduce their mobility. Although women have less propensity than men to remain in rural areas, among those who do remain the rate of unemployment is much higher than it is for men.

The particularly high unemployment rate for rural women (but not for men) is not a consequence of the low labor absorption of Colombian agriculture, which affects both men and women. It is partly a consequence of social norms that discourage participation in off-farm work, except in specific operations such as the coffee harvest. However, the unemployment of women may be aggravated by inadequate access to land. This hits relatively immobile women more severely than more mobile males. The usual response to slack labor demand is to increase the use of unpaid family labor on the farm. For families with insufficient farmland, either rented or owned, this option does not exist.

This brief review suggests that rural labor markets or labor market policies are not a major cause of the misallocation of land and labor in Colombia. Instead, the disquieting feature of extremely high female unemployment may be a consequence of agricultural and land market policies that cause low labor absorption and inefficient land use.

Temporary Shocks in Rural Labor Markets

Although the rural labor market appears reasonably flexible, temporary shocks to the farm sector can still have major adverse effects on employment. In 1992–93, the combination of severe drought and a sharp downturn in world prices seriously depressed the sector. The employment impact was probably greater than it would have been if the rural demographic transition had already been fully negotiated. Although the rural population has declined in absolute as well as relative terms since 1985, the population of working age increased at an annual average of 2.9 percent between 1985 and 1990, much higher than in previous periods. This reflects the relatively high rural birth rates of the late 1960s.

Between 1990 and 1993, the number employed in agriculture fell from 2.3 million to 2.2 million. The loss of agricultural employment (54,000 jobs lost) was more than offset by the growth of employment in the "rural nonfarm" sector (163,000 jobs created; see Reyes and Martínez 1994). Within the farm sector, employment trends varied significantly by type of crop. Between 1990 and 1993, nonperennial crops contracted 19 percent, most of this related to the declining profitability of cotton, rice, and vegetables (and, to a lesser extent, maize, sorghum, soybeans, and wheat). Coffee contracted 2.4 percent. Other perennials grew 9 percent: falls in the employment generated by cassava, *panela*, and cocoa were more than offset by increases in bananas, plantains, oil palm, and sugarcane (this refers to changes in the number of man-days—*jornales*—worked in each crop per year; see Suárez and others 1993).

The impact of contraction would have been felt most strongly by households that depend significantly on rural wage incomes. In 1993, 57 percent of rural household heads had worked for a wage, and wage earnings made up 46 percent of rural household income. A larger proportion of the rural labor force was self-employed (58 percent), but incomes from these activities accounted for only 37 percent of rural household income. Wage employment was a much more significant source of income in the Caribbean region than in other rural areas (Reyes and Martínez 1994).

This brief overview suggests that labor market rigidities are not a major problem in Colombia. However, there is an urgent need to find adequate ways of responding to sharp downturns in rural labor demand occasioned by the periodic agricultural crises associated with import surges and price slumps. The second issue is how to ease female unemployment. Policy interventions could usefully stress the creation of an enabling environment for rural nonfarm enterprises, including broader opportunities for female employment, and the promotion of public works programs to alleviate temporary downturns in the agricultural demand for labor.

The Case for Land Reform

Government interventions to drive up the demand for labor in periods of a temporary downturn will do little to cure the fundamental misallocation of land and labor in Colombian agriculture. Nor will interventions to improve the productivity of hillside agriculture of poor farmers be sufficient. More radical approaches are required. A strong case can be made for reorienting and revitalizing the Colombian land reform. Three reasons are paramount. First, migration away from the countryside has not—contrary to the initial expectations of Colombian policymakers—eliminated rural poverty, nor has it reversed the tendency of small farmers to cultivate steep slopes in an unsustainable manner. Small farmers need to get access to some of the more fertile bottomland currently used for extensive livestock grazing. Second, putting land into the hands of the rural poor will help to defuse one of the causes of rural violence and may help to promote investment in agriculture. Third, the evidence suggests that small farmers in Colombia have high productivity if given access to land, markets, and agricultural services. They are capable of making significant productivity gains. In other words, putting more land in the hands of small farmers is likely to generate efficiency as well as equity gains, making Colombian agriculture more competitive.

Land reform is a highly appropriate response to this problem of segmentation. An alternative would be to create incentives for larger estates to hire in more labor—for example, by reducing capital investment and livestock subsidies. But rural violence would still reduce the propensity for large farmers to hire labor, and, even in the absence of violence, the supervision costs associated with using hired labor would make these estates less efficient than small farms using family labor. Therefore, there are good grounds for facilitating the subdivision and sale of estates to small farmers. Leasing to small farmers could also be encouraged, but incentive problems will make these farms somewhat less efficient than small owner-operated farms.

Land reform will only work if complementary measures are designed to reduce the "nonfarm" incentives for holding large estates—measures to reduce the wedge between the price of land and the capitalized value of farm profits.

In addition, INCORA, the parastatal land reform agency, has been inefficient in carrying out much land reform. This ineffectiveness can be attributed to several causes:

- The major episode of redistributive land reform occurred in the 1960s, initiated at the end of the civil war known as *La Violencia* in 1958. After that initial spurt, budget allocations for the purpose of land reform dried up.

- Land acquisition was based on expropriation of insufficiently used farms, a divisive process fraught with endless litigation. As a consequence, INCORA was a slow and poor buyer of land, acquiring mostly already invaded and marginal lands, often at inflated prices.
- Selection of land reform beneficiaries, land allocation, and farm development were carried out by a centralized bureaucracy in paternalistic ways. Beneficiaries received incomplete rights to land and poor services.
- INCORA continued to allocate frontier land to favored individuals in large units, rather than as family-farm units, thus reproducing the unfavorable agrarian structure at the frontier. Its settlements of smallholders on the frontier suffered from centralized and paternalistic bureaucratic approaches similar to those of its land reform program.

Given this less-than-encouraging record, future land reform efforts clearly must be based on radically new approaches that are cheaper, faster, more decentralized, and more participatory.

The government of Colombia has recently taken steps to revitalize land reform and to change how it is administered. It has introduced new legislation (Law 160 of 1994) that proposes to use the land market for land reform.[6] Beneficiaries buy land directly from large farmers, using grant and loan resources provided by the government. Land allocation on the frontier is restricted to family-farm plots for poorer groups. The legislation also decentralizes land reform administration and makes it much more participatory. Thus the legislation has many positive features. However, little consideration has so far been given to the policy issues that affect the efficiency with which land markets work. Next, we briefly discuss the strengths and shortcomings of the new legislation.

Evaluating Law 160

The new law provides for the poor to be given grants to buy land. Persons deemed eligible for grants are allowed to identify land that they want to buy and to negotiate a price with the owner. This marks a clear break from the previous arrangement whereby INCORA bought land and transferred it to beneficiaries. The centerpiece of the new initiative is the provision for direct negotiation between buyer and seller: INCORA facilitates these negotiations but does not buy land, except as a last resort. INCORA may compulsorily purchase land (a) on behalf of indigenous or other designated target groups, (b) to protect the environment, and (c) when beneficiaries are unable to reach agreement with landowners who previously had agreed to sell to them.

The grant for land purchase amounts to 70 percent of the cost of buying a family-farm holding; beneficiaries cover the remaining 30 percent of the purchase price through loans that INCORA helps them to negotiate. A substantial grant element seems justified in order to offset the wedge between the market price of land and the capitalized value of farm profits. The size of the wedge is a function of the severity of policy distortions; complementary measures are needed to reduce these distortions. Under the new arrangement, the loan component is not so large as to saddle beneficiaries with a high debt-asset ratio that would compromise their ability to secure credit for working capital.

However, a 70 percent subsidy may be too large. First, it entails a major fiscal cost, which the state may be poorly placed to bear. Second, it could be argued that if the policy distortions are removed, a smaller purchase subsidy would be needed. Also, the size of the subsidy may encourage persons to apply for it, even when they have no intention of continuing to farm.

In the past, the agrarian reform ceded ownership of large tracts of public lands on the frontier, thus failing to redress the concentration of rural property. Under the new law, land on the frontier is parceled out in family-farm units. Persons whose net worth exceeds the Col$1,000 minimum wage will not qualify for these homesteads; neither will agrarian reform officials. These provisions represent a significant break with the past, holding out the prospect of a more equitable distribution of land on the frontier.

There are a number of weaknesses in the new law. First, holdings that are sold off under the auspices of the land reform may not be sold or rented for 12 years after the transfer: this moratorium seems excessive. The stipulation is unwise because it is unlikely to be enforceable. To the extent that it is enforced, it will needlessly restrict the flexibility of land and labor markets. Some of the newly created farm enterprises will inevitably fail, and farmers who fail must be allowed to exit.

Second, it is not clear that government needs to make the award of grants conditional on purchase of a "family-farm-size" holding or to restrict the scope for subsequent subdivision of such holdings. The proposal denies the option to the poor of buying a "subfamily farm" and supplementing farm revenue with income from off-farm sources. Arguably, there is no need to fix a minimum acceptable holding size. INCORA could simply pay 70 percent of the cost of land purchase up to a maximum amount rather than fix a target farm size. This would incidentally remove the need for time-consuming and costly surveys to determine, for each locality, what area of land is needed to support a farm family.

Third, the present law reopens the possibility of regularizing the tenure status of persons who have invaded land. This is valid: past invasions need to be resolved. But, in regulating the law, it will be important to create incentives to acquire land through the market rather than invasion, to ensure that title is granted more quickly if land is not invaded, and to use the provision of infrastructure as leverage—concentrating investment on areas where land has been transferred through market negotiation, not invasion.

Fourth, private property rights are deemed to lapse if land is abandoned, unsustainably exploited, or used for illicit purposes (cultivation of narcotics). The new law indicates, for example, that "mere tree felling" is not a sufficient indicator of "adequate exploitation." However, there is much room for discretionary interpretation of what constitutes proper use of the land. The best way to get around this problem is to remove the policy distortions that encourage people to hang on to large tracts of land: changing the incentive structure will work better than trying to regulate land use.

Fifth, the new law makes no attempt to alleviate the severe restrictions on leasing. These restrictions entail substantial equity and efficiency losses and have resulted in so much premature eviction of labor from the sector. Opposition to leasing is based on the perception that, in the past, such arrangements (particularly sharecropping) were exploitative. Evidence from around the world suggests that sharecropping is often a way for differently endowed enterprises to pool their resources to mutual benefit, overcoming credit constraints and helping to manage risk. Safeguards could be adopted to protect small farmers from the possibility of exploitation. Minimum standards for tenancy contracts may be appropriate. These contracts could be enforced by municipal arbitration councils whose judgment would, in the last resort, allow for appeal through the courts. Arbitration councils would provide statutory representation to all parties involved in disputes concerning labor, tenancy, mortgage, and land sales contracts.

Although the law promises a new wave of land reform, significant policy problems in the law itself, and in other agricultural policies that favor large-scale farms, will have to be addressed. Only then will land reform's potential to reduce both poverty and land degradation be fully realized.

Conclusions

This article has argued that rural poverty, inefficient resource allocation, and natural resource degradation are joint phenomena, often induced by a common nexus of policy failures that favor the modernization of large-scale farming at the expense of more efficient and employment-intensive family farms. The policy nexus has as an additional adverse consequence the concentration of impoverished populations with few investment resources on marginal lands, at tropical forest frontiers, and on erodible hillsides. Eliminating the adverse policy nexus should be given higher priority than reducing population growth, relocating the impoverished population to the cities, or improving farming techniques on marginal lands.

Of course, eliminating the privileges of rural elites who are embedded in these policies is an extremely difficult task. These policies did not come about by accident, but as the consequence of a historical evolution that involved bargaining among politically strong groups who protected their interests at each point in time. Analyzing the joint consequences of the policies may help in the reform effort but is unlikely to be sufficient. Peasants have been poorly articulated in the Colombian political process, and steps must be taken to provide them with a greater input into policymaking.

There are very few checks and balances when it comes to agricultural policymaking. The power of large farmers—articulated through the commodity associations (*gremios*) and through congress—is in no way offset by countervailing peasant or consumer lobbies. It is particularly striking that government gives peasant organizations very little scope to intervene in the policy dialogue. The Ministry of Agriculture is dominated by the *gremios* and by congress and spends much of its time responding to demands from commercial farmers—primarily, producers of importable goods—for greater protection and subsidies. Policymaking, therefore, tends to focus on short-term issues of crisis management; there is little pressure on the government to address issues of long-term strategy or poverty reduction. Unless some political actors see that it is in their interest to bring poor peasants more strongly into the policy process, land and policy reforms will continue to be a slow process.

Notes

This chapter was originally published in 1996 as John Heath and Hans Binswanger, "Natural Resource Degradation Effects of Poverty and Population Growth Are Largely Policy-Induced: The Case of Colombia," *Environment and Development Economics* 1(1):65–83. The authors are grateful to Cambridge University Press for permission to reprint this edited version.

1. Specifically, Grepperud (1994) states that "This paper tests the population pressure hypothesis for the Ethiopian highlands using quantitative methods. The hypothesis posits that, under comparable physical conditions, heavily eroded areas occur in highly populated regions. A soil erosion severity index, a proxy variable for soil erosion, was chosen as the dependent variable. Because the dependent variable is categorical and ordinal, an ordinal cumulative logit model was chosen for the analysis. Two alternative variables were applied to reflect population pressure: the ratio of rural population to arable land and the ratio of the population-support capacity to rural population. For the first ratio, the effect on the level of soil erosion seemed linear and weak, because substantial increases in population density were needed to increase the odds of an area being classified as more seriously eroded. For the second ratio, a reciprocal transformation improved the explanatory power of the model, suggesting a hyperbolic relation between the ratio and the soil erosion index. The probability of an area being classified above any given level of soil erosion increases rapidly as a rural population exceeds the population-supporting capacity of its region."

2. Taking Mexico, Central America, the Andean countries, and the Caribbean as a whole, Posner (1981) estimates that 30 percent of peasant households farm steep slopes, generating 20–40 percent of the output of annual crops.

3. Law 101 of 1993 offers a mandate for continuing credit subsidies and low interest rates for agriculture indefinitely: the policies do not facilitate small farmer access to credit, most of the subsidies being captured by larger farmers. Because it is mostly large farmers who qualify for working capital loans, they will reap the lion's share of the subsidy; the law implicitly discriminates against small farmers. Chronic farmer indebtedness is a major problem, one that was exacerbated by the drought and downturn in world prices of 1992. There has been very little success in recovering overdue loans from large producers who have borrowed money from Caja Agraria, the state-owned agricultural credit agency: of the total value of overdue loans (Col$150 billion), half is attributable to large producers; most of the money recovered has come from small producers. Refinancing schemes have been poorly targeted and may have helped to postpone necessary adjustment in the sector, setting a precedent that might discourage larger producers form repaying loans.

4. Recent studies confirm the earlier findings and provide a theoretical rationale for the observed relationship based on incentives issues and missing markets. This literature is summarized by Binswanger, Deininger, and Feder (1995).

5. In Colombia, about 5 percent of farmland was sold in 1990 and 1991 (Suárez and others 1993). The percentage of farmland transferred on average each year is 3 percent of the total in the United States, 1.5 percent in Great Britain and in the white sector of South Africa, and 0.5 percent in Ireland and Kenya (Moll 1988: 354).

6. Between 1995 and 1998, the intention is to facilitate the purchase of 1 million hectares by 69,900 families and to title 5 million hectares of frontier land, benefiting 178,600 families. The total number of beneficiaries is estimated at 1.2 million, or 13 percent of the rural population.

References

The word "processed" describes informally reproduced works that may not be commonly available through libraries.

Anderson, Jock R., and Jesuthason Thampapillai. 1990. *Soil Conservation in Developing Countries: Project and Policy Intervention.* Policy and Research Series 8. Washington, D.C.: World Bank.

Ashby, J. A. 1985. "The Social Ecology of Soil Erosion in a Colombian Farming System." *Rural Sociology* 50(3):377–96.

Barbier, Edward B., and Joshua T. Bishop. 1995. "Economic Values and Incentives Affecting Soil and Water Conservation in Developing Countries." *Journal of Soil and Water Conservation* 50(2, March-April):133–37.

Bejerano, J. A. 1988. "Efectos de la violencia en la producción agropecuaria." *Coyuntura Económica* 18(3, September):25–45.

Berry, R. A. 1992. "Agriculture during the Eighties' Recession in Colombia." In A. Cohen and F. R. Gunter, eds., *The Colombian Economy: Issues of Trade and Development.* Boulder, Colo.: Westview Press.

Berry, A., and W. Cline. 1979. *Agrarian Structure and Productivity in Developing Countries.* Baltimore, Md.: Johns Hopkins University Press.

Binswanger, Hans. 1989. "Brazilian Policies That Encourage Deforestation in the Amazon." *World Development* 19(7):821–29.

Binswanger, Hans P., Klaus Deininger, and Gershon Feder. 1995. "Power, Distortions, Revolt, and Reform in Agricultural Land Relations." In J. Behrman and T. N. Srinivasan, eds., *Handbook of Development Economics.* Vol. 3B. Amsterdam: Elsevier.

Blakemore, H., and C. T. Smith. 1971. *Latin America: Geographical Perspectives.* London: Methuen.

Boserup, Ester. 1965. *The Conditions of Agricultural Growth: The Economics of Agrarian Change under Population Pressure.* New York: Aldine.

Cleaver, Kevin M., and Gotz A. Schreiber. 1994. *Reversing the Spiral: The Population, Agriculture, and Environment Nexus in Sub-Saharan Africa.* Washington, D.C.: World Bank.

Currie, Lauchlin. 1965. *El manejo de cuencas en Colombia: Estudio sobre el uso de las tierras.* Bogotá: Tercer Mundo.

Grepperud, Sverre. 1994. "Population-Environment Links: Testing a Soil Degradation Model for Ethiopia." Divisional Working Paper 1994-46. Pollution and Environmental Economics Division, World Bank, Washington, D.C. Processed.

IGAC (Instituto Geográfico Colombiano). 1988. *Suelos y bosques de Colombia.* Bogotá.

Misión de Estudios del Sector Agropecuario. 1990. *El desarrollo agropecuario en Colombia.* Vol. 1. Bogotá: Ministerio de Agricultura, Departamento Nacional de Planeación.

Moll, P. G. 1988. "Transition to Freehold in the South African Reserves." *World Development* 16(3): 349–60.

Pingali, Prabhu, Yves Bigot, and Hans Binswanger. 1987. *Agricultural Mechanization and the Evolution of Farming Systems in Sub-Saharan Africa.* Baltimore, Md.: Johns Hopkins University Press.

Posner, J. L. 1981. *The Densely Populated Steep Slopes of Tropical America: Profile of a Fragile Environment.* New York: Rockefeller Foundation.

Reyes, Alvaro, and Jaime Martínez. 1994. "Funcionamiento de los mercados de trabajo rurales en Colombia." In Clara González and Carlos Felipe Jaramillo, eds., *Competitividad sin pobreza.* Bogotá: Tercer Mundo Editores.

Ruthenberg, Hans. 1980. *Farming Systems in the Tropics.* New York: Oxford University Press.

Suárez, R., G. Hurtado, L. A. Pacheco, and E. Segura. 1993. *El mercado de tierras y la formación de propietarios en Colombia.* Bogotá: Centro de Estudios Ganaderos y Agrícolas.

Syrquin, Moshe, and H. Chenery. 1989. *Patterns of Development, 1950–83.* Washington, D.C.: World Bank; Cambridge, Mass.: Harvard International Institute for Development.

Tiffen, Mary, Michael Mortimore, and Francis Gichuki. 1994. *More People, Less Erosion: Environmental Recovery in Kenya.* Chichester, U.K.: John Wiley.

World Bank. 1995a. "Colombia: Poverty Assessment." Country Department III, Natural Resources Management and Rural Poverty Division, Latin American and the Caribbean Region, Washington, D.C. Processed.

———. 1995b. "Colombia: Review of Agricultural and Rural Development Strategy." Report 13437. Natural Resources Management and Rural Poverty Division, Latin America and the Caribbean Region, World Bank, Washington, D.C. Processed.

World Bank, LACTAD (Latin America and the Caribbean Technical Advisory Group). 1994. "Trade Surveillance Study: Colombia." Washington, D.C. Processed.

4

Toward More Conducive Policies for Sustainable Agriculture

Jules N. Pretty

As this century draws to a close, agricultural development faces unprecedented challenges. By the year 2020, the world may have to support some 8.4 billion people. Even though enough food is being produced in aggregate to feed everyone, some 800 million people still do not have access to sufficient food. This includes 180 million underweight children suffering from malnutrition. The gap between the wealthy and the poor has widened. Despite a doubling in global income in the past three decades, the number of people living in poverty continued to rise, from 944 million to 1,300 million.

Recent models constructed to investigate changes in agricultural production and food security over the next quarter to half century all conclude that food production will have to increase substantially (IFPRI 1995, Crosson and Anderson 1995, Leach 1995, CGIAR 1995, and FAO 1993, 1995). But the views on how to proceed vary enormously. Some are optimistic, even complacent; others are darkly pessimistic. Some indicate that not much needs to change; others argue that agricultural and food systems need fundamental reforms. Some indicate that a significant growth in food production will occur only if new lands are taken under the plow; others suggest that there are sufficient social and technical solutions for increasing yields on existing farmland.

There are five distinct schools of thought for future options in agricultural development (see also McCalla 1994, Hazell 1995, Hewitt and Smith 1995, Pretty 1995b, and Hinchcliffe, Thompson, and Pretty 1996).

The *business-as-usual optimists*, with a strong belief in the power of the market, say that supply will always meet increasing demand and that world food production will continue to grow alongside expected reductions in population growth (Rosegrant and Agcaolli 1994, Mitchell and Ingco 1993, and FAO 1993). They argue that food prices are falling (down 50 percent in the past decade for most commodities), indicating that there is no current crunch in demand. Food production will continue to expand as the fruits of biotechnology research ripen, boosting plant and animal productivity, and as the area under cultivation expands, probably some 20–40 percent by 2020 (this may mean that an extra 79 million hectares of uncultivated land are converted to agriculture in Sub-Saharan Africa alone). They expect that population growth will slow and that developing countries will substantially increase food imports from industrialized countries (perhaps as much as fivefold by 2050).

The *environmental pessimists* contend that ecological limits to growth are being approached or have already been reached (Harris 1995, Brown 1994, CGIAR 1995, Kendall and Pimentel 1994, Brown and Kane 1994, and Ehrlich 1968). Following a neo-Malthusian line, these pessimists claim that populations continue to grow too rapidly, while growth in the yield of major cereals will slow or even fall, particularly because of growing production constraints in the form of resource degradation (soil erosion, land degradation, forest loss, excessive use of pesticides, and excessive exploitation of fisheries). Dietary shifts, especially increasing consumption of livestock products, are an emerging threat, because this results in the consumption by livestock of an even greater share of cereal products. They do not believe that new technological breakthroughs are likely. Solving these problems means that seeking some form of population control should be the first priority.

The *industrialized-world-to-the-rescue* lobby believes that developing countries will never be able to feed themselves, for a wide range of ecological, institutional, and infrastructural reasons, and that the looming food gap will have to be filled by modernized agriculture in the industrialized countries (Avery 1995, Wirth 1995, DowElanco 1994, Carruthers 1993, and Knutson and others 1990). Increasing production in large, mechanized operations will force smaller and more "marginal" farmers to go out of business, taking the pressure off natural resources. These resources can then be conserved in protected areas and wildernesses. The larger producers will then be able to trade their food with those who need it or have it distributed by famine relief or food aid. They vigorously argue that any adverse health and environmental consequences of chemically based agricultural systems are minor in comparison with those wrought by the expansion of agriculture into new lands. External inputs (especially pesticides and fertilizers) and free trade are said to represent a crucial part of any strategy for feeding the world (see Avery 1995, in particular).

One group, what we might call the *new modernists*, argues that increases in biological yield are possible on existing lands, but that this food growth can only come from "modern" high-external-input farming (Borlaug 1992, 1994a, 1994b, Sasakawa Global 2000 1993–95, World Bank 1993, Paarlberg 1994, Winrock International 1994, and Crosson and Anderson 1995). This group argues that farmers simply use too few fertilizers and pesticides, which are said to be the only way to improve yields and keep the pressure off natural habitats. This repeat of the green revolution model is widely termed science-based agriculture, the objective being to increase farmers' use of fertilizers and pesticides. It also argues that high-input agriculture is more environmentally sustainable than low-input agriculture, which makes intensive use of local resources that may be degraded in the process.

Another group, though, is making the case for the benefits of *sustainable intensification*, on the grounds that substantial growth is possible in currently unimproved or degraded areas, while at the same time protecting or even regenerating natural resources (Pretty 1995a, 1995b, Hazell 1995, McCalla 1994, 1995, Scoones and Thompson 1994, Northwest Area Foundation 1994, and Hewitt and Smith 1995). Proponents argue that empirical evidence now indicates that regenerative and low-input (but not necessarily zero-input) agriculture can be highly productive, provided farmers participate fully in all stages of technology development and extension. This evidence also suggests that the productivity of agricultural and pastoral lands is as much a function of human capacity and ingenuity as it is of biological and physical processes. Such sustainable agriculture seeks the integrated use of a wide range of pest, nutrient, soil, and water management technologies. It aims for an increased diversity of enterprises within farms combined with increased linkages and flows between them. By-products or wastes from one component or enterprise become inputs to another. As natural processes increasingly substitute for external inputs, the impact on the environment is reduced.

What Is and What Is Not Sustainable Agriculture?

Sustainability is a word that has entered common use in recent years. Since the Brundtland Commission put "sustainable development" on the map in the middle to late 1980s, close to 100 definitions of sustainability have been published. Each emphasizes different values, priorities, and practices. Clearly no reasonable person is opposed to the idea. But what does it mean? To some it implies the capacity of something to continue unchanged for a long time. To others, it implies not damaging natural resources. To others still, it is just accounting for the environment while continuing on a business-as-usual track. Does any of this help in the context of farming? We all know that sustainability represents something good, but what exactly? And, more important, has the notion of sustainable agriculture contributed to better farm practices, or is the term too easily hijacked?

In any discussion of sustainability, it is important to clarify what is being sustained, for how long, for whose benefit and at whose cost, over what area, and measured by what criteria. Answering these questions is difficult, because it means assessing and trading off values and beliefs (Pretty 1995b, Viederman 1994, Steer and Lutz 1993). It is critical, therefore, that sustainable agriculture not prescribe a concretely defined set of technologies, practices, or policies. This would serve only to restrict the future options of farmers. As conditions and knowledge change, so must farmers and communities be encouraged and allowed to change and adapt. Sustainable agriculture is, therefore, not a simple model or package to be imposed. It is more a process for learning (Pretty 1995a and Röling 1994).

During the past 50 years, agricultural and rural development policies have successfully emphasized external inputs as the means to increase food production. This has produced remarkable growth in global consumption of pesticides, inorganic fertilizer, animal foodstuffs, and tractors and other machinery. However, these external inputs have substituted for natural control processes and resources, rendering them more vulnerable. Pesticides have replaced biological, cultural, and mechanical methods for controlling pests, weeds, and diseases; inorganic fertilizers have substituted for livestock manure, compost, nitrogen-fixing crops, and fertile soils; information for management decisions comes from input suppliers, researchers, and extensionists rather than from local sources; and fossil fuels have substituted for locally generated energy sources. What were once valued local resources often now become waste products.

The basic challenge for sustainable agriculture is to make better use of available physical and human resources. This can be done by minimizing the use of external inputs, by regenerating internal resources more effectively, or by combining the two in various ways. This ensures the efficient and effective use of what is available and keeps any dependencies on external systems to a reasonable minimum.

A more sustainable agriculture is any food production system that systematically pursues the following goals:

- A thorough integration of natural processes such as nutrient cycling, nitrogen fixation, and pest-predator relationships into agricultural production processes, so ensuring profitable and efficient food production
- A minimization of the use of those external and nonrenewable inputs with the potential to damage the environment or harm the health of farmers and consumers and targeted use of the remaining inputs used with a view to minimizing costs
- The full participation of farmers and other rural people in all processes of problem analysis and technology development, adaptation, and extension, leading to an increase in self-reliance among farmers and rural communities
- A greater productive use of local knowledge and practices, including innovative approaches not yet fully understood by scientists or widely adopted by farmers
- The enhancement of wildlife and other public goods of the countryside.

Sustainable agriculture seeks the integrated use of a wide range of pest, nutrient, soil, and water management technologies. It aims for an increased diversity of enterprises within farms combined with increased linkages and flows between them. By-products or wastes from one component or enterprise become inputs to another. As natural processes increasingly substitute for external inputs, the negative impacts on the environment are reduced, and positive contributions are made to regenerate natural resources.

Current Extent and Impact of Sustainable Agriculture

Increasing evidence shows that regenerative and resource-conserving technologies and practices can bring both environmental and economic benefits for farmers, communities, and nations. The best evidence comes from countries of Africa, Asia, and Latin America, where the concern is to increase food production in the areas where farming has been largely untouched by the modern packages of externally supplied technologies. In these lands, farming communities adopting regenerative technologies have substantially improved agricultural yields, often using only a few or no external inputs (Bunch 1990, 1993, GTZ 1992, UNDP 1992, Krishna 1994, Shah 1994, SWCB 1994, Balbarino and Alcober 1994, de Freitas 1994, and Pretty 1995b).

But these are not the only sites for successful sustainable agriculture. In the high-input and generally irrigated lands, farmers adopting regenerative technologies have maintained or improved yields while substantially reducing their use of inputs (Bagadion and Korten 1991, Kenmore 1991, van der Werf and de Jager 1992, UNDP 1992, Kamp, Gregory, and Chowhan 1993, and Pretty 1995b). And in the industrialized countries, farmers have been able to maintain profitability, even though input use has been cut dramatically, such as in the United States (Liebhardt and others 1989, National Research Council 1989, Hanson and others 1990, Faeth 1993, Northwest Area Foundation 1994, and Hewitt and Smith 1995) and in Europe (El Titi and Landes 1990, Vereijken 1990, Jordan, Hutcheon, and Glen 1993, Pretty and Howes 1993, Reus, Weckseler, and Pak 1994, Somers 1997, and Pretty 1998).

Current Extent

The International Institute for Environment and Development has examined the extent and impact of sustainable agriculture in a select number of countries (Pretty, Thompson, and Hinchcliffe 1996). The government and nongovernmental programs and projects included in this analysis share important characteristics. They have:

- Made use of resource-conserving technologies in conjunction with group or collective approaches to agricultural improvement and natural resource management
- Put participatory approaches and farmer-centered activities at the center of their agenda, so that these activities are occurring on local people's terms and are more likely to persist after the projects and programs have ended
- Not used subsidies or food-for-work to "buy" the participation of local people or to encourage them to adopt particular technologies, so that improvements are unlikely to fade away or simply disappear at the end of the project or program
- Supported the active involvement of women as key producers and facilitators
- Emphasized "adding value" to agricultural products through agroprocessing, marketing, and other off-farm activities, thus creating employment, income-generating opportunities, and surplus retention in the rural economy.

Two types of transition to sustainable agriculture were assessed: from modern or conventional high-external-input agriculture (such as farming in green revolution lands or in the industrialized countries) and from traditional, rainfed agriculture where cereal yields have largely remained constant over centuries. Because these transitions are recent (within the past 5–10 years), they provide evidence that similar improvements could occur elsewhere and could be repeated on a larger scale.

In the 20 countries of the South (and the total of 63 projects) examined and analyzed, some 2.38 million households are farming 4.37 million hectares with sustainable agriculture technologies and practices (see table 4.1). The data in table 4.1 do not represent a comprehensive survey of sustainable agriculture in each of the countries. They do illustrate, however, what has been achieved by specific projects and what could be replicated elsewhere. Most of these improvements have occurred in the past 10 years (many in the past two to five years). The assumption is that these represent what is possible on a wider scale. It could be argued, however, that they are only successful because they have occurred where there is a combination of the least resistance and most opportunity, although the sheer diversity of approaches and contexts represented undermines such an assertion. Moreover, many of the improvements are occurring in difficult, remote, and resource-poor areas that have commonly been assumed to be incapable of developing innovations or producing food surpluses.

Contested Views

This empirical evidence is still contested. In the United States, some 82 percent of conventional farmers believe that low-input agriculture will always be low output, even though the top quarter of sustainable agriculture farmers obtain better yields and gross margins than conventional farmers (Hewitt and Smith 1995 and Northwest Area Foundation 1994). Influential politicians continue to reinforce these beliefs. In 1991 the former U.S. secretary of agriculture, Earl Butz, said,

> We can go back to organic agriculture in this country if we must—we once farmed that way 75 years ago. However, before we move in that direction, someone must decide which 50 million of our people will starve. We simply cannot feed, even at subsistence levels, our 250 million Americans without a large production input of chemicals, antibiotics, and growth hormones. (Quoted in Schaller 1993.)

Yet a selection of recent evidence shows that at least 40,000 farmers in 32 states are using sustainable agriculture technologies and have cut their use of external inputs substantially. This includes 2,800 sustainable agriculture farmers in the northwestern states, who grow twice as many crops as conventional farmers, use 60–70 percent less fertilizer, pesticide, and energy, and obtain roughly comparable yields; they also spend more money on local goods and services (Northwest Area Foundation 1994).

Despite the growing number of successful sustainable agriculture initiatives in different parts of the world, it is clear that most of these are still only "islands of success." There remains a huge challenge to find ways to spread or scale up the processes that have brought about these transitions.

Table 4.1. Examples of the Extent and Impact of Sustainable and People-Centered Agriculture in Different Agricultural Systems

Countries	*Number of farming households reported*	*Number of hectares reported*	*Dominant cereal crop*	*Cereal yield improvement factor (percent)*
Rainfed systems				
Brazil	223,000	1,330,000	Maize, wheat	198–246
Burkina Faso	22,500	37,360	Sorghum, millet	250
Ethiopia	24,175	21,850	Maize	154
Guatemala	17,000	17,000	Maize	250
Honduras	27,000	42,000	Maize	250
India	307,910	993,410	Sorghum, millet	288
Kenya	222,550	250,000	Maize	200
Mexico	7,400	23,500	Coffee	140
Nepal	3,000	1,300	Maize, wheat	164–307
Philippines	850	920	Upland rice	214
Senegal	200,000	400,000	Sorghum, millet	300
Uganda	9,426	21,379	Maize	150
Zambia	6,300	6,300	Sorghum, millet	200
Total	1,146,111	3,257,519		
Irrigated systems				
Bangladesh	11,025	4,772	Rice	110
China	47,000	12,000	Rice	111
India	50,000	71,300	Rice	108
Indonesia	400,000	267,000	Rice	107
Malaysia	2,500	3,925	Rice	108
Philippines	175,000	385,000	Rice	112
Sri Lanka	100,000	95,350	Rice	117
Thailand	500	2,040	Rice	109
Vietnam	6,600	3,540	Rice	108
Total	792,625	844,927		
Industrialized systems				
Germany (integrated)	75,000	200,000	Wheat, barley	90
Netherlands (integrated)	500	—	Wheat, barley	85
United States (integrated)	40,000	632,000	Wheat, barley	95
European Union (organic)	50,000	1,200,000	Wheat, barley	80
Total	165,500	1,832,000		

— Not available.

Note: Improvements are measured against nonsustainable farming equivalents, which are taken to be 100 percent. Thus an improvement of 200 percent implies a doubling of yields; one of 90 percent implies a fall in yields of 10 percent. The time frame for these improvements is the life of program activities, usually less than five years. Some improvements are expected to occur in the season following the adoption of sustainable agriculture, and these tend to increase over time.

Source: See Pretty, Thompson, and Hinchcliffe 1996 for details of data sources and surveys.

The Record of Conventional Agricultural Projects

When the recent record of development assistance is considered, it is clear that sustainability has been poor. There is a widespread perception among both multilateral and bilateral organizations that agricultural development is difficult, that agricultural projects perform badly, and that resources may best be spent in other sectors. Reviews by the World Bank, the European Commission, the Danish International Development Agency, and the British Department for International Development (formerly the Overseas Development Administration) have all shown that agricultural and natural resource projects both performed worse in the 1990s than in the 1970s–80s and worse than projects from other sectors (World Bank 1993, Pohl and Mihaljek 1992, European Commission 1994, DANIDA 1994, and Dyer and Bartholomew 1995). They are also less likely to continue achievements beyond the provision of aid inputs.

A recent analysis of the evaluations of 95 agricultural projects recorded on the database of the Development Assistance Committee of the Organisation for Economic Co-operation and Development (OECD) shows a disturbing rate of failure, with at least 27 percent of projects having nonsustainable structures, practices, or institutions and 10 percent causing significantly negative environmental impact (Pretty and Thompson 1996). The reasons given for failure include an emphasis on external technologies alone, lack of participation by local people, ineffective training of professionals, and institutions that were working with no view toward the diversity of local conditions and needs of local people.

This evidence from completed agricultural development projects suggests three reasons for the lack of sustainability:

- If coercion or financial incentives are used to encourage adoption, sustainable agriculture technologies, such as soil conservation, alley cropping, and integrated pest management, are not likely to persist.
- If imposed, new institutional structures, such as cooperatives or other groups at the local level or project management units and other institutions at the project level, rarely persist beyond the project.
- If introduced with no thought to how they will be paid for, expensive external inputs, including subsidized inputs, machinery, or high-technology hardware, generally do not persist beyond the project.

The Problems with Comprehensive Technology Packages

Modernist agricultural development has begun with the notion that there are "on-the-shelf" technologies that work, and it is just a matter of inducing or persuading farmers to adopt them. Yet few farmers are able to adopt whole packages of conservation technologies without adjusting their own practices and livelihood systems. For example, alley cropping, an agroforestry system comprising rows of nitrogen-fixing trees or bushes separated by rows of cereals, has long been the focus of research (Kang, Wilson, and Lawson 1984 and Attah-Krah and Francis 1987). Many productive and sustainable systems, needing few or no external inputs, have been developed. They stop erosion, produce food and wood, and can be cropped over long periods. But the problem is that very few, if any, farmers have adopted these alley cropping systems as designed. Despite millions of dollars of research expenditure over many years, many of the systems produced are suitable only for research stations (Carter 1995). There has been some success, however, where farmers have been able to adapt one or two components of alley cropping to their own farms. In Kenya, for example, farmers planted multiple rows of leguminous trees next to field boundaries or single rows through their fields; and in Rwanda, extension workers planted alleys that soon became dispersed through fields (Kerkhof 1990).

But the prevailing view tends to be that farmers should adapt to the technology. Of the Agroforestry Outreach Project in Haiti, the evaluators said that

> Farmer management of hedgerows does not conform to the extension program ... Some farmers prune the hedgerows too early, others too late. Some hedges are not yet pruned by two years of age, when they have already reached heights of 4–5 meters. Other hedges are pruned too early, mainly because animals are let in, or the tops are cut and carried to animals ... Finally, it is very common for farmers to allow some of the trees in the hedgerow to grow to pole size. (Bannister and Nair 1990.)

This could be read as a great success—farmers were clearly adapting the technology to their own special needs. But it was not. The language of the evaluators is quite clear: this was considered a failure.

In Laos, one project used food for work to encourage shifting agriculturalists to settle and adopt contour farming with bench terraces (Fujisaka 1989). But these fields became so infested with weeds that farmers were forced to shift to new lands, and the structures were so unstable in the face of seasonal rains that they led to worsened gully erosion. Farmers then refused to do further work when the incentives were gone.

What are the implications for sustainable agriculture? How should we proceed so as to ensure that farmers are fully involved in developing and adapting these sustainable and productive technologies?

Enhancing Farmers' Capacity to Innovate

Important evidence comes from a variety of soil conservation and agricultural regeneration programs in Central America (Bunch and López 1996). The Guinope (1981–89) and Cantarranas (1987–91) programs in Honduras and the San Martín Jilotepeque program in Guatemala (1972–79) were collaborative efforts between World Neighbors and other local agencies. All began with a focus on soil conservation in areas where maize yields were very low (400 to 660 kilograms per hectare) and where shifting cultivation, malnutrition, and outmigration prevailed. All show the importance of developing resource-conserving practices in partnership with local people.

There were several common elements. All forms of paternalism were avoided, including giving things away, subsidizing farmer activities or inputs, or doing anything for local people. Each started slowly and on a small scale, so that local people could meaningfully participate in planning and implementation. They used technologies, such as green manure, cover crops, contour grass strips, in-row tillage, rock bunds, and animal manure, that were appropriate to the local area and were finely tuned through experimentation by and with farmers. Extension and training were done largely by villager farmers who had already experienced success with the technologies on their own farms.

Each program substantially improved agricultural yields, increasing output per area of land from some 400–600 kilograms per hectare to 2,000–2,500 kilograms per hectare. Altogether improvements have been made in some 120 villages. Over time, soils have not been simply conserved but regenerated, with depth increases from 0.1 meter to 0.4–1.3 meters not uncommon. These programs have also helped to regenerate local economies. Land prices and labor rates are higher inside the project areas than outside. There are housing booms, and families have moved back from capital cities. There are also benefits to the forests. Farmers say they no longer need to cut the forests, because they have the technologies to farm permanently the same piece of land. Before the programs, national park authorities sought to keep villagers out of the forests; now there is no such concern because the forests are no longer threatened.

Few published studies give evidence of impacts years after outside interventions have ended. In 1994, however, staff of the Honduran organization COSECHA (Asociación de Consejeros para una Agricultura Sostenible, Ecológica y Humana) returned to the three program areas and used participatory methods with local communities to evaluate subsequent changes (Bunch and López 1996).

They first divided all 121 villages into three categories, according to where they felt there had been good, moderate, and poor impact. Twelve villages were sampled from these—four from each program comprising one of the best, two of the moderate, and one poor. These villages had some 1,000 families (with a range of 30 to 180 per village). The first major finding was that crop yields and adoption of conserving technologies continued to grow after the project ended (see table 4.2).

Surprisingly, though, many of the technologies known to be successful during the project had been superseded by new practices. Had the original technologies been poorly selected? It would appear not, as many that had been dropped by farmers are still very successful elsewhere. The explanation appears to be that changing external and internal circumstances had reduced or eliminated their usefulness, such as changing markets, droughts, diseases, insect pests, land tenure, labor availability, and political disruptions.

Altogether, some 80–90 successful innovations were documented in these 12 villages. In one Honduran village, Pacayas, there had been 16 innovations, including four new crops, two new types of green manure, two new species of grass for contour barriers in vegetables, chicken pens made of king grass, marigolds for nematode control, lablab and velvet bean as cattle and chicken feed, nutrient recycling into fish ponds, human wastes placed in composting latrines, napier grass to stabilize cliffs, and homemade sprinklers for irrigation.

Technologies had been developed, adopted, adapted, and dropped. The study concluded that the half-life of a successful technology in these project areas is six years. Quite clearly the technologies themselves are not sustainable. As Bunch and López (personal

Table 4.2. Changes in the Adoption of Resource-Conserving Technologies, Maize Yields, and Migration Patterns in Three Programs in Central America during and after Projects

Technology	*At initiation*	*At termination*[a]	*In 1994*
Number of farmers with the technology			
Contour grass barriers	1	192	280
Contour drainage ditches	1	253	239
Contour rows	0	100	245
Green manure	0	35	52
Crop rotation	12	209	254
No burning fields or forests	2	160	235
Organic matter as fertilizer	44	195	397
Yields of maize (kilograms per hectare)			
1. San Martín, Guatemala (1972–79)	400	2,500	4,500
2. Guinope, Honduras (1981–89)	600	2,400	2,730
3. Cantarranas, Honduras (1987–91)	660	2,000	2,050
Migration (number of households)			
1. San Martin: 2 villages			
San Antonio Correjo	65	n.d.	4
Las Venturas	85	n.d.	4
2. Guinope: 3 villages	38	0	(2)[b]
3. Cantarranas: 3 villages	n.d.	10	(6)[b]

n.d. No data.
a. Program termination dates were San Martín, 1979; Guinope, 1989; Cantarranas, 1991.
b. Numbers in parentheses refer to negative outmigration, that is, families returning to their villages.
Source: Bunch and López 1996.

communication, 1996) have put it, "What needs to be made sustainable is the social process of innovation itself." Sustainability does not equal fossilization or continuation of a thing or practice forever: rather it implies an enhanced capacity to adapt in the face of unexpected changes and emerging uncertainties.

A similar picture has emerged in Gujarat, where many farmers have developed new technical innovations after support from the Aga Khan Rural Support Program for undertaking simple conservation measures. Farmers have introduced planting of grafted mango trees and bamboo near embankments, making full use of residual moisture near gully traps. They have also introduced cultivation of vegetables, such as brinjal and lady's finger, other leguminous crops, and tobacco in the newly created silt traps. This has increased production substantially, particularly in years of poor rainfall. Most of these innovations and adaptations have been introduced and sustained with support from the local network of village extensionists (Shah 1994).

In South Queensland, Australia, extensionists from the Department of Primary Industry, using very simple learning tools that enabled farmers to investigate the impact of rainfall on their soil, have encouraged more than 80 percent of farmers to adopt conservation technologies. Many of these have gone on to develop and adopt new and different technologies for their own farms, and they now fully support the values and principles they once would have opposed (Hamilton 1995).

Another example comes from Thailand, where the four phases of the Thai-German Highland Development Project clearly illustrate the importance of genuine participation with local people (Thai-German Highland Development Project 1995 and Steve Carson, personal communication, 1996). The project has been working with upland communities in northern Thailand to support the transition toward sustainable agriculture. The resource-conserving technologies developed and adapted for local use include hedgerows on contours, buffer strips, new crop rotations, integrated pest management, crop diversification, and livestock integration. The approach, however, has changed significantly since the mid-1980s (see table 4.3). In the first phase, cash incentives and free inputs were used to encourage adoption of these technologies; as a result adoption rates were high, although there was little or no adaptation of the technologies by the farmers. In 1990 all the incentives were stopped when the project adopted a participatory approach; adoption rates fell immediately, and withdrawal increased threefold. By 1993–94, the participatory village planning had fully involved communities, and the ratio of adopters to withdrawers was equal. Most recently, the number of farmers using sustainable technologies has grown rapidly, and, crucially, farmers are now actively adapting those technologies and innovating new ones to satisfy their particular needs (Steve Carson, personal communication, 1996).

The Policy Environment in Industrial Countries

Current policies often do not reflect the long-term social and environmental costs of resource use. The external costs of modern farming, such as soil erosion, health damage, or polluted ecosystems, generally are not incorporated into individual decisionmaking by farmers. In this way, resource-degrading farmers bear neither the costs of damage to the environment or economy nor those incurred in controlling the polluting or damaging activity (Pretty 1996). In principle, it is possible to imagine pricing the free input to farming of the clean, unpolluted environment. If charges were levied in some way, then degraders or polluters would have higher costs, would be forced to pass them on to consumers, and would be forced to switch to more resource-conserving technologies. This notion is captured in the polluter-pays principle, a concept used for many years in the nonfarm sector (OECD 1989). However, beyond the notion of encouraging some internalization of costs, it has not been of practical use for policy formulation in agriculture.

In general, farmers are entirely rational to continue using high-input degrading practices under current policies. High prices for particular commodities, such as key cereals, have discouraged mixed farming practices, replacing them with monocultures. This is strikingly clear in the United States, where commodity programs inhibit the adoption of sustainable practices by artificially making them less profitable to farmers. In Pennsylvania, the financial returns to monocropped and continuous maize are about the same as those to sustainable agriculture involving mixed rotations (Faeth 1993). But continuous maize attracts about twice as much direct support in the form of deficiency payments. In addition, continuous maize farms require much more nitrogen fertilizer, erode more soil, and cause three to six times as much damage to off-site resources. Quite clearly, a transition to resource-conserving rotations would substantially benefit both farmers and the national economy.

In this context of systemic support for high-input agriculture, many countries have sought to "bolt" conservation goals onto these policies. These countries have tended to rely on conditionality, such as cross-compliance, whereby farmers receive support only if they adopt certain types of resource-conserving technologies and practices. Such cross-compliance

Table 4.3. Changing Phases in the Thai-German Highland Development Project: The Case of 113 Villages in Nam Lang, Northern Thailand

Phase and date	*Characteristics*	*Ratio of adoption to withdrawal*
1. 1987–90	Cash incentives and free inputs; high adoption, but little or no adaptation of technologies	5:1
2. 1991–92	All incentives stopped; beginning of participatory work; adoption rates fell to 25 percent of phase 1; withdrawal immediately increased threefold	1:2
3. 1993–94	Participatory village planning; communities fully involved; adopters and withdrawers now equal	1:1
4. 1995–96	Adopters increasing; farmers adapting technologies and diversifying (pineapple strips, lemongrass, cash crops, soil and water conservation)	3:1

Source: Steve Carson, personal communication, 1996.

occurs widely in the South too. The process of agricultural modernization has widely involved encouraging farmers to adopt modern practices through the linkage of credit or other benefits. If farmers wish to receive one type of support, they must adopt a particular set of technologies and practices. In some cases, coercion has been used to achieve certain levels of adoption. The problem with these cross-compliances is that they can create long-term resentment and lead to a reversal of practices when policies change or the money runs out.

Although a growing number of policy initiatives are oriented specifically toward improving the sustainability of agriculture, most have focused on input reduction strategies. Only a few as yet represent coherent plans and processes that clearly demonstrate the value of integrating policy goals. A thriving and sustainable agricultural sector requires both integrated action by farmers and communities and integrated action by policymakers and planners. This implies both horizontal integration with better linkages between sectors and vertical integration with better linkages between the micro and macro level.

Integration has been the policy buzz of the 1990s. But putting this desired integration into practice has been much more difficult. There have been substantial differences in the views of major policy actors, such as those representing the interests of farmers, environmentalists, and treasuries. Other problems include market failure, such as nonpayment or underpayment for resource degradation, or intervention failure, such as undersupply of public goods. Nonetheless, small steps are occasionally being taken both to penalize polluters and to encourage resource conservers, particularly in industrial countries where levels of facilities and pesticide applications are much higher than in developing nations.

Policy Initiatives in the South

The Group Farming Initiative of the Kerala State government in India is a good example of how coordinated action within the agricultural sector can have a significant impact on farming practices (Sherief 1991). Land reform in the 1970s led to the formation of a new class of small farmers, with some 70 percent owning less than 2 hectares. But as the costs of inputs and pest and disease control spiraled, the area under rice fell from 810,000 to 570,000 hectares between 1975 and 1988. Small farmers were unable to adopt the whole technological package.

In 1989 the Group Farming for Rice Program was launched. Local committees comprising all rice farmers were formed to chart a detailed plan of farming. In this group, activities, such as water management and labor operations, are agreed jointly. Costs are reduced through community nursery raising of rice, fertilizer applications on the basis of soil testing, the introduction of integrated pest management and minimum use of pesticides, and the formation of plant protection squads. The average cost reduction to farmers has been Rs1,000 per hectare, and rice yields have improved 500 kilograms per hectare.

In China, agricultural policy is encouraging farmers to grow green manure in the rice fields (Yixian 1991). During the 1980s, continuous and monocropping of rice caused widespread soil fertility and pest problems. The agricultural ministry set up multiplication bases for green manure, which are expected to produce 5.5 million kilograms of seed each year. In some regions, farmers selling green manure seed to state-run farm cooperatives receive fertilizers at lower prices. Green manure and plant residues are now used on 68 percent of the 22 million hectares of rice fields.

Similar successes have been observed in Indonesia where, during the late 1980s, pesticide subsidies were cut from 85 percent to zero by January 1989 and farmer field schools were established for integrated pest management (Kenmore 1991, Matteson 1995, FAO 1994, and Röling and van der Fliert 1997). The country saves some $130 million–$160 million each year, and pesticide production fell nearly 60 percent between 1985 and 1990. Rice yields have continued to improve, despite the cut in inputs. This raises a crucial policy issue when it comes to cutting inputs. Farmers who depend on external inputs need support to make the transition to a more sustainable agriculture. As this program demonstrates, if this support increases farmers' capacity to learn and act on their own farms, the substitute is more than adequate. The FAO is supporting similar integrated pest management programs for rice in eight other countries of South and Southeast Asia (FAO 1994). Together, these programs are training hundreds of thousands of farmers and have saved many millions of dollars in pesticides.

Policy Processes: Conditionality or Participation?

Policy reform has been under way in many countries, with some new initiatives supporting elements of a more sustainable agriculture. Most of these have focused on input reduction strategies, because of concerns over foreign exchange expenditure or environmental damage. Only a few as yet represent coherent plans and processes that clearly demonstrate the value of integrating policy goals.

Governments can do much with existing resources to encourage and nurture the transition from modernized systems toward more sustainable alternatives (see table 4.4). The first action that governments can take is to declare a national policy for sustainable agriculture. This would help to raise the profile of these processes and needs as well as give explicit value to alternative societal goals. It would also establish the necessary framework within which the more specific actions listed in table 4.4 can fit and be supported.

It is important to be clear about just how policies should be trying to address issues of sustainability. As suggested at the beginning of this chapter, precise and absolute definitions of sustainability, and therefore of sustainable agriculture, are impossible. Sustainability itself is a complex and contested concept. Sustainable agriculture should not, therefore, be seen as a set of practices to be fixed in time and space. It implies the capacity to adapt and change as external and internal conditions change. Yet there is a danger that policy, as it has tended to do in the past, will prescribe the practices that farmers should use rather than create the enabling conditions for locally generated and adapted technologies.

Environmental policy has sometimes tended to take the view that rural people mismanage natural resources. The history of soil and water conservation, rangeland management, protected area management, irrigation development, and modern crop dissemination shows a common pattern: technical prescriptions are derived from controlled and uniform conditions, supported by limited cases of success, and then applied widely with little or no regard for diverse

Table 4.4. Polices That Work toward Sustainable Agriculture

Policy	*Intent*
Policy 1	Declare a national policy for sustainable agriculture
	Encourage resource-conserving technologies and practices
Policy 2	Establish a national strategy for integrated pest management
Policy 3	Prioritize research into sustainable agriculture
Policy 4	Grant farmers appropriate property rights
Policy 5	Promote farmer-to-farmer exchanges
Policy 6	Direct limited grants toward sustainable technologies
Policy 7	Link support payments to resource-conserving practices
Policy 8	Provide better information for consumers and the public
Policy 9	Adopt natural resource accounting
Policy 10	Establish appropriate standards and licensing for pesticides
	Support local groups for community action
Policy 11	Encourage the formation of local groups
Policy 12	Foster rural partnerships
Policy 13	Support training and field schools for farmers
Policy 14	Provide incentives for on-farm employment
Policy 15	Permit groups to have access to credit
	Reform external institutions and professional approaches
Policy 16	Encourage the formal adoption of participatory methods and processes
Policy 17	Support information systems to link research, extension, and farmers
Policy 18	Rethink the project culture
Policy 19	Strengthen the capacities of NGOs to scale up
Policy 20	Foster stronger NGO-government partnerships
Policy 21	Reform teaching and training establishments
Policy 22	Develop capacity in planning for conflict resolution and mediation

Source: Adapted from Pretty 1995b.

local needs and conditions (Pretty and Shah 1994, Benhke and Scoones 1992, and Pimbert and Pretty 1995). Differences in receiving environments and livelihoods often make the technologies unworkable and unacceptable. When they are rejected locally, policies shift to seeking success through the manipulation of social, economic, and ecological environments and, in some cases, through outright enforcement.

For sustainable agriculture to spread widely, policy formulation must not repeat these mistakes. Policies will have to arise in a new way. They must be enabling and create the conditions for sustainable development based on locally available resources and local skills and knowledge. Achieving this will be difficult. In practice, policy is the net result of the actions of different interest groups pulling in complementary and opposing directions. It is not just the normative expression of governments. Effective policy will have to recognize this and seek to bring together a range of actors and institutions for creative interaction and joint learning.

References

The word "processed" describes informally reproduced works that may not be commonly available through libraries.

Attah-Krah, A. N., and P. A. Francis. 1987. "The Role of On-Farm Trials in the Evaluation of Composite Technologies: The Case of Alley Farming in Southern Nigeria." *Agricultural Systems* 23:133–52.

Avery, Denis. 1995. *Saving the Planet with Pesticides and Plastic*. Indianapolis, Ind.: The Hudson Institute.

Bagadion, B. U., and F. F. Korten. 1991. "Developing Irrigators' Organisations: A Learning Process Approach." In Michael M. Cernea, ed., *Putting People First*, 2d ed. New York: Oxford University Press.

Balbarino, Edwin A., and Doris L. Alcober. 1994. "Participatory Watershed Management in Leyte, Philippines: Experience and Impacts after Three Years." Paper prepared for the International Institute for Environment and Development/ActionAid conference new horizons: the social, economic, and environmental impacts of participatory watershed development, Bangalore, India, 28 November–2 December. Processed.

Bannister, M. E., and P. K. R. Nair. 1990. "Alley Cropping as a Sustainable Agricultural Technology for the Hillsides of Haiti: Experience of an Agroforestry Outreach Project." *American Journal of Alternative Agriculture* 5(2):51–59.

Beaumont, Peter. 1993. *Pesticides, Policies, and People*. London: The Pesticides Trust.

Benhke, Roy, and Ian Scoones. 1992. *Rethinking Range Ecology: Implications for Rangeland Management in Africa.* Drylands Program Issues Paper 33. London: International Institute for Environment and Development.

Borlaug, Norman. 1992. "Small-Scale Agriculture in Africa: The Myths and Realities." *Feeding the Future* (newsletter of the Sasakawa Africa Association) 4:2.

———. 1994a. "Agricultural Research for Sustainable Development." Testimony before the U.S. House of Representatives Committee on Agriculture, Washington, D.C., 1 March. Processed.

———. 1994b. "Chemical Fertilizer 'Essential [Letter].'" *International Agricultural Development* (November-December):23.

Brown, Lester R. 1994. "The World Food Prospect: Entering a New Era." In *Assisting Sustainable Food Production: Apathy or Action?* Arlington, Va.: Winrock International.

Brown, Lester R., and Hal Kane. 1994. *Full House: Reassessing the Earth's Population Carrying Capacity.* New York: W. W. Norton and Co.

Bunch, Roland. 1990. *Low-Input Soil Restoration in Honduras: The Cantarranas Farmer-to-Farmer Extension Program.* Gatekeeper Series SA23. London: International Institute for Environment and Development, Sustainable Agriculture Program.

———. 1993. "EPAGRI's Work in the State of Santa Catarina, Brazil: Major New Possibilities for Resource-Poor Farmers." Asociación de Consejeros para una Agricultura Sostenible, Ecológica y Humana, Tegucigalpa. Processed.

Bunch, Roland, and Gabino López. 1996. *Soil Recuperation in Central America: Sustaining Innovation after Intervention.* Gatekeeper Series SA 55. London: International Institute for Environment and Development, Sustainable Agriculture Program.

Carruthers, Ian. 1993. "Going, Going, Gone! Tropical Agriculture as We Knew It." *Tropical Agriculture Association Newsletter* 13(3):1–5.

Carter, Jane. 1995. *Alley Cropping: Have Resource-Poor Farmers Benefited?* Natural Resource Perspectives 3. London: Overseas Development

Institute.

CGIAR (Consultative Group on International Agricultural Research). 1995. "Sustainable Agriculture for a Food Secure World: A Vision for International Agricultural Research." Expert panel of CGIAR, Washington, D.C., and Swedish International Development Authority, Stockholm. Processed.

Crosson, Pierre, and J. R. Anderson. 1995. "Achieving a Sustainable Agricultural System in Sub-Saharan Africa." Building Block for Africa Paper 2. AFTES Africa Technical Department, Environment and Social Division, World Bank, Washington, D.C. Processed.

DANIDA (Danish International Development Agency). 1994. *Agricultural Sector Evaluation: Lessons Learned.* Copenhagen: Ministry of Foreign Affairs.

de Freitas, Hercilio V. 1994. "EPAGRI in Santa Catarina, Brazil: The Micro-Catchment Approach." Paper prepared for the new horizons conference, International Institute for Environment and Development, Bangalore, India, November. Processed.

DowElanco. 1994. *The Bottom Line.* Indianapolis, Ind.: DowElanco.

Dyer, Nick, and A. Bartholomew. 1995. "Project Completion Reports: Evaluation Synthesis Study." Evaluation Report Ev583. Overseas Development Administration, London. Processed.

Ehrlich, Paul. 1968. *The Population Bomb.* New York: Ballantine.

El Titi, Adel, and H. Landes. 1990. "Integrated Farming System of Lautenbach: A Practical Contribution toward Sustainable Agriculture." In C. A. Edwards, R. Lal, P. Madden, R. H. Miller, and G. House, eds., *Sustainable Agricultural Systems.* Ankeny, Iowa: Soil and Water Conservation Society.

European Commission. 1994. *Evaluation des projets de développement rural finances durant les conventions de Lomé I, II, et III.* Brussels.

Eveleens, Kees G., Richard Chisholm, Elske van der Fliert, M. Kato, Pham Thi Nhat, and Peter Schmidt. 1996. *Mid-Term Review of Phase III FAO Intercountry Programme for the Development and Application of Integrated Pest Control in South and South East Asia.* GCP/RAS/145-147/NET. Rome: Food and Agriculture Organization.

Faeth, Paul, ed. 1993. *Agricultural Policy and Sustainability: Case Studies from India, Chile, the Philippines, and the United States.* Washington, D.C.: World Resources Institute.

FAO (Food and Agriculture Organization of the United Nations). 1993. *Strategies for Sustainable Agriculture and Rural Development (SARD): The Role of Agriculture, Forestry, and Fisheries.* Rome.

———. 1994. "Intercountry Program for the Development and Application of Integrated Pest Control in Rice in South and Southeast Asia, Phase I and II. Project Findings and Recommendations." Terminal report. Rome. Processed.

———. 1995. *World Agriculture: Towards 2010.* Rome.

Fujisaka, Sam. 1989. "The Need to Build on Farmer Practice and Knowledge: Reminders from Selected Upland Conservation Projects and Policies." *Agroforestry Systems* 9:141–53.

GTZ (German Agency for Technical Cooperation). 1992. *The Spark Has Jumped the Gap.* Eschborn.

Hamilton, Gus. 1995. "Learning to Learn with Farmers." Ph.D. diss., Wageningen Agricultural University, Wageningen, Netherlands.

Hanson, J. C., D. M. Johnson, S. E. Peters, and R. R. Janke. 1990. "The Profitability of Sustainable Agriculture on a Representative Grain Farm in the Mid-Atlantic Region, 1981–1989." *Northeastern Journal of Agriculture and Resource Economics* 19(2):90–98.

Harris, J. M. 1995. *World Agriculture: Regional Sustainability and Ecological Limits.* Discussion Paper 1. Medford, Mass.: Tufts University, Center for Agriculture, Food, and Environment.

Hazell, Peter. 1995. "Managing Agricultural Intensification." IFPRI 2020, Brief 11. International Food Policy Research Institute, Washington, D.C. Processed.

Hewitt, Tracy I., and Katherine R. Smith. 1995. *Intensive Agriculture and Environmental Quality: Examining the Newest Agricultural Myth.* Greenbelt, Md.: Henry Wallace Institute for Alternative Agriculture.

Hinchcliffe, Fiona, John Thompson, and Jules N. Pretty. 1996. "Sustainable Agriculture and Food Security in East and Southern Africa." A report prepared for the Swedish International Development Cooperation Agency, Stockholm, and International Institute for Environment and Development, London. Processed.

IFPRI (International Food Policy Research Institute). 1995. *A 2020 Vision for Food, Agriculture, and the*

Environment. Washington, D.C.

Jordan, Vic W. L., J. A. Hutcheon, and D. M. Glen. 1993. *Studies in Technology Transfer of Integrated Farming Systems: Considerations and Principles for Development.* Bristol: Institute of Arable Crops Research, Long Ashton Research Station.

Kamp, K., R. Gregory, and G. Chowhan. 1993. "Fish Cutting Pesticide Use." ILEIA *Newsletter* 2(93):22–23.

Kang, B. T., G. F. Wilson, and T. L. Lawson. 1984. *Alley Cropping: A Stable Alternative to Shifting Agriculture.* Ibadan: International Institute for Tropical Agriculture.

Kendall, H. W., and D. Pimentel. 1994. "Constraints on the Expansion of the Global Food Supply." *Ambio* 23:198–205.

Kenmore, Peter. 1991. *How Rice Farmers Clean up the Environment, Conserve Biodiversity, Raise More Food, Make Higher Profits. Indonesia's IPM—A Model for Asia.* Manila: Food and Agriculture Organization of the United Nations.

Kerkhof, Paul. 1990. *Agroforestry in Africa: A Survey of Project Experience.* London: Panos Institute.

Knutson, R. D., J. B. Taylor, J. B. Penson, and E. G. Smith. 1990. *Economic Impacts of Reduced Chemical Use.* College Station: Texas A&M University.

Krishna, Anirudh. 1994. "Large-Scale Government Programs: Watershed Development in Rajasthan, India." Paper prepared for the new horizons conference, International Institute for Environment and Development, Bangalore, India, November. Processed.

Leach, Gerry. 1995. *Global Land and Food in the 21st Century.* Polestar Series Report 5. Stockholm: Stockholm Environment Institute.

Liebhardt, W., R. W. Andrews, M. N. Culik, R. R. Harwood, R. R. Janke, J K. Radke, and S. L. Rieger-Schwartz. 1989. "Crop Production during Conversion from Conventional to Low-Input Methods." *Agronomy Journal* 81(2):150–59.

Matteson, Patricia. 1995. "The 50 Percent Pesticide Cuts in Europe: A Glimpse of Our Future?" *American Entomologist* 41(4):210–20.

McCalla, Alex. 1994. "Agriculture and Food Needs to 2025: Why We Should be Concerned." Sir John Crawford Memorial Lecture, October 27. CGIAR Secretariat, World Bank, Washington, D.C. Processed.

———. 1995. "Towards a Strategic Vision for the Rural/Agricultural/Natural Resource Sector Activities of the World Bank." Paper prepared for the fifteenth annual agricultural symposium, World Bank, Washington, D.C., 4–6 January. Processed.

Mitchell, D. O., and M. D. Ingco. 1993. "The World Food Outlook." International Economics Department, World Bank, Washington, D.C. Processed.

National Research Council. 1989. *Alternative Agriculture.* Washington, D.C.: National Academy Press.

Northwest Area Foundation. 1994. *A Better Row to Hoe: The Economic, Environmental, and Social Impact of Sustainable Agriculture.* St. Paul, Minn.

OECD (Organisation for Economic Co-operation and Development). 1989. *Agricultural and Environmental Policies.* Paris.

Paarlberg, Roy L. 1994. "Sustainable Farming: A Political Geography." IFPRI 2020, Brief 4. International Food Policy Research Institute, Washington, D.C. Processed.

Pimbert, Michel A., and Jules N. Pretty. 1995. *Parks, People, and Professionals: Putting 'Participation' into Protected Area Management.* UNRISD Discussion Paper 57. Geneva: United Nations Research Institute for Social Development and Worldwide Fund for Nature International.

Pohl, G., and D. Mihaljek. 1992. "Project Evaluation and Uncertainty in Practice: A Statistical Analysis of Rate-of-Return Divergences of 1,015 World Bank Projects." *The World Bank Economic Review* 6(2):255–77.

Pretty, Jules N. 1995a. "Participatory Learning for Sustainable Agriculture." *World Development* 23(8):1247–63.

———. 1995b. *Regenerating Agriculture: Policies and Practice for Sustainability and Self-Reliance.* London: Earthscan Publications; Washington, D.C.: National Academy Press; Bangalore: ActionAid.

———. 1996. "A Three-Step Framework for Agricultural Change." *Pesticides News* 32(June): 6–8.

———. 1998. *The Living Land: Agriculture, Food, and Community Regeneration for Rural Europe.* London: Earthscan Publications, Ltd.

Pretty, Jules N., and Rupert Howes. 1993. "Sustainable Agriculture in Britain: Recent Achievements and New Policy Challenges." IIED Research Series. International Institute for Environment and Development, London. Processed.

Pretty, Jules N., and Parmesh Shah. 1994. *Soil and Water Conservation in the 20th Century: A History of Coercion and Control.* Rural History Centre Paper 1. Reading, U.K.: University of Reading.

Pretty, Jules N., and John Thompson. 1996. "Sustainable Agriculture at the Overseas Development Administration." Report for Natural Resources Policy Advisory Department, Overseas Development Administration, London. Processed.

Pretty, Jules N., John Thompson, and Fiona Hinchcliffe. 1996. *Sustainable Agriculture: Impacts on Food Production and Challenges for Food Security.* Gatekeeper Series SA 60. London: International Institute for Environment and Development.

Reus, J. A. W. A., H. J. Weckseler, and G. A. Pak. 1994. *Towards a Future EC Pesticide Policy.* Utrecht: Centre for Agriculture and Environment (CLM).

Röling, Niels. 1994. "Platforms for Decisionmaking about Ecosystems." In L. Fresco, ed., *The Future of the Land.* Chichester, U.K.: John Wiley and Sons.

Röling, Niels, and Elske van der Fliert. 1997. "Transforming Agricultural Extension for Sustainable Agriculture: The Case of Integrated Pest Management in Rice in Indonesia." In N. G. Röling and W. A. E. Wagemakers, eds., *Social Learning for Sustainable Agriculture.* Cambridge, U.K.: Cambridge University Press.

Rosegrant, M. W., and M. Agcaolli. 1994. *Global and Regional Food Demand, Supply, and Trade Prospects to 2010.* Washington, D.C.: International Food Policy Research Institute.

Sasakawa Global 2000. 1993–95. *Annual Reports.* Tokyo: Sasakawa Africa Association.

Schaller, N. 1993. "Sustainable Agriculture: Where Do We Go Now?" *Forum for Applied Research and Public Policy* (fall):78–82.

Scoones, Ian, and John Thompson, eds. 1994. *Beyond Farmer First: Rural People's Knowledge, Agricultural Research, and Extension Practice.* London: Intermediate Technology Publications.

Shah, Parmesh. 1994. "Village-Managed Extension Systems in India: Implications for Policy and Practice." In Ian Scoones and John Thompson, eds., *Beyond Farmer First.* London: Intermediate Technology Publications.

Sherief, A. K. 1991. "Kerala, India: Group Farming." *AERDD Bulletin* (University of Reading) 32:14–17.

Somers, B. M. 1997. "Learning about Sustainable Agriculture: The Case of Dutch Arable Farmers." In Niels Röling and W. A. E. Wagemakers, eds., *Social Learning for Sustainable Agriculture.* Cambridge, U.K.: Cambridge University Press.

Steer, A., and Ernst Lutz. 1993. "Measuring Environmentally Sustainable Development." *Finance and Development* (December):20–23.

SWCB (Soil and Water Conservation Branch). 1994. *The Impact of the Catchment Approach to Soil and Water Conservation: A Study of Six Catchments in Western, Rift Valley, and Central Provinces, Kenya.* Nairobi: Ministry of Agriculture.

Thai-German Highland Development Project. 1995. "Thai-German Highland Development Project Annual Report." Chiang Mai, Thailand. Processed.

UNDP (United Nations Development Programme). 1992. *The Benefits of Diversity: An Incentive toward Sustainable Agriculture.* New York.

van der Werf, E., and A. de Jager. 1992. *Ecological Agriculture in South India: An Agro-Economic Comparison and Study of Transition.* The Hague: Landbouw-Economisch Institut; Leusden: ETC Foundation.

Vereijken, Peter. 1990. "Research on Integrated Arable Farming and Organic Mixed Farming in the Netherlands." In C. A. Edwards, R. Lal, P. Madden, R. H. Miller, and G. House, eds., *Sustainable Agricultural Systems.* Ankeny, Iowa: Soil and Water Conservation Society.

Viederman, S. 1994. "Ecological Literacy: Can Colleges Save the World?" Keynote address to Associated Colleges of the Midwest Conference on Ecological Education, Beloit College, March 11. Jessie Smith Noyes Foundation, New York. Processed.

Winrock International. 1994. *Assisting Sustainable Food Production: Apathy or Action?* Arlington, Va.

Wirth, Tim. 1995. "U.S. Policy, Food Security, and Developing Countries." Presentation of the under-secretary of state for global affairs to the Committee on Agricultural Sustainability for Developing Countries, Washington, D.C. Processed.

World Bank. 1993. "Agricultural Sector Review." Agriculture and Natural Resources Department, Washington, D.C. Processed.

Yixian, G. 1991. "Improving China's Rice Cropping Systems." *Shell Agriculture* 10:28–30.

5

Small-Farmer Decisionmaking, Market Imperfections, and Natural Resource Management in Developing Countries

Stein T. Holden and Hans P. Binswanger

This chapter analyzes the decisionmaking of small farmers in the context of imperfect markets and its implications for efficiency and for sustainability of natural resource management. We examine the basis for market imperfections and their structure, pervasiveness, and implications for behavior (and vice versa). We discuss policies in relation to market imperfections with some historical hindsight, complemented by empirical studies and some theoretical and applied model studies. We end by discussing promising approaches to promoting sustainable intensification in small-farm agriculture. More research is required to develop and test these approaches.

Small farmers represent the majority of the population in developing countries. They represent even a larger share of the population below the poverty line because rural poverty is more extensive than urban poverty. They also represent a major link between the economy and the environment because their livelihoods depend so directly on the use of land resources (soil and vegetation). Small farmers are usually only partly integrated into markets. Typical market imperfections include missing markets, partly missing markets (rationing, seasonality), thin markets (imperfect competition), and interlinked markets. In a world with such market imperfections, incorrect or missing price signals may accrue from society's perspective and possibly result in inefficiencies. Possible outcomes include the too rapid extraction of and too low investment in natural resources. However, interventions or corrective measures to deal with these problems have to be defended on efficiency, equity, or sustainability grounds that incorporate the costs of intervention.

A world with transaction costs and imperfect information is almost always constrained Pareto-inefficient (Greenwald and Stiglitz 1986). Unfortunately, it is not obvious which interventions are most effective in reducing transaction costs, eliminating or reducing market failures (inefficiencies), and solving equity and sustainability problems. We prefer to distinguish between market imperfections and market failures and define market failures as a subcategory of market imperfections that implies efficiency losses. In a world with transaction costs, not all market imperfections may imply inefficiencies because the costs of correcting the imperfections may be higher than the benefits of doing so. Efficiency and equity concerns can no longer be separated, because redistributive policies may sometimes be defended on efficiency grounds alone. We relate this to the debate on the relative efficiency of small versus large farmers. We ask whether small farmers are too poor to be efficient and to manage natural resources carefully.

Policy failures in the form of neglect and severe taxation of the agricultural sector are largely to blame for poverty, economic stagnation, and decline in many low-income countries. New polices that stimulate rural development and promote sustainable management of natural resources are badly needed. To succeed, these policies have to build on an understanding of the decisionmaking environment of small farmers.

Policymakers have to experiment with the behavioral responses of these rural households, but they have a limited number of policy instruments. The fact that small farmers are rural households that are both production and consumption units complicates the analysis, particularly when market imperfections cause their production and consumption decisions to be nonseparable (Singh, Squire, and Strauss 1986 and de Janvry, Fafchamps, and Sadoulet 1991). Nonseparability implies that consumption needs and asset distribution may have significant impacts on production decisions and thus the management of natural resources.

Schultz (1964) has characterized small farmers as "poor but efficient." Research that has followed has shown that small farmers largely are rational and respond to changes in the set of constraints and opportunities they face. (The transaction cost and imperfect information school sees agents as acting in a way that advances their objectives, given the information and opportunities that they have; see Hoff, Braverman, and Stiglitz 1993.) Inefficiencies may still accrue due to economies of scale, poverty, and subsistence constraints, unequal access to markets, and various policies disfavoring small farmers. It is thus possible that small farmers are too poor to be efficient from society's perspective. These inefficiencies may also lead to underinvestments in conservation. An important question is therefore whether small farms are operated less sustainably than large farms. If small farmers are poor and poverty leads to myopic behavior, this could be a threat to sustainability. There may thus be grounds for interventions from the viewpoints of efficiency, equity, as well as sustainability.

The population-poverty-agriculture-environment nexus is complex and subject to much debate (Malthus 1987, Boserup 1965, Lele and Stone 1989, Cleaver and Schreiber 1994, Tiffen, Mortimore, and Gichuki 1994, Kates, Hyden, and Turner 1993, Heath and Binswanger 1996, and Holden and Sankhayan 1997). It appears that population growth under certain conditions may stimulate sustainable intensification, while in others it may lead to land degradation. The incentive structures and how these are affected by policies appear to have a strong impact on whether small farmers are able to choose a sustainable and welfare-improving development path or are forced onto a nonsustainable and welfare-reducing one. Population growth and higher population densities may reduce transaction costs and facilitate better market integration when intensification paths are open, and market integration may further stimulate intensification and investment in agricultural development. Particularly in Africa, where policy environments have been adverse to agriculture, and in areas that are poorly integrated into markets, small farmers appear to degrade their environment. Population growth is seen as fragmenting scarce land resources and leading to nonsustainable intensification under stagnant technology conditions (Grepperud 1994, World Bank 1996, Lele and Stone 1989, and Cleaver and Schreiber 1994). These rural pockets appear to be in a poverty-environment trap resembling a neo-Malthusian scenario. They represent one of the largest challenges to the developing and industrial world because they can lead to large-scale famines, forced migration, enormous refugee problems, and war. Induced technological and institutional innovation appear to be the best means, together with stable macroeconomic conditions, of switching from a Malthusian to a Boserupian development path (Hayami and Ruttan 1985). Agricultural research and extension, infrastructure development, basic education, and family planning are important elements, together with land tenure policies and other measures to provide secure livelihoods. Environmental degradation necessitates a better integration of agricultural and environmental policies and calls for a new and greener green revolution.

Covariate risk, asymmetric information, and transaction costs lead to pervasive market imperfections. These problems are even more acute in areas with low population density where the population derives a living from marginal land resources (Binswanger and Rosenzweig 1986a, Binswanger, McIntire, and Udry 1989, and McIntire 1993). Market imperfections in these types of areas may not be due primarily to market-distorting policy failures because they will persist after the market-distorting policies have been removed.

Interlinkage of markets may reduce the costs of information and transaction costs associated with such imperfections (Cheung 1969, Stiglitz 1974, Binswanger and Rosenzweig 1984, Williamson 1985, and Bardhan 1989). Braverman and Stiglitz (1986) have argued that interlinkages may affect the adoption of new technologies. We may thus look at interlinkages as endogenous institutional response that may have the potential to be powerful policy tools for influencing the adoption and development of technology (Holden and Shanmugaratnam 1995). Further research is required to investigate the potential of interlinkage, cross-compliance, or cost-sharing policies as tools for promoting sustainable natural resource management. We discuss some of the issues here.

Market Imperfections in Rural Economies

In this section, we provide an overview of market structures and their determinants in rural economies dominated by rainfed agriculture, drawing heavily on Binswanger and Rosenzweig (1986a), Binswanger, McIntire, and Udry (1989), and Hoff, Braverman, and Stiglitz (1993). The focus is both on material and behavioral determinants of institutional structures and on particular markets. We start out by using a rough typology of village economies (Holden, Taylor, and Hampton, forthcoming), which is illustrated in figures 5.1 and 5.2. Village economies are characterized along two main dimensions, in relation to transaction costs or isolation from the outside world and to the degree of internal differentiation of access to resources. This diagram gives four extreme corner situations, with the neoclassical model as the best representation when transaction costs are zero or minimal. Local trade is only relevant when there is an internal differentiation in access to resources and when local transaction costs do not prohibit local trade. With isolation, due to long distances, poor infrastructure, and poor access to transportation technologies, external markets may become missing or only seasonal in character. The result may be endogenous village prices that may be observed in local markets or be shadow prices that may not be observed and are internal to households and various contractual arrangements. Dynamic processes as well as policy interventions may change the positions of villages in the diagram. Villages in land-abundant economies (low population density) may be located close to the upper-left-hand corner, while population growth tends to create a move in the southeastern direction in the diagram. Wealth accumulation and specialization usually lead to a movement from left to right (increased differentiation), while a land reform or a progressive tax may cause a move in the opposite direction. Infrastructure investments and pan-territorial pricing may reduce transportation costs and the degree of isolation from the outside world for remote villages (downward movement), while the removal of pan-territorial prices and reduced investments in infrastructure may lead to an upward movement (Holden 1997).

The following material conditions contribute to market imperfections (Binswanger, McIntire, and Udry 1989):

- Dispersed, low-density population
- High transportation costs
- Seasonal rainfall causing seasonal demand for agricultural labor
- Simple technologies without significant economies of scale.

Figure 5.1. Village Economy Typology

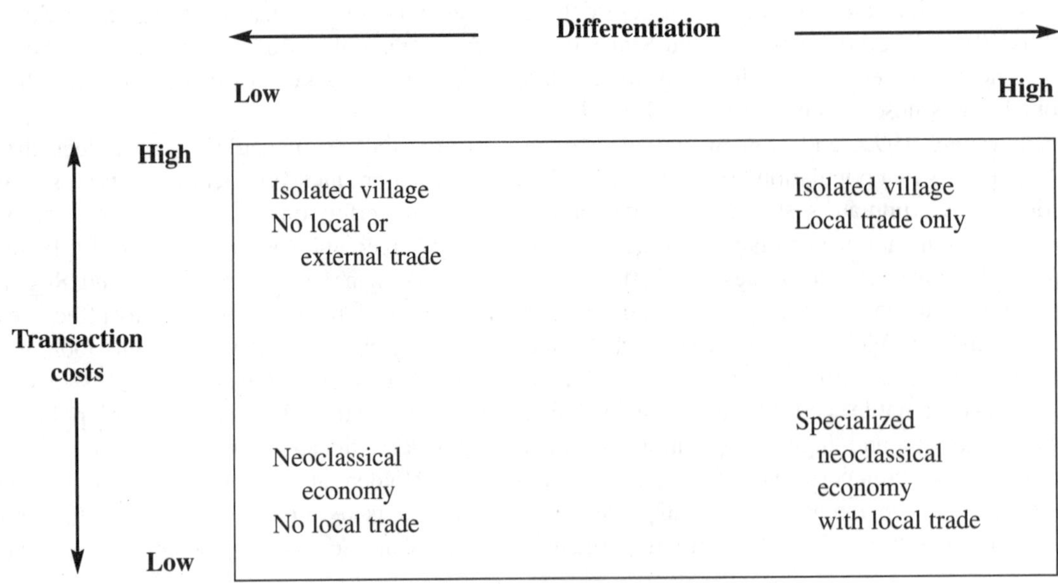

The following behavioral and ability assumptions are important:

- Individuals are interested in their own well-being and value consumption.
- Individuals have limited access to information about their environment, and obtaining information is costly.
- Individuals are averse to drudgery above a certain level of effort.
- Individuals face risks and are in general averse to risk. The degree of risk aversion varies among individuals and between levels of wealth.
- Individuals have minimum subsistence requirements.

The following implications of the behavioral and ability assumptions are the most important (Hoff, Braverman, and Stiglitz 1993; Binswanger and Rosenzweig 1986a):

- Asymmetric information leads to problems of adverse selection and moral hazard, which increase the costs of contract formation, monitoring, and enforcement.
- Economies in general are not constrained Pareto-efficient. Individual rational behavior does not necessarily lead to socially optimal outcomes.
- Allocational efficiency itself depends on the distribution of resources.
- Asymmetric information and transaction costs lead to new types of market failures (Pareto-relevant externalities) and institutional responses to reduce their costs.
- New sorts of market failures may imply new sorts of policy solutions.
- Local institutions may play a significant role in implementing policies aimed at eliminating market failures.
- High weather risk causes covariate crop yields and the failure of intertemporal markets.
- Poverty and imperfections in intertemporal markets may lead to intertemporal externalities that require remedial action on sustainability grounds.

The government may have a central role to play in stimulating social innovation and cooperative solutions to natural resource management in rural areas and in imposing sustainability constraints to protect the interests of future generations.

In the following section, we outline typical market imperfections that are found in rural areas with low- and high-density populations in developing countries, interactions among imperfect markets, nonmarket institutions, and dynamic processes affecting the distribution of resources and human welfare.

How Imperfect Are Intertemporal Markets?

There are sound a priori reasons for supposing that, in a low-income rural setting, credit and crop insurance markets will be either highly imperfect or simply

Figure 5.2. Typology of Village Economy Models

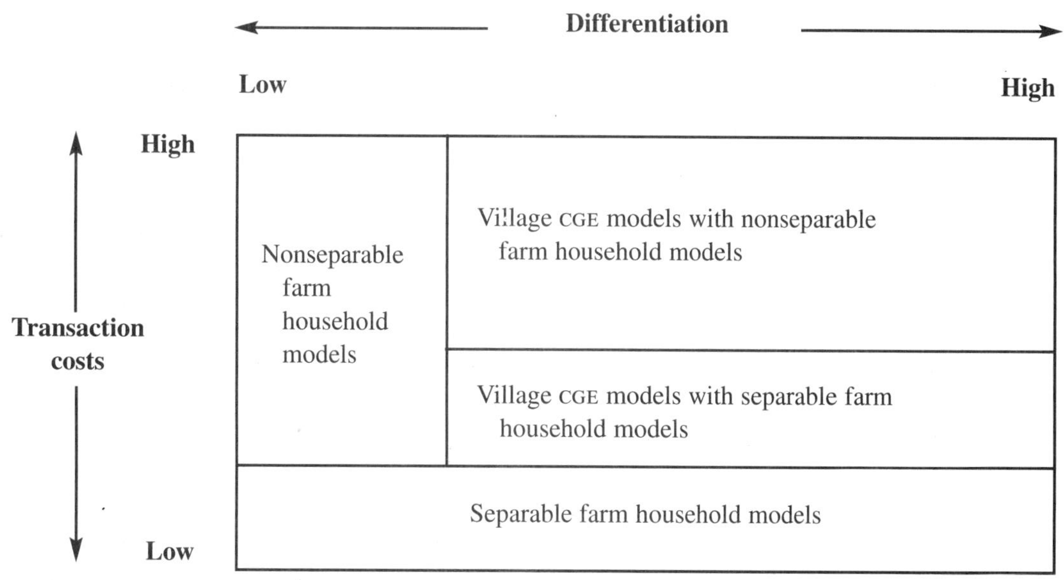

nonexistent (Binswanger and Rosenzweig 1986a). To begin with, there is a problem of moral hazard. It is hard for providers of credit or insurance to screen applicants for creditworthiness or moral integrity and costly to administer credit and insurance extended to a multitude of small producers. Information is asymmetric: clients are able to mislead providers about key indicators such as crop yields, thus altering in their favor the terms of contracts or the amount of compensation paid.

Collateral can be used to circumvent moral hazard. However, farmers may lack assets or their assets may be hard to collateralize (such as movable goods), making it less likely that lenders will deem them creditworthy.

Second, *high covariance discourages the provision of both credit and insurance:* spatially in relation to yield risk and temporally in relation to the sequence of depositing and borrowing. In the event of drought, a large number of farmers in the affected region will simultaneously default on their loans and submit insurance claims. Also, because all farmers are bound by the seasons to a fairly uniform cropping calendar, deposits and withdrawals from the credit fund will be synchronized. To deal with the resulting covariance of deposits, withdrawals, and claims, credit and insurance providers must set aside larger cash reserves than they would if there were lower covariance between the borrowing and claiming activities of their clients.

Crop insurance (other than for very specific risks) is therefore not commercially viable (Hazell, Pomareda, and Valdes 1986), and rural credit markets are usually poorly developed. Empirical evidence assembled by village-level studies of the International Crop Research Institute for the Semi-Arid Tropics (ICRISAT) lends additional support to the a priori arguments (Binswanger and others 1985 and Binswanger and Rosenzweig 1986b). The research examines longitudinal data from village households that ICRISAT monitored between 1978 and 1985. The survey villages were selected on the basis of agroclimatic risk to allow for controlled comparison of the impact of shallow soils and unreliable rainfall on the provision of credit. Formal credit facilities were found to be more developed in low-risk than in high-risk villages. This finding was confirmed independently by analyzing district-level data from 80 districts in India (Binswanger and Khandker 1989).

In the ICRISAT villages, the growth of bank lending was associated with a reduction in the relative importance of moneylenders after the most creditworthy clients gravitated toward formal sector loans. However, both traditional and modern credit arrangements failed to serve producers with few assets. Moneylenders were only prepared to lend for the duration of a crop season or less, offering small amounts (much smaller than the expected value of output). The takeup of poor farmers by the formal system was impeded by the unsatisfactory repayment record of bank clients (inferior to that of the moneylenders' clients), threatening the solvency of these institutions. The high default rate had many causes, including inadequate screening procedures, scarcity of adequate collateral, political impediments to the seizure of assets, and the high level of agroclimatic risk in certain regions.

Another indication that intertemperal markets in these villages are highly imperfect comes from an experimental study of individual-specific internal rates of discount (Pender 1996). Farmers were given the option of choosing bags of rice today or larger bags of rice 12 months from today. About one-third of the sample always chose rice today, no matter what the tradeoff, thereby exhibiting internal rates of discount in excess of 100 percent. The mean value of internal rates of discount for the other two-thirds ranged from 0.33 to 0.41, depending on the precise nature of the games offered. These internal rates of discount were well in excess of the highest interest rates paid by each of the farmers. The results indicate severe credit rationing. Internal rates of discount were much higher for the poor than for the wealthier farmers.

These credit and insurance markets imply that inherited endowments are central to the ability of farmers to acquire productive assets out of equity and to engage in additional borrowing for investments and for working capital. Therefore, inherited endowments, not just abilities, determine the capital stock with which an individual has to work, *and inefficiency ensues.*

This inefficiency is consistent with the results of an investigation of whether poverty is transmitted across generations via imperfect markets (Singh and Binswanger 1988). A comparison was made of the assets, income, and inherited goods of two generations (valued at 1983 prices), controlling for the level of schooling. The panel format (unlike cross-sectional data) makes it possible to estimate the policy effect of providing individuals with additional schooling or assets, such as land: unobservable ability variables do not contaminate the results. The research findings demonstrate that the level of inherited wealth and schooling strongly influences the individual's capacity to earn and accumulate capital. In contrast, with perfect markets, the capacity to earn and accumulate should depend on ability, not inherited wealth.

Having established the absence or imperfection of intertemporal markets, the next step is to ask whether producers adopt risk-diffusing strategies and whether such strategies offer full compensation for the inadequacy of credit and insurance markets, in the sense that they enable producers to maximize expected profits.

Do informal mechanisms for diffusing risk significantly reduce consumption risk? ICRISAT village-level studies and other studies have described many of the ex ante and ex post risk diffusion mechanisms used by farmers. These include crop diversification, intercropping, storage of fodder and grain, and sharecropping (Jodha 1978, Walker and Jodha 1986, and Walker and Ryan 1990). In addition to these practices, there are other methods of diffusing risk involving capital accumulation and social relations.

First, storage and savings are significant forms of risk diffusion. The ownership of bullocks emerged as the preferred means of accumulating "buffer stock," presenting low risks because these animals are fairly resistant to disease and drought. However, the transportability of livestock reduces its value as collateral, so the ownership of bullocks does not greatly facilitate borrowing. Also, the absence of a rental market for bullocks (owing to the synchronicity of demand for their services) prevents the owner from realizing cash through leasing. These factors account for the frequent sale of bullocks, which tends to coincide with periods of low income and enables households to smooth consumption over time (Rosenzweig and Wolpin 1989).

Second, it is important to consider the impact of female marriage contracts (Rosenzweig and Stark 1989). The researchers hypothesized that marrying daughters to locationally distant but kinship-related households, as is typical in South India, is an attempt to mitigate income risks and smooth consumption in an environment characterized by high information costs and spatially covariant risks. The ICRISAT village data indicate that marriage and migration significantly reduce the variability of a household's food consumption because they pool income between households in different regions and the risks attached to the various sources of income are not highly covariant. Households exposed to higher income risks are more likely to invest in longer-distance migration and marriage arrangements. The research findings suggest that agricultural technical change may significantly affect spatial marriage patterns by altering the spatial covariance and level of risk, rendering the assessment of risk and the establishment of implicit risk arrangements more difficult. If formal credit markets are improved so that they facilitate consumption smoothing, risk will play a smaller role in marital arrangements and rural migration, perhaps diminishing rural mobility.

Third, the membership, size, and composition of households and cross-household kinship ties contribute further to smoothing consumption and mitigating risk. The membership of household units typically spans the generations, allowing knowledge of optimal productive packages to be passed down from young to old. Households entering into new, implicit contracts with other households reduce their participation in sharecropping, the previous form of risk protection, and increase the extent to which transfers, most of which originate outside the village, mitigate the volatility of intertemporal consumption. The intergenerational and spatial extension of households helps to reduce their risk ex post, thereby making them more able to bear risk ex ante through the organization of productive resources (Rosenzweig 1988a, 1988b).

Does the undiffused risk that remains affect consumption, farm investment, and its efficiency? The research found that although farmers adopt these multiple risk-mitigating strategies, risk is not sufficiently diffused to permit farmers to maintain consumption steady over time and to maximize profits. In other words, the strategies fail to compensate fully for the efficiency-inhibiting impact of nonexistent or imperfect markets for credit and insurance. If risk diffusion is not sufficient, farmers tend to invest too heavily in inputs that reduce risk and to invest too little in high-risk inputs. This produces a factor mix that is inefficient from the standpoint of profit maximization.

The ICRISAT data provide a unique opportunity to test these hypotheses. Rosenzweig and Binswanger (1994) first show that individual-specific income shocks do not affect food consumption of households. However, covariate village-specific income shocks do affect household food consumption. This suggests that the informal mechanisms for risk diffusion are insufficient to fully diffuse risks, which therefore are likely to spill over into investment behavior.

To test this hypothesis the authors estimate a profit function, using a flexible functional form, incorporating nine capital inputs, two rainfall measures, and personal characteristics such as family composition, age, and education. The estimation procedure exploits the panel data and allows for the possibility that unmeasured farmer attributes and abilities influence the composition and level of investment. For each of the investment inputs, the marginal contributions to profits and riskiness of the portfolio are computed.

The test statistics (Binswanger and Rosenzweig 1989: table 3) show that the *investment portfolios of poor farmers are not allocated to maximize profits.* Indeed, uninsured weather risk is a significant cause of lower efficiency and lower average incomes: a decrease of 1 standard deviation in the timing of the rainy season would raise average profits by up to 35 percent among farmers in the poorest quintile.

Wealthier farmers, in contrast, are able to absorb risk, and the profitability of their investment is not affected by an increase in riskiness.

The results indicate that improving the ability of poor farmers to smooth consumption, perhaps via increased consumption credit, would increase the profitability of their agricultural investments. Similarly, providing insurance against failure of the monsoon, if feasible, would raise the overall level of profit and decrease the inequality of earnings in high-risk areas. The research also demonstrates that differences in investment portfolios across wealth classes do not arise from technical scale economies. Thus differences in the composition of investment principally reflect differences in the capacity of farmers to absorb risk and differences in objective risk (variable weather).

The investment portfolio also reflects the liquidity of the assets, that is, their usefulness in permitting the farmer to compensate for unforeseen losses. In particular, farmers frequently sell bullocks in the event of drought-induced crop failure. In South India, the frequent selling of bullocks (to permit consumption smoothing) means that *average bullock stocks are typically below the level necessary to ensure maximum profitability and efficiency of farms.* Simulations based on parameters estimated from a dynamic structural model applied to the ICRISAT data demonstrate that augmenting sources of off-farm income (such as welfare payments, rural industry, and public works programs) reduces the farmer's need to sell bullocks and boosts the stock of bullocks, which could raise agricultural productivity (Rosenzweig and Wolpin 1989).

To sum up, poor farmers are less likely to diffuse risk successfully because they have fewer assets for collateral and fewer "buffer" stocks. The research in India concludes that risk aversion varies inversely with wealth. Poor farmers are less able to cope with uncertain weather. One of the most important findings is that, contrary to Schultz (1964), poor farmers are not able to maximize profits: specifically, in the ICRISAT villages, for every rupee invested, poor farmers obtain 30 percent less than they would if they had been able to adopt a profit-maximizing portfolio (Binswanger and Rosenzweig 1984). *Thus poor farmers are too poor to be efficient.*

Small Farms, Efficiency, and Sustainability

Efficiency and equity discussions have been at the heart of the analysis of interlinked markets or contracts, while sustainability issues rarely have been addressed.

Interlinkage and Efficiency

The theories of interlinkage emerged from efforts to understand the persistence of share tenancy, which Marshall (1920) claims is an inefficient institutional arrangement because the tenant receives only a share of his marginal benefit but has to cover the entire marginal cost. The works of Cheung (1969) and Stiglitz (1974) claim that share tenancy may be a superior solution to fixed-rent and owner-operator (with hired labor) contracts because it provides a combination of risk-sharing and incentive effects that represent a compromise solution.

Interlinked contracts may implicitly provide credit and insurance to tenants in environments where credit and insurance markets are imperfect (Otsuka, Chuma, and Hayami 1992). Cost-sharing arrangements may relieve the working capital constraints of tenants. Nevertheless, adjusting for quality differences in plots, farmers in the ICRISAT villages had a 16 percent lower efficiency on their sharecropped plots than their own plots. For further review of the empirical literature, see Otsuka, Chuma and Hayami, 1992.

Share tenancy represents primarily the interlinkage of labor and land markets (and the risk and insurance market) but may also include credit and input markets (with a choice of technology implications). Other examples of interlinked markets include credit and land (collateral) markets; credit, input, and output markets (often government-sponsored programs); labor and output markets (food-for-work contracts), which often provide insurance as well (food security); and simple barter trade (linking two commodity markets).

Small Farms and Efficiency

A common belief is that small farmers eventually will become either landless workers or commercial farmers and thus only represent a temporary mode of production (Ellis 1988 and Sadoulet and de Janvry 1995). This belief has led to an overemphasis on the development of large mechanized farms and collective farms. However, these attempts to benefit from economies of scale frequently have been countered by large moral hazard problems.

Even though we have seen in the last section that small farmers have great difficulties to diffuse covariate risk in poorly developed intertemporal markets, many studies indicate that small farms may be more efficient than large farms. Few of these studies have used a comprehensive total factor productivity approach for the estimation. Some studies in developing countries hold a higher methodological quality (Berry and Cline 1979 and Prosterman and Riedinger 1987). They also show an inverse relation between farm size and productivity. Differences in land quality and use of capital cannot explain all of the negative relationship. Moral hazard problems favor family farms over large farms depending on hired labor. Berry and Cline (1979) find that the second smallest class of farm, which includes the smallest full-time farmers, has the highest output per unit of area.

The optimal size of family farms may have increased as a result of economies of scale due to mechanization, an increase in the acceptable level of family income, and a fall in the price of farm output, which could be avoided by engaging in some off-farm activities. Cash scarcity, lack of insurance against production and market risks, poorer access to credit markets, higher interest rates, and indivisibility of input and capital investments may decrease the use of inputs and increase the use of less-efficient technologies on small farms. Imperfections in more than one market are required to obtain the inverse relationship between farm size and productivity. Srinivasan (1972) has shown that a combination of missing land (rental) and insurance markets under conditions of risk and risk-averse producers (with nonincreasing absolute risk aversion and nondecreasing relative risk aversion in wealth) will yield the type of inverse relationship observed. Similarly, Feder (1985) and Carter and Kalfayan (1989) have shown that imperfections in credit and labor markets may yield similar results.

Economies of scale in processing and the need for close coordination of harvesting, processing, and marketing have led to the development of plantations for sugarcane, tea, oil palm, and banana. However, contract farming may eliminate the link of farm production with the processing plant (where the economies of scale are located) via joint ownership in a plantation. Examples are sugarcane in India and Thailand, tea in Sri Lanka, and oil palm in Indonesia. The plantation mode of production has therefore declined sharply at the expense of contract farming with smallholders (Binswanger, Deininger, and Feder 1995). Rosenzweig and Binswanger (1994) showed that in the ICRISAT villages there are no technical economies of scale at the farm level. They also showed that, despite their inferior risk diffusion capacity discussed in the last section, small farmers have a higher profit rate than large farmers, indicating no productivity advantages. Indeed, *together with the inefficiencies associated with sharecropping, these results indicate that a redistribution of assets (mainly land) could increase productivity.*

However, the empirical literature also suggests that there may be economies of scale at very small farm sizes and diseconomies of scale at larger farm sizes. Thus excessive fragmentation could be harmful for efficiency. If population growth is pushed as far as to lead to excessive fragmentation and poverty, population growth may be bad for efficiency, leading to a population-poverty trap and reinforcing the Malthusian scenario. This outcome could be prevented by further intensification, diversification to high-income crops, off-farm opportunities (market integration), migration, income-increasing technologies, and family planning.

Small Farms and Sustainability

Missing credit markets in combination with poverty and risk may cause poor farmers to have very high discount rates, as shown by Pender (1996) in India and Holden, Shiferaw, and Wik (1998) in Ethiopia, Indonesia, and Zambia. Pender finds that as much as one-third of the population has discount rates above 100 percent. Holden, Shiferaw, and Wik find average rates of 93 percent among transmigration settlers in Seberida, Sumatra, 105 percent among peasant households in northern Zambia, and 53 percent among small farmers in a high-potential cereal zone in the highlands of Ethiopia. Average incomes per capita are $107 in Sumatra, $108 in Zambia, and $196 in Ethiopia. Shortage of cash appears to have a strong impact on the discount rates in Sumatra and Zambia, where the average income levels are lowest. Wealth in the form of livestock and total income is most strongly correlated (negatively) with the discount rate in Ethiopia.

These studies have shown a remarkable consistency across sites, which supports the hypothesis that poverty combined with liquidity constraints leads to high discount rates. This implies that poverty reduction itself or profit-increasing reform may reduce the intertemporal externality, because the discount rates of the poor will decline as incomes increase. However, in locations where severe environmental degradation is taking place, the high discount rates found in these studies point to the need for additional interventions to stimulate conservation.

What are the implications of high discount rates for investment in conservation technologies? This intertemporal externality may create incentives for nonsustainable resource extraction as a short-term survival strategy.

Under conditions of poor social organization and/or adverse policies, population pressure may also lead to encroachment into and production of food crops on fragile steep slopes, which may be rapidly degraded. Livestock may serve as insurance, causing farmers to build up their herds beyond the carrying capacity of the land, leading to overgrazing and inefficient use of grazing lands.

Are small farms more efficient in some cases because they are more intensively run, making higher efficiency no more than a rapid mining of the soil and vegetation? The Indonesian rubber farmers with small plots and few trees have managed to increase their rubber production per tree significantly in the short run by tapping their rubber trees as frequently as every day despite the recommendation to tap young trees only twice a week (Holden, Hvoslef, and Simanjuntak 1995). However, in the *chitemene* shifting cultivation system in northern Zambia, the level of income or material welfare depends directly on the extent of cultivation by the individual household. Labor-rich households therefore have higher incomes but also cause the most deforestation when the carrying capacity of the system has been exceeded. Female-headed households and labor-poor households require only one-third to half of the area per capita required by labor-rich households (Holden 1991). In this case, better market access or higher income could imply more rapid deforestation.

Does improved market access make small farm production more sustainable? The vast literature on Boserup effects clearly shows that this is not generally the case. Higher profitability from market access tends to induce more land investment. For, example, improved market access may create incentives for intensification by encouraging farmers to switch from systems with low to systems with high external inputs. An example is the promotion of maize production in Zambia, which reduced the extent of shifting cultivation (Holden 1991, 1997).

However, in some cases, improved market access leads to more rapid deforestation and extraction of forest products due to higher output prices. Under conditions of uncertain or missing tenure rights, improved market access may initiate more clearing of land as a way to obtain property rights to land (homesteading; Angelsen 1997). The immediate effect may thus be more deforestation.

Models of Farm Households

Further insight on these issues can be gained by drawing on modeling efforts. Chayanov (1966) was the first to develop farm household models. He did this for Russian rural society, which was characterized by land abundance and no labor market during the 1920s. His models later were formalized into a neoclassical theoretical framework by Nakajima (1986), who has also developed models for a number of specific market structure settings.

The neoclassical farm household model developed by Barnum and Squire (1979) is a separable model that contains only one missing market (for land) and ignores intertemporal markets (credit, risk and insurance, and investment decisions). Production decisions can still be treated as independent of consumption decisions.

The farm household model becomes nonseparable in the presence of a missing labor market, significant risk (and lack of insurance), credit market imperfections, and/or binding subsistence requirements. With nonseparability, a farm household approach becomes essential for policy analysis because production decisions depend on the consumption side of the model and vice versa. Endogenous shadow prices link the sides, and fewer qualitative results may be drawn unambiguously from theoretical models.

De Janvry, Fafchamps, and Sadoulet (1991) have developed a general farm household model with missing markets. They have developed a numerical model for a typical African peasant household that simulates situations with missing markets for food and labor, labor only, food only, and no missing markets. They illustrate the effects of a price change for a cash crop in each of these four cases. They show that the supply response for the cash crop is substantially reduced in the case with market imperfections. The reduction is greatest in the absence of both labor and food crop markets. The reduction is less when only the labor market is missing because households can substitute from the production of food crops to the production of cash crops and purchase more of their food. With a missing food crop market, this substitution is much less possible, while hired labor may substitute for family labor in the production of cash crops.

De Janvry and others (1992) use nonseparable household models for small and medium farms in Morocco to simulate the effects of an agricultural structural adjustment program in a case with a missing market for child labor and milk and a credit constraint. Children are used to herd animals in communal grazing lands. The results show that the credit constraint severely limits the ability of farmers to hire labor, rent machinery, and buy fertilizer and thus lowers the supply response in agricultural production. The adjustment program favored the production of crops over livestock, which reduced the production of milk and beef. Rising forage prices, however, caused a substitution into more communal grazing and more use of child labor, with increased overgrazing and school absenteeism as negative side effects. This is an example of how market imperfections can contribute to unpredictable and paradoxical effects.

Holden, Taylor, and Hampton (forthcoming) have developed a village economy model for a typical remote village in northern Zambia, consisting of two independent nonseparable farm household models representing two groups of households, one more market oriented than the other. Unlike the model of de Janvry and others, these models have a credit constraint, the cash crop (maize) is partly consumed by the households, and the food crops (millet, groundnuts, cassava, and beans) are produced in an extensive system of shifting cultivation and are partly marketed. Credit institutions provide formal credit for the purchase of fertilizer and improved seeds of maize and require that the maize be marketed through them (interlinkage of input, credit, and output markets). The model is used to analyze the responses to changes in cash crop price, fertilizer price, and provision of credit. Historically, peasants in northern Zambia have been very responsive to the possibility of producing this cash crop if they are given access to input and credit markets. The introduction of structural adjustment policies removed the input, output, and credit subsidies and the implicit transportation subsidies of the pan-territorial pricing system. This resulted in a less favorable ratio of fertilizer/output prices for peasants in remote locations in northern Zambia, reducing the price of the cash crop. The model is able to trace the consequences of any policy change to the expansion of shifting cultivation of food crops, which can lead to deforestation and land degradation, if carrying capacity has been exceeded.

The model is then used to simulate this price response. A 10 percent reduction in the price of maize reduces the profit in maize production by 16–19 percent, reduces marketed surplus of maize by 2–3 percent, increases shifting cultivation by 2–6 percent, increases marketed surplus of food crops by 14–44 percent, and reduces full income by 1–4 percent. This occurs under the assumption that there is no change in the rationed supply of credit, implying no change in fertilizer use. The privatization of transportation costs may thus result in more extensive cultivation practices and more deforestation and land degradation in remote regions. These model results compare well with empirical findings (Holden 1997).

Some recent theoretical developments have significance for how risk may influence the behavior of small farmers. Sandmo (1971) has established that a risk-averse firm facing output risk will produce less output than a risk-neutral firm. Small farmers have been found to be risk-averse (Binswanger 1980 and Binswanger and Sillers 1983). Credit market imperfections and the inability of poor farmers to pool risks across time may be complementary explanations for high discount rates and lower use of cash inputs (Binswanger 1980, Binswanger and Sillers 1983, Eswaran and Kotwal 1990, and Morduch 1995).

Risk has also been used to explain the heterogeneity of production (crop diversification) that may reduce overall production risk. However, other explanations, like the need for seasonally more even distribution of labor, heterogeneity of soils (Bellon and Taylor 1993), and commodity market imperfections combined with heterogeneous demand for food (complementary commodities), may be important as well (Holden 1991 and Wik and Holden 1996).

Finkelshtain and Chalfant (1991) have argued that a peasant household that consumes part of its production and faces output price risk really faces multivariate risk because the price of a consumption good is random. They show that such a household, which is risk-averse, may even choose to produce more than a risk-neutral profit-maximizer when the price is random. Sandmo's qualitative result therefore does not always hold for marketed surplus-producing households. Sandmo's result does hold if the good produced is a normal good. However, if the good is an inferior good, the household may increase its production with increasing risk. For a net buyer of food, an inferior good will be produced at higher quantities under risk than under certainty. For a normal good, the Sandmo result may or not hold for net buyers.

Wik and Holden (1996) have tested the relevance of the Finkelshtain and Chalfant result for peasant households in northern Zambia. Adjustment policies imply a removal of price controls for maize, the main cash crop there, and this results in more variable prices. Because maize is found to be a normal good and most producers are net sellers and are risk-averse, the Sandmo result should hold; the increase in risk as a consequence of the introduction of the adjustment policy (assuming no change in average relative prices) should result in lower production of maize. They also show that the increase in price risk is likely to lead to more credit rationing because it will lead to higher default rates and lower expected returns to the lenders. This has been confirmed by Tviland (1996), who finds that, in a sample of 161 households, the share of peasants who received formal credit dropped from 56 percent in 1993 to 14 percent in 1995. This is not explained by a change in credit demand, as approximately 60 percent of the peasants applied for credit in both years. The default rate for those who received credit increased from 15 to 35 percent from 1994 to 1995. (However, this may also be a result of a collapsed credit system that did not consist of rational lenders.)

Models of Deforestation and Land Degradation or Conservation

Much of tropical deforestation is caused by agents other than small farmers as shown by FAO (1996, cited in chapter 22 of this volume). Nevertheless, labor constraints, cash constraints, limited access to input, credit, and output markets, and subsistence requirements may be highly relevant for the decisions of small farmers to clear forests. Models that ignore the resulting nonseparabilities may fail to give good predictions about behavioral responses to policy interventions.

Most dynamic models of soil degradation and conservation have assumed separability (McConnel 1983, Barbier 1990, LaFranche 1992, and Shiferaw 1996). As noted by McConnel, when markets are perfect, the private intertemporal path for soil use will mimic the socially efficient path. Thus there are no reasons to be concerned with land degradation unless some market imperfections are Pareto-relevant or relevant to social welfare for other reasons.

Pagiola (1995) has developed a dynamic farm household model with subsistence requirements and uses it to explore the conditions under which subsistence requirements prevent the adoption of sustainable land use practices. He discusses four cases and shows that under certain conditions (severe poverty), *poor farmers may have more incentives to adopt sustainable practices than other farmers* because the future disutility of degrading the resource is potentially unbounded.

Conversely, he also looks at the case where subsistence constraints prevent farmers from adopting sustainable practices. *Farmers may be too poor and lack the access to credit markets that would enable them to make the necessary investments.* When sustainable practices do not exist or are unattainable, poor farmers without migration options degrade the soil more slowly than households with migration options.

Holden, Taylor, and Hampton (forthcoming) have also shown that households with high (constant) discount rates may save and invest more to secure future survival as they become poorer but that their level of savings and investment is low at higher income levels. However, if the discount rate is increasing with falling income, this pulls in the opposite direction. Households may reach a switch point at which they see no hope for survival in the future.

Grepperud (1997) has developed a dynamic model of soil depletion choices under production and price uncertainty. It is a model with one market imperfection only (risk and insurance). The model has three control variables; productivity increasing but degrading input, conservation but productivity-reducing input, and "win-win" input, which both increases productivity and conserves the soil. The model illustrates why risk and risk aversion have unclear effects on conservation (Anderson and Thampapillai 1990 and Ardilla and Innes 1993). Detailed knowledge of the farming systems and sources of risk is required to predict the direction of effects. One of the conclusions of the model is that if output price is uncertain, risk aversion may induce farmers to use less of all inputs. Another is that uncertain production may induce farmers to use fewer degrading inputs and more conservation inputs. Thus production uncertainty may be more favorable for conservation than price uncertainty.

Pender and Kerr (1996) have developed a theoretical model for farm household conservation decisions with market imperfections. It is a two-period model incorporating several categories of assets. They analyze the four cases: perfect markets, a missing labor market, a missing credit market, and both labor and credit markets missing. They also have a category of nontradable assets. They assume a strongly concave production function with all cross-partial derivatives as non-negative and a strictly concave utility function. They derive comparative statics results and show that conservation investments are independent of labor endowments and savings but may be influenced by the level of fixed assets and the initial level of conservation investments. With a missing labor market, investment in conservation will increase with the labor endowment of households. The impact of other categories of assets is negative or ambiguous. With a missing credit market, investment in conservation will increase with the household's amount of saving. Effects of labor endowments and other assets are ambiguous. If both labor and credit markets are missing, conservation investment will still increase with savings, while the impact of labor endowments and other categories is ambiguous.

Pender and Kerr (1996) have conducted an econometric study of conservation investment decisions in three (ICRISAT) villages in India. They test for the possible impact of market imperfections using their own theoretical model and find that imperfect land markets cause farmers to invest less in conservation on leased land in two of the study villages (Aurpalle and Shirapur). Owners invest more in their land than tenants do. Indian land tenure laws, which undermine long-term leases, may explain this effect, because a long-term lease would give the tenant right to the land. In two of the villages (Aurpalle and Kanzara), conservation investments decrease with farm size, while in the third village (Shirapur) conservation investments increase with farm size. These effects may also be signs of land and other market imperfections, leading to endogenous land prices at the farm level. Conservation investment increases with the adult male work force but, strangely, decreases with the adult female work force, nevertheless indicating imperfections in labor markets. The study does not analyze how investment per unit of area changes with farm size, however. The study finds that conservation investment increases with debt, which may be a sign that those with low debt are credit constrained. Savings affect conservation investments positively in only one of the villages, indicating some credit market imperfections.

Shiferaw and Holden (1997b) have developed an applied farm household model incorporating the user's cost of land and the conservation decisions of farm households in the Ethiopian highlands. The model includes subsistence requirements, liquidity, and credit market constraints and is thus a nonseparable farm household model. Short- and long-term responses are analyzed for cases when conservation technologies reduce, make no difference, and increase yields. Low (10 percent) and average (53 percent) discount rates are used to estimate user costs of land and then are compared with results when user costs are ignored. *The results indicate that conservation technologies will be adopted when they do not negatively affect yields.* (See also chapter 16.) However, if they have a negative effect on yields in the short run, which is usually assumed to be the case in the Ethiopian highlands, these technologies will not be adopted unless the farmer has a low discount rate (10 percent) or is provided incentives to reduce his short-term costs or increase his short-term benefits. If the social rate of discount is low (less than 10 percent) compared with the private rate (53 percent), this is an intertemporal externality that calls for intervention.

Shiferaw and Holden (1997a) have used the same model as a baseline for testing the efficiency of various policy measures to reduce land degradation and increase conservation investments. We return to the results of this exercise when we discuss various policy measures to promote sustainable natural resource management.

Policy and Market Imperfections

Ideology had a considerable impact on policy design in the past. Leftists thought that markets were "evil capitalistic creatures" that should be condemned and replaced with more fair systems of exchange and sharing. The influence of leftist ideologies or of rent-seeking behavior in developing countries caused the implementation of policies aimed at replacing markets to ensure a more fair distribution but neglecting the implications for efficiency. The neglect of behavioral conditions undermined not only efficiency but also equity because the policies frequently had a strong urban bias and reinforced rural underdevelopment. Parallel "black markets" developed and allowed the elite and corrupt persons to make fortunes at the expense of the poor.

There are also many examples of well-intended but unsuccessful interventions to promote agricultural development, including price, credit, input, marketing, and research and extension policies. These unfortunate developments were partly due to a lack of understanding of the functioning of public institutions. Improper incentives and control mechanisms resulted in problems related to moral hazard, corruption, and growing rather than declining transaction costs. With the economic decline and debt crisis in the early 1980s, the role and functioning of the state were assessed more critically, and state activities were reduced and reorganized.

Some policies were aimed at correcting market imperfections, which were believed to represent market failures, but more frequently interventions were made on grounds other than efficiency. Interventions to establish credit markets where they were missing or where interest rates in informal credit markets were very high may have been implemented on efficiency as well as equity grounds. In many cases, they were not successful. Price regulation in food crop markets was another key arena for interventions with limited success, often causing severe consequences such as stagnant agricultural and rural development, increased rural-urban migration, and huge efficiency losses. This, in addition to the debt crisis, economic stagnation, and decline facing many developing countries, is the context in which adjustment policies were implemented beginning in the early 1980s.

State property rights to land have frequently resulted in de facto open access and degradation of natural resources by undermining customary rights and informal cooperative management systems. These traditional systems often provide security of access to land. Institutionalization of private property rights has not succeeded in many developing countries due to the high costs of implementation and maintenance. The introduction of formal title has not improved land productivity in Africa (Migot-Adholla and others 1993), and the absence of formal title is therefore not considered a constraint to development. Rather, there is evidence that customary rights regimes in Africa are becoming more privatized as population pressure and commercialization grow. The titling program in Kenya has been expensive and had no significant positive effects on security of ownership or access to credit. Feder (1993), however, has found that in Thailand land titles have a positive effect on access to credit, since land can be used as collateral in a well-developed credit system. By facilitating access to credit, land titles have had a positive effect on the level of investment and output and on market values of land.

Can we trust that removing policy distortions will solve the problems? Will first-best policies work in a second-best world? There is no straightforward answer to this. The need for complementary policies has to be assessed in each case. High transportation costs, poor infrastructure, inelastic demand, droughts, and poorly developed information systems have added to the difficulties of designing good food policies. Removing price controls may thus increase the output price risk, and this may have adverse effects on efficiency and the functioning of credit markets (Wik and Holden 1996). Investments in infrastructure and systems for storage may improve both food and credit markets.

Policy and Project Design Implications

The research in India suggests that the concern of policymakers in developing countries, the World Bank, and the rural development community at-large with providing access to crop insurance and credit for poor farmers not only has an equity basis but also may be justified on efficiency grounds, at least in the context of semiarid India. The problem is that almost all past public sector programs failed (Hazell, Pomareda, and Valdes 1986). For small-farmer credit, the number of successful cases is very limited as well. But failing to provide poor farmers with access to more efficient forms of savings may have a cost. Paying financial intermediaries a fee for serving these groups may be amply justified to restore efficiency. Such a scheme should, of course, not subsidize interest rates because that would provide perverse incentives to serve large farmers. Instead, it might be better to pay institutions for rendering specific services to small farmers, irrespective of the size of transaction. Further experimentation and research are clearly warranted.

Although failures in the credit and insurance markets appear to be at the root of many inefficiencies, this does not imply that the second-best intervention is to be found in these markets. It could be that the moral hazard and covariance problems that led to the market failure in the first place cannot be overcome directly with a public policy intervention, however well designed and well intended. Instead, research on other policy instruments should take account of the possible impact of more stable income or consumption on investment efficiency.

Alternatively, with respect to credit and insurance, there may be a case for redistributing assets to enable the poor to obtain loans and to insure against income fluctuations. However, land reform poses severe financing problems of its own (Binswanger and Elgin 1990).

The research results suggest that it often makes sense to provide poor producers with transfers (food subsidies) or publicly sponsored employment (food-for-work programs). So far, such measures have been evaluated by looking at the consumption and nutrition benefits derived by beneficiaries. Much concern has been expressed about the disincentives to work created by such transfers and the diversion from more remunerative employment that may follow from job creation schemes. No consideration has been given to the possibility that providing the poor with more secure incomes may enable them to make better decisions regarding the formation of both physical and human capital.

As much of the research surveyed in this chapter has amply demonstrated, questions concerning the role of financial markets and risk in influencing the efficiency of agricultural investment and productivity can best be addressed using panel data. It is necessary to follow the same household over a relatively long period in order to detect fluctuations in risk, in asset holdings, and in the pattern of investment. It is therefore regrettable that the ICRISAT data set (which surveyed the same households frequently over a 10-year period) should remain the only one of its kind.

Interlinkage and Induced Innovation

Agricultural development has been promoted through the provision of interlinked input (fertilizer and seeds) and credit packages, which efficiently promoted the adoption of technology (green revolution) in many cases. These efforts were efficiency-enhancing elements of induced technological and institutional innovations (Hayami and Ruttan 1985). They followed or were accompanied by public investments in agricultural research and extension, infrastructure, and marketing development. However, they were not based on a thorough analysis and understanding of rural market and nonmarket institutions because they were introduced before the major developments in the economics of rural organization.

By drawing on these new theoretical developments, it may be possible to identify good ideas for successful institutional innovation and to reduce the number of counterproductive interventions. For example, the use of group lending and peer monitoring to control moral hazard problems in credit markets is shown to be justified by this literature. The successes of the Grameen Bank and Aga Khan Rural Support Programme have shown that it may be possible to reach poor people with credit and retain low default rates. Other lessons include the desirability of communal management of local commons. There often are good grounds for cooperative solutions when the group is fairly small and homogeneous, mutual trust exists, communication and monitoring opportunities are good, a system of responsibilities and sanctions is in place, and cooperation is like a repeated game (Dutta, Ray, and Sengupta 1989, Ostrom 1990, and Balland and Plateau 1996).

Interlinked Markets and Contracts

Holden and Shanmugaratnam (1995) have forwarded the idea that interlinkage mechanisms may be powerful tools for promoting sustainable agricultural development. They use the example of land degradation linked to poverty and missing or imperfect markets to argue that this may represent a Pareto-relevant externality. Poverty may prevent poor farmers from investing in land conservation due to imperfections in credit markets and high subsistence requirements. Poor farmers may use agricultural practices that cause large net losses in the stock of soil nutrients. Under such conditions, one may argue for a Pigouvian subsidy on fertilizers to reduce the nutrient loss externality that would negatively affect future productivity. However, a fertilizer subsidy alone may not be sufficient to internalize the externality because cheap fertilizers may become a substitute for good management (Repetto 1987 and Low 1986). But if a subsidy is used in relation to an interlinked contract where the farmers receive a combination of fertilizer, high-yielding seeds, and eventually credit, *if* they implement certain conservation practices (terracing), it may be possible to internalize a larger share of the externality because this package will ensure better nutrient recycling and more efficient use of nutrients.

The problem may not be solved by introducing a

credit market, however, because improved access to credit may not lead to more sustainable resource extraction and investments in conservation. Credit may also be used to increase resource extraction and consumption in the short run. Targeting credit to investments in conservation may be preferable, but not without problems (Holden and Shanmugaratnam 1995).

Shiferaw and Holden (1997b) use an applied nonseparable household model from the Ethiopian highlands to investigate the plausible responses to various policy measures, including seed, fertilizer, and labor subsidies (cost sharing) linked to the installation of conservation structures on erodible land. These packages were designed for cases in which the conservation technology alone had a negative effect on yields (because it took up space and increased pest problems) or had no effect on yields (but required additional investments, primarily labor). It has been found that input subsidies linked to conservation investments can increase the adoption of conservation technologies even when conservation technologies reduce yields by 20 percent. However, the subsidy rates have to be very high. The benefit/cost ratio is less than 1 for a fertilizer subsidy (75 and 90 percent) linked to conservation, while it is higher than 1 (1.61 and 1.66) for a seed subsidy (90 and 100 percent) for improved seeds of *teff*, assuming that the organizational costs of intervention are 10 percent of the costs of direct subsidy. However, the seed subsidy initiates the conservation of only a small share (12 percent) of the erodible uplands.

If the conservation technologies have no negative impact on yields in the short run, lower subsidy rates are sufficient to create incentives for adoption, and both fertilizer subsidies and seed subsidies may give positive benefit/cost ratios although the seed subsidies give the highest ratios.

Linking fertilizer and seed subsidies with conservation requirements yields positive adoption responses and benefit/cost ratios above 1 at lower subsidy levels when conservation investments reduce yields by 20 percent. The model is also used to test the response to taxation (20 percent) of the erosive crop (*teff*) and subsidization of soil-improving crops (pulses). Taxation of *teff* reduces the *teff* area considerably (more than 50 percent) and reduces soil erosion by 15 percent, but this policy instrument does not create incentives to implement other conservation techniques and has a severe, negative effect on household income (–11 percent).

Targeted Credit in Surplus-Producing Areas

Targeting of credit for productivity-increasing inputs and conservation investments is likely to function best in surplus-producing areas because the activity financed must produce a surplus that has to be used in part to pay back the loan. It is more likely to succeed where production and market risks, probability of leakages, and monitoring costs are low. This suggests it may be difficult to reach poor, deficit-producing farmers, living in risky environments with poor market access. The state or implementing agency may have to be prepared to absorb a large share of the costs in such areas. If successful, such programs may, however, change farmers and societies from subsistence and deficit-producing to surplus-producing, thus reducing the need for outside support. Where policies are not the main reason for the poverty and degradation trap, a temporary cost-sharing scheme may thus break the vicious spiral of poverty and environmental degradation and lift society onto a sustainable development path. Family planning and slower population growth may be necessary to sustain the improvement.

It is usually assumed that making preventive investments to avoid land degradation is cheaper than rehabilitating degraded lands. Yet it is possible for rehabilitation to be more realistic if no interventions to promote conservation have been made. Population pressure increases the value of land and may thus induce incentives to rehabilitate degraded lands where lower population pressure (and land values) does not provide sufficient incentives for undertaking preventive conservation investments. Credit constraints, high private discount rates, and other land market imperfections (insecure tenure) may be other explanations.

Food-for-Work Programs in Deficit-Producing Areas

Food for work is a common policy instrument. It may be seen as linking labor and food markets and as providing poor unemployed people with employment and food security, while contributing to economic growth, if the work represents productive investments. If such investments are made to improve the productivity or the resource base in low-resource economies, they may contribute to efficiency as well as equity (if the returns to investment are higher than for alternative uses). The negative environmental externalities of not intervening may strengthen the basis for intervention. So do negative externalities in the form of famines, forced migration, social conflicts, and human suffering, which may be reduced by providing food assistance and employment to drought- and poverty-stricken societies (von Braun 1995). These programs work best if they set the wage rate (or payment in kind) low to ensure targeting of the poor (self-selection).

The Importance of Local Participation

Local participation may have several meanings and levels, ranging from forced participation to self-mobilization with many intermediate forms. Active participation and broad collective mobilization provide a good basis for successful programs for natural resource management. The local leadership and commitment can have a large impact on the opportunities for participatory organization of activities. Food-for-work programs may not have the intended investment (conservation) effects if participation is primarily for employment and food security reasons and the program does not represent local priorities. The ownership and future responsibility for maintenance and derivation of benefits must be clear and have local support. Education, information, and communication systems are crucial to improving the basis for local participation. One important strategy may be to facilitate learning across communities by supporting networking and exchange programs. Collective action may have more chances to succeed if it is led by relatively young, literate persons who have been exposed to the outside world, know the traditional society, and have integrity (Balland and Plateau 1996).

The Importance of Local Institutions and Decentralization

Governments, donors, and nongovernmental organizations should cooperate to strengthen local institutions. Decentralization of power is often a prerequisite for success. Likewise, if the benefits are short term, local people should be willing to invest their time and energy because they tend to have high discount rates. Sometimes, local institutions may take on new responsibilities; other times, it may be necessary to establish new institutions because existing ones are unsuitable and difficult to modify. Forming new institutions will require time to build a platform for cooperation as well as knowledge and managerial skills. It may be wise to start with the tasks that are easiest and most likely to succeed and to expand to more difficult tasks as experience is gained.

Conclusions

Standard tools or rules of thumb from environmental economics or microeconomics in general may not always hold when applied to small farmers. Taxing the output from their land-degrading activities (the polluter-pays principle) may induce poor small farmers to increase their land-degrading activities. Such taxation may worsen their poverty and cash liquidity constraints and reduce their ability to invest in conservation. Policies aimed at the input side are more likely to have the expected effects. Interlinkage and targeting policies may be useful in resource-poor remote areas and where poverty causes nonsustainable resource management. More empirical research is needed to investigate the relationships among poverty, farm size, market imperfections, and natural resource management.

Market imperfections are more likely to hurt small and poor farmers than large and rich farmers because poor farmers are more likely to be rationed out of credit markets, for example, and to be less able to solve their problems through consumption smoothing (coping strategies). They may therefore have to go for ex ante income smoothing. This explains why the most vulnerable households are the most likely to diversify their production (Morduch 1995). A case can therefore be made for intervening on the basis of efficiency as well as equity. The possibility that poverty reinforces environmental degradation (and vice versa) strengthens the case for intervention.

The big and difficult question is how best to intervene in such cases. What is the optimal mix of incentive-based and command-and-control mechanisms? And what role should markets have? The difference between "market" and "nonmarket" institutions becomes blurred because various forms of interlinked contracts or markets may be perceived as either one. We believe that there are good hopes for improving market performance and that there are many win-win-win gains from doing so. However, particularly in remote and marginal regions and regions with severe poverty, we need to look for supplementary policy measures. Interlinked contracts between local communities, which have better local information (adverse selection, moral hazard, and enforcement problems), and outside institutions, which face less covariate risk problems, may provide win-win-win solutions. Such contracts should be formed on the basis of local participation and local priorities that are compatible with efficiency, equity, and sustainability objectives. A stepwise procedure should be chosen in the design and expansion of such contracts to allow for learning on both sides, to achieve cooperative solutions, and to reduce information asymmetries. This may only be achieved if the principal is perceived to be benevolent.

The disadvantage of this approach is that it demands knowledge on both sides, requires continuity and stability in leadership, and is easily undermined by power struggles. Contracts should be designed to safeguard against this. Group responsibility may be extended to areas other than shared responsibility for paying back loans. It could also be linked to the management of local resources, where a share of the costs or insurance against weather risk, for example, could be provided from the outside (principal). Food security for sustainable management of local resources may be one option. Family planning may have to be introduced in situations where the ratio between resources and population is unfavorable. Support from the outside should be made conditional on local performance, creating incentives for cooperation.

Agency theory, contract theory, and mechanism design theory, combined with solid knowledge of local conditions, may provide a basis for institutional innovation by building on local priorities and participation. The complexity of the issues points toward a strategy with pilot projects and flexible project design accompanied by systematic research to allow for maximum learning and consequent adjustment and upscaling as confidence is gained.

References

The word "processed" describes informally reproduced works that may not be commonly available through libraries.

Anderson, Jock R., and Jesuthason Thampapillai. 1990. *Soil Conservation in Developing Countries: Projects and Policy Intervention*. Policy and Research Report 8. Washington, D.C.: World Bank.

Angelsen, A. 1997. "Tropical Agriculture and Deforestation: Economic Theories and a Study from Indonesia." Ph.D. diss. Norwegian School of Economics and Business Administration, Bergen.

Ardilla, S., and R. Innes. 1993. "Risk, Risk Aversion, and On-Farm Soil Depletion." *Journal of Environmental Economics and Management* 25:27–45.

Balland, J.-M., and J.-P. Plateau. 1996. *Halting Degradation of Natural Resources. Is There a Role for Rural Communities?* Oxford: Clarendon Press.

Barbier, Edward B. 1990. "The Farm-Level Economics of Soil Conservation: The Uplands of Java." *Land Economics* 66(2):199–211.

Bardhan, P., ed. 1989. *The Economic Theory of Agrarian Institutions*. Oxford: Clarendon Press.

Barnum, H. N., and Lyn Squire. 1979. *A Model of an Agricultural Household*. Washington, D.C.: World Bank.

Bellon, M., and E. Taylor. 1993. "'Folk' Soil Taxonomy and the Partial Adoption of New Seed Varieties." *Economic Development and Cultural Change* 41(4):763–86.

Berry, R. A., and W. R. Cline. 1979. *Agrarian Structure and Productivity in Developing Countries*. Geneva: International Labor Organization.

Binswanger, Hans P. 1980. "Attitudes towards Risk: Experimental Evidence from Rural India." *American Journal of Agricultural Economics* 62:395–407.

Binswanger, Hans P., V. Balaramaiah, V. B. Rao, M. J. Bhende, and K. V. Kashirsagar. 1985. "Credit Markets in Rural India: Theoretical Issues and Empirical Analysis." Discussion Paper ARU 45. Agriculture and Rural Development Department, World Bank, Washington, D.C. Processed.

Binswanger, Hans P., K. Deininger, and Gershon Feder. 1995. "Power, Distortions, Revolt, and Reform in Agricultural Land Relations." In J. Behrman and T. N. Srinivasan, eds., *Handbook of Development Economics*. Vol. 3B. Amsterdam: North-Holland.

Binswanger, Hans P., and M. Elgin. 1990. "What Are the Prospects for Land Reform?" In Allen Maunder and Alberto Valdes, eds., *Agriculture and Governments in an Interdependent World*. Proceedings of the twentieth international conference of agricultural economists, 24–31 August 1988. Buenos Aires: Dartmouth Publishing Co.

Binswanger, Hans P., and S. R. Khandker. 1989. "Agroclimatic Potential, Infrastructure, and the Expansion of the Financial System in Rural India." Latin America and the Caribbean Country Department II, World Bank, Washington, D.C. Processed.

Binswanger, Hans P., John McIntire, and C. Udry. 1989. "Production Relations in Semi-Arid African Agriculture." In P. Bardan, ed., *The Economic Theory of Agrarian Institutions*. Oxford: Clarendon Press.

Binswanger, Hans P., and M. R. Rosenzweig. 1984. *Contractual Arrangements, Employment, and Wages in Rural Labor Markets in Asia.* New Haven, Conn.: Yale University Press.

———. 1986a. "Behavioral and Material Determinants of Production Relations in Agriculture." *Journal of Development Studies* 22(3):503–39.

———. 1986b. "Credit Markets, Wealth, and Endowments in Rural South India." Discussion Paper ARU 59. Agriculture and Rural Development Department, World Bank, Washington, D.C. Processed.

———. 1989. "Wealth, Weather Risk, and the Composition of Agricultural Investments." Agriculture Department, World Bank, Washington, D.C. Processed.

Binswanger, Hans P., and D. Sillers. 1983. "Risk-Aversion and Credit Constraints in Farmers' Decisionmaking: A Reinterpretation." *Journal of Development Studies* 20:5–21.

Boserup, Ester. 1965. *The Conditions of Agricultural Growth: The Economics of Agrarian Change under Population Pressure.* London: Allen and Unwin.

Braverman, Avishay, and Joseph E. Stiglitz. 1986. "Landlords, Tenants, and Technological Innovations." *Journal of Development Economics* 23:313–32.

Carter, M. R., and J. Kalfayan. 1989. "A General Equilibrium Exploration of the Agrarian Question." Madison, Wis. Processed.

Chayanov, A. V. 1966. *The Theory of Peasant Economy.* ed. by Daniel Thorner, Basile Kerblay, and R. E. F. Smith. Homewood, Ill.: Richard Irwin for American Economic Association.

Cheung, S. N. S. 1969. *The Theory of Share Tenancy.* Chicago: University of Chicago Press.

Cleaver, K. M., and G. A. Schreiber. 1994. *Reversing the Spiral: The Population, Agriculture, and Environment Nexus in Sub-Saharan Africa.* Washington, D.C.: World Bank.

de Janvry, Alain, M. Fafchamps, M. Raki, and E. Sadoulet. 1992. "Structural Adjustment and the Peasantry in Morocco: A Computable Household Model." *European Review of Agricultural Economics* 19(4):427–53.

de Janvry, Alain, M. Fafchamps, and E. Sadoulet. 1991. "Peasant Household Behavior with Missing Markets: Some Paradoxes Explained." *Economic Journal* 101:1400–07.

Dutta, B., D. Ray, and K. Sengupta. 1989. "Contracts with Eviction in Infinitely Repeated Principal-Agent Relationships." In P. Bardhan, ed., *The Economic Theory of Agrarian Institutions.* Oxford: Clarendon Press.

Ellis, F. 1988. *Peasant Economics: Farm Households and Agrarian Development.* Cambridge, U.K.: Cambridge University Press.

Eswaran, M., and A. Kotwal. 1990. "Implications of Credit Constraints for Risk Behavior in Less Developed Economies." *Oxford Economic Papers* 42(April):473–82.

FAO (Food and Agriculture Organization of the United Nations). 1996. *Forest Resources Assessment 1990: Survey of Tropical Forest Cover and Study of Change Processes.* FAO Forestry Paper 130. Rome.

Feder, Gershon. 1985. "The Relation between Farm Size and Farm Productivity: The Role of Family Labor, Supervision, and Credit Constraints." *Journal of Development Economics* 18(August): 297–313.

———. 1993. "The Economics of Land and Titling in Thailand." In Karla Hoff, Avishay Braverman, and Joseph E. Stiglitz, eds., *The Economics of Rural Organization: Theory, Practice, and Policy.* New York: Oxford University Press.

Finkelshtain, Israel, and J. A. Chalfant. 1991. "Marketed Surplus under Risk: Do Peasants Agree with Sandmo?" *American Journal of Agricultural Economics* 73(August):557–67.

Greenwald, Bruce C., and Joseph E. Stiglitz. 1986. "Externalities in Economies with Imperfect Information and Incomplete Markets." *Quarterly Journal of Economics* 101(May):229–64.

Grepperud, Sverre 1994. "Population-Environment Links: Testing a Soil Degradation Model for Ethiopia." Divisional Working Paper 1994-46. Pollution and Environment Economics Division, World Bank, Washington, D.C. Processed.

———. 1997. *Soil Depletion Choices under Production and Price Uncertainty.* Discussion Paper 186. Oslo: Statistics Norway.

Hayami, Yujiro, and V. W. Ruttan. 1985. *Agricultural Development: An International Perspective.* Baltimore, Md.: Johns Hopkins University Press.

Hazell, Peter, C. Pomareda, and Antonio Valdes, eds. 1986. *Crop Insurance for Agricultural Development: Issues and Experience.* Baltimore, Md.: Johns Hopkins University Press.

Heath, J., and Hans P. Binswanger. 1996. "Natural

Resource Degradation Effects of Poverty and Population Growth Are Largely Policy-Induced: The Case of Colombia." *Environment and Development Economics* 1(1):65–83.

Hoff, Karla, Avishay Braverman, and Joseph E. Stiglitz, eds. 1993. *The Economics of Rural Organization: Theory, Practice, and Policy.* New York: Oxford University Press.

Holden, S. T. 1991. "Peasants and Sustainable Development: The Chitemene Region of Zambia. Theory, Evidence, and Models." Ph.D. diss. Department of Economics and Social Sciences, Agricultural University of Norway, Ås.

———. 1997. "Adjustment Policies, Peasant Household Resource Allocation, and Deforestation in Northern Zambia: An Overview and Some Policy Recommendations." *Forum for Development Studies* 1:117–34.

Holden, S. T., H. Hvoslef, and R. Simanjuntak. 1995. "Transmigration Settlements in Seberida, Sumatra: Deterioration of Farming Systems in a Rainforest Environment." *Agricultural Systems* 49:237–58.

Holden, S. T., and P. Sankhayan. 1997. "Population Pressure, Agricultural Change, and Environmental Degradation in the Himalayan Region: A Conceptual, Historical, and Methodological Basis." Discussion Paper D-05/1997. Department of Economics and Social Sciences, Agricultural University of Norway, Ås. Processed.

Holden, S. T., and N. Shanmugaratnam. 1995. "Structural Adjustment, Production Subsidies, and Sustainable Land Use." *Forum for Development Studies* 2:247–66.

Holden, S. T., B. Shiferaw, and M. Wik. 1998. "Poverty, Market Imperfections, and Time Preferences: Of Relevance for Environmental Policy?" *Environment and Development Economics* 3(1):83–104.

Holden, S. T., J. E. Taylor, and S. Hampton. Forthcoming. "Structural Adjustment and Market Imperfections: A Stylized Village Economy-Wide Model with Nonseparable Farm Households." *Environment and Development Economics.*

Jodha, N. 1978. "Effectiveness of Farmers' Adjustment to Risk." *Economic and Political Weekly* 13(25):38–48.

Kates, R. W., G. Hyden, and B. L. Turner II, eds. 1993. *Population Growth and Agricultural Change in Africa.* Gainesville: University Press of Florida.

LaFranche, J. T. 1992. "Do Increased Commodity Prices Lead to More or Less Soil Degradation?" *Australian Journal of Agricultural Economics* 36(1):57–82.

Lele, U., and S. W. Stone. 1989. "Population Pressure, the Environment, and Agricultural Intensification: Variations on the Boserupian Hypothesis." MADIA Discussion Paper 4. Agriculture Department, World Bank, Washington, D.C. Processed.

Low, A. 1986. *Agricultural Development in Southern Africa: Farm Household Economics and the Food Crisis.* London: James Currey.

Malthus, T. R. 1987. *An Essay on the Principles of Population.* Vol. 1, ed. by Patricia James. Cambridge, U.K.: Cambridge University Press.

Marshall, Alfred. 1920. *Principles of Economics.* 8th ed. London: Macmillan.

McConnel, K. E. 1983. "An Economic Model of Soil Conservation." *American Journal of Agricultural Economics* 65(1):83–89.

McIntire, John. 1993. "Markets and Contracts in African Pastoralism." In Karla Hoff, Avishay Braverman, and Joseph E. Stiglitz, eds., *The Economics of Rural Organization: Theory, Practice, and Policy.* New York: Oxford University Press.

Migot-Adholla, S., Peter B. Hazell, B. Blarel, and F. Place. 1993. "Indigenous Land Right Systems in Sub-Saharan Africa: A Constraint on Productivity?" In Karla Hoff, Avishay Braverman, and Joseph E. Stiglitz, eds., *The Economics of Rural Organization: Theory, Practice, and Policy.* New York: Oxford University Press.

Morduch, J. 1995. "Income Smoothing and Consumption Smoothing." *Journal of Economic Perspectives* 9(3):103–14.

Nakajima, C. 1986. *Subjective Equilibrium Theory of the Farm Household.* Amsterdam: Elsevier.

Ostrom, Elinor. 1990. *Governing the Commons: The Evolution of Institutions for Collective Action.* Cambridge, U.K.: Cambridge University Press.

Otsuka, Keijiro, H. Chuma, and Y. Hayami. 1992. "Theories of Share Tenancy: A Critical Survey." *Economic Development and Cultural Change* 37 (October):31–68.

Pagiola, Stefano. 1995. "The Effect of Subsistence Requirements on Sustainable Land Use Practices." Paper presented at meetings of the American Agricultural Economics Association, Indianapolis, Ind., 6–9 August. Processed.

Pender, John L. 1996. "Discount Rates and Credit Markets: Theory and Evidence from Rural India."

Journal of Development Economics 50(2):257–96.

Pender, John L., and J. Kerr. 1996. *Determinants of Farmers' Indigenous Soil and Water Conservation Investments in India's Semi-Arid Tropics.* EPTD Discussion Paper 17. Washington, D.C.: International Food Policy Research Institute.

Prosterman, R. L., and J. M. Riedinger. 1987. *Land Reform and Democratic Development.* Baltimore, Md.: Johns Hopkins University Press.

Repetto, Robert. 1987. "Economic Incentives for Sustainable Production." *Annals of Regional Science,* special edition on environment management and economic development 21(3):44–59.

Rosenzweig, Mark R. 1988a. "Risk, Implicit Contracts, and the Family in Rural Areas of Low-Income Countries." *The Economic Journal* 98 (December):1148–70.

———. 1988b. "Risk, Private Information, and the Family." *American Economic Review* 78(2): 245–50.

Rosenzweig, Mark R., and Hans P. Binswanger. 1994. "Wealth, Weather Risk, and the Profitability of Agricultural Investment." *Economic Journal* 103:56–78.

Rosenzweig, Mark R., and O. Stark. 1989. "Consumption Smoothing, Migration, and Marriage: Evidence from Rural India." *Journal of Political Economy* 97(4):905–26.

Rosenzweig, Mark R., and K. I. Wolpin. 1989. "Credit Market Constraints, Consumption Smoothing, and the Accumulation of Durable Production Assets in Low-Income Countries: Investment in Bullocks in India." Department of Agricultural Economics, University of Minnesota, St. Paul. Processed.

Sadoulet, E., and Alain de Janvry. 1995. *Quantitative Development Policy Analysis.* Baltimore, Md.: Johns Hopkins University Press.

Sandmo, A. 1971. "On the Theory of the Competitive Firm under Price Uncertainty." *American Economic Review* 6:65–73.

Schultz, T. W. 1964. *Transforming Traditional Agriculture.* New Haven, Conn.: Yale University Press.

Shiferaw, B. 1996. "Economic Analysis of Land Degradation and Incentives for Soil Conservation in Smallholder Farming: A Theoretical Development." Discussion Paper D-30/1996. Department of Economics and Social Sciences, Agricultural University of Norway, Ås. Processed.

Shiferaw, B., and S. T. Holden. 1997a. "Analysis of Economic Incentives for Soil Conservation: The Case of Highland Smallholders in Ethiopia." Discussion Paper D-09/1997. Department of Economics and Social Sciences, Agricultural University of Norway, Ås. Processed.

———. 1997b. "A Farm Household Analysis of Resource Use and Conservation Decisions of Smallholders: An Application to Highland Farmers in Ethiopia." Discussion Paper D-03/1997. Department of Economics and Social Sciences, Agricultural University of Norway, Ås. Processed.

Singh, Inderjit, Lyn Squire, and John A. Stauss, eds. 1986. *Agricultural Household Models: Extensions, Applications, and Policy.* Baltimore, Md.: Johns Hopkins University Press.

Singh, R. P., and Hans P. Binswanger. 1988. "Income Growth in Poor Dryland Areas of India." Agriculture and Rural Development Department, World Bank, Washington, D.C. Processed.

Srinivasan, T. N. 1972. "Farm Size and Productivity: Implications of Choice under Uncertainty." *Sankhya, The Indian Journal of Statistics* (series B) 34(4):409–20.

Stiglitz, Joseph E. 1974. "Incentives and Risk-Sharing in Sharecropping." *Review of Economic Studies* 41(2):219–55.

Tiffen, M., M. Mortimore, and F. Gichuki. 1994. *More People, Less Erosion: Environmental Recovery in Kenya.* New York: John Wiley and Sons.

Tviland, M. 1996. "Credit to Small-Scale Farmers in Northern Zambia: Reasons for High Default Rates and Means of Reducing Them." MSc. thesis. Department of Economics and Social Sciences, Agricultural University of Norway, Ås. Processed.

von Braun, Joachim, ed. 1995. *Employment for Poverty Reduction and Food Security.* Washington, D.C.: International Food Policy Research Institute.

Walker, T. S., and N. S. Jodha. 1986. "How Small Farm Households Adapt to Risk." In Peter Hazell, C. Pomareda, and Antonio Valdes, eds., *Crop Insurance for Agricultural Development: Issues and Experience.* Baltimore, Md.: Johns Hopkins University Press.

Walker, T. W., and J. G. Ryan. 1990. *Village and Household Economies in India's Semi-Arid Tropics.* Baltimore, Md.: Johns Hopkins University Press.

Wik, M., and S. T. Holden. 1996. "Risk and Peasant Adaptation: Do Peasants Behave According to the Sandmo Model?" Discussion Paper D-24/1996. Department of Economics and Social Sciences,

Agricultural University of Norway, Ås. Processed.

Williamson, O. 1985. *The Economic Institutions of Capitalism.* New York: Free Press.

World Bank. 1996. *Towards Environmentally Sustainable Development in Sub-Saharan Africa: A World Bank Agenda.* Washington, D.C.

6

Agricultural Trade Reforms, Research Incentives, and the Environment

Kym Anderson

In recent years the world has, for the first time this century, begun reducing distortions to agricultural incentives in significant ways. The long-term growth in agricultural protection in high-income countries (see McCalla 1969 and Lindert 1991) has slowed, and the heavy direct and indirect taxation of agriculture is being reduced in numerous lower-income countries. Economists have hailed this as a breakthrough that will lead to more efficient use of global resources, and they have provided many empirical estimates of the comparative static effects of these reforms on prices, production, trade, and economic welfare (Tyers and Anderson 1992 and Martin and Winters 1996). Others, however, have been less positive. Some environmental groups, for example, claim that economic modelers give too little attention to the (in their view) adverse effects that trade liberalization may have on our rural and natural environments, on agriculture's resource base, and on food safety. The aid community is another example: it is worried that the reduction in government attention to agriculture in high-income countries may be contributing to a loss of interest in funding agricultural development assistance programs, including international agricultural research. To date, comparative static economic modelers have paid little attention to the effects of reform on the environment or on research and development (R&D).

This chapter looks at both of these concerns. It begins by exploring the claim that the income growth and resource reallocations that follow farm price and trade policy reforms will adversely affect the environment. It takes as an example the General Agreement on Tariffs and Trade (GATT) / World Trade Organization (WTO) Uruguay Round agricultural agreement to reduce high-income countries' import protection and producer and export subsidies for farm products. The opposite type of change in developing countries (that is, the unilateral removal of policies that keep domestic food prices artificially low) is reinforcing the relocation brought on by the Uruguay Round of farm production away from protected high-income countries. An important feature of those reforms, though, is their impact on incentives to invest in agricultural research. This is the subject of the second section of the chapter. The concluding section draws out some implications of the analysis for developing countries and for aid agencies. It also mentions areas where further empirical economic research is needed. Economists are confident that the removal of trade distortions will in almost all situations boost economic welfare and will in many situations also improve the environment, particularly if governments have in place and enforce optimal environmental policies. Nonetheless, the chapter makes clear that the environmental effects of many policy changes can have both positive and negative elements. Hence quantifying the main environmental effects of economic reform is needed to convince skeptics that on balance trade liberalization does indeed, as espoused in the preamble to the agreement establishing the WTO, contribute to sustainable global development.

The Environmental Consequences of Reducing Agricultural Protection in Industrial Countries

Concerns about the environmental effects of reducing agricultural protection stem in large part from the fact that reducing the distortions to international trade increases real incomes in the countries whose exports expand and changes the mix and international location of production and consumption. Both of these effects worry some environmentalists in high-income countries. That, in turn, concerns people in poorer countries, who fear that environmental groups may use their political power to convince industrial-country governments to impose trade or other sanctions on poorer countries deemed to have inadequate environmental and food safety policies.

Effects of Boosting Incomes

With respect to the effect of trade reform on raising incomes in developing countries, many environmentalists assume that the consequent increase in spending will place greater demands on the environment. That assumption, however, is questionable, because income growth brings with it at least three important changes in behavior. One is that population growth tends to decline as incomes rise, which reduces one important source of pressure on both urban and rural environments of developing countries. Another is that education investment expands with income, and with it comes more skillful management of all resources, including private use of the natural environment. And third, the governments of modernizing communities with rising incomes and improving education tend eventually to improve private property rights and to put more-stringent environmental policies in place (Radetzki 1992, Grossman 1995, and Grossman and Krueger 1995). Not only does the demand for pollution abatement policies appear to be quite income elastic (at least beyond a certain threshold), but also the cost of compliance falls, because trade liberalization expands the opportunities to acquire more environmentally benign production technologies, inputs, and consumer products.

Environmentalists may be disappointed that governments adopt less-stringent environmental standards and charges than they would like; but the appropriate response in most circumstances is for them to advocate tougher domestic environmental standards in their own and other countries as incomes rise, rather than to argue against price and trade reforms that can raise those incomes. The main exceptional circumstance is when the environmental effects of greater spending spill over national boundaries. The spillovers could be physical, as with carbon emissions and large-scale deforestation (climate change, reduced biodiversity) and with chlorofluorocarbons (CFCs) and halons (ozone depletion). Or the spillovers could be of an emotional kind, as concern for the welfare of animals. One possible solution when there are international spillovers is to seek international environmental agreements (the Montreal Protocol on CFC phaseout and the Convention on International Trade in Endangered Species). But such agreements typically are very difficult to conclude, not least because of large differences across countries in incomes and hence in the ability of people to place a high value on the goods and services of the natural environment. In those circumstances, wealthier and more-concerned environmentalists will look for other ways to influence the environmental damage of other countries (as they perceive it), and using their own country's trade policy as a stick or carrot to influence behavior in other countries may be one of the few options they have (Anderson 1997b).

Effects of Relocating Production

With respect to changes in the mix and international location of production that would accompany agricultural trade liberalization, environmentalists have expressed at least two concerns. They fear that in the highly protected countries of Western Europe the rural countryside and villages will be less visually attractive and less populated as farmers respond to lower domestic food prices by "getting bigger or getting out." And they fear that higher food prices in international markets, following reduced exports and increased imports by the highly protectionist economies, will raise agricultural land prices in tropical and Southern Hemisphere countries. That, in turn, will stimulate greater net deforestation to expand the area of agricultural land (thereby reducing the forest's absorption of carbon dioxide from the atmosphere and its contribution to biodiversity) and the use of heavier doses of agricultural chemicals (which not only degrade the local environment but also increase the amount of chemical residues in food that the reforming countries might import). These concerns are understandable, but they are based on only a small number of direct and indirect environmental effects involved in the relocation of production that would be induced by trade reform. It is possible to examine a few additional influences by using some estimates of the production effects of a multilateral reform as simulated by a model of world food markets (for some early studies, see Abler and Shortle 1992, Barbier and Burgess 1992, Lutz 1992, and Runge 1994).

The estimates of Tyers and Anderson's (1992) model are from an extreme simulation: it assumes complete removal of all farmer support policies in all industrial countries and U.S. land set-asides as of 1990 and full adjustment within a year. In fact, the Uruguay Round cuts will be less than one-fifth of that and will be phased in over a six-year period for industrial countries and up to 10 years (to 2004) for poorer countries (Martin and Winters 1996). Even assuming such a huge liberalization and instantaneous adjustment, the estimated impact on world food output in aggregate is estimated to be negligible and the relocation of production only minor: grain and meat production would be 5 or 6 percent lower in industrial countries and 3 to 8 percent higher in developing countries. The big declines in output would occur in Japan and Western Europe, but they would be partly offset by increases in North

Figure 6.1. Shares of Various Countries in Global Changes in Food Production Following Full Liberalization of Industrial-Country Food Policies
(percent)

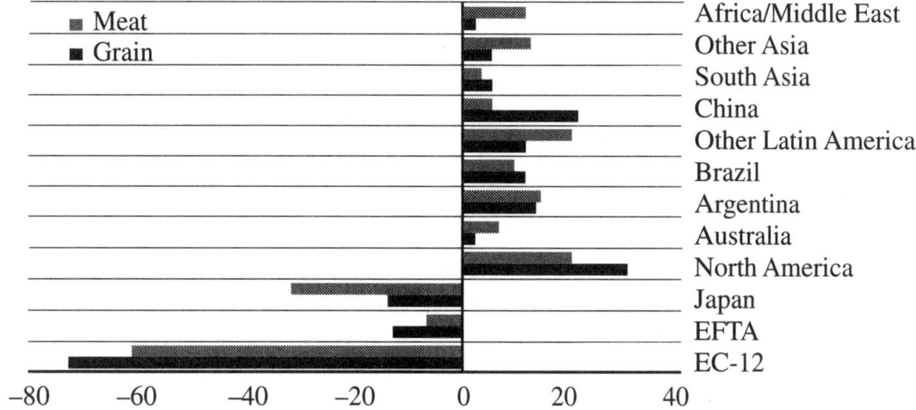

Source: Anderson and Strutt 1996.

Figure 6.2. Proportional Changes in Food Production of Various Countries Following Full Liberalization of Industrial-Country Food Policies
(percent)

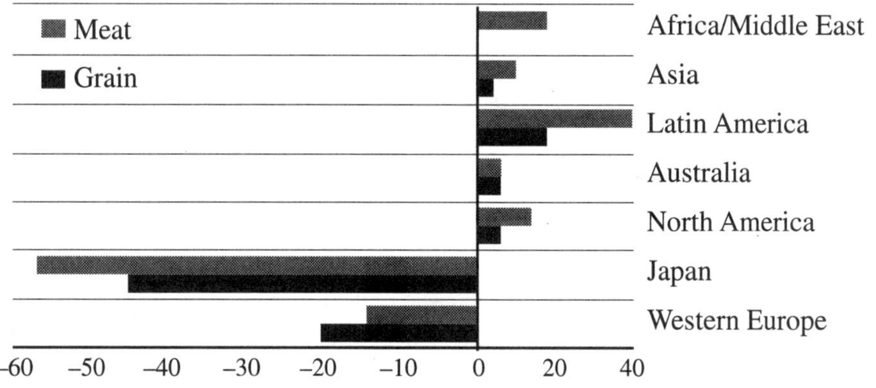

Source: Anderson and Strutt 1996.

Figure 6.3. Relationship between Agricultural Producer Subsidy Equivalent (PSE) for 1979–89 and the Use of Chemical Fertilizers and Pesticides per Hectare of Arable and Pasture Land

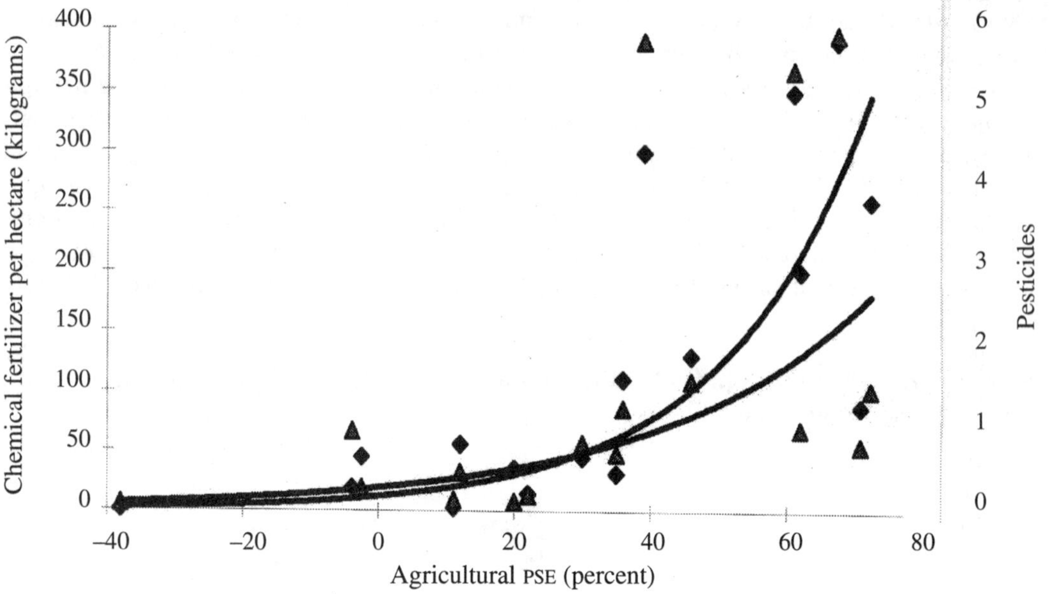

Source: Anderson and Strutt 1996, based on World Resources Institute 1990: table 18.2.

America and Australasia. The rich regions in North America and Australasia are estimated to account for more than a quarter of the increases, with developing countries providing the balance (see figures 6.1 and 6.2).

What would such a relocation do for the use of pollutive farm inputs, for land use, and for the use of labor and capital? We consider each in turn.

Changes in the use of environmentally unfriendly farm inputs. Figure 6.2 suggests that for regions where production would fall, the reductions would constitute a fairly large proportion of output. But for regions where production would expand, the increases would constitute a relatively small proportion of current output, especially for grain. The difference is because the price decreases in the former regions are much larger than the price increases in the latter regions. This has important implications for environmental degradation and chemical residues in food, because the contracting regions are relatively densely populated with high land prices and so use farm chemicals and intensive livestock methods much more than the expanding regions. In the case of fertilizer and pesticides, for example, the highly protected countries use more than 10 times as much per agricultural hectare as Australasia and most developing countries (figure 6.3). Of course, the demand for farm chemicals depends on more than just the level of distortion in output price, but previous econometric work suggests that a thorough multivariate regression analysis that includes additional variables such as wage rates is still likely to find a strong positive correlation between chemical use and output prices.

Furthermore, land-scarce Western Europe and Japan crop twice as much of their total land area as does the rest of the world on average, so the extent of contamination of their soil, water, and air from the use of farm chemicals is even greater than figure 6.3 suggests, relative to other countries. That is, the relocation of crop production from densely populated protectionist countries to the rest of the world would cause a much larger reduction in degradation in the former compared with any increased degradation in the latter, where chemical use would expand from a low base and to still-modest levels. Should the expanding region have or choose to introduce an optimal tax on the sale of farm chemicals at the time of the rise in food prices following the reform abroad, that tax would not eliminate externalities from chemical use but would at least ensure that at the margin the benefit to farmers from applying more farm chemicals would just equal the external cost that those chemicals impose on others. Rather than *taxing* the pollution caused by irrigation and fertilizer and pesticide use in farming, many developing-country governments have contributed to environmental damage by *subsidizing* the use of those inputs in the past. Reforms are occurring though. For example, Bangladesh and Indonesia have chosen to phase out such subsidies in recent years, even if they have yet to add pollution taxes; and already the reductions in environmental degradation are evident (World Bank 1997).

What about the effect of trade policy reform on the use of inputs in the livestock sector? The relocation of meat and milk production from the most densely populated rich countries to relatively lightly populated poorer countries would be associated with a decline in the extent to which the world's livestock is fed grain and supplements rather than pasture. With this would come a decline in the use of growth hormones and veterinary medicines, partly because animals are less valuable in less-protected economies and partly because the risk of the spread of disease is lower in range feeding than in intensively housed conditions. The greater use of less-intensive methods would reduce not only air, soil, and water contamination associated with the disposal of animal effluent but also the chemical additives in the food produced. Moreover, insofar as this relocation leads to greater use of crop and leguminous pasture rotation methods, there will be less need for chemical fertilizer and hence less water pollution from nitrates.

Changes in the use of land. What about land degradation? Think first of the nonsubsidizing countries, where product prices would rise slightly following reforms abroad. The lightly populated countries such as Argentina, Australia, and New Zealand, where most of the potential farming land is already cleared for agriculture, would see a more-intensive use of that land, but from its current relatively low base. In tropical countries, some land might be attracted from crop plantation areas (coffee, palm oil, and so forth), which may cause more or less degradation than the now-more-profitable alternative. But conservationists are concerned that more tropical forests would be felled. How much is an empirical question, about which there is relatively little evidence at present. It is worth noting, however, that during the period 1961–91 cereal production in developing countries increased 165 percent, but the area of arable land increased only 17 percent (Alexandratos 1995), and that area is projected to expand only 13 percent during the subsequent three decades, even though per capita food supplies will expand and cereal prices in international markets are expected to be 20 percent lower by 2020 (Rosegrant and Ringler 1997).

In rich countries where price supports are lowered and there are no land set-aside policies, land values would fall, which would probably further reduce the use of land substitutes such as farm chemicals, irrigation water, and feed concentrates. (The word "probably" is necessary because some offsetting forces make it uncertain whether reducing output prices leads to more or less care of farmland, particularly because the altered incentives to avoid on-site damage also affect off-site damage; see, for example, Barrett 1991, Clarke 1992, and LeFrance 1992.) Much of that would happen immediately, but over time their use would fall even more as the decline in land price reduces the incentive to seek land-saving technologies (Hayami and Ruttan 1985). The use of some land would switch to products whose domestic prices have not fallen or have fallen less, such as the subset of fresh fruits and vegetables that are not easily tradable internationally because of high transport costs. Such products may generate higher or lower negative externalities than the ones previously produced. But other land might revert to recreational and other nonfarm uses (golf courses, for example), whose positive externalities may be more or less for nonfarmers than the aesthetic pleasure they derive from seeing the land being farmed.

Changes in the use of labor and capital. The fall in farm profitability in industrial countries would encourage more labor and capital to be employed in nonfarm sectors, most likely the dominant services sector, which is relatively unpollutive. By contrast, in the developing countries the greater profitability of farming because of higher international and domestic farm product prices would encourage more labor and capital to be employed in commercial agriculture. Those productive factors would come from other activities where they may well be contributing more environmental damage than they would in an expanded commercial agricultural sector. One possibility is manufacturing, which in many newly industrializing countries can be quite pollutive until incomes rise sufficiently for people to demand stricter enforcement of environmental policies. The other possibility, particularly in less-advanced economies, is that underemployed labor will be attracted to commercial farming. Whether such workers come from the urban slums or from rural areas, they are likely to do less environmental damage in their new job. In the case of workers who would otherwise be eking out a subsistence income by squatting on marginal hillsides, the result would be less deforestation and soil degradation on those hillsides. As well, the increased value of rural labor would raise the opportunity cost of collecting and chopping wood for fuel. Cleaner fuels such as kerosene would then be used instead, and forests would be depleted less as a result. This positive effect on the environment could be substantial, given that four-fifths of logs felled in developing countries are used as fuel (World Resources Institute 1996: table 9.3).

Effects of Policy Reinstrumentation

What about other policy responses to trade policy reform under the Uruguay Round agreement on agriculture? Because the agreement bans the use or requires less use of some but not all policy instruments, there is the possibility that some protection reinstrumentation will take place as governments of industrial countries attempt to reduce the economic loss to powerful producer groups. Four examples serve to illustrate the extra uncertainty that this can introduce to predicting the environmental consequences of trade policy reforms: output-limiting instruments, environmental measures, R&D investments, and quarantine import restrictions.

Production-limiting policy instruments. In the case of the Uruguay Round agreement on agriculture, price supports for products subject to acreage set-asides and other production-limiting instruments were not counted in the aggregate measure of support (AMS). Instead they were classified as "blue box" instruments, whereby U.S. and European Union governments, in trying to minimize the political cost of meeting commitments to reduce the AMS along with cuts in export subsidies and increases in import market access, were allowed to maintain price supports for products with an output-limiting instrument attached. However, there is a considerable degree of leakage in such instruments. For instance, farmers may be allowed to set aside their worst-yielding acres, which reduces their output by only a fraction of the reduction in acreage. Yet the fact that land is set aside makes the remaining land used more valuable, which encourages higher doses of land-augmenting farm chemicals per hectare—not an obvious improvement from an environmental viewpoint. If more price support programs were to use this type of policy instrumentation, this could reduce future environmental gains in farm assistance–reducing countries.

Environmental instruments in lieu of price supports. The second example has to do with environmental measures per se. An important consequence of making traditional barriers to international trade less acceptable is that industry assistance in politically correct green garb is likely to be used more. Hence it is not surprising that there are already European Union Council Regulations (for example, no. 2078/92) allowing environmental subsidies to agriculture, including subsidies to reduce the use of fertilizer and pesticides, to promote environmentally sound methods of production, to encourage extensive agricultural techniques, to maintain practices that are already compatible with the environment, and to assist organic farming. These are allowable ("green box") measures in the Uruguay Round agreement on agriculture and hence are likely to become more common in Europe and elsewhere over time, substituting somewhat for traditional economic protectionism. These measures may well improve the environment but are likely to be excessive if used as substitutes for farm price supports—and they are not necessarily the first-best ways to achieve particular environmental outcomes. Hence monitoring their rise may be just as important as monitoring the promised declines in traditional border protection and other means of farm price support, with a view to arguing against excessive or inappropriate interventions.

Agricultural research investments. A decline in the profitability of farming in industrial countries because of cuts in protection would lead one to expect the profitability of agricultural research to fall as well. However, because much of that research is supported by government and because public agricultural research funding appears in the WTO's "green box" and so is not counted as part of a country's AMS, public research investment may rise rather than fall. Conceivably that rise could more than offset the likely fall in private investment in agricultural research. And with environmental policies in these countries altering farmers' incentives in ways that discourage the use of environmentally damaging practices, new investments in public agricultural research are likely to reinforce that trend.

Quarantine import restrictions. Quarantine import restrictions have always been a potential substitute for more direct measures of assistance to farmers. Agricultural exporting countries were well aware of the possibility that the Uruguay Round agreement on agriculture could encourage greater use of quarantine import restrictions as a new form of protectionism and, for that reason, ensured that an agreement on the application of sanitary and phytosanitary (SPS) measures (to protect human, animal, or plant life or health) was part of the Uruguay Round package. Under that agreement, members of the WTO have a right to apply their own standards, provided they are not unduly trade-restrictive. The SPS agreement establishes detailed rights and obligations for environmental health and food safety as well as measures to prevent the spread of pests or diseases among animals and plants. It outlines procedures for product inspection, treatment, and processing, risk assessment, and allowable maximum levels of pesticide residues and certain food additives based on agreed international standards (the Codex Alimentarius Commission for food safety, the International Office of Epizootics for animal health, and the International Plant Protection Convention for plant health). Members can adopt higher standards if they wish, but only with appropriate scientific justification. The SPS agreement allows governments to challenge, via the WTO's dispute settlement body, another country's food safety or other technical requirements based on evidence showing whether the measure is justified. Canada, for instance, is challenging Australia's justification for sanitary measures banning salmon imports. Another important food case before the WTO in 1997 had to do with the European Union's ban on the import of beef produced with the use of growth hormones. The European Union claimed that the ban applies to domestic producers and so does not contravene the national treatment provision of the GATT, whereas the United States argued (successfully, as it happened) that the hormones are not harmful to humans and hence the European Union standard is excessive and labeling should be sufficient protection for consumers. For the first time, the WTO called on scientific experts to help resolve that case.

This is clearly an area where trade tensions could escalate. It is telling that during the 47 years of the GATT, virtually no formal trade disputes arose over SPS measures, because a country's import restrictions to protect human, health, and plant life were difficult to challenge under previous GATT agreements. In the first 18 months of the WTO's formation, by contrast, seven formal complaints were lodged under WTO dispute settlement procedures. The SPS agreement is thus likely to help those agricultural exporters who have been facing unduly restrictive barriers in potential export markets abroad and to reduce the returns to producers who have enjoyed protection from import bans on quarantine grounds that cannot be justified scientifically. In the latter cases, removal of unjustified import barriers could boost the welfare of domestic consumers by more than it would harm the welfare of domestic producers (James and Anderson 1998). Only time will tell whether the SPS agreement is tough enough to prevent reinstrumentation toward a greater use (abuse) of quarantine measures as a means of supporting farm incomes.

The nonagricultural Uruguay Round agreements. Trade reforms are typically not confined to just one sector. The multilateral Uruguay Round involved more than a dozen agreements, of which agriculture was just one. Likewise, regional agreements tend to involve many, if not all, sectors. And unilateral economic reforms are often across-the-board or at least not focused on the farm sector alone. Hence relative prices of farm products are affected by the change in nominal price not only of those goods but also of other goods affected by the reform package.

The Uruguay Round, for example, not only goes only a small fraction of the way toward freeing farm trade (less than one-fifth) but also frees up manufacturing and other trade as well. Hence real international food prices are expected to be raised by no more than 1 percent, mainly because the international price of some manufactures also would rise (Goldin and van der Mensbrugghe 1996). For countries in East Asia, in particular, the Uruguay Round agreement on textiles and clothing ensures that domestic prices will rise more for those products than for farm products over the Uruguay Round's implementation period. As a result, the share of agriculture in the Association of Southeast Asian Nation's (ASEAN) gross domestic product (GDP) is projected to fall by 27 percent instead of 21 percent between 1992 and 2005, the share of other primary products is projected to fall by 17 instead of 13 percent, while the share of light manufactures in ASEAN's GDP is expected to rise by 42 percent instead of 16 percent, according to the Global Trade Analysis Project (GTAP) modeling results reported in Anderson and others (1997). The estimated environmental implications of the Uruguay Round for that group of countries clearly would be very different if the focus were only on the agriculture agreement rather than on the full package of agreements.

With this in mind, Strutt (forthcoming) uses an economywide approach to examine the environmental effects of the Uruguay Round and of the proposed Asia-Pacific Economic Community trade liberalizations for the case of Indonesia. The global GTAP model is enhanced with environmental modules to examine air, soil, and water pollution over the next two decades without and with those trade reforms. It produces striking results in several respects. First, that computable general equilibrium (CGE) model suggests that only some rather than all of Indonesia's agricultural subsectors will expand as a result of the Uruguay Round. Second, it suggests that while the expansion in aggregate output and the shift in composition of that output because of the Uruguay Round would add slightly to soil erosion, the cost of the damage caused by that increased erosion is minuscule, amounting to less than 0.2 percent of the national economic welfare gain (as traditionally measured) from Uruguay Round liberalization. Third, the trade policy reforms slated for the next two decades are shown to reduce many forms of air and water pollution, water use, and deforestation in Indonesia, and even the forms of pollution that are projected to increase would add only very slightly to environmental degradation—even without toughening the enforcement of existing environmental regulations or adding new ones. And fourth, for the pollution levels that do rise with trade reform, the extent of the rise in all cases is a tiny fraction (less than 2 percent) of the rise projected to accompany Indonesia's economic growth (again assuming no toughening of environmental regulations) over the period of implementation (see Strutt and Anderson 1998).

Conclusion

In short, the environmental consequences of farm price and trade policy reforms are very complex, with lots of offsetting forces at work, of which only a subset will ever be easily quantified. The reforms' effects, both on boosting income and hence encouraging more output and on relocating production globally, could increase or decrease global environmental damage. But what this makes clear is that (a) there is much more likelihood that the reforms will have a positive net effect on the environment than is often admitted by environmental groups and (b) where the environmental effect is negative, the extra damage caused by trade reforms is only a minuscule fraction of that which would occur as a result of normal economic growth, whose adverse welfare effects are likely to be far more than offset by the welfare gains from trade reform. Indeed, a recent empirical study that attempted to place a monetary value on estimated changes in pollution following the Uruguay Round finds that the changes reduced developing countries' estimated economic welfare gains from the Uruguay Round by less than 2 percent while *raising* the welfare gains to some advanced economies (Cole, Rayner, and Bates 1997). And of course if the trade reform were accompanied by reform that brought environmental policies closer to optimal regulations, the likelihood of an even better outcome would be enhanced, as Beghin and others (1997) demonstrate quantitatively in a case study of Mexico.

This review focuses only on the comparative static effects of farm policy reform (that is, the one-off increase in income, resource adjustment, and production relocation). Yet such policy reforms can also affect incentives to invest, including R&D efforts, that can permanently boost economic growth rates (Grossman and Helpman 1991). Because faster rates of economic growth tend to speed up improvements in the management of environmental policies, they are desirable both from an economic and an environmental viewpoint. The next section of this chapter therefore examines the impact of a rise in farm product prices in developing countries on the incentive to invest in agricultural research, because such research is arguably one of the fastest ways to boost the welfare of the world's poorest people (Schultz 1980).

The Impact of Higher Real Farm Product Prices in Developing Countries on Returns to Agricultural Research

Traditionally, many (especially food-surplus) developing countries have adopted price, trade, and exchange rate policies that have discriminated heavily against the rural sector (Krueger, Schiff, and Valdes 1988 and Tyers and Anderson 1992). Independent of the Uruguay Round, those distortionary policies have been coming under scrutiny in recent years and are beginning to be dismantled unilaterally in a growing number of developing countries. Such unilateral reforms could have a much larger impact—by raising agricultural prices in these countries—than the reductions in agricultural protection resulting from the Uruguay Round. They will reinforce the global income growth and relocation of farm production away from protected high-income countries brought on by the Uruguay Round and so tend to amplify the effects on the environment outlined in the previous section. But because the rise in domestic farm prices could be very substantial in many cases (in China, prices rose at least 50 percent in the late 1970s, early 1980s, for example), effects on economic welfare and the environment, through altered agricultural research incentives and rewards, also are worthy of consideration.

A simple way to explore this issue is to consider a market for a single commodity (say a staple such as maize) in a small country. Assume that, for the benefit of poor net buyers of this food staple (albeit at the expense of net sellers, even though they may be the majority of the poor population), the country has kept the domestic price low via a ban on exports of this good. If the ban were lifted, this would become a bigger industry in terms of gross output or value added, and hence the scope for levying the income of producers to fund national agricultural research would rise. In a numerical example, Anderson (1997a) shows that even if the export ban depressed the domestic price by just one-third (not unusual in developing countries; see Krueger, Schiff, and Valdes 1988), in the case where the supply and demand curves had 45° slopes, it would reduce:

- The scope for raising national agricultural research funds via a grower levy by between two-thirds and three-fourths (depending on whether funds are to be levied as a percentage of farm income gross or net of farm input costs)
- The benefit to producers from (and hence their incentive to seek) the new technology by between 75 and 88 percent (depending on whether the supply curve shift is parallel or proportional)
- The net economic benefit from the new technology (from which national agricultural research costs have to be subtracted) by between 50 and 63 percent and therefore
- The internal rate of return from national agricultural research investment by as much as 63 percent.

In short, reforming the policies that depress food prices, other things being equal, is going to raise much more than proportionately the rates of return from agricultural research for that reforming economy. It is also going to alter the relative returns to research among the various farm products to the extent that they face different degrees of distortion or reform. Should that policy reform also involve raising the prices of environmentally damaging farm inputs, it would induce more of a search for new precision technologies that use those inputs less intensively (Hayami and Ruttan 1985).

If developing countries were to reduce their policy discrimination against agriculture, there would be larger national agricultural research systems to adapt and promote technologies developed at international agricultural research centers. Hence the measured returns to investment in those development assistance agencies would also rise following such reforms. The latter should help to offset the aid community's concern that the reduction in government support for agriculture in high-income countries, following the Uruguay Round, will contribute to a loss of interest in funding agricultural development assistance programs, including those at international agricultural research centers.

The boost to agricultural research in developing countries that would be stimulated by unilateral price and trade policy reforms, together with a possible reduction in national agricultural research in those high-income countries that reduce their high food prices to meet their Uruguay Round commitments, would tend to amplify the relocation of agricultural production discussed in connection with figures 6.1 and 6.2. One might therefore expect it to amplify the environmental consequences of that relocation. However, two caveats should be kept in mind.

One is that the research itself may lead to the adoption of new technologies that have different environmental effects than present practices. Indeed there have been calls for more development and dissemination of precision technologies to reduce environmental degradation, but that is unlikely to be heeded until more developing-country governments have removed subsidies on farm inputs and taxed their pollutive content (Zilberman, Khanna, and Lipper 1997).

The other caveat is that any stimulus to international agricultural research centers may well have a positive effect on advanced economies also. Byerlee and Traxler (1996) find that a substantial proportion of the productivity gains associated with past investments in international agricultural research centers have accrued in the donor countries, because of their well-developed capacity to import and adapt new technologies. The more that happens, the more world output will grow, putting downward pressure on agricultural prices in international markets. Those developing countries that are net food importers will enjoy a terms-of-trade gain, while the net food exporters will benefit so long as their growth in farm productivity more than offsets the downturn in prices of their exports.

Policy Implications and Areas for Further Quantitative Economic Research

Economists are confident that the removal of trade distortions will in almost all situations boost economic welfare. The discussion in this chapter suggests that in the case of agricultural policy reform it would in many situations also make a net improvement in the environment. Where it would not, the reason is commonly that governments do not have appropriate environmental policies in place. Having said that, it is clear that the environmental effects of many policy changes can have some negative elements even when the net welfare effect might be positive. Hence before it is possible to convince skeptics that trade liberalization does indeed—as espoused in the preamble to the agreement establishing the WTO—contribute to sustainable global development, more attempts will need to be made to quantify the environmental changes likely to be associated with trade and other economic reforms. The papers by Beghin and others (1997), Cole, Rayner, and Bates (1997), and Strutt and Anderson (1998) are a small start to that process.

For developing countries, several policy implications follow from the analysis in this chapter. One is that raising domestic prices of farm products up toward those in international markets is highly desirable but needs to be accompanied by other policy changes to ensure that the country's agricultural development is sustainable. Specifically, at those higher output prices, the cost of not reducing distortions in markets for farm chemicals and irrigation water—degradation of the farm sector's natural resource base—will be much greater than when product prices and hence input intensities are lower. If the country seeks to export some of that extra output to high-income countries, the cost of not reducing those distortions might be high in two other senses as well. One is that the food safety standards of the importing countries would need to be met, including limits on the level of chemical additives in food. The other is that production deemed to be environmentally irresponsible could trigger trade sanctions, as Mexico found when it continued to catch dolphin in tuna nets and sought to export the tuna to the United States, which had banned that form of netting. Such sanctions are much less likely in the future, however, following the GATT dispute settlement finding in favor of Mexico on that issue.

Another implication for developing countries that allow their low domestic prices of farm products to rise is that the returns from investing in agricultural research will be much higher in the future. Available empirical studies suggest that developing countries have underinvested in this activity even at low prices (Alston, Craig, and Pardey 1996). In recent years, governments have sought to downsize, so the degree of underinvestment will be even greater at less-depressed price levels.

The international agricultural research centers too will enjoy higher rates of return to their research on behalf of reforming developing countries. Insofar as those countries' reforms involve raising prices not only for farm products but also for chemical inputs and water, the rewards will be especially high from generating environmentally friendlier precision technologies as distinct from ones using heavy doses of chemicals and water (Zilberman, Khanna, and Lipper 1997). These changes should make it easier for international agricultural research centers to entice financial support from donors, not least because the precision technologies are also becoming more sought-after in the advanced economies supplying development assistance funds. This suggests yet another topic for empirical modeling work, namely, projecting the effects of new research on the relocation of the world's agricultural production, and the associated impact on the environment, under various assumptions about the degree of spillover adoption by advanced economies.

Finally, what might be in store for agriculture as the World Trade Organization approaches the next century? Clearly much remains to be done before agricultural trade is as fully disciplined as under the WTO or as free as world trade in manufactures. The Cairns Group of agriculture-exporting countries is planning simultaneous action on numerous fronts (Anderson 1998). Their first priority is to secure an early commitment to begin a new round of multilateral trade negotiations at the turn of the century, one that is comprehensive enough to allow intersectoral tradeoffs. The second priority is to ensure that all possible areas for opening agricultural markets are on the table. In addition to reductions in production and export subsidies and bound tariffs, this includes removing the "blue box" in the Uruguay Round agreement on agriculture, which has allowed the United States and European Union to continue subsidizing industries subject to set-aside and other production-limiting measures, expanding tariff rate quotas, demonopolizing state trading enterprises, and phasing out special safeguard provisions. It also includes tightening up the provisions in the SPS agreement to reduce further the scope for countries to use (misuse) quarantine restrictions to raise their farmers' incomes and striving to get socialist economies in transition (most notably China, Russia, Ukraine, and Vietnam) to reform sufficiently to be welcomed into WTO membership.

On the specific issue of trade and environment linkages, there will be pressure from environmental groups to have the matter on the agenda of the next WTO round of talks. It is too early to tell what form that discussion might take, but it is unlikely to result in agreement to do anything as extreme as allow countries to use trade measures against other countries with less-stringent environmental policies. Even calls to have an environmental impact assessment of any agreements that further reduce trade restrictions will be seen as impractical, for if any such assessment would lead to required amendments to the negotiated agreement, the whole agreement would unravel. Such calls will be less likely, however, as more empirical research fails to demonstrate the significant adverse environmental consequences of trade liberalization.

References

The word "processed" describes informally reproduced works that may not be commonly available through libraries.

Abler, David G., and John S. Shortle. 1992. "Environmental and Farm Commodity Policy Linkages in the U.S. and the E.C." *European Review of Agricultural Economics* 19(2):197–217.

Alexandratos, Nikos, ed. 1995. *World Agriculture: Towards 2010.* New York: John Wiley and Sons for the Food and Agriculture Organization of the United Nations.

Alston, Julian, Barbara Craig, and Philip Pardey. 1996. "Research Lags and Research Returns." Paper presented to the global agricultural science policy conference, 26–28 August, Melbourne. Processed.

Anderson, Kym. 1997a. *The Changing International Context of Australian Aid.* CIES Policy Discussion Paper 97/06. Adelaide: University of Adelaide.

———. 1997b. "Environmental Standards and International Trade." In M. Bruno and B. Pleskovic, eds., *Annual World Bank Conference on Development Economics 1996*, pp. 317–38. Washington, D.C.: World Bank.

———. 1998. *Agriculture and the WTO into the 21st Century.* CIES Policy Discussion Paper 97/06. Adelaide: University of Adelaide.

Anderson, Kym, Betina Dimaranan, Tom Hertel, and Will Martin. 1997. "Economic Growth and Policy Reforms in the APEC Region: Trade and Welfare Implications by 2005." *Asia-Pacific Economic Review* 3(1, April):1–21.

Anderson, Kym, and Anna Strutt. 1996. "On Measuring the Environmental Impact of Agricultural Trade Liberalization." In M. E. Bredahl, N. Ballenger, J. C. Dunmore, and T. L. Roe, eds., *Agriculture, Trade, and the Environment: Discovering and Measuring the Critical Linkages*, ch. 11. Boulder, Colo.: Westview Press.

Barbier, Edward B., and Joanne C. Burgess. 1992. "Agricultural Pricing and Environmental Degradation." Background paper for the *World Development Report 1992.* World Bank, Washington, D.C. Processed.

Barrett, Scott. 1991. "Optimal Soil Conservation and the Reform of Agricultural Pricing Policies." *Journal of Development Economics* 36(2):167–87.

Beghin, John, Sebastian Dessus, David Roland-Holst, and Dominique van der Mennsbrugghe. 1997. "The Trade and Environment Nexus in Mexican Agriculture: A General Equilibrium Analysis." *Agricultural Economics* 17(2-3, December): 115–31.

Byerlee, Derek, and Greg Traxler. 1996. "The Role of Technology Spillovers and Economies of Size in the Efficient Design of Agricultural Research Systems." Paper presented to the global agricultural science policy conference, 26–28 August, Melbourne. Processed.

Clarke, Harry. 1992. "The Supply of Nondegraded Agricultural Land." *Australian Journal of Agricultural Economics* 36(1, April):31–56.

Cole, M. A., Anthony J. Rayner, and J. M. Bates. 1997. *The Environmental Impact of the Uruguay Round.* CREDIT Research Paper 97/20. Nottingham: University of Nottingham.

Goldin, Ian, and Dominique van der Mensbrugghe. 1996. "Assessing Agricultural Tariffication under the Uruguay Round." In Will Martin and L. Alan Winters, eds., *The Uruguay Round and the Developing Countries.* Cambridge, U.K.: Cambridge University Press.

Grossman, Gene M. 1995. "Pollution and Growth: What Do We Know?" In I. Goldin and L. A. Winters, eds., *The Economics of Sustainable Development.* Cambridge, U.K.: Cambridge University Press.

Grossman, Gene M., and Elhanan Helpman. 1991. *Innovation and Growth in the Global Economy.* Cambridge, Mass.: MIT Press.

Grossman, Gene M., and Alan B. Krueger. 1995. "Economic Growth and the Environment." *Quarterly Journal of Economics* 110(2, May):353–78.

Hayami, Yujiro, and Vernan W. Ruttan. 1985. *Agricultural Development: An International Perspective*, rev. ed. Baltimore, Md.: Johns Hopkins University Press.

James, Sallie, and Kym Anderson. 1998. "On the Need for More Economic Assessment of Quarantine/SPS Policies." *Australian Journal of Agricultural and Resource Economics* 42(4, December), forthcoming.

Krueger, Anne O., Maurice Schiff, and Alberto Valdes. 1988. "Measuring the Impact of Sector-Specific and Economywide Policies on Agricultural Incentives in LDCs." *The World Bank Economic Review* 2(3, September):255–72.

LeFrance, Jeffrey T. 1992. "Do Increased Commodity Prices Lead to More or Less Soil Degradation?" *Australian Journal of Agricultural Economics* 36(1, April):57–82.

Lindert, Peter H. 1991. "Historical Patterns of Agricultural Policy." In P. Timmer, ed., *Agriculture and the State: Growth, Employment, and Poverty*. Ithaca, N.Y.: Cornell University Press.

Lutz, Ernst. 1992. "Agricultural Trade Liberalization, Price Changes, and Environmental Effects." *Environmental and Resource Economics* 2(1):79–89.

Martin, Will, and L. Alan Winters, eds. 1996. *The Uruguay Round and the Developing Countries.* Cambridge, U.K.: Cambridge University Press.

McCalla, Alex F. 1969. "Protectionism in International Agricultural Trade, 1850–1968." *Agricultural History* 43(3, July):329–44.

Radetzki, Marian. 1992. "Economic Growth and Environment." In Patrick Low, ed., *International Trade and the Environment.* Discussion Paper 159. Washington, D.C.: World Bank.

Rosegrant, Mark W., and Claudia Ringler. 1997. "World Food Markets into the Twenty-First Century: Environmental and Resource Constraints and Policies." *Australian Journal of Agricultural and Resource Economics* 41(3, September): 401–28.

Runge, C. Forde. 1994. "The Environmental Effects of Trade in the Agricultural Sector." In *The Environmental Effects of Trade*, pp. 19–54. Paris: Organisation for Economic Co-operation and Development.

Schultz, Theodore W. 1980. "Nobel Lecture: The Economics of Being Poor." *Journal of Political Economy* 88(4, August):639–51.

Strutt, Anna. Forthcoming. "Economic Growth, Trade Policy, and the Environment in Indonesia." Ph.D. diss. University of Adelaide.

Strutt, Anna, and Kym Anderson. 1998. "Will Trade Liberalization Harm the Environment? The Case of Indonesia to 2020." Paper prepared for a World Bank conference on trade, global policy, and the environment, Environment Department, World Bank, Washington, D.C., 22–23 April. Processed.

Tyers, Rod, and Kym Anderson. 1992. *Disarray in World Food Markets: A Quantitative Assessment.* Cambridge, U.K.: Cambridge University Press.

World Bank. 1997. "Subsidy Policies and the Environment." In *Expanding the Measure of Wealth: Indicators of Environmentally Sustainable Development*, ch. 4. Washington, D.C.: World Bank.

World Resources Institute. 1996. *World Resources 1996–97*. New York: Oxford University Press.

Zilberman, David, Mahdu Khanna, and Leslie Lipper. 1997. "Economics of New Technologies for Sustainable Agriculture." *Australian Journal of Agricultural and Resource Economics* 41(1, March):63–80.

7

Property Regimes in Economic Development: Lessons and Policy Implications

Daniel W. Bromley

This chapter is concerned with property regimes and property relations as they pertain to development projects undertaken in the interest of improving the management of natural resources. Development projects represent new streams of income and associated benefits to a local community and sometimes to the nation-state in which the community exists. But most development projects bestow the bulk of their beneficial (and harmful) effects on those in closest proximity to the development intervention. Sometimes the local community will seek these new circumstances. Sometimes the development project will appear without local initiative.

The major difference in the long-run performance of these two types of projects can be profound indeed. One thing we know with almost striking certitude is that when local citizens play a role in the conception of a development project, the chances for success are usually better than when projects simply appear from outside and are seen either as gifts or as impositions. Although not central to the issues raised here, the point bears mention because it addresses a fundamental issue in the domain of "ownership"—whether we think of ownership in relation to land and related natural resources or in relation to alterations in the social and economic circumstances among local people arising from development projects.

Perceptions of ownership arise in the domain of property relations in and around the community in which development projects appear. These property relations will dominate the probability of success of all projects concerned with land and related natural resources. The centrality of property relations arises not because property relations connect people to land and other physical objects. Rather, property relations are central in development because they connect people to one another *with respect to land and related natural resources*. Notice that the emphasis is not on people and objects, but on people *in relation to objects or circumstances*. Property relations are simply socially constructed contractual arrangements among a group of people with respect to objects and circumstances of value to them. Property relations are created by human communities to mediate individual and collective behaviors regarding objects and circumstances of value to members of the community.

Development projects often are less successful than they would otherwise be precisely because they ignore existing property relations or, just as frequently, because they misunderstand existing property relations.

This chapter addresses property relations in the context of developmental efforts. Some of these developmental efforts seek to mitigate natural resource degradation. Other such efforts are concerned primarily with increasing agricultural production and other income-earning possibilities. And of course some projects seek to do both. In the discussion, I summarize what we have learned over the years in the domain of property relations in the developing world. I call attention to what has worked and what has not worked with respect to project design and implementation and draw implications for policy reform and institutional design. In essence, I emphasize the necessary components of a development program focused on land use and land management in the developing world. I begin with a section on the lessons learned before elaborating why these particular findings are—or seem to be—true.

Property Regimes: Lessons for Development Planning

In this chapter, the term property relations and property regimes are used interchangeably. Both terms are much preferred to the more common term property rights because they focus on the full panoply of property structures in a particular community (see box 7.1). To talk of property rights is to jump immediately to a more legalistic and judicial idea as opposed to the social dimension of property regimes and property relations. To deal with property rights is to enter the domain of contention and disputes, whereas to deal with property relations is to address structures of contractual relations among economic agents.

Property Regimes Are Part of the Larger Institutional Structure

Property relations over land and related natural resources must be understood as part of the larger institutional structure of a society (Bromley 1989). In other words, property regimes are a social construct and therefore cannot be seen as apart from the society in which we address development problems. It is curious that one must make this point, but some development planners clearly imagine that property regimes stand apart from local culture. They are often likely to suppose, as well, that only one possible property structure exists—or ought to exist—throughout the world. Their favorite candidates are (1) the structure into which they (development planners) were socialized or (2) the structure that was implicit (or explicit) in the particular discipline in which they were trained.

As social scientists, we are acutely aware that the institutional setup of a Muslim society is very different from one that is Hindu or Christian. We readily accommodate those differences in our analysis of development projects. A rural health program must be

---

Box 7.1. Property Rights

A property right is the capacity of the holder to compel the authority system to come to his or her defense. To have a property right is to have the capacity to require the authorities to defend your interests in an object or a circumstance.

Common property. A group of co-owners as a "corporate" entity has a right to exclude nonowners, and that right is recognized by an authority system in which the common property regime is situated (say, the nation-state). Individual co-owners are bound by both rights and duties regarding behavior in respect of the asset.

Open access. An open-access resource is one in which there are no property rights for anyone. Under open-access resource settings, the first individual or group of individuals to appropriate resources becomes, by default, their "owner." But ownership under this situation arises from capture, not from prior legitimization by the state. Under open access, individuals are free to use a resource without regard for the implications accruing to others. We say that the individual in such settings has privilege to use the resource but also has no right to prevent others from using it. An open-access resource is a free-for-all in which the rule of capture drives all users to take as much as possible, as quickly as possible. There are no property rights in open-access regimes.

State property. The nation-state maintains control of the area and will usually manage it through the activities of a government agency—forestry service, parks service, and so forth. Individuals in the nation-state may have the right to enter and use the resource, with a clear duty to observe certain strictures on use.

Private property. Individuals (or groups) have a right to undertake socially acceptable uses and a duty to refrain from socially unacceptable uses. All others ("nonowners") have a duty to respect the boundaries of the property regime and have a right to expect that only socially acceptable uses will occur.

---

structured differently if it is intended for Pakistan than if it is being implemented in Paraguay. A livestock project must look very different in northern India than in eastern Africa. A program to improve primary schools should be quite different in Yemen than it is in the Philippines. That is, we have learned that development programs and projects must be crafted to different cultures and institutional structures.

Despite this cultural sensitivity to programs and projects, there is often a degree of institutional singularity when it comes to property regimes. For some reason, economists often find it difficult to admit that property regimes can and do show as much variability across cultures as do other social arrangements. The reason for this difficulty, no doubt, is that property relations are incorporated into economic models and economic analysis such that there is but one way in which to organize behavior. Private (individualized) property is implicit in a theory of economic behavior that reifies individual autonomy predicated on rational self-interest. Simply put, no other property regime is consistent with a theory of behavior founded on methodological individualism. Methodological individualism is the philosophical position that believes the single agent to be the *sufficient* unit of analysis. Contemporary economics is built on the precepts of methodological individualism.

But the institutional arrangements of a society—including property relations—must be understood to reflect prior values and expectations regarding future opportunities. Just as there is no "right" culture, there is no "right" property regime. Rather, there are property regimes responding to, indeed reflecting, manifold interests and priorities. As Dahlman (1980: 85) observes:

> In the process of defining property rights, the economic system must make two interrelated decisions ... The first is to decide on the distribution of wealth; who shall have the rights to ownership of the scarce economic resources even before, as it were, trading and contracting begin. The second refers to the allocative function of property rights; they confer incentives on the decisionmakers within the economic system ... one set of decisions must be treated as endogenous for the system and constitute the exogenous conditions for each trading agent in the resulting set of trades; the second set of decisions is made in the context of the making of these trades.

Property relations—property regimes—reflect this *preallocative* function of any economic system. And, by being preallocative, property relations must be understood as social constructs whose nature and existence are precisely antecedent to what we ordinarily regard as "economic" behavior.

It is indeed encouraging to see institutional arrangements receiving more attention in economics. However, intellectual incoherence is inevitable if this recognition of the importance of institutions is allowed to proceed under the patently false notion that institutions are *constraints* on the otherwise "efficient" workings of an economy. Institutions—property relations—are simply structural attributes of an economy that provide agents with domains of choice within which they may act. They are, at once, constraints and liberation. A property right for Alpha (a liberation) is a "duty" for Beta (a constraint). Property relations exist for just this reason—to liberate Alpha and to constrain Beta. The social problem, therefore, is how it shall be decided that Alpha deserves liberation and therefore that Beta warrants constraint. That question cannot be decided from within economic models that are themselves predicated on that institutional setup. Nor can that choice be decided on grounds of "efficiency" for the simple reason that judgments of efficiency are themselves artifacts of the prices and income distribution (and hence structures of "effective" demand) that are the products of a particular institutional structure (Bromley 1989, 1990).

We are left with the realization that property relations—property structures—must be seen as logically and socially prior to the economizing behavior we attempt to analyze and explain in our economic models. The first lesson, therefore, is that we must understand property regimes as particularistic structures that gain their rationale because they are seen to address fundamental questions about which members of society deserve the protection from the state that property relations entail. Property relations are manifestations of whose interests count in a particular social setting.

Property Regimes Require External Legitimacy

The previous discussion provides a logical precursor to a second lesson about property relations. One of the very clearest lessons from the past several decades of economic assistance to the developing world is that customary (and traditional) property regimes have fallen prey to the priorities and imperatives of influential forces at the center of the nation-state. Against these pressures—which can manifest themselves as large-scale logging of valuable timber, the encouragement of extraction of minerals, the subsidization of commercial fishing fleets, the promotion of massive resettlement projects, or the development of commercial agriculture in pastoral or forest regions—traditional property regimes are nothing. Communities of resource users cannot be expected to resist the economic and political power from the national capital in order to stand up to such pressures for resource extraction.

Sometimes, pressure on local property regimes is more subtle. In particular, perhaps crop agriculture is given preference over pastoral activities. In such instances, the political and economic vulnerability of pastoralists means that the property regimes central to pastoralism will be undermined. If a development project is aimed at improving range and forest conditions for pastoral peoples, then the property regimes central to this particular economic activity must be recognized by the authority system (the nation-state) as a precondition of the project. It is not enough that pastoralists have traditional grazing "rights" in a particular area. If the property regime central to pastoralism is to survive incursion by outsiders, then pastoralists cannot be expected to mobilize their own defense of the assets central to their survival. After all, holders of private property are not expected to defend their own claims to ownership; the authority system does that for them. Although we know that in most urban places in the developing world, the inattention and unreliability of official enforcement structures (local police) mean that protection of private property (homes and businesses) has necessarily been "privatized" by those who can afford it. The rest, largely without assets worth serious protection, take their chances. Indeed, to have a property right is to know that the authority system will come to your defense against the predatory behavior of others interested in your situation or circumstances.

But successful development projects and programs require more than willingness of the authority system to defend the property regimes in a particular location. Members of the community will also have to have developed local structures of authority to manage natural resources in a sustainable manner. Very often development projects come to those areas in which resource degradation is rather advanced. And just as often, part of the reason for that degradation is the breakdown of authority over the resource—by internal conflict or by an intrusion of outsiders into the area.

The second lesson, therefore, is that any structure of property relations requires a commitment from the recognized system of authority that enforcement will be collectively assured instead of privately required (Bromley 1991). After all, when individuals must enforce their own property rights, the concept of a property "right" becomes a contradiction in terms.

Ambiguous Rights Regimes Have Ambiguous Efficiency and Distributional Consequences

The record of development projects around the world is somewhat mixed with regard to the distributional implications of ambiguous property rights structures. In some places we know that the rich will exploit the institutional vacuum that characterizes open-access resources and even some common property resources (Blaikie, Harriss, and Pain 1992 and Jodha 1986). Yet we also know that some ambiguity in property regimes can work to the advantage of persons who do not control private property resources, which are rarely, if ever, ambiguous. That is, some natural resources over which rights are ambiguous are available to individuals who would be excluded if the resources were more valuable and therefore vulnerable to expropriation by the powerful elements in a community. The very meagerness of the bounty from such assets means that the better-off elements in society do not bother with them (Cordell and McKean 1992).

However, when development projects increase the income flow from natural resources over which rights had previously been ambiguous, powerful individuals at the local level will usually figure out a way to expropriate at least a part—if not the majority—of the new income streams. Hence, although ambiguity can often work to the advantage of the dispossessed, it is unlikely to do so when development projects are introduced into this kind of institutional environment. Therefore, the third lesson suggests that development projects introduced into settings in which property relations are highly ambiguous will most probably have economic and social impacts quite different from those postulated in the proposals and feasibility analyses that led to the project's acceptance and implementation in the first place.

Property Relations Must Be Specified Prior to the Implementation of Development Projects

A development project is a "relational contract" involving three entities: (1) the donor agency, (2) the nation-state (the co-sponsor) into which the project is to be placed, and (3) the local individuals whose lives we hope will be altered for the better. We must recognize that each of these entities will always have its own agenda and that only in the rarest of circumstances will those agendas be entirely coincident and mutually enforcing. (This is offered not as an assertion of impropriety but as an empirical observation of great import to the ultimate success of development efforts.) But disparate agendas suggest disparate perceptions of ends and means—of objectives and of instruments.

Although the authority structure of donors and nation-states assures that the implementation of a project will appear to be unified and in harmony, we encounter no such hegemony at the local level. Indeed, we place projects in areas that—as a general rule—warrant outside assistance precisely on account of the incoherence and contentious nature of local institutional arrangements—of which property relations are of paramount importance. Those conditions of institutional incoherence are, after all, often related to situations of chronic poverty, serious natural resource degradation, and the inability of local people to undertake and to sustain adequate material provisioning. That is the "development problem."

Institutional dissonance and ambiguous property relations almost certainly favor the already advantaged when the benefits of a development project begin to appear. And the record of development interventions over the past several decades suggests that new income streams arising from development projects have a difficult time finding their way into the hands and pockets of those most in need of help. The puzzlement about persistent poverty in the wake of decades (and billions of dollars) of "development" projects is really no puzzle at all. The poor remain poor because the institutional arrangements rendered them poor prior to the development intervention, and there are durable pressures—and nontrivial individuals—to make sure that the mere advent of a "development project" does not somehow upset the institutional arrangements that created the current structure of economic advantage in the first instance.

The fourth lesson, therefore, is that *every development intervention* must be preceded by a concerted effort to ensure that the institutional arrangements have been modified so as to make sure the benefits go to the intended beneficiaries. This modification of the "working rules" of the local economy must occur *before the new benefit stream begins to materialize*.

This respecification of the working rules (including property regimes) need not entail every possible detail of future circumstances associated with the development project. But if certain general principles are not spelled out, problems will certainly surface as the benefits of the development intervention begin to appear. To prevent this, it is good to think in terms of a general framework out of which greater detail will emerge. For instance, if a reforestation project is planned, it is important to have in place a general program of protection and management before the trees are planted. Such a framework needs to specify who is responsible for monitoring use of the new plantings, what uses are acceptable during which times of the year (and when they may start), and how breakdowns in compliance will be handled.

For a project to rehabilitate an abandoned irrigation system, a "constitution" should be worked out that specifies the order in which irrigators spread out along the major and minor distribution arteries shall receive water in any particular cycle of water availability. It will also be necessary to make sure that rules of maintenance are spelled out, in a general way, before the first water flows through the system.

Of course we know that this is easier said than done. Or, more correctly, it is more easily done than enforced. This brings us to the issue of the design and implementation of sustainable interventions (in contrast to sustainable development). Issues of sustainable interventions take us, inevitably, back to a discussion of the larger institutional context within which the need for development projects arises.

Resource Degradation Is the Result of Problems That Precede Property Regimes

There is a genre of literature that seeks to blame certain property regimes for widespread degradation of natural resources. Most often, resource degradation is laid at the feet of so-called "common property." Although there are indeed common property regimes in which resource degradation is prevalent, the more probable institutional setup is one of "open access" in which no property rights exist. But situations of open access—either in a narrow legalistic sense or in an operational sense—are themselves manifestations of larger institutional problems.

Hence, a more realistic assessment and diagnosis would indicate that resource degradation arises for many reasons—only one of which may relate to the particular property regime in place (Deacon 1994, 1995). Much of the soil erosion from agricultural lands occurs on private land; this is as true in the developing world as it is in the Corn Belt of the United States. It is curious indeed that private property is not blamed for soil erosion in Nebraska, while common property is blamed for soil erosion in Namibia. When Nebraska farmers allow 15 tons of soil per hectare annually to wash into the Niobrara River, the blame is said to lie with the wrong technology (moldboard plowing), with a rate of time preference that is too high, or with ignorance of the future yield-reducing effects of such erosion over the long run. But when Namibian farmers allow 15 tons per hectare of soil annually to wash into the Konkiep River (or to blow into the South Atlantic), the blame goes to the fact that they are farming under a "primitive and quaint" common property regime. This asymmetry reveals more about the ideological predisposition of the commentator or analyst than it does about the causal structure in each setting.

These particular resource management "outcomes" are largely explained by existing institutional structures, of which property regimes are but a part. The outcomes are explained, as well, by existing customs and norms—and economic imperatives—that are often much more important than the particular property regime in which both farmers operate.

The lesson we draw from this is that before offering development advice, we must build structurally coherent models that gain their legitimacy from their capacity to *explain observed phenomena* as opposed to being *merely consistent with observed phenomena*. For example, constructing a simple model showing that "common property" will result in more erosion than "private property" explains very little. More seriously, such models—when employed in the service of prediction—quickly become the basis for normative prescriptions without the benefit of empirical evidence or conceptual logic. Once someone has built a model "proving" that common property is more conducive to erosion than private property, it is not a long reach to predict that all common property regimes will have erosion in excess of the erosion from ecologically comparable private property regimes. And from there, the normative element slips in to claim that all common property regimes should be converted to private property in order to reduce resource degradation in general and soil erosion in particular.

The lesson here is that we must be careful to avoid building flawed causal models that become the basis for suggesting false solutions to land management problems.

Ecological Variability Demands Flexible Institutions and Actions

The instrumental nature of institutional arrangements, including property regimes, is best demonstrated by thinking about the intersection of the realms of ecological variability and the holding of social and economic capital. In temperate climates, fixed assets such as land hold great economic potential and social status. Small wonder that we find in such places elaborate institutional and technological structures to define and control access to, and control over, land. Most residents of temperate climates regard it as quite normal to view a fixed land base as both an economic asset and a source of social status. These same individuals have been known to offer a bemused smile when told that African pastoralists gain social status from holding not land but cattle. But of course this quaint practice is a product of their "primitive stage of development," it is said.

However, the fact that cattle are privately owned and land is not says less about stages of "development" than it does about the ecological reality within which Africans must make a living. This institutional structure reflects an adaptive response to the reality of provisioning where soils are poor, rainfall is fickle, and irrigation is not feasible. Pastoralism is a response to these circumstances, and the institutions of resource management are a constructed overlay that allows herding to function in such settings. It is not surprising that this structure of flexibility does, at times, bump up against an alternative property rights structure—private property—constructed for rigidity. But institutional flexibility is a necessary attribute of certain economic systems (Cousins 1995, Behnke 1994, and Behnke and Scoones 1992).

Conflicts between pastoralists and sedentary agriculture in the arid climes must be understood as a conflict over property regimes. According to van den Brink, Bromley, and Chavas (1995: 374):

> In agriculture as well as livestock production, property rights emerge to secure income streams generated by production activities. The nature of the income stream, then, may affect the type of property right that is likely to be established. The crucial difference between sedentary farming and nomadic livestock production is that they differ in their ability to react ex post to temporal uncertainty; in other words, they differ in flexibility.

Unfortunately, property rights essential for livestock production in the Sahel have been eroded by a long history of conflicts. More recently, a number of state interventions have expropriated pastoralist property rights crucial to their economic systems and clearly favored farmers over pastoralists in the allocation of private property rights. These changes have created general uncertainty over property rights to natural resources, thereby inducing a de facto open-access situation. The resulting tragedy of open access, induced by public policy, has substantially increased the (transaction) costs of running the pastoralist economy and adversely affected the pastoralists' ability to overcome periods of drought (van den Brink, Bromley, and Chavas 1995).

The lesson for sustainable resource management is that ecological settings that exhibit great variability require property regimes that allow quick and low-cost adaptations to new circumstances. Livestock as an asset, because of their mobility, provide this flexibility in a way that the immobility of land cannot possibly equal. Small wonder that the institutions over livestock and land differ so profoundly between the temperate climates of the middle latitudes and the arid and semiarid reaches of Africa and Asia.

Resource Degradation Is Contextual

Once we recognize that the social and economic meaning of various assets differs across ecological and social circumstance, then it follows automatically that the social and economic meaning of what is a "resource" also differs. The social and economic meaning of a resource is something that brings value; a resource is an "input" into something from which benefits flow. But this means that "resource degradation" cannot be understood without reference to the prior notion of what, exactly, is the set of resources to which a particular society is predisposed to pay close attention.

Overgrazing is a term that clearly derives its operational content from a larger social and economic context. Unless one treasures particular desert plants for their own sake, their relative paucity in particular locations is without normative significance in a development context. Of course more and better plants would allow more and better livestock, that would then allow—and the chain is getting more tenuous as we proceed—more and richer herders, that would allow more economic surplus to flow to urban areas, that would allow "lift-off" to some state-defined "development." Absent that teleological sequence, more or fewer plants in some corner of the Sahel, or robust as opposed to scrawny plants, carry no policy message whatsoever.

The general "condition" of a specific natural resource in a particular place is a socially constructed concept. The policy relevance enters when different constructions become associated with the same ecological reality.

But the lesson we have learned over the past several decades of development interventions in the domain of land and related natural resources is that the only coherent social construction is that which derives from those whose existence is most directly connected to the resources. Of course, technical experts can help to ensure that local people grasp the prevailing scientific explanations (or conjectures) for particular ecological outcomes. But given the large number of scientific "truths" that no longer command even minimal assent, such "technology transfer" must be quite cautious and circumspect.

Land Titles Are Not Necessary for Efficient Investments in Productivity

I now wish to discuss the apparently self-evident proposition that a "secure title" to land is a necessary condition for investment in, or the "wise" management of, land and related natural resources (Feder and Noronha 1987, Mendelsohn 1994, and Southgate 1988). Interestingly, this "truth" is less true than many imagine (Place and Hazell 1993). In fact, the contrary proposition—that investment in land is a necessary condition to secure "title" to that land—is equally probable. Rather than investment requiring security, security requires investment (Sjaastad and Bromley 1997).

The confusion in this domain has to do with the precise meaning of "security" and "title." It seems to be the experience in some parts of the world that the titling and registration of land were the immediate precursor to the dispossession of individuals who imagined that this step would ensure their longevity on the land. Instead, such title became a means whereby moneylenders and others with some measure of political influence were able to acquire what the poor once thought was "their" land. Rather than enhancing security, title had quite the opposite effect. But, of course, title served to establish security for those who managed, by whatever means, to gain control of land that had heretofore been controlled by chiefs and other local notables. So the issue is not one of mere title and security, but of *security for whom.*

To those trained in the modern scheme of things, it seems self-evident that individuals will only invest in some asset if their control over that asset is recognized and secure. However, to return to the very first point about property regimes being part of a larger institutional structure, in some settings, Alpha may wish to invest in land for the benefit of Beta. When this is the case, it is not Alpha's future that must be secured, but Beta's.

The lesson here is that we must be careful about the direction of causality when discussing fundamental economic behavior among disparate peoples in widely scattered corners of the developing world. To suppose that investment is stifled because bankers cannot see a clear title against which a loan might be offered is to impose a particular constellation of assumptions on a setting that has an entirely different structural makeup.

Mobilizing Local Interests Can Improve the Chances of Program Success

The importance of ownership was suggested at the outset. Many development programs and projects require that established patterns of human interaction—and traditional resource uses—be modified. We know that efforts to impose such changes from outside are not sustainable, and this means that the benefit streams intended from such interventions will not materialize over the long term.

In natural resource projects, there is growing recognition that "community-based conservation" is an important innovation in creating the conditions for sustaining both natural resources and the local commitment to sustainability (Western, Wright, and Strum 1994). The essence of community-based conservation is to reconstitute the incentives at the local level so that those closest to the resource are given a greater stake in its long-run viability (Bromley 1994). In effect, those closest to the resource are given an "ownership" interest in its future value. That approach generally gets the incentives right.

Of course, this is necessary but not sufficient to ensure improved behaviors with respect to resources. For instance, it is possible that an ownership interest in the future viability of particular natural resources simply allows for its more efficient degradation. After all, we have seen that an ownership interest in topsoil is no guarantee that it will not be washed away. So although creating a sense of ownership among locals in particular resources can solve much of the incentive problem, such arrangements may still require an oversight role for some higher authority if there is a national or international interest in the particular resource (Arya and Samra 1995).

Still, an important lesson learned over the years is that the chances of sustainability are enhanced as a direct function of the extent to which local people acquire "ownership" in the future benefit stream associated with a particular natural resource.

Implications for Development Policy

I have spelled out nine lessons for persons working in the general area of land rights and resource management. In essence, these lessons pertain to (1) the social embeddedness of property regimes, (2) the concept of a "resource" and resource "degradation," (3) the social instrumentality of institutions such as property, and (4) those institutional arrangements that seem to enhance the success and sustainability of development interventions.

We must keep in mind that development endeavors are both threatening to certain interests and also very rewarding to others. Property regimes play the central role in directing benefits and costs to different groups at the local level. This means that the outcomes of development activities depend fundamentally on institutional arrangements in general and on property relations in particular.

This should not be surprising. After all, the institutional structure is largely responsible for the conditions that lead to the need for development assistance in the first instance. Ameliorative activities in the way of development programs and projects cannot possibly succeed without careful analysis of these institutional precursors of current problems and without a clear program of correction as a precursor to project implementation.

References

The word "processed" describes informally reproduced works that may not be commonly available through libraries.

Arya, S. L., and J. S. Samra. 1995. "Participatory Process and Watershed Management: A Study of the Shiwalik Foothill Villages in Northern India." *Asia-Pacific Journal of Rural Development* 5(2, December):35–57.

Behnke, R. H. 1994. "Natural Resource Management in Pastoral Africa." *Development Policy Review*

12(1):5–27.

Behnke, R. H., and Ian Scoones. 1992. "Rethinking Range Ecology: Implications for Rangeland Management in Africa." Environment Department Working Paper 53. World Bank, Washington, D.C. Processed.

Blaikie, Piers, John Harriss, and Adam Pain. 1992. "The Management and Use of Common- Property Resources in Tamil Nadu, India." In Daniel W. Bromley and others, eds., *Making the Commons Work: Theory, Practice, and Policy,* ch. 11. San Francisco: ICS Press.

Bromley, Daniel W. 1989. *Economic Interests and Institutions: The Conceptual Foundations of Public Policy.* Oxford: Blackwell Publishers.

———. 1990. "The Ideology of Efficiency: Searching for a Theory of Policy Analysis." *Journal of Environmental Economics and Management* 19(1):86–107.

———. 1991. *Environment and Economy: Property Rights and Public Policy.* Oxford: Blackwell Publishers.

———. 1994. "Economic Dimensions of Community-Based Conservation." In David Western, R. Michael Wright, and Shirley C. Strum, eds., *Natural Connections: Perspectives in Community-Based Conservation,* ch. 19. Washington, D.C.: Island Press.

Cordell, John, and Margaret A. McKean. 1992. "Sea Tenure in Bahia, Brazil." In Daniel W. Bromley and others, eds., *Making the Commons Work: Theory, Practice, and Policy,* ch. 8. San Francisco: ICS Press.

Cousins, Ben. 1995. *A Role for Common Property Institutions in Land Redistribution Programmes in South Africa.* Gatekeeper Series 53. London: International Institute for Environment and Development.

Dahlman, Carl. 1980. *The Open-Field System and Beyond.* Cambridge, U.K.: Cambridge University Press.

Deacon, Robert T. 1994. "Deforestation and the Rule of Law in a Cross-Section of Countries." *Land Economics* 70(4, November):414–30.

———. 1995. "Assessing the Relationship between Government Policy and Deforestation." *Journal of Environmental Economics and Management* 28(1):1–18.

Feder, Gershon, and Raymond Noronha. 1987. "Land Rights Systems and Agricultural Development in Sub-Saharan Africa." *The World Bank Research Observer* 2(2, July):143–69.

Jodha, N. S. 1986. "Common Property Resources and Rural Poor in Dry Regions of India." *Economic and Political Weekly,* 5 July, 21(27):1169–81.

Mendelsohn, Robert. 1994. "Property Rights and Tropical Deforestation." *Oxford Economic Papers,* supplement, 46(5, October):750–56.

Place, Frank, and Peter Hazell. 1993. "Productivity Effects of Indigenous Land Tenure Systems in Sub-Saharan Africa." *American Journal of Agricultural Economics* 75(February):10–19.

Sjaastad, Espen, and Daniel W. Bromley. 1997. "Indigenous Land Rights in Sub-Saharan Africa: Appropriation, Security, and Investment Demand." *World Development* 25(4):549–62.

Southgate, Douglas. 1988. "The Economics of Land Degradation in the Third World." Environment Department Working Paper 2. World Bank, Washington, D.C. Processed.

van den Brink, Rogier, Daniel W. Bromley, and Jean-Paul Chavas. 1995. "The Economics of Cain and Abel: Agro-Pastoral Property Rights in the Sahel." *Journal of Development Studies* 31(3):373–99.

Western, David, R. Michael Wright, and Shirley C. Strum, eds. 1994. *Natural Connections: Perspectives in Community-Based Conservation.* Washington, D.C.: Island Press.

8

Equity and Environmental "Modifiers" for Rural Development Policies

Peter Hazell

Agriculture is different from other sectors in many developing countries. In addition to being an important production sector, it is the principal means of livelihood for most of the poor, it plays a key role in ensuring food security, and, because a nation's farmers are the custodians of many of its natural resources, it has a direct bearing on how these resources are managed. For these reasons, agricultural development should simultaneously contribute to four principal goals: growth, poverty reduction, food security, and sustainable natural resources management.

Experience suggests that there can be a high degree of complementarity between these four goals if agricultural development (a) is broadly based and involves small and medium-size farms, (b) is market-driven, (c) is participatory and decentralized, and (d) is driven by productivity-enhancing technological change that does not degrade the resource base. Such growth reduces food prices and increases the effective demand for nonfood goods and services, particularly in rural towns and market centers. By reducing poverty and promoting economic diversification in rural areas, it also relieves pressures on the natural resource base.

The requirements for broadly based agricultural development are reasonably well understood and should not be forgotten in our contemporary quest for "sustainability." Because they are so important, I briefly review them before discussing sustainability concerns.

Back in the 1950s and 1960s, agricultural development experts tended to be interested in agricultural growth, and the lessons that emerged from their experience can be summarized as the 5 "I"s for agricultural growth (or for getting agriculture "moving," as Mosher aptly stated it):

- *Innovation*, the need for strong national agricultural research and extension systems (public and private) to generate and disseminate productivity-enhancing technologies
- *Infrastructure*, the need for good rural infrastructure, particularly roads and transport systems
- *Inputs*, the need for efficient systems for delivering agricultural services, especially farm inputs, agro-processing, irrigation water, and credit
- *Institutions*, the need for efficient markets that provide farmers with unfettered access to domestic and international markets and the need for effective public institutions to provide key public services
- *Incentives*, the need for conducive government policies that do not penalize agriculture.

It later became evident that governments also need to promote the role of the private sector in agricultural production, marketing, and service provision.

In the 1970s and 1980s, attention focused on using agricultural development to reduce poverty and food insecurity as well as to contribute to growth. The lessons that emerged from that area can be summarized as follows. In addition to the 5 "I"s, the following six "equity modifiers" are needed for agricultural growth:

- *Broadly based agricultural development.* There are few economies of scale in agriculture, hence targeting family farms is attractive on efficiency and equity grounds. Small and medium-size farms need to be given priority in agricultural research and extension and in marketing, credit, and input supplies.
- *Participation.* All rural stakeholders (not just the rich and powerful) need to participate in the prioritization of public investments from which they are expected to benefit or help finance.
- *Land reforms.* Market-assisted redistribution programs, in particular, may be needed where too much land is concentrated among large farms.
- *Investments in human capital.* Investments in rural education, clean water, health, family planning, and nutrition programs are needed to improve the productivity of poor people and increase their opportunities for gainful employment.
- *Women.* Because women play a key role in farming and ancillary activities, targeted programs are warranted in agricultural extension and education as well as credit and small nonfarm business assistance programs.
- *The rural nonfarm economy.* The nonfarm economy in rural areas should be actively encouraged because it is an important source of income and employment, especially for the poor, and because it benefits from powerful income and employment multipliers when agriculture prospers. In many countries, these potential multiplier effects are constrained by investment codes and related legislation that discriminate against small, rural nonfarm firms.

The new priority for environmental sustainability that has emerged in the 1990s does not negate the need for agriculture to continue contributing to growth, poverty alleviation, and increased food security; it is just that agriculture is now required to do this in ways that do not degrade the environment. To achieve this, we will still need the five "I"s and the six equity modifiers. There are no shortcuts here. But some new environmental modifiers for sustainable agricultural development are now required. These modifiers have yet to be worked out and tested through development experience, a process that in many ways is still at the research and design stage.

Two fundamentally different types of environmental problems are associated with agriculture. The first type is associated with the misuse of modern inputs (irrigation water, fertilizers, and pesticides) in intensive farming systems. This misuse is related to poor management of inputs by public agencies (some irrigation departments), inappropriate polices that encourage farmers to overuse inputs (fertilizer and pesticide subsidies, underpricing of water), and externality problems that lead farmers to undervalue environmental costs and benefits (water pollution, destruction of beneficial insects).

The second type of environmental problems arises in extensive farming systems and is associated with rapid population growth, poverty, and growth rates in agricultural productivity that is insufficient to meet increasing need for food livelihood. Farmers in these areas typically use low levels of modern inputs (and have limited opportunities to misuse inputs), and, because yield grows too slowly, they reduce fallows and expand cultivation into marginal and environmentally fragile areas. Poverty and insufficient agricultural intensification are the fundamental problems, but resource degradation is often worsened by insecure property rights, externality problems, high costs of collective action for managing common properties or undertaking resource-improving investments, and inappropriate government policies.

Recognizing these differences, I propose the following eight environmental modifiers for "sustainable" agricultural development, these have yet to be fully worked out:

- Give greater priority to backward regions in agricultural development, even though many of these may be resource-poor. Given rapid population growth and limited nonfarm opportunities, agricultural growth often is the only viable means of meeting the food and livelihood needs of growing populations in these areas for the next few decades without excessive outmigration that adds to already overloaded urban slums. Failure to do so will lead to worsening poverty and further degradation of hillsides, forests, and soils. This will require providing additional resources for agricultural development rather than diverting resources from favorably endowed agricultural regions.
- Give greater attention in agricultural research to sustainability features of recommended technologies, to broader aspects of natural resource management at the watershed and landscape levels, and to the problems of resource-poor areas.
- Ensure that farmers have secure property rights over their resources. This does not necessarily imply that government should invest in ambitious land registration programs. In many cases (such as

Sub-Saharan Africa), indigenous tenure systems still work surprisingly well and are better able to meet equity needs and recognize multiple-user rights than fully privatized property rights systems.

- Privatize common property resources or, where this is not desirable (because of externality benefits or for equity reasons), strengthen community management systems.
- Resolve externality problems through optimal taxes on polluters and degraders, regulation, empowerment of local organizations, or appropriate changes in property rights. Free market prices are not always best; externalities may require optimal tax or subsidy interventions.
- Improve the performance of relevant public institutions that manage and regulate natural resources (irrigation and forestry department). Devolve management decisions to resource users, or groups of users, wherever possible. This also requires the transfer of secure property or use rights.
- Correct price distortions that encourage excessive use of modern inputs in intensive agriculture (remove subsidies on fertilizers and pesticides, charge full costs for irrigation water and electricity). It may still be necessary to subsidize fertilizers in backward regions where current use is low and soil fertility is being mined.
- Establish resource monitoring systems to track changes in the condition of key resources, educate farmers about the environmental effects of their actions, and delineate and protect sites of particular environmental value.

Part II
Institutional and Social Perspectives

In part I, we take mainly a policy perspective and argue that a sound policy environment is essential for sustainable agricultural development because, among other things, it creates an incentive framework that induces farmers to innovate. Here in part II, we take a more institutional and social focus and make the point that creation of appropriate institutional frameworks and consideration of the social dimensions are equally important because they provide the structure or fabric within which policies are applied and within which farmers work.

Limited success of central, top-down approaches to development has led to recognition of the need for participatory community-driven development (Narayan, chapter 9). Community-driven development is a process in which community groups initiate, organize, and take action to achieve common interests and goals. No single model is appropriate for all places and times. But successful community-driven development generally is characterized by five factors: local organizational capacity or the existence of viable community groups, an appropriate fit of technology to community capacity, effective agency outreach strategies, client-responsive agencies, and enabling policies.

Local organizational capacity is embedded in the norms that enable groups of people to trust one another, work together, organize, solve problems, mobilize resources, resolve conflicts, and network with others to achieve agreed goals. Viable community groups are key to the success of community-driven development. Getting local groups and organizations to become self-managing organizations can extend over several years and does not happen without investment in capacity building. Many programs have assumed that geographic community is synonymous with community of interest. But communities are rarely homogeneous entities; nevertheless, the need to solve urgent priority problems may bring different class and power groups together. The problem with existing social organization is that it is generally invisible and often excludes women and the poor from the most important production and decisionmaking networks and associations. It is important to understand the existing social system so as to draw on its strengths and embed new organizational forms in the existing social milieu. Local elites often take leadership roles, and although this is not necessarily bad, it can result in the hijacking of resources unless transparency and accountability are somehow enforced.

The most important criteria determining *appropriateness of technology* are local management capacity and the influence that the introduction of a technology has on social cohesion and benefit flows. Thus technology needs to be viewed in a holistic way to determine whether it fits into the existing social system and the direction of change that its introduction will set into motion. In natural resource management and agricultural projects, technological considerations are important when considering what to plant, how to harvest, and how to monitor.

Outreach mechanisms can be classified broadly

into two basic approaches, the extension approach and the *empowerment* approach. The two often overlap, and an effective outreach strategy contains elements of both. The extension approach is characterized more by information dissemination, demand promotion, and delivery of particular services or inputs. The empowerment approach has the main goal of empowering local groups or creating local organizational capacity for self-management through involvement in decisionmaking. The empowerment approach uses staff more intensively and usually requires greater investment in local capacity building. In the extension approach, the agency's desired outcomes are the use of inputs and services delivered and increased effectiveness and efficiency of programs. In the empowerment approach, the key outcomes are that households or groups at the community level have organized for self-management and show increased capacity for coordinated action, including management of local services or resources.

Client-centered agencies are needed to ensure that demand is met and community self-management is adequately supported by community fieldworkers. To become client centered, agencies must be convinced to change their rules and incentive structures so that the benefits from client responsiveness are more important to the agency than the costs. Experience demonstrates that trying to support community-based approaches without fundamentally changing institutional structures, rules, and incentives does not work. Experience also shows that strong political and government support at the highest levels generally is needed for public agencies to become client centered.

A critical question is how to bring about government commitment to change. World Bank staff have used two main strategies to generate interest and commitment: pilot projects as instruments to demonstrate alternative strategies and a variety of participatory techniques, including field visits to other regions and countries, to generate interest and a commitment to new ideas and action strategies. Indicators of success are also important because they send important signals to staff about program priorities. If community involvement, number of women or the poor being reached, and number of functioning systems are not reflected in indicators of success, there is little incentive for staff to change their way of doing things to reach these goals. Almost all projects that adjust as they are implemented invest in numerous small studies that provide feedback on how different approaches are working.

Community-driven development on a large scale requires *enabling policies and political support* to protect agencies so they can initiate the reform process and give it time to take root. It is much easier to support in the context of decentralization strategies. Although individuals play key roles in starting or protecting the reform process while it is still in the early stages, change can be sustained only if the rules and regulations at different levels provide incentives to work and organize.

For *designing participatory community-based development projects*, the following 10 lessons from worldwide experience should be reflected in the design of new projects, subject to testing with pilot activities: (1) develop clearly stated objectives logically linked to strategies, outputs, indicators of success, and physical or capacity-building outcomes, (2) identify the key actors and their capacity and interests (through social assessment), (3) assess demand (as measured, for example, by commitment before construction and substantial financial investment up-front in capital cost), (4) craft a self-selection process (with a significant financial or organizational contribution up-front), (5) structure subsidies that do not distort demand (for example, by asking the poor to contribute substantial amounts through their labor, (6) restructure the release of funds to support demand (such as by fiscal decentralization), (7) plan for learning and for a plurality of models (the project must be seen as likely to adjust and change over time as local priorities change and as local organizations mature), (8) invest in outreach mechanisms and social organization (investments are needed, and there are no shortcuts to strengthening local social organization for collective action), (9) use participatory monitoring and evaluation (listening to people's suggestions is empowering and encourages innovation and responsibility), and (10) redefine procurement rules, where necessary (so that they do not hinder community initiative).

Ashby and others (chapter 10) illustrate these participatory principles by examining the experience involving local organizations in watershed management. Their discussion is oriented toward providing guidelines rather than a theoretical analysis. Also, it is based on experience of working with local organizations from the perspective of development practitioners. This experience is evolving and accumulating rapidly, and many practitioners are promoting a learning process in which their approach to local-level watershed management is adapted and reoriented swiftly based on analysis of what has gone before. Thus, although this discussion focuses on some of the lessons learned and principles that can be derived from this experience, Ashby and others believe that the last word on what works and what does not work is by no means decided. Indeed, the dynamic nature of the process of involving local organizations in watershed management is one of its most important characteristics and one that requires a focus on managing the process rather than on defining a blueprint for how to go about it.

Watershed management, especially in low-income countries, usually involves a combination of conservation and development of natural resources. Watershed inhabitants depend on these resources for their livelihoods, and one of the basic challenges is to establish and maintain management practices that reconcile their economic needs with the long-run conservation of soil, water, and vegetation, especially where resource degradation has significant off-site effects, such as the silting up of watercourses or reservoirs of interest to influential stakeholders downstream. On-site and off-site degradation of watersheds is widespread, and the management practices for conservation and regeneration are technically well understood: land use planning to designate certain areas for conservation, grazing, annual or perennial crops, and natural or managed forest; use of contour barriers on steep slopes; use of ground cover; use of agroforestry and agrosilvopastoral systems; creation of buffer zones around watercourses and springs; and minimum tillage, green manuring, and other soil improvement practices that regenerate soils and permit the intensification of agriculture where suitable. However technically desirable these practices, their adoption by resource-poor farmers is very uneven, in many cases because the short-term economic returns to investing in these practices exceed the costs for the individual farmer. As a result, top-down efforts to enforce mandatory use of these practices or to promote them using temporary subsidies have by and large failed to achieve lasting adoption.

Involving local organizations in designing technical innovations for soil and water conservation and in planning where to locate them in the landscape of a watershed is now recognized as a key element of successful adoption. There are many examples of enforced use of soil and water conservation practices leading to accelerating erosion once the local people stop maintaining the structures, because they have no commitment to keeping the structures in place. In other instances, entire structures are leveled as soon as the outside agencies imposing them leave the village. In contrast, innovative management practices are rapidly developed and adopted when local organizations involve farmers in experimenting with and adapting the principles of known conservation techniques to their own needs and constraints and in combining this information with local knowledge of the landscape, soils, and vegetation.

Few local organizations at the watershed scale have arisen completely spontaneously without some intervention from outside agencies to catalyze their formation and action. It appears to be a characteristic of local organization on a watershed scale that outsiders play a significant role in enabling stakeholders to arrive at a joint plan of action that takes into account transboundary effects not readily perceived or measured by stakeholders other than those directly affected.

By providing critical information about transboundary effects, external agencies can enhance the capacity of local organizations to mobilize local cooperation in watershed management. Another important point of connection between local organizations and external agencies in watershed management is the need for external intervention in identifying the relevant stakeholders (which often have conflicting objectives) and bringing them "to the table," whether this refers to a negotiating table or to participation in collective labor, monitoring, or enforcement of sanctions. The involvement of local organizations in watershed management does not necessarily ensure equitable participation of all relevant stakeholders in a resource management initiative (Ashby and others, chapter 10; a point also made by Narayan, chapter 9). Case studies show that women, the landless, marginal ethnic groups, and the laboring poor are all likely to be unrepresented in participatory watershed management. Leadership in local organizations is generally captured by the higher status, more well-to-do members of dominant elites, and the priorities of local organizations tend to reflect their interests. Watershed development projects should deliberately seek to overcome this bias by setting up catchment committees with representation of underprivileged groups. Ashby and others (chapter 10) highlight these important principles with information from a case study tracing the evolution over time of local-level watershed management in the Río Cabuyal in Colombia.

Participation is also one of the keys to successful technology generation among low-income farmers

(Farrington and Thiele, chapter 11). In many areas of rainfed agriculture, where agroecological conditions are complex, diverse, and risk-prone and where standards of physical (communications, transport) and social (education, health) infrastructure are low, specific efforts will have to be made to elicit farmers' perceptions of needs and opportunities. In these areas the private sector undertakes little research and development activity even though this is needed the most. Attention has therefore turned to ways of improving the effectiveness of the kinds of research institutions likely to dominate these areas in the coming decades, namely those in the public sector. A radical view is that public sector research services are so ineffective that they should be shut down and replaced with a low-cost but efficient capacity to "borrow" ideas. But even then, capacity is still needed to adapt and test "imported" technologies under local conditions.

The public sector has generally performed poorly in difficult areas for several reasons: (a) inadequate orientation toward clients and participation by them in the design and implementation of research and (b) insufficient interaction with local organizations having close contact with farmers (nongovernmental organizations, membership organizations, development projects). Local organizations can create a "demand pull" on the public sector by helping to articulate farmers' requirements.

Multiagency approaches—not replacement by the private sector or complete abolition—are one way of overcoming the limitations of the public sector. Innovative multiagency approaches are now being tested in various ways. They seek synergistic interactions between different types of organizations, in which all work together to analyze particular problems, contribute jointly to their solution, review progress in an iterative fashion, and make course corrections by mutual agreement as necessary.

Field-based efforts to establish multiagency approaches to technology generation and dissemination in agriculture and natural resources have proliferated in the past five years. These have involved varying combinations of government organizations, nongovernmental organizations (NGOs), farmers organizations, and, at times, the private commercial sector. Although some success has been achieved in introducing participatory approaches at the diagnostic stage of the research cycle (for instance, participatory rural appraisal is now widely used, although often badly), much less attention has been paid to ensuring continued interaction between researchers and their clients during the identification, testing, and dissemination of technologies. Significant organizational and management challenges must be overcome if multiagency approaches and participation are to become part of the mainstream of government research and extension strategy (Farrington and Thiele, chapter 11).

NGOs providing development-related services tend to follow intensive face-to-face approaches with small, local groups, often over a period of years. This empowering approach to participation allows NGOs to take the moral high ground and to dismiss efforts that do not reach these standards. However, from a wider policy perspective, government organizations cannot mandate themselves to concentrate resources in a few villages as NGOs do, and they have to make the best of the difficult job of achieving whatever level of participation is feasible through a thin spread of resources over a wide area. This leads government organizations to seek functional participation to perform technology generation and dissemination more efficiently. Clearly, at times there are gray areas between functional and empowering forms of participation, which represent different ends of a spectrum rather than uniquely distinguishable categories. Nevertheless, the distinction is important in policy contexts, not least because each approach tends to be associated with very different kinds of organizations.

Farrington and Thiele (chapter 11) present three case studies to illustrate innovative, participatory technology generation. The first case study describes scaling up participatory approaches in an unstructured context over three years (1994–97) in the Udaipur district in Rajasthan, India. These have sought to identify how NGOs and government research and extension services might work more closely together. Some successes are noted, but a number of difficulties remain. The principal constraints are deeply rooted in philosophies and operational procedures on both sides: government organization programs remain largely centrally determined. Problems on the NGO side include a reluctance of some to think in terms of strategic development beyond the confines of a few villages in which they are working. This case study finds that the efforts to scale up participatory approaches will, for the future, have to be organized on a more structured basis.

A second case study describes scaling up participatory approaches in a structured context and uses an example from microwatershed rehabilitation. Microwatersheds of 500–1,500 hectares containing one or more villages have become the focus of policy attention in several South Asian countries as a natural unit for implementing efforts to intensify agricultural and natural resource productivity. Improved natural resource management potentially allows a quantum leap in agricultural production technology and in productivity.

A third case study describes changing the organization and management of government services to scale up participation. It reports on the evolution in the eastern lowlands of Bolivia of a model for strengthening links between research, extension, and farmers. Central to the model is the highly innovative Research-Extension Liaison Unit (RELU) explicitly charged with linking the state agricultural research center with a broad range of intermediate institutions, including NGOs, that carry out extension. When the RELU was created, its job was seen principally as that of transferring technology placed on the shelf by the existing commodity-based structure of Centro de Investigación en Agricultura Tropical (CIAT), but applied research and testing soon became the principal activity for some RELU staff.

Participatory principles are also key to successful extension with long-run impact (Bunch, chapter 12). One such approach—participatory people-centered agricultural development—has been developed and tested extensively in Central America by World Neighbors and has spread to NGOs in many nations around the world.

Government extension services tend to neglect farmers in resource-poor areas. One way to overcome this is to enlist farmers in solving their problems of low productivity. A participatory process teaches villagers how to experiment and teach one another through the most efficient teaching process: *learning by doing*.

People-centered agricultural development consists of a series of principles with much room for adaptation to local circumstances, farmer needs, and the institutional context. Principle 1 is to motivate and teach farmers to experiment with new technologies on a small scale, thereby reducing the risk of adoption and providing a means for them to continue to develop, adopt, and adapt new technologies in a permanent scientific process of innovation. This principle is frequently referred to as participatory technology development. Farmers who have very little or no formal education are not accustomed to learning through the written word or presentations; they learn much more efficiently through their own experience. Thus when they themselves experiment with new technologies, they learn in the best way they can. Being able to validate and modify technologies for their own use, as well as to create entirely new technologies, makes the farmer much less dependent on outside sources of information, which is very important in situations where infrastructure and information services are lacking.

Principle 2 suggests the use of rapid, recognizable success of selected experiments to motivate farmers to innovate, rather than artificial incentives or subsidies. Principle 3 stresses the use of technologies that rely primarily on inexpensive, locally available resources. Principle 4 recommends beginning the process with a very limited number of technologies, so that the program is focused and can achieve the maximum possible success right from the start, and involving even the poorest farmers in the process. Finally, principle 5 recommends training village leaders as extensionists and supporting them as they teach additional farmers, thereby creating and nurturing a community-based multiplier effect. Now called farmer-to-farmer extension, this principle is effective and keeps costs down. To be selected as an extensionist, farmers must have introduced innovations successfully on their own farm (which provides credibility!).

In many instances where people-centered agricultural development has been used successfully, yields of major crops have tripled and new crops have been introduced. But success has not been uniform. According to one survey, in the 25 to 30 percent of villages where the response was best, yields continued to increase after program termination, the number of organizations increased, land prices increased dramatically, incomes increased, and outmigration either diminished drastically or was reversed. In 40 to 50 percent of the average villages, yields more or less maintained or increased only marginally, whereas in the 20 to 35 percent of the villages where impact was poorest, yields decreased, although they continued to be better than the yields at program initiation.

One of the major lessons of people-centered agricultural development is that sustainability does not reside in technologies, which have a half-life of perhaps five years. Markets change, input prices increase, new technological opportunities appear, varieties degenerate, pests spread, and competition becomes stiffer. The hope for sustainability of agricultural development is in the nature of an ongoing social process of widespread experimentation, sharing of information, and innovation and group problem-solving. In people-centered agricultural development, it is more important *how* things are taught than *what* particular technologies are taught. Thereby the process creates a sustainable agricultural development process. It also, by its very nature, empowers people in ways useful for rural development efforts in general.

Giving farmers access to land and security of tenure is essential to creating the right incentives for sustainable rural development. Theoretical reasons and empirical evidence suggest that land reform may provide benefits in terms of both equity and efficiency (Deininger, chapter 13). First, a large body of research has demonstrated the existence of a robustly negative relationship between farm size and productivity due to the supervision cost associated with employing hired labor. This implies that redistribution of land from large to small farms can increase productivity. Second, in many situations, landownership is associated with improved access to credit markets, providing benefits as an insurance substitute to smooth consumption intertemporally. By enabling the poor to undertake indivisible productive investments (or preventing them from depleting their base of assets), this could lead to higher aggregate growth. Third, even in aggregate cross-country regression, the distribution of productive assets—more than the distribution of income—seems to have an impact on aggregate growth. Finally, land reform is expected to increase the environmental sustainability of agricultural production and to reduce rural violence.

Notwithstanding this apparent potential, actual experience with land reform has been disappointing in all but a few exceptional cases, such as Korea, Japan, Taiwan (China), and to some extent Kenya. Despite—or because of—this, land reform remains a hotly debated issue in a number of countries (Brazil, Colombia, El Salvador, Guatemala, South Africa, Zimbabwe).

There is a new type of negotiated land reform in which land transfers are based on voluntary negotiation and agreement between buyers and sellers and in which the government's role is restricted to the provision of a land purchase grant to eligible beneficiaries. This is being tried in Brazil, Colombia, and South Africa and is Deininger's main focus in chapter 13.

As Heath and Binswanger in chapter 3 indicate, small farmers in Colombia were often driven off their traditional lands and forced to eke out a living in marginal and environmentally fragile areas, and much of the best agricultural land continued to be devoted to extensive livestock grazing or was not farmed at all due to violence. Only 25 percent of the land suitable for crop production was actually devoted to this use, while the rest was left to pasture. This suggests that there are indeed large tracts of unused or underused land that could be subjected to land reform in order to increase agricultural productivity—a notion in line with available empirical evidence.

Because centralized land reform has not had much success, negotiated, market-based land reform is being attempted. It is characterized by three main principles. First, potential beneficiaries (with assets below a certain minimum level) can negotiate independently with landowners and, once a deal has been struck, are eligible to receive a grant of up to 70 percent of the land's purchase price (subject to an upper limit). Second, the government's role is limited to that of regulatory oversight and grant disbursement. Third, a decentralized institutional structure is needed to ensure that land reform is demand driven and coordinated with other government programs. Despite favorable preconditions, and the government's expressed determination to distribute 1 million hectares within four years, this new type of land reform has had a disappointingly slow start.

Brazil and South Africa also have, under different conditions, recently initiated programs of negotiated land reform. With an institutional background very similar to that of Colombia (presence of land reform legislation and a central land reform institute dating from the early 1960s), negotiated land reform in Brazil has been initiated by individual states. The purpose of the Brazilian interventions is to establish cheaper, more agile policy alternatives to centralized land reform in an environment where the issue of land reform is high on the political agenda and potential beneficiaries have at least some idea of what to do with the land. By contrast, negotiated land reform in South Africa has been adopted in the context of the national reconstruction program, in an environment in which productive small-scale agriculture was eradicated almost a century ago. There, much greater effort is required to establish the decentralized infrastructure necessary to implement land reform, to provide complementary services such as marketing and technical assistance, and to increase beneficiaries' agricultural and entrepreneurial capacity.

Although it is too early to judge whether negotiated land reform can rise to the challenges that administrative land reform has failed to solve, the experience unfolding in Brazil, Colombia, and South Africa (as described in chapter 13) can at least provide the basis for a first assessment. Comparison of the approaches taken in different countries suggests that (a) land reform through negotiation can only succeed if measures are taken to make the market for land sales and rental more transparent and fluid, (b) productive projects are a core element of market-assisted land reform (designed to establish economically viable and productive projects at a socially justifiable cost rather than to transfer assets), (c) the only way to achieve effective coordination of the various entities involved in this process is through demand-driven and decentralized implementation, and (d) the long-run success of land reform is likely to depend critically on getting the private sector involved in implementation and the ability to use the land purchase grant to "crowd in" private money.

Several of the chapters in part II (particularly chapters 9–12) stress the importance of participation for successful and sustainable rural development. Some also touch on the need to pursue decentralized approaches, which are important in and of themselves. Yet a certain degree of decentralization also represents a precondition or facilitating role for participation.

Decentralization is one aspect of the broader issue of good governance and is, along with participation, a key element in sustainable rural development. In chapter 14, Caldecott and Lutz focus on whether or not decentralization promotes the conservation of biological diversity and, if so, under what conditions. Many past attempts to conserve biodiversity failed. One reason was an overly centralized approach, often involving top-down planning by technicians and bureaucrats without concern for the opinions or well-being of the people affected by their decisions. In many other countries too, the viability of centrally planned fortress reserves was undermined by their cost and also by the democratic deficit built into them. Nevertheless, complete decentralization can also be counterproductive to conservation. Under those conditions, local people, perhaps in cooperation with outside entrepreneurs, may simply degrade and deplete the resources faster and more efficiently. So central governments clearly continue to have a role. The challenge is to find the appropriate degree of decentralization.

The most potent form of decentralization involves fiscal decentralization, that is, the transfer of authority over spending decisions. These particular powers are retained by elite groups far more diligently than those over other administrative functions. Thus the most complete cases of decentralization are those where local people collect revenues and decide locally how they will be spent. But not all functions of government can usefully be decentralized, and nonlocal groups may be in a better position than local ones to appreciate long-term or large-scale issues and to act as disinterested arbiters of local disputes that cannot be solved locally. This nonlocal perspective is vital in conservation, which fundamentally concerns avoiding or managing interspecies, intergenerational, interregional, and international conflicts of interest. Empowerment of local groups should therefore be balanced by a continuing role for central government to deal with market failures and to ensure both social equity and environmental protection.

After stressing participation and decentralization as necessary elements of sustainable rural development throughout part II, we would be remiss if we failed to stress more explicitly the need to integrate gender. For several decades, food, agricultural, and natural resource management policies have been designed without acknowledging that rural men and women may have different preferences, face different constraints, and respond differently to incentives. Indeed, until Ester Boserup's path-breaking work in 1970, women were largely "invisible" within agriculture, and the predominant assumption was that the male household head made most, if not all, farm allocation and production decisions.

Quisumbing and others, in chapter 15, show how neglecting the gender dimension in the design of development policy has led to failures in project implementation, failure to adopt new agricultural technologies, or the adoption of new technology with negative unanticipated impacts. Therefore, policymakers would do well to take gender issues into account when formulating food, agricultural, and natural resource management policies. Based on a review of empirical evidence, reducing inequalities between men and women in human and physical capital can lead to major efficiency and productivity increases in agriculture. Moreover, addressing inequalities in the underlying distribution of property rights between men and women may be essential not only for increased agricultural productivity but also for environmental and social sustainability. The second-round effects are also important: when women control the increases in income, those increases have a greater impact both on household food security and on investments in child health and schooling.

Although participatory approaches to project design have been recognized as a way of ensuring community involvement, commitment, and ownership of development projects, they are not synonymous with incorporating gender analysis. Participatory approaches to project design by themselves do not guarantee that women will participate in either assessment of their needs or design of the proposed project. This means that a special effort needs to be made to have gender properly addressed, such as by means of gender analysis (Quisumbing and others, chapter 15). This goes beyond a women-in-development approach, which identifies the differences in resource access, ownership, and control between men and women. Rather a gender-and-development approach recognizes the determinants of these asymmetries and their consequences for individuals, households, communities, and economic development. In essence, the women-in-development approach recognizes an outcome, whereas a gender-and-development approach recognizes both the outcome and its consequences as well as the process leading to those outcomes. The understanding gained from the gender-and-development approach facilitates understanding not only how a planned project or policy will affect men and women but also how it will affect the underlying processes that condition the allocation of rights, resources, and responsibilities within communities and households.

9
Participatory Rural Development

Deepa Narayan

In many countries, limited government success in managing natural resources, providing basic infrastructure, and ensuring primary social services has led to the search for alternative institutional options. In recent years, a shift has occurred away from supply-driven toward demand-driven approaches and from central command-and-control to local management or coproduction of resources and services in an effort to increase effectiveness, efficiency, and equity.

One of these institutional options is participatory community-driven development, a process in which community groups initiate, organize, and take action to achieve common interests and goals. Substantial experience, both about what works and what does not, shows that no single model is appropriate for all places and times. Although community groups have organized themselves from time immemorial to meet their needs, for the government or other outsiders to induce community-driven development on a large scale requires agencies to invest in local organizational capacity and support community control in decisionmaking. This reorientation, in turn, requires agencies to shift their internal rules, incentives, and skills to focus on matching what the agency supplies with what community groups want, are willing to pay for, and can manage.

Experience also shows that community-driven development does not automatically include marginalized groups, the poor, women, or ethnic minorities unless their inclusion is specifically highlighted as a goal both at the agency and community levels.

In order to induce community-driven development on a large scale, it is important first to understand the dynamics at the household, group, or community level. Based on this understanding, it is important to define what needs to happen at successively higher and more distant levels to support community action.

Key Features of Successful Community-Driven Development

Successful community-driven development is characterized by five main factors: local organizational capacity or the existence of viable community groups, the appropriate fit of technology to community capacity, effective agency outreach strategies, client-responsive agencies, and enabling policies.

Local Organizational Capacity

Local organizational capacity or local-level *social capital* is embedded in the norms that enable groups of people to trust one another, to work together, and to organize to solve problems, mobilize resources, resolve conflicts, and network with others to achieve agreed goals. Participatory decisionmaking processes help build confidence and capacity. When people cooperate and work together, they can overcome problems of coordination, risk, and limits to their information and individual skills. Viable community groups are key to the success of community-driven development. Development projects can be instrumental in facilitating the emergence of local groups that organize themselves to solve their own problems and network with others, including support agencies, to mobilize resources and design solutions. However, for collective action to succeed, certain conditions have to be met and some common mistakes avoided.

Quick, local benefits. Collective action is easier to stimulate when benefits are quick, visible, and local, when they accrue only to those who participate, and when they are felt to be proportionate to costs. If benefits from conservation measures such as forestry, fisheries, or wildlife protection accrue only at the district or national level, local community groups have little incentive to participate. For this reason, cleanup of rivers and lakes is difficult to induce through collective action. The long gestation period before benefits accrue in forestry and tree-planting projects is a disincentive to investment, particularly when combined with insecurity over tenure. Projects that provide access to fodder, nontimber forest products, and firewood while timber is actually maturing are generally more successful than those that rely solely on promised timber benefits.

However, when quick visible results are induced artificially by agency-initiated shortcuts, they may be counterproductive to collective action. For example, in an attempt to motivate the formation of pastoral associations, the Mali Pastoral Associations Project changed its strategy and constructed water points prior to mobilization; an evaluation of the project found that the change had the opposite effect because it violated the principle of local contribution and involvement in decisionmaking (Shanmugaratnam and others 1992).

Another shortcut with negative long-run consequences is to lower or eliminate a community's financial contributions to induce quick action. Once eliminated, contributions are difficult to introduce later in the same area; the result is lack of local ownership and financial viability. In a rural water supply project in Nigeria, required community contributions were dramatically lowered to induce quick action and accelerate the pace of implementation. This lowered the stake that community groups had in the success of the boreholes (Boerma 1993).

A more useful approach to inducing quick benefits is to *sequence* tasks and break them up into small "doable" bits. Success in small tasks generates mutual trust and satisfaction. This is also one of the reasons why nongovernmental organizations (NGOs) that are not limited to activities in a particular sector can start with activities that produce quick results and then move gradually to activities for which there is a time lag. The Agha Khan Rural Support Program in northwestern Pakistan starts by first building one small productive infrastructure that is identified as the community's first priority. The work is undertaken in partnership with the community (World Bank 1996).

The FUNDASAL (Fundación Salvadoreña de Vivienda Mínima) housing project for the poor in El Salvador builds on the principle of *incremental development.* Houses are constructed in stages through community self-help; a high priority is assigned to securing tenure and services, completing the basic structure of the house quickly, and letting families expand their houses over time. Each house is constructed by 20 to 25 families working together; once completed, the house is assigned by lottery to a participating family. The group then collectively starts constructing the next house (Bamberger, Gonzalez-Polio, and Sae-Hau 1982).

Common interest. When people experience problems that they consider serious and cannot solve through their own individual efforts, they are more likely to come together to act collectively. Many programs have assumed that geographic community is synonymous with "community of interest." This is not necessarily the case and is a common source of problems in mobilizing collective action. The community woodlots movement in India, for example, largely failed in its early days because it was assumed that the community was the appropriate unit of "common interest" to manage woodlots (Cernea 1989). Communities are rarely homogeneous entities. In irrigation associations, the interests of those at the tail end invariably clash with those at the head end; formal village councils, such as the *panchayats* of India, may have different interests than the poor in the community.

However, the need to solve urgent high-priority problems may bring different class and power groups together. In South India, Robert Wade (1994) has documented how an entire village manages community-based irrigation systems and has developed a monitoring system to discourage water theft. Nonetheless, as groups grow or spread over large areas, the influence of social cohesion begins to break down, and it becomes more difficult to control and monitor the behavior of individuals. For this reason, as groups become larger, either they formalize regulations and delegate decisionmaking to smaller working groups or they join in a federated structure that leaves decisionmaking at the local level.

Authority over resources. In the past, agencies have been reluctant to shift the control and authority over resource management to community groups. The assumption has been that community groups will not make rational decisions and will misuse financial resources. The experience has been to the contrary. In the Philippines, when farmers groups were given authority to set and collect irrigation fees, farmers actually raised fees and improved their collection, dramatically increasing the returns to the National Irrigation Agency (NIA). In Orangi, an urban neighborhood in Karachi, Pakistan, the cost of sewers declined as the incentive for government engineers to overdesign sewers was removed. In several social funds (the Village Infrastructure Project in Indonesia and the social fund in Malawi and Eritrea), as community groups were given the authority to manage financial resources, they carefully monitored all expenses and often reduced the officially sanctioned labor costs paid to themselves to extend the money available for physical construction. To encourage accountability, more and more projects are supporting the public posting of all project agreements with communities, including financial information, unit costs, labor costs, and so forth, to create a well-informed public and thus maximize the protection of project resources.

Social embeddedness. Experience demonstrates the importance of nurturing institutions at the local level that have their roots in the local social context (Evans 1995). The Balochistan Primary Education Project in Pakistan shows how quickly community action spreads when it is embedded in the local social organization, when tasks are clearly defined, and when time is spent developing the local organization.

The problem with existing social organization is that it is generally invisible and often excludes women and the poor from the most important production and decisionmaking networks and associations. It is important to understand the existing social system so as to draw on its strengths and embed new organizational forms in the existing social milieu.

In Nepal, for example, when a government policy was promulgated to create farmers associations, assistant overseers found many informal groups of farmers organized around irrigation systems. Rather than create new organizations, existing groups were encouraged to register themselves as official farmers irrigation associations. In a Livestock Development Project in Mauritania, pastoral associations based partly on traditional organization and partly on new organizations were created on a pilot basis, using environmental and socioeconomic data. The average association covered 50 villages, 2,500 square kilometers, and a population of 14,000 people. These groups proved too large to work effectively and were quickly subdivided (Shanmugaratnam and others 1992).

Group leadership and capacity. The capacity of groups to organize themselves to undertake coordinated action is important to their success. Local elites often take leadership roles, and although this is not necessarily bad, it can result in the hijacking of resources unless transparency and accountability are somehow enforced. For example, in community-based rural water projects in Indonesia and Pakistan, success in community organizing was closely linked to the presence of strong leaders interested in changing the water supply situation. However, in Zimbabwe, when local leaders implemented the policy regarding communal grazing schemes, they often grabbed the external resources for themselves or obstructed the formation of grazing groups (Scoones and Matose 1993). Charismatic leadership may be necessary to seed collective action, but managerial leadership is called for once the period of creation and innovation is complete.

When collective action is induced from the outside, the focus is often on creating the structure of committees, without matching the task to the capacity, knowledge, or technical skills of groups. Groups also fail because too much is expected of them too soon without any supportive training in management or specific skills. Getting local groups and organizations to become self-managing organizations can extend over several years and does not happen without investment in capacity building.

Group ownership and enforcement. To function effectively, community groups have to set the rules that define membership requirements, responsibilities and benefits, accountability, punishments, and the resolution of disputes. This may be done formally, with written texts, fees, and fines, or informally through practice. Often, rules evolve in the course of the work.

Entry rules define who belongs and who does not

as well as the obligations of members. Membership can be based on a variety of factors: ownership of land, participation in farmers groups, gender, or age. To limit the benefits to those who do the work, successful groups often impose membership or user fees.

Allocational rules define responsibilities, contributions, and benefits. Unless members and managing committees know what their responsibility is and how they will benefit, they cannot be expected to perform their functions. A 1987 evaluation of the Orissa Social Forestry Project in India found that 82 percent of the villagers, all of whom were supposedly members, did not know how produce from the village woodlot would be distributed. Most did not expect to receive any share from the final output and looked on the community woodlot as another type of reserve forest. Not surprisingly, they had no interest in contributing to its establishment or maintenance.

Mechanisms to ensure accountability are also important. The fact that group members know that an effective monitoring mechanism is in place can deter violators and encourage others to report violations. Monitoring can focus on payment of tariffs, extraction and use of resources, and performance of group management committees, agency staff, and contractors. In small groups that live in physical proximity, social pressure through peer monitoring is an important low-cost and effective technique. The Grameen Bank peer monitoring system is an important part of why the group lending system works. If an individual defaults on a loan, the whole group is accountable. This creates incentives among members to monitor and support one another.

The greater the resolution of disputes at the local level, the less burdened and expensive the overall system. When conflicts cannot be solved quickly, group-based schemes fall apart. In Zimbabwe, many grazing schemes have not been able to resolve basic boundary conflicts or differences of opinion about technical soundness. The presence of physical fences to mark boundaries has made no difference, and fences have been removed, ignored, or not maintained.

Access by women and the poor. In most societies, there are important differences between men and women as regards their roles, needs, networks, skills, and knowledge. It cannot be assumed that what is appropriate for men is appropriate for women at the community level. A study of rural water supply projects, for example, showed that only 17 percent of the projects involved women in decisionmaking, although most project documents stated that a goal was to reach women (Narayan 1994).

Nor do community-based programs necessarily reach the poor. Work with pastoral associations reveals that as many as 80 percent of the herd in Mali and 40 percent in Niger belong to absentee herd owners. The capture of benefits and resources by the rich and powerful is ever present when the poor are bound to the rich in client/patron relationships and the resource under consideration is scarce and therefore has great economic value (Shanmugaratnam and others 1992).

Thus, in Pakistan, village elites play a positive role in the northern provinces in initiating simple drinking water supply schemes, but in the southern provinces, particularly Sind, elites capture irrigation water, often blatantly and disproportionately (Byrnes 1992). In Kenya, where community leaders are sympathetic to the plight of the poor, communities have instituted a sliding scale of fees for the poor. In Tanzania, sliding fees based on poverty appear universal in community-managed systems. In other countries and settings, however, the poor are excluded when communities impose user charges to achieve financial viability.

The poor are most easily reached when program services and structures are targeted specifically to their needs. In the FUNDASAL Project in El Salvador, the poor constructed each other's houses, and this mutual help eliminated the need for a 10 percent downpayment. Families were required to work in construction teams for 30 weekends and to participate in meetings and training in organizational techniques (Bamberger, Gonzalez-Polio, and Sae-Hau 1982).

The Appropriate Fit of Technology to Community Capacity

The most important criteria determining appropriateness of technology are the level of local management capacity and the influence that the introduction of a technology has on social cohesion and benefit flows. Thus technology needs to be viewed in a holistic way to determine its fit with the existing social system and the direction of change that its introduction will set into motion.

In the drinking water sector, poor maintenance led to the search for village-level operations and maintenance technologies, including hand pumps. Although cost is important, the least-cost option is not necessarily the best. Experience has established that people desire house connections and are often more willing to pay a relatively high charge for piped water than a small fee for access to communally managed hand pumps, which are less convenient and reliable.

In natural resource management and agricultural projects, technological considerations are important when considering what to plant, how to harvest, and how to monitor. Experience shows that community response is greatest when projects adopt approaches that provide a continuous flow of benefits, especially where indigenous people or women are dependent on nontimber products, fodder or thatch grass, agricultural intercrops, and seed.

In Nepal, where indigenous farmers organizations manage the majority of irrigated land, agency-led and donor-financed construction of permanent headworks based on efficiency concerns has been less than satisfactory (Ostrom 1993). The Kamala Irrigation Project was meant to serve 25,000 hectares but has never reached that goal. The system includes large, permanent, concrete headworks and a fully lined canal. No fees have ever been imposed or collected, and agency staff spend most of their time operating and maintaining the huge concrete headworks and very little time maintaining the rest of the system. No farmer organization was created at construction nor were farmers involved in decisionmaking. Because the headworks increased the control that nearby farmers had over water, these farmers had little incentive to bargain with the tail-end farmers. As a result, farmers located at a distance from the headworks have broken through the branch canals to obtain water, and armed conflict for water often occurs between the tail-enders and head-enders.

By contrast, in the nearby Pithuwas Irrigation Project, which is not dominated by permanent headworks, the need to pool labor every year to construct the intakes and maintain the canal creates incentives for head-end farmers to collaborate and negotiate water distribution. In sharp contrast to the Kamala project, in the Pithuwas project, farmers have managed to irrigate 1,300 hectares using a system designed to serve 600 hectares. Although no farmer groups were created at construction, a branch committee was later formed; this model has since spread through the entire system (Laitos and others 1986).

If the tasks to be performed by the community-level actors are complex and on a recurrent basis, such as long-term operations and maintenance, then the investment needed in capacity building is generally higher.

Effective Outreach

Outreach mechanisms can be broadly classified into two basic approaches, the *extension* approach and the *empowerment* approach. In reality, the two approaches often overlap; effective outreach should combine aspects of both. The extension approach is characterized by information dissemination, demand promotion, and delivery of particular services or inputs. In the empowerment approach, the main goal is to empower local groups or create local organizational capacity for self-management through involvement in decisionmaking. The key challenge is to match the outreach system with the desired outcomes to be achieved at the community level. The empowerment approach is more staff-intensive and usually requires greater investment in local capacity building. In an irrigation project in Pakistan where success depended on autonomously functioning water users groups, the project provided for only 20 days of work per group per community organizer. This resulted in community organizers rushing from group to group, creating groups without even allowing time for them to assess their own needs (Byrnes 1992). The groups collapsed soon after project completion.

The institutional forms within both systems vary. Outreach can be done by restructuring agencies to add client outreach personnel (NIA in the Philippines); by contracting outreach services to NGOs (Kenya Water and Health Organization in rural water supply), the private sector (in agriculture in Malaysia), other government agencies, or other bilateral agencies; or by using existing local groups.

A recent study of several community-based programs in Mexico found that a key issue in reaching the poor was investment in social organization or group formation (Carrasco, Esmail, and Piriou-Sall 1997). In FONAES, the national Solidarity Fund for Social Enterprises, which provides assistance for productive activities to households below the poverty line, agency staff encourage the formation of groups of 15 people who then collectively develop detailed plans including financial details. FONAES staff also help groups to register legally. The work is concentrated in a microregion and then spreads outward, with members of groups training new groups and raising awareness about the program in nearby communities.

Because community-level workers are often the first contact through which community people learn about agency programs, they are of critical importance. Unfortunately, the design of community outreach systems is usually not given careful thought, often with disastrous consequences. It does not help that community-level workers themselves have little voice within their own agencies. Merely adding more community workers without encouraging higher-level staff to support community-level workers makes no difference.

Nature of the task. The extension approach is useful in increasing the use of inputs not highly dependent on coordinated action by individuals (such as agricultural inputs, seeds, fertilizer, nutritional supplements). The dissemination of information is often made more efficient if groups are organized and ready to receive information or other inputs on predetermined days. This includes, for example, the training and visit (T&V) system, use of contact farmer groups, use of visiting health teams to monitor the growth of babies on fixed days in a month (being done successfully in India and Indonesia), and use of the veterinary department to vaccinate livestock (in Pakistan). It is also efficient when communities or individuals are only required to perform one-shot tasks such as collecting money, contributing labor for construction over a very short period of time, immunizing children, or eliminating a particular pest.

The empowerment approach is essential when community groups need to be involved in decisionmaking and take responsibility for long-term management or when the tasks involve coordination at different levels.

Role of field agents. In the extension approach, field agents are often message focused. They channel information, provide technical expertise, and deliver inputs, seeds, fertilizer, health or nutrition education, and so forth. The extension approach is weak in supporting local institutional development for self-management. In the empowerment approach, field agents are first facilitators, catalysts, and organizers for empowerment. Technical information is provided by specialists or the field agent to help local groups make informed decisions after weighing the costs and benefits of various options.

To manage natural resources, in the Department of Agricultural, Technical, and Extension Services (AGRITEX) system of extension in Zimbabwe the extension worker promotes a fixed and limited technical package through a range of T&V group approaches, demonstration plots, master farmer certificate training, and field days. Participation of farmers in decisionmaking is limited by the structure of the extension approach, which emphasizes technical instruction (Scoones and others 1993).

In contrast, in Mali the role of agricultural extension agents working on natural resources has been to enhance the capacity of village groups to manage natural resources. In the 1992 Natural Resources Management Project, extension agents supported communities in carrying out their own needs analysis, in developing a plan of action, and in liaising with external support agencies. This is a good example of an extension approach that is empowering.

Skills and characteristics of field agents. Extension agents have technical skills, are information specialists, and are often outsiders, typically with loyalty to the agency. Empowerment agents, in contrast, must be highly trusted if they are to motivate community groups to collective action and must have some technical information and skills. In India and Korea, the difference in the performance and characteristics of irrigation patrollers, the lowest-level worker involved in canal operations, provides a graphic example (Wade 1994). The Indian system, with its hierarchical organizational system based on centralized control and public works orientation, minimizes identification between patrollers and local farmers and maximizes orientation and accountability to the Irrigation Department. Patrollers are full-time employees of the agency and, hence, are only marginally involved in farming themselves. They are selected by the agency engineer and become permanent employees after a period of probation. They are not posted near their own village and must move within six years to another area. The relationship is one of institutionalized mistrust and control by superiors.

The incentives, skills, and rules in the Korean system are almost the inverse and maximize the identification of patrollers with the existing social network of other farmers and village chiefs. The patrollers are paid part-time; the rest of the time they are farmers. They are selected by the village chiefs and approved by the irrigation hierarchy; they must be renominated every year by the village chiefs. They must have land and reside in their area of work, and they are not posted from one place to another. In the Korean system, if the irrigation system does not work well, the irrigation patroller himself suffers, together with other farmers with whom he has an ongoing relationship.

Another important example is from the Tamil Nadu Nutrition Project, where the process of selecting community-level health workers is an important contributor to success. Preference is given to poor, married women from within the village, who are of scheduled caste with primary-level education and two healthy children. They are hired as part-time workers. Because the women have deep roots within the community and are poor but have thriving children themselves, they are credible nutrition workers to families with malnourished children. The health workers have a limited number of specific tasks and work with a local group of women to whom they are accountable (Heaver 1989).

By contrast, in the Integrated Child Development Services Project in India, the *anganwadi* (child development) workers are without clear priorities and are overloaded with a wide range of tasks including feeding preschool children, conducting home visits, and paying attention to pregnant and lactating mothers. Studies have found that almost half of the women are not recruited from the village in which they work and are often of high caste. In the Indian context, it is very unlikely that these women will try to reach needy children from scheduled castes (Subbarao 1989 and Heaver 1989).

Hiring women becomes particularly important in reaching women. In Pakistan and Yemen (the 1993 Balochistan Primary Education Program and the 1992 Basic Education Project), female teachers had to be hired before parents were willing to send female children to school. In Nigeria, hiring female agricultural extension workers tripled the number of female farmers who were in contact with extension workers.

Control over decisions. In the typical extension approach, control of and authority over decisions are retained by the agency, and the field agent is the messenger. In the empowerment approach, the agency sets the parameters or starting conditions of partnership, but control over management details is left with the community.

Some social funds use the empowerment approach in which fieldworkers offer a menu of options from which the community chooses those that best fit its needs and financial and management capacity. The agency defines the rules for partnership (for example, the amount of financial assistance available).

The role of information. In the extension approach, where the primary purpose is dissemination of information, social marketing and the use of mass media are of great importance. In the empowerment approach, the primary activity initially is building confidence by tapping into people's knowledge and involving people in tasks; specialized information is introduced only as needed. Mass media are used to publicize the availability and rules of the program and to establish transparency.

Accountability. Within the extension approach, fieldworkers typically have their heads turned toward the agency, which rewards, punishes, promotes, or ignores. In the empowerment approach, agents are accountable to community groups even when they are paid by agencies. To increase accountability, some programs require communities to certify that staff have completed work satisfactorily; in others, agency staff are replaced by local people who are sometimes also paid directly by communities.

Outcomes desired at the community level. In the extension approach, the agency's desired outcomes are that the inputs and services delivered are used and that programs are more effective and efficient. In the empowerment approach, the key outcomes are that households or groups at the community level have organized for self-management and show increased capacity for coordinated action including management of local services or resources.

Client-Responsive Agencies

Client-centered agencies are needed to ensure that demand is met and community self-management is adequately supported by community fieldworkers. To become client centered, agencies must be convinced to change their rules and incentive structures so that the benefits from client responsiveness are more important to the agency than the costs. Experience demonstrates that trying to support community-based approaches without fundamentally changing institutional structures, rules, and incentives does not work. Key design features of client-responsive agencies are summarized in table 9.1.

Experience shows that strong political and government support for making public agencies client centered is needed at the highest levels. Such support has occurred, for example, with the Programa Integral para el Desarrollo Rural (PIDER) in Mexico, NIA in the Philippines, public works programs in Indonesia, local government in Paraguay, the Ministry of Forestry in West Bengal in India, and in Zimbabwe. The performance of the irrigation sector in Mexico is

Table 9.1. Reinventing Agencies to Support Community-Driven Development

Lessons from the community	*Agency characteristics*	*Agency mechanisms*
Group must have a felt need for service; there is "commonality of interest"	Agency responds to demand and knows what people want	Use demand assessment, institute outreach mechanisms, and employ social analysis to identify and understand key actors, their power, interests, and needs
Group controls and has authority over resources, decisions, and rule making	Agency "lets go" of control over implementation details and spells out framework of rules for interaction and negotiation with communities	Define objectives and indicators of success to support achievement of local control and reorient staff and performance criteria
Local group has the needed capacity and skills and can mobilize financial resources for long-term survival	Agency puts local empowerment and capacity building high on its agenda	Institute the capacity-building process, invest time and resources in training, devise new funding mechanisms that reach communities quickly, focus on strategies for groups to achieve financial self-sufficiency, and phase in payment of outreach workers by community groups
Every group or community is unique	Agency plans for diversity and encourages local adaptation	Use a learning process approach, employ short planning horizons, modify implementation plans with feedback from monitoring and evaluation
In many cases, the poor and marginal, including women and indigenous groups, are left out	Agency focuses on reaching the poor and marginal	Reflect in objectives, institutional mechanisms, targeting strategies, and indicators of success the focus on poor and marginal groups

a notable example of rapid and successful institutional reform.

The critical question is how to bring about government commitment to change. World Bank staff have used two main strategies to generate interest and commitment: (a) pilot projects as instruments to demonstrate alternative strategies and (b) a variety of participatory techniques, including field visits to other regions and countries, to generate interest in and commitment to new ideas and strategies for action.

Indicators of success send important signals to staff about program priorities. If community involvement, the number of women or the poor reached, and the number of systems functioning are not reflected in indicators of success, there is little incentive for staff to change their way of doing things to reach these goals.

Almost all projects that adjust as they implement invest in numerous small studies that provide feedback on how different approaches are working. The Women-in-Development (WID) Project in The Gambia created a monitoring and evaluation unit during project preparation that honed its skills by carrying out several small studies. The Indonesia Rural Water Project has developed a national system of monitoring key process and output indicators focusing on changes at the community level.

The most graphic examples of the importance of defining the parameters while supporting the evolution of local rules have emerged from the water and natural resource management sectors. Projects with standardized forms of participation and universal applicability of implementation rules formulated by the agency have not worked and, despite stated objectives, have become agency-led rather than user-led programs. Programs that have created the overall framework of rules of interaction between users groups and agencies, while letting users groups work out the details of how to manage resources, have emerged with numerous variations in local rules that are managed, monitored, and enforced by local people.

Elinor Ostrom (1990) highlights this fact in comparing irrigation systems across and within countries. For example, a study of a farmer-led Karjahi irrigation system in Nepal found a diversity of rules even within one small self-organized system, yet the system functioned well (Hilton 1990). In contrast, Frances Cleaver (1990) in a national sample survey of more than 400 hand pumps in Zimbabwe found that applying microscopic rules (such as how deep the community was required to dig) deterred community participation, local initiative, and problem solving. Similarly, uniform application of rules in grazing schemes, tree plantations, and woodlot management has proven counterproductive.

The importance of social organization is increasingly recognized, and hence most agencies involve institutional and social science (noneconomic) expertise early in program development. Understanding the social organization includes understanding local leadership and power systems; who has control and access to resources; who are the key social actors; what are the local patterns of landownership, tenure, and use rights; and what are the number, structure, and functioning of existing informal and formal groups. In the experience of NIA in the Philippines, baseline sociocultural profiles and process documentation played important roles in pointing the direction for institutional change. In Mexico, the National Water Authority has an in-house group of senior social scientists and communication specialists who design the strategy for community outreach, applied research, and communications. This work is then subcontracted to the private sector.

Enabling Policies

Community-driven development on a large scale requires enabling policies and political support to protect agencies so that they can initiate the reform process and give it time to take root. It is much easier to support reform in the context of decentralization strategies. Although individuals do play key roles in starting or protecting the reform process while it is still in the early stages, change can be sustained only if the rules and regulations at different levels provide incentives to work and organize. Three broad categories of rules are important: legislation, titles, and rights; control and use of funds and local fees; and other rules of the game. Also important is gaining political support or patronage in the right quarters.

Legislative issues that govern the functioning of agencies and the relationship of agencies with other agencies and communities should be identified during sectoral reviews. Key issues to examine include the mandate of agencies, redefining the role of government agencies as facilitators and regulators, civil service reform, new funding mechanisms, new systems of accountability for performance, legal status of community groups, simplification of legal registration requirements, individual, group. or community ownership, and use and tenure rights over assets (particularly water, land, and other natural resources).

Agencies and communities have little incentive to mobilize resources if they do not realize any immediate benefits. At the agency level, when fees collected by agencies revert to central treasuries with no linkage to the agency's subsequent share of resources, agencies have little incentive to pay attention to tariff payment or collection. Dramatic improvements can occur when agencies are required to be financially autonomous and have access to user fees.

It should not be assumed that having a centralized agency manage the collection of user fees is always the most effective way to proceed. One of the key problems that had to be addressed in the Bank-supported NIA participatory irrigation projects in the Philippines was poor maintenance of irrigation canals. At the outset, it was assumed that the best way to solve the problem was to impose targeted increases both in the annual funding levels of operations and maintenance expenses that NIA had to incur and in the total amount of irrigation fees it had to collect annually. However, this undermined farmer-managed irrigation, including fee collection, which was more efficient than NIA fee collection. Eventually NIA changed its rules and gave control to farmers. Farmer-managed irrigation schemes worked so well that NIA subsidies were eliminated in two years (NIACONSULT 1993).

Many rules and regulations may need to be changed, from required qualifications of community workers, teachers, and health educators to procurement rules. In the Balochistan Primary Education Project, the quick spread of community-managed schools for girls required several policy changes. Some of the key policy changes to ensure recruitment of and support for female teachers from within the village included lowering the minimum age to 14 years and raising the maximum age to 40 years, legalizing a mobile teacher training program, and "sanctioning" the school and teacher post through the Education Department Community Coordination Unit. Without these changes, no amount of community participation and organization would have helped spread primary education to girls quickly. In changing the rules, attention must be paid to protecting accountability and performance.

When change is initiated on a large scale, there are usually losers and winners. Change is a political process that rearranges power structures and alters people's access to and use of resources. Most of the large-scale community-based approaches, especially those that have worked through government and instituted rapid, radical change, have had strong political support or patronage.

Judith Tendler (1993) highlights the positive and important roles played by "godfathers." Behind many stories of successful implementation has stood a "good governor" or a state secretary or a reform-minded senior civil servant with access to and support of top political leaders. The good governors have provided protection from the pressure to hire mediocre staff or to fire excellent staff on political grounds, to make technically undesirable choices, and to delay the transfer of funds from the central government. They have pressured agencies to produce results and chosen one or two components of a project as "signature activities" with which to leave their mark. Thus in Sergipe, the governor fashioned the project around rural water supply; in Piaui, the governor supported the land purchase component; and in Ceara, the small-scale riverine irrigation component won the governor's support. Good governors have also mobilized their own resources, sometimes as much as three times the amount provided in the Bank loan, to support their favorite component within the time period of their tenure in office.

In Nepal, when the government decided in 1987 to promote farmer associations in irrigation management through radical policy changes, the Bank responded with a $20 million irrigation line of credit with technical assistance to support the project through the United Nations Development Programme. The policy changes devolved power from a large central bureaucracy to lower levels, including farmers. These changes were made possible by support at the highest levels, including the king, the minister of finance, the minister of water resources, and the director general of the water directorate. In Mexico, the *solidaridad* program, which devolves fiscal power to the municipalities, had the blessing of the president.

The Gambia WID Project survived the preparation process because of the strong interest of the president. At later stages, when it appeared that the project might not proceed, the president flew to Washington, D.C., to meet with then-president of the World Bank, Barber Conable, to persuade him of the importance of the project to his country. Since then, the project has been designated one of the Bank's most successful, and strategies developed to reach women in agriculture have been incorporated in mainstream agricultural projects (Schmidt 1993).

Godmothers play important roles as well. In Indonesia, wives of governors often are involved in getting service delivery agencies to take on gender issues in a serious way. In Colombia, the community-based day care program movement was actively supported by the wife of the president.

Designing Participatory Community-Based Development Programs

The lessons from the worldwide experience in community-driven development must be reflected in the design of new projects. This section discusses 10 features that are particularly important in the design of participatory community-driven development projects. In situations of many unknowns, pilot activities using the new learning and innovation loans are the most appropriate.

Set Objectives: Clarity, Priority, Links to Outputs

A surprising number of projects do not have clearly stated objectives logically linked to strategies, outputs, indicators of success, and physical or capacity-building outcomes. Goal-oriented planning tools such as the logical framework analysis are aids in achieving clarity. These tools help to clarify and prioritize objectives and articulate underlying assumptions, activities, and indicators of success. Brainstorming meetings and discussions can also be useful.

Establishing the priority of objectives is particularly important in community development projects because as the pressure to produce tangible results builds, shortcuts are taken, resulting in "build first, listen-dialogue-and-organize later." This is most likely to happen when there are no godfathers, when technical agencies are in charge, or when success is based on construction completed and inputs distributed rather than on services functioning, being operated and maintained, and being used effectively.

Project experience establishes that women and the poor are not automatically reached through community-driven projects. If this is a goal, it must be reflected in the objectives, strategies, and indicators of success. In such projects, targeting strategies that have low transaction costs become very important. Gender and poverty analysis tools are particularly useful.

Early in project preparation, staff have found it useful to be open-minded about overriding objectives. These become increasingly clear as project preparation proceeds. They may change radically, as happened in the Matruh Natural Resource Management Project, where objectives shifted from livestock to natural resource management. To avoid designing complex projects with multiple and competing objectives, a hierarchy of objectives that can change in importance over time can be developed, thus making the project manageable.

Identify the Key Actors, Their Capacity, and Interests

Identifying the key actors at both the community and agency levels is critical. It should not be assumed that the community is the logical unit of interaction with support agencies. Households or groups within a community, individuals within a household, children, women, or those in particular occupations (such as farmers, informal sector workers, and the landless) may prove more appropriate.

Social assessment is an important tool for identifying key actors and their interests, the existing social organization, and aspects that need to be strengthened or changed. At the community level, data from social analysis are helpful in identifying local laws, rules, and regulations governing interaction and access to resources. At the agency level, social assessment becomes an institutional tool for identifying the incentives, interest, and capacity of various service delivery agencies to support community-driven development. Based on this analysis, either other agencies are drawn in or fundamental reform is initiated. This may encompass redefining agency functions, restructuring funding mechanisms, or adding a new cadre of staff to the organization.

Task teams find it useful to include national and international expertise in teams. Social scientists, including NGO or community development specialists, have been particularly helpful in identifying and sketching the workings of existing social organization from which much can be learned about what does and does not work in a particular context. A range of participatory and nonparticipatory techniques and tools are now available (Jacobs 1998, Narayan and Srinivasan 1995, and Rietbergen-McCracken and Narayan 1998).

Assess Demand

Many problems have arisen because programs have been developed on needs assumed by planners rather than effective demand of community groups. A review of 121 rural water supply projects shows that demand, as measured by commitment before construction and substantial financial investment up-front in capital cost, contributes significantly toward both project effectiveness and overall beneficiary participation in decisionmaking (Narayan 1994).

Demand can be assessed using a variety of methodologies: participatory techniques (community self-diagnosis, ranking of priority problems), beneficiary assessment or "listening to the people" techniques (informal interviews, community meetings, participant observation), or contingency valuation techniques that gauge what people are willing to pay for different levels of service. In assessing demand, data should be disaggregated by gender, wealth level, and other relevant characteristics.

Craft a Self-Selection Process

If collective action and some degree of self-management are the goal, designing a self-selecting process is important. This includes community groups taking the initiative to become part of a project rather than agencies initiating the selection of communities. Self-selection is not a key strategy if delivery and use of inputs to a targeted group is the prime objective.

The single most important self-selection strategy is to institute a significant financial or organizational contribution up-front before any project outputs are delivered. Many projects now require community groups to enter contracts and sign documents that detail mutual responsibility. These are useful only to the extent that people understand what they are signing and when enforcement mechanisms exist on both sides (community and agency) to keep the other party accountable.

As a gauge of interest, agencies may stipulate organizational tasks that community groups must perform before receiving project assistance. These may include mobilization of local resources and cash, formation of committees, demonstration of consultation with everyone in the community, and submission of a proposal. Some projects require legal registration, an important factor if laws require legally constituted bodies before governments or banks can transfer assets to the group or involve the groups in decisionmaking.

Structure Subsidies That Do Not Distort Demand

To ensure inclusion of the poor and to retain a demand orientation, it is important to structure subsidies so that they do not violate the principle of demand. Project experience illustrates that even the poor are willing to contribute substantial amounts by offering their labor or by accessing credit, where available, if the project meets a need and the service provider is perceived to be reliable and trustworthy.

Five approaches to structuring subsidies are common:

- The subsidy is indirect and is invested in strengthening community organization or capacity through outreach and training (Grameen Bank outreach systems).
- Upper limits are set on the base amount; if higher levels of service are desired, the community or user must pay (Indonesia rural water supply project for low-income communities).
- Where initial investments are high and beyond the capacity of the community, organizational tasks completed by community groups are used to gauge commitment rather than requiring significant community contribution to capital investment (Sri Lanka rural water systems in areas where hydrogeological systems are difficult; Senegal livestock centers).
- Agencies defer or spread out large capital outlays until the community has proven its management ability and interest in the service (Pakistan Balochistan Primary Education Project; the government constructs a school only after a community has managed a community-based girls school for three years; the government trains the teacher, provides school supplies, and pays the teacher after the first three months).
- Agencies invest in the "trunk" or main network structures, and communities invest in secondary and tertiary distribution networks (irrigation, sewerage, piped water systems).

Restructure Fund Release to Support Demand

A large-scale project may be demand oriented, but if the flow of funds is not responsive, it is impossible to maintain a demand focus. Depending on the national and sectoral context, projects have used different strategies.

One approach is to decentralize the programming of funds closer to where they will be used. In Indonesia rural water supply, programming of funds has been devolved to different levels. The higher the cost, the further the project proposal has to travel for approval. However, authorities cannot tamper with the basic design of the water system. In Ethiopia's social fund, regional offices have been opened.

Another strategy is to create new funding mechanisms inside or outside the formal government structure. The structure and management of social funds falls in this category. Some have been established with independent boards (managed with a majority of NGO and private sector representatives) outside a particular ministry; others are semiautonomous, but with high-level protection, often under the wing of the president or prime minister. Funding may also be channeled through intermediaries, NGOs, banks, and other multilaterals, particularly the United Nations Children's Fund (UNICEF) and International Fund for Agricultural Development (World Bank 1997 and Narayan and Ebbe 1997).

Plan for Learning and Plurality of Models

Community development, especially the end of the continuum that aims for community empowerment, is by definition dependent on hundreds of discrete decisions made by individuals in communities. The review of experiences across sectors reveals that no single form of project is successful in all settings; projects have worked in some areas at certain times and not at others.

Hence, it is important for the basic planning assumption to be one of learning embedded in local knowledge systems. The project must be seen as likely to evolve, adjust, and change over time as local priorities change and as local organizations mature. This, in turn, means that project components may change over time in order to maintain a fit with community needs. The new adaptable lending instrument actively supports such a learning process approach.

It is also important not to search for one universal answer or model for all times and all places, but to plan for a multiplicity of implementation models. This is particularly important for projects that require communities to perform complex long-range tasks.

An increasing number of infrastructure (Tanzania, Brazil, Indonesia) and natural resource management (Burkina Faso, Mali, Egypt) projects are adopting structured learning as a specific strategy. These large-scale projects are based on a learning process approach rather than on a blueprint design. The focus is on a framework of principles and processes rather than on implementation blueprints. Planning is on a yearly basis, with plans for subsequent years dependent on previous performance. This allows for the evolution of plans based on experience and fine-tuning with time. Procurement of most materials is done on a yearly basis, budgets are indicative and flexible, and the project is conceptualized as a series of subprojects that are not implemented in a uniform way.

Invest in Outreach Mechanisms and Social Organization

The level of investment needed in community social organization and intermediation is a function of the tasks expected of communities and the existing strength of community organizations. In general, strengthening indigenous or existing organizations is easier than starting from scratch, although in some cases this may be justified. If existing groups are dominated by the powerful and the elite, they may or may not reach women and the poor. Hence, many projects that specifically target women or the poor invest in special strategies to reach them. These may include forming special organizations of the poor such as the Grameen Bank, the Self-Employed Women's Association of India, and women farmers groups in Nigeria and The Gambia.

Perhaps the most important lesson that emerges is that there are no shortcuts to strengthening local social organization for collective action. All shortcuts for speedy implementation that circumvent local involvement in decisionmaking backfire sooner or later.

Use Participatory Monitoring and Evaluation

There is a great temptation to freeze project designs once implementation starts, even though much is written in project staff appraisal reports about learning, conducting studies, and performing monitoring and evaluation. Learning is more effective and efficient when feedback is listened to and when changes are put in place. Listening to people's suggestions is empowering and encourages innovation and responsibility. Seeing information systems actually being used rewards the filling out of forms, which otherwise can degenerate into meaningless busy-work.

The Gambia project used a very simple method to involve village women in identifying priority indicators for monitoring and evaluation. A consultant was hired to develop a list of indicators. These were taken to the villages, and well-publicized consultative meetings were held with women's groups who identified the key indicators from their perspective. These indicators were then monitored by the project monitoring and evaluation unit.

Redefine Procurement Rules

Procurement rules can hinder community initiative. When appropriate, these should support local-level procurement procedures. International bidding and bulk purchase are hardly appropriate when projects consist of many subprojects, whose nature and timing will be determined by community needs and readiness.

Three strategies have been found to be useful. First, maximum unit costs are set, as in Bolivia's social fund. As long as the costs of local procurement do not exceed the maximum unit costs, local procurement is allowed, with a system put into place for ensuring completion of quality work. Second, a cost threshold can be set, below which communities procure goods locally based on market principles and above which procurement follows standard local bidding or international bidding as appropriate. Third, as in India, community projects are clustered to allow for economies of scale and to attract a larger pool of local contractors and NGOs.

References

The word "processed" describes informally reproduced works that may not be commonly available through libraries.

Bamberger, Michael, Edgardo Gonzalez-Polio, and Umnuay Sae-Hau. 1982. *Evaluation of Sites and Services Projects: The Evidence from El Salvador.* Staff Working Paper 549. Washington, D.C.: World Bank.

Boerma, Pauline. 1993. "Lessons Learnt from the

Implementation of the RUSAFIYA Project in a Number of Selected Communities in Nigeria." United Nations Development Programme–World Bank Water and Sanitation Program, World Bank, Washington, D.C. Processed.

Byrnes, K. J. 1992. *Water User's Associations in World Bank–Assisted Irrigation Projects in Pakistan.* Technical Paper 173. Washington, D.C.: World Bank.

Carrasco, Tanya, Talib Esmail, and Suzanne Piriou-Sall. 1997. "Decentralization and the Promotion of Productive Projects: Case Studies of Community-Based Natural Resource Management and Income-Generating Projects in Oaxaca and Hidalgo, Mexico." Report prepared for the Decentralization, Fiscal Systems, and Rural Development Group, World Bank, Washington, D.C. Processed.

Cernea, Michael M. 1989. *User Groups as Producers in Participatory Afforestation Strategies.* Discussion Paper 70. Washington, D.C.: World Bank.

Cleaver, Frances. 1990. "Community Maintenance of Hand Pumps." Department of Rural and Urban Planning, University of Zimbabwe, Herare. Processed.

Evans, Peter. 1995. "Government Action, Social Capital, and Development: Creating Synergy across the Public-Private Divide." University of California, Berkeley. Processed.

Heaver, Richard. 1989. *Improving Family Planning, Health, and Nutrition in India: Experience from Some World Bank–Assisted Programs.* Discussion Paper 59. Washington, D.C.: World Bank.

Hilton, Rita. 1990. "Cost Recovery and Local Resource Mobilization: An Examination of Incentives in Irrigation Systems in Nepal." Report prepared for the Decentralization, Finance, and Management Project, U.S. Agency for International Development, Washington, D.C. Processed.

Jacobs, Sue. 1998. *A Manual for Participatory Methods for Social Assessment.* Washington, D.C.: World Bank.

Laitos, Robby, and others. 1986. *Rapid Appraisal of Nepal Irrigation Systems.* Water Management Synthesis Report 43. Fort Collins: Colorado State University.

Narayan, Deepa. 1994. *The Contribution of People's Participation: Evidence from 121 Rural Water Supply Projects.* Environmentally Sustainable Development Occasional Paper Series 1. Washington, D.C.: World Bank.

Narayan, Deepa, and Katrinka Ebbe. 1997. *Design of Social Funds, Participation, Demand Orientation, and Local Organizational Capacity.* Discussion Paper 375. Washington, D.C.: World Bank.

Narayan, Deepa, and Lyra Srinivasan. 1995. *Participatory Development Tool Kit.* Washington, D.C.: World Bank.

NIACONSULT. 1993. "An Evaluation of the Impact of Farmer's Participation on the National Irrigation System's Performance." Report commissioned by the Participation Learning Group, Social Development Department, World Bank, Washington, D.C. Processed.

Ostrom, Elinor. 1990. "Crafting Irrigation Institutions: Social Capital and Development." Paper prepared for the workshop on political theory and policy analysis, Indiana University, Bloomington. Processed.

———. 1993. "The Evolution of Norms, Rules, and Rights." Paper prepared for the workshop on property rights and the performance of natural resource systems, Royal Swedish Academy of Science, Stockholm, 2–4 September. Processed.

Rietbergen-McCracken, Jennifer, and Deepa Narayan. 1998. "Participation and Social Assessment: Tools and Techniques." Poverty Group, Social Development Department, World Bank, Washington, D.C. Processed.

Schmidt, Mary. 1993. "Case Study in Popular Participation: WID in Gambia." Report prepared for the Learning Group on Participation, Social Development Department, World Bank, Washington, D.C. Processed.

Scoones, Ian, Jeanette Clark, Frank Matose, Colin Phiri, Ole Hofstad, Isaac Makoni, and Sara Mvududu. 1993. "Future Directions for Forestry Extension." In P. N. Bradley and K. McNamara, eds., *Living with Trees: Policies for Forestry Management in Zimbabwe.* Technical Paper 210. Washington, D.C.: World Bank.

Scoones, Ian, and Frank Matose. 1993. "Local Woodland Management: Constraints and Opportunities for Sustainable Resource Use." In P. N. Bradley and K. McNamara, eds., *Living with Trees: Policies for Forestry Management in Zimbabwe.* Technical Paper 210. Washington, D.C.: World Bank.

Shanmugaratnam, Nadarajah, Trond Vedeld, Anne Mossige, and Mette Bovin. 1992. *Resource Management and Pastoral Institution Building in the West African Sahel.* Discussion Paper 175. Washington, D.C.: World Bank.

Subbarao, K. 1989. *Improving Nutrition in India: Policies and Programs and Their Impact.* Discussion Paper 49. Washington, D.C.: World Bank.

Tendler, Judith. 1993. *New Lessons from Old Projects: The Workings of Rural Development in Northeast Brazil.* Operations Evaluation Study. Washington, D.C.: World Bank.

Wade, Robert. 1994. "Public Bureaucracy and the Incentive Problem: Organizational Determinants of a 'High-Quality Civil Service,' India and Korea." Institute of Development Studies, Sussex University. Processed.

World Bank. 1996. *The Aga Khan Rural Support Program, A Third Evaluation.* Operations Evaluation Study. Washington, D.C.: World Bank.

———. 1997. "Portfolio Improvement Program Review of Social Funds Portfolio." Washington, D.C. Processed.

10

Involving Local Organizations in Watershed Management

Jacqueline A. Ashby, Edwin B. Knapp, and Helle Munk Ravnborg

Involving local organizations in watershed management is widely recognized as advantageous for three reasons. First, local organizations can often be very effective in generating and securing compliance with rules for the use of common property such as water, common grazing land, or forest and with management of buffer zones around conservation areas, all of which may be important features of watershed management. This is in contrast to nonlocal agencies, especially in low-income countries, which have difficulty imposing sanctions on undesirable management practices or providing incentives that are lasting. Second, organizations that involve local stakeholders in the development of management practices and the selection of technologies aimed at improving watershed resource conservation often promote innovation by identifying locally appropriate technologies and securing their adoption more effectively than external agencies. A third reason why central government agencies find it attractive to involve local organizations in watershed management is that devolution of responsibility can both externalize some costs of enforcing conservation from the state to local communities and reduce costs overall by creating conditions in which nonlocal agencies become more efficient and effective through collaboration with local organizations.

In watershed management, therefore, local organizations regulate resource use, provide a forum for resolving conflicts among local stakeholders, and provide a channel for the representation of on-site and off-site stakeholder groups in the negotiation of common property resource use. In this respect, local organizations may have an important role to play in making micropolicy within the watershed as well as in defining enabling policies that may be formulated and enforced by the state. They also have a role in promoting technological innovation and the adoption of conservation practices, whether these require collective or individual action. And they may mobilize and deploy significant resources, both in cash and in kind, that are needed to implement sustainable management of watershed resources.

This chapter examines experience with involving local organizations in watershed management. The discussion seeks to provide guidelines, rather than a theoretical analysis, based on experience of working with local organizations from the perspective of development practitioners. This experience is evolving and accumulating rapidly, and many practitioners are promoting a learning process in which their approach to local-level watershed management is swiftly adapted and reoriented based on analysis of what has gone before. Although this discussion focuses on some of the lessons learned and principles that can be derived from experience, the last word on what works and what does not work is by no means decided. Indeed, the dynamic nature of the process of involving local organizations in watershed management is one of its most important characteristics, and one that requires a focus on managing the process rather than on defining a blueprint for how to go about it. To highlight this important principle, we illustrate some of the dynamics of process management with information from a case study that traces the evolution over time of local-level watershed management.

The Social Ecology of Watersheds

In order to analyze the role of local organizations in watershed management, it is necessary to define both the physical and the social terms used. The perspective of social ecology can be useful because it focuses on the interaction between a physical landscape, including changes in the natural resource base, and the adaptive strategies people use to obtain livelihoods and to organize socially within that landscape. This interaction between the possibilities for resource use and ways of organizing is central to understanding the circumstances under which local organizations turn out to be effective managers of watershed resources.

Watersheds have been defined as "physically defined subsets of rural society" and watershed management as "a question of social relations and coordination among individuals with vested common interests in the watershed" (White 1995: 3) This definition recognizes that a watershed is a geohydrological unit comprised of all land and water within the confines of a drainage divide as well as an ecosystem where people, livestock, vegetation, land, and water interact (Jensen 1996).

In the watershed landscape, the flow of water creates a set of interlinked environmental effects from upstream to downstream and involves multiple resources (water, soil, vegetation) with multiple uses: for example, water for irrigation, rural domestic use, or downstream urban industrial use; forest for firewood, forage, construction, timber, or ground cover; and land for buffer zones, arable cropping, or grazing. Watersheds often include a mix of privately owned and common property resources. The latter may be open access or may have self-organized or externally imposed rules that govern access and use. The transboundary effects typical of resource flows in watersheds, whether they originate from common or private property, mean that proper management requires collective, rather than solely individual, action. The need for cooperation, negotiation, and collective agreements to manage transboundary effects is a basis for involving local organizations.

However, cooperation in watershed management is not necessarily easy to initiate or sustain, even when local organizations are involved. The difficulty of establishing definitive usufruct rights and obligations over resources with multiple uses is one reason for skepticism about the efficacy of local organizations for watershed management (Uphoff 1986). Multiple uses for any given watershed resource create multiple stakeholders, often with competing priorities. Typically, use by one stakeholder will generate negative effects or externalities for others, as in the case of upstream users who clear forests for arable cropping, which causes soil erosion that silts downstream reservoirs and may damage the hydrological cycle so that seasonal flow is affected. For this reason, conflict among stakeholders over the rights and conditions for use of a given resource is a common feature of watershed management.

A useful approach to assessing the potential role of local organizations in watershed management is to distinguish resources by their physical characteristics, which affect the incentives and strategies users pursue and which for this reason require different institutional arrangements. Two physical distinctions are whether a resource is stationary (grazing areas, groundwater basins) or mobile (flowing streams) and whether the resource has storage capacity, either natural or man-made (reservoirs or tanks). These two physical characteristics affect the ease with which users can obtain or generate information about the resource and thus their ability to coordinate resource management (Schlager, Blomquist, and Tang 1994). For example, in common property resources with mobile flows that fluctuate unpredictably, such as stream flow in a watershed, it is very difficult for users to assess the effects of use by one stakeholder on the amount or quality of water available to another stakeholder or the benefits to either user from a conservation intervention. The extent to which local organizations can monitor change in a mobile resource in a watershed and can obtain information about cause-and-effect relationships in the management of a mobile resource will determine how easy it is for them to play a role in that management.

When cause and effect cannot be determined readily, conflict over usufruct rights is more likely to occur, and it is easier to cheat on regulations about use. Even more problematic is when a mobile resource cuts across multiple locations, as in the case of upstream and downstream locations in a watershed: coordinating rules of use across widely dispersed locations increases the costs of management to local people (Schlager, Blomquist, and Tang 1994).

The existence or creation of storage capacity in a common property resource helps users to establish some control over use of the resource. It is then easier for local organizations to devise rules and sanctions governing use, and, because users have a better chance of capturing the benefits of conserving and developing a resource with storage capacity, they are more willing to work together to maintain storage and preserve the resource without the intervention of external authorities. Locally defined institutional arrangements for managing irrigation vary depending on whether or not the resource has storage capacity (Schlager, Blomquist, and Tang 1994), but no empirical test of this relationship has been published with reference to management of watershed resources.

In summary, the physical characteristics of a given resource targeted for regulation in a watershed are likely to affect the type of institutional arrangements that are workable and the role of local organizations in these arrangements. The role of local organizations in watershed management will be strengthened by improving their access to information about the effects of alternative management interventions on mobile resources or by creating storage capacity in resources that are a focus of conflict.

Local Organizations and Social Capital

With reference to natural resource management, local organizations are often understood to be self-organizing groups or groups of users voluntarily cooperating in the regulation of common property resources. There are numerous examples of these and extensive evidence that, all over the world, these self-organizing groups have been weakened and undermined by the growth of government intervention. As a result, natural resources degrade). However, very few of these groups have organized spontaneously for watershed management. In most documented cases from developing countries, external agencies have played an important role in catalyzing the formation of local watershed management organizations.

The special nature of local organizations is that they provide a basis for collective action, which includes building consensus about problems and needed solutions, seeking and disseminating information about these, coordinating action, designating rights and responsibilities, as well as ensuring accountability (Cernea 1993 and Uphoff 1992). Self-organizing groups for common property management are only one subset of the local organizations that use watershed resources and form the tapestry of local institutional arrangements in a watershed. Many other types of local organizations are operative in watersheds: exchange labor groups, which may be sporadic or long term, credit groups that manage rotating funds or community banks, farmer-to-farmer agricultural research or extension groups, marketing associations and cooperatives, religious groupings, as well as groups involved with local government. Although many of these are not involved directly in the governance of resource management, they are usually stakeholders with interests in one or more watershed resources.

The presence of such organizations, and the degree to which participation in them is widespread among the local inhabitants of a watershed, is important because the social dynamics of cooperation on which local organization for natural resource management often depends are fundamentally driven by the extent to which the people involved trust one another and are therefore able to predict how other users of a common property resource are likely to behave. This interpersonal trust is generated and sustained by repeated face-to-face participation in small groups, and the more people interact in this way, the greater the level of trust and cooperation that is possible (Glance and Huberman 1994). Institutions, and specifically the rules, norms, and conventions that local people develop to govern voluntary cooperation for resource management, depend on each individual's assurance that other people can be expected to cooperate and adhere to these rules. Field studies in villages show that people in small communities are commonly motivated by the idea of fairness, as well as zeal—the opposite of free-ridership—driven by the desire to obtain approval from the community (Bardhan 1993). All local organizations are a channel for positive sanctions, group approval, and status acquired for altruistic behavior that spills over into natural resource management.

Organizations are a structured sets of roles, which themselves are defined by bundles of norms and rules about what social behavior a particular role involves (Uphoff 1992). Local organizations may define roles in natural resource management for individuals or groups, such as monitoring the state of forests, irrigation channels, or burning for land clearance. Such roles involve applying sanctions for noncompliance with the locally defined rules for forest, water, or land management. The wider universe of local organizations found in watersheds have a number of functions that, although they may not be concerned directly with watershed resource management, have significance for the efficacy of local organizations that do have this responsibility. These other functions include mobilizing resources, such as labor or credit, lobbying external agencies, managing information, and mediating and resolving conflicts (Pretty 1995).

Organizational density, measured by the number of local organizations and the degree to which participation and overlapping membership is widespread in a population (Cernea 1993), can be used as an indicator of social capital or the capacity to work together. The concept of social capital refers to reciprocity based on interpersonal trust, which is the foundation of cooperative behavior. A high level of social capital, or organization, that builds experience in working together is an important ingredient in the capacity of local people to manage their affairs successfully without dependence on outside agencies. For example, a study of 25 agricultural projects several years after completion found that those which had made an effort to build local organizations, such as water user associations, had been able to sustain high rates of return; projects that had not done this had rates of return 5–10 times lower than anticipated (Cernea 1987).

Social capital is fostered by local organizations that reinforce norms of reciprocity (favors done now will be returned later), create channels of communication that enable people to verify the trustworthiness of others, provide a local institutional memory about experiences of successful collective action, and implement sanctions against those who act opportunistically. Social capital can grow, when local organizations are active, and can be depleted when they are not (Sirianni and Friedland 1995).

Thus one of the key functions of local organizations in watershed management is building the social capital required for people to build consensus about and enforce the agreed use of watershed resources for the diverse, multiple, and often conflicting possibilities described in the previous section. A study of the development of collective action groups to control transboundary erosion in 22 small, multiowner watersheds in Maissade, Haiti, illustrates this phenomenon. The watersheds averaged 9 hectares in size and ranged between 2 and 34 hectares. After two years of activity, 590 check dams were built by the groups, each of which retained an average of 39 tons of soil per hectare per year, providing financial benefits to farmers from increased and diversified production. After examining a number of factors in order to explain what motivated individuals to cooperate, the study concluded that short-term economic gain was not the primary motivation. Participation in the cooperative resource conservation activities was most strongly predicted by membership in preexisting, self-organized precooperatives (averaging eight members), which in turn was highly correlated with membership in labor exchange groups. White (1995: 1691) concludes that

> Participation in these groups apparently provides the necessary experience of sharing information, building trust, constructing rules, monitoring relations, and sanctioning. This experience and the assurance of solidarity and reciprocity allow members to share risk, leverage resources, extend payback periods, test innovations, and make leading contributions to collective actions even when they know that these ventures might not be rewarded.

An example from Kenya illustrates how the degree of self-organization and decisionmaking (or social capital formation) that local organizations achieve in planning and implementing small catchment resource conservation is correlated with their capacity to sustain locally managed conservation. In 1988 the Kenyan Ministry of Agriculture adopted the catchment approach, which concentrates soil conservation efforts in a specified catchment (usually 200–500 hectares) for a limited time, during which all farms are planned and conservation practices laid out. It also involves the formation of a local catchment conservation committee consisting of 8–15 people, sometimes all newly elected by the inhabitants of a catchment and sometimes based on existing traditional organizations such as groups of elders or cooperatives. The catchment committee helps other farmers to plan and carry out soil and water conservation activities, after receiving training from the ministry. In some cases, catchment committees have continued their activities after the ministry staff have moved on to new catchments, expanding their work to include marketing and road maintenance. A comparison was made of six catchments to assess the impact of the approach. The study found that when the catchment committees are freely elected and actively involved in planning the conservation measures, they tend to be active after ministry staff have moved on and to develop a higher degree of self-organization, formulating their own management rules. These active committees are also associated with independent replication of the new practices by neighboring communities without payment or subsidy (Pretty, Thompson, and Kiara 1995).

Local Organizations and Technical Innovation in Watershed Management

Watershed management, especially in low-income countries, usually involves a combination of conservation and development of natural resources. Watershed inhabitants depend on these resources for their livelihoods, and one of the basic challenges is to establish and maintain management practices that reconcile their economic needs with the long-run conservation of soil, water, and vegetation, especially where there are significant off-site effects of resource degradation, such as the silting up of watercourses or reservoirs of interest to influential downstream stakeholders. On-site and off-site degradation of watersheds is widespread, and the management practices for conservation and regeneration are technically well understood: land use planning to designate certain areas for conservation, grazing, annual or perennial crops, and natural or managed forest; use of contour barriers on steep slopes; use of ground covers; creation of agroforestry and agrosilvopastoral systems; creation of buffer zones around watercourses and springs; and use of minimum tillage, green manuring, and other soil improvement practices that regenerate soils and permit the intensification of agriculture where suitable. However technically desirable these practices, their adoption by resource-poor farmers is very uneven, in many cases because the short-term economic returns to investing in these practices exceed the costs for the individual farmer. As a result, top-down efforts to enforce mandatory use of these practices or to promote them using temporary subsidies have by and large failed to achieve lasting adoption.

Involving local organizations in designing technical innovations for soil and water conservation and in planning where to locate them in the landscape of a watershed is now recognized as a key element of successful adoption. There are many examples of enforced use of soil and water conservation practices that have accelerated erosion once the structures stop being maintained, because the local people have no commitment to keeping the structures in place. In other instances, entire structures have been leveled as soon as outside agencies imposing them leave the village (Hinchcliffe and others 1995). In contrast, when local organizations involve farmers in experimenting with and adapting the principles of known conservation techniques to their own needs and constraints and in combining this information with local knowledge of the landscape, soils, and vegetation, then innovative management practices are developed and adopted rapidly. One example is the Empresa de Pesquisa Agropecuaria e Difusão de Tecnologia (EPAGRI) microwatershed program in Santa Catarina, Brazil. Catchment committees were formed in microwatersheds of about 150 families, and these promoted collective action in conjunction with the agronomists of the state extension and research service. Extensive testing with farmers of more than 60 species of green manure and cover crops, no-till agriculture, and recycling as manure the effluent from piggeries stabilized erosion, improved yields, and improved water quality. Similarly, the impact study of six villages with catchment committees in Kenya referred to earlier found that the more active, self-organizing committees were associated with a higher percentage of farms adopting conservation practices and that farmers associated this with higher yields, increased availability of fodder, and groundwater recharge (Pretty, Thompson, and Kiara 1995).

Local organizations can initiate institutional change by creating new rules for watershed resource management that catalyze or induce technical change of various kinds. An example is the Agha Khan Rural Support Program in India. This program forms a village institution to take on planning, conflict resolution, extension, monitoring, and evaluation activities in small catchments. Collective watershed management carried out through village institutions achieved widespread adoption of new practices in an area covering more than 3,000 hectares from 1988 to 1991. This not only increased crop and livestock productivity but also stimulated the collective use of farm equipment as well as innovation in plant protection practices, postharvest processing, and pooled marketing of produce and credit (Shah 1994).

An important feature of their contribution to improved watershed management is that local institutions can institutionalize local capacity for innovation through collectively organized experimentation with new practices and can provide a conduit for receiving and exchanging information about technical innovations, which brings down the individual's cost of experimenting. The Agha Khan project is an example of how progressing toward collective action to implement watershed development reduces the costs of implementation: for example, comparing two periods—1988–89 and 1990–91—costs of planning watershed management have declined from Rs325 to Rs25 per acre and the cost of arranging community plowing decreased from Rs125 to Rs13 per acre

(Shah 1994: table 1). A detailed case study of 22 participatory watershed development projects found that one of the main results of involving local organizations in changing farmers' soil and water conservation practices was the capacity building that enabled a process of technical innovation to be sustained (Hinchcliffe and others 1995).

In summary, local institutions can play an important role in promoting technical change in watershed management. Collective organization can reduce the costs of acquiring information and experimenting with and learning about potential changes that require action on a collective rather than an individual basis and can build local capacity to innovate. Self-organizing groups may also reduce the costs of conservation to the individual by making it more likely that environmentally damaging opportunistic behavior will be sanctioned, so that the individual who innovates does not bear all the costs of getting his neighbors to do so. Finally, institutional change—new rules and regulations about watershed management that are created by local organizations—can induce technical change by creating conditions that make it more economically feasible to undertake new land use systems and production or marketing innovations that increase the benefits of conservation management.

The Interface between Local Organizations and External Agencies

Local organizations are referred to throughout the preceding discussion as self-organizing groups voluntarily cooperating in the regulation of watershed resources. Few groups organized at the watershed level have arisen completely spontaneously without some intervention from outside agencies to catalyze their formation and action. At the watershed level, outsiders appear to play a significant role in enabling stakeholders to arrive at a joint plan of action that takes into account transboundary effects not readily perceived or measured by stakeholders other than those who are directly affected.

This may be especially the case when mobile as opposed to stationary resource flows are the focus of watershed management or when multiple, competing uses exist for a given resource, such as forest, that are not easily perceived by stakeholders in different locations. When, for example, watershed management is concerned with stream flow, a mobile resource that can often fluctuate unpredictably, it is difficult for users to inventory, monitor, and assess causes and effects of variations in flow and to gain consensus on what are reasonable limits to use. Moreover, regulation of use may have to be coordinated among geographically disperse users in upstream and downstream locations, which involves a high transaction cost for local organizations built on face-to-face interaction. External agencies have a critical role to play in helping local organizations obtain and manage information about cross-scale or transboundary effects.

Some authors have interpreted the need for expert knowledge from outside agencies to mean that the role of local organizations will therefore be diminished (Schlager, Blomquist, and Tang 1994 and Uphoff 1986). An alternative interpretation is that by providing critical information about transboundary effects, external agencies can enhance the capacity of local organizations to mobilize local cooperation in watershed management (Knapp and others 1997). This approach draws on research on the dynamics of cooperative behavior that shows that when a large group containing several factions (or stakeholders) begins cooperating, it does so through progressive transitions, with cooperation spreading from small group to small group. The subgroup with the greatest tendency to cooperate at first will probably be the one with the lowest average costs for cooperation or the longest time horizon; their decision triggers cooperation in another group with the next lowest costs and so on until there is a "cascade of further cooperation" as information spreads about the benefits of cooperation from small group to small group, reducing uncertainty within them (Glance and Huberman 1994: 80). The relationship of external agencies to self-organizing, voluntary cooperation for watershed management is one of facilitating the access of stakeholder groups to information, supporting their skills in processing information, and promoting the cascade-like sharing of information among these groups in a watershed.

The Indo-German Watershed Program in Maharashtra, India, provides an illustration of how technical information about resource flows that are not readily perceived by local stakeholders can, when provided by the outside agency to local people, fundamentally alter the resource conservation strategy. The program sought to enhance percolation over the whole microwatershed so that it acts as a large underground reservoir, creating underground flows that reach the mid-slopes some two months earlier than the lower slopes and providing additional water to the generally poorer farmers located higher up the slope. At the outset, farmers preferred to build expensive check dams on streams to improve irrigation and to invest first in improving the lower slopes (where the better-off villagers live). Specialized survey and land use planning teams provided technical information to a village watershed committee and integrated that information with local knowledge of the watershed landscape. In a planning meeting, all landowners in the watershed were involved in the development of detailed action plans, which the village committee was responsible for implementing (Farrington and Lobo 1997). This blending of local knowledge, mobilized through the village watershed committees, with externally supplied technical information considerably relaxed the rigor with which technical information is usually applied in the planning process in order to define a type and location for conservation interventions that farmers found acceptable. An important function of the local organizations involved in watershed management in this program was to provide a forum for blending local knowledge with externally supplied technical information and to mobilize local people together with outsiders in a process that enables them to adapt technical standards for planning and monitoring resource use to meet local needs.

The process of sharing information reduces each group's uncertainty about whether other groups will cooperate to implement conservation, clarifies expectations, and achieves consensus about the action to be planned and implemented by local organizations. In many experiences of participatory watershed development, the appearance of local organizations like catchment committees is catalyzed by participatory rapid appraisals, which use techniques of group analysis to facilitate the collective diagnosis by watershed stakeholders of the status of physical and social resources and the causal relationships leading to degradation (for an overview of these techniques, see Pretty 1995). It is critical that this group analysis result in a plan of action formulated by local people (in contrast to a diagnosis used by the outside agency to formulate the plan of action). Otherwise the level of voluntary cooperation and local commitment to the plan is likely to be weak and of short duration.

In most cases, local organizations, such as catchment committees, with an active role in managing watershed development operate at a relatively small scale, in catchments of a few hundred hectares at most. There are as yet few reported experiences where local organizations coordinate actions dealing with transboundary effects in watersheds of several thousand hectares.

The Landcare experience in Australia provides some insight into the roles of local and nonlocal organizations in a large-scale process. Landcare began in 1989 when an important farmers union and conservation lobby obtained political support for a 10-year funding program of community-based, conservation extension groups. These groups involve a broad cross-section of rural people with a stake in catchment planning, as a framework for individual property plans: farmers, schools, scientists, state agencies, and agribusiness, for example. The Landcare groups foster an ethic of land stewardship, fueled by extensive voluntary participation in innovative environmental monitoring (land literacy). Users employ techniques and information that were largely the domain of specialists a few years ago, such as geographic information system (GIS) and aerial surveys. Nationally funded facilitators and coordinators work with Landcare groups, and one of their main functions is to assist groups with process management—how to manage participatory planning, conflict resolution, and information sharing, for example. This has, in turn, changed the role of external agencies from being regulators of resource use to assisting the Landcare groups in the process of learning and capacity building (Campbell 1994).

Another important point of interface between local organizations and external agencies in watershed management is the need for external intervention in identifying the relevant stakeholders and bringing them to the table, whether this is literally a negotiating table or participation in collective labor, monitoring, or enforcement of sanctions. The involvement of local organizations in watershed management does not ensure equitable participation of all relevant stakeholders in a resource management initiative. Case studies show that women, the landless, marginal ethnic groups, and the laboring poor are all unlikely to be represented in participatory watershed management (Hinchcliffe and others 1995).

Leadership in local organizations is generally captured by the higher status, more well-to-do members of dominant elites, and the priorities of local organizations tend to reflect the interests of these elites. Some watershed development projects deliberately seek to overcome this bias by setting up catchment committees with representation of underpriviledged groups (Jensen and others 1996); the effectiveness of this approach has yet to be evaluated. In the Indo-German Watershed Development Program, the village committee is nominated by the village assembly (*gram sabha*)—the gathering of all persons in a village boundary with voting rights. The program guidelines stipulate that the committee should include representatives of all social groups and that at least 30 percent should be women. It is important to ensure that stakeholder groups represent the different interests in use of a given watershed resource, even if these groups do not already exist or have representation within existing local organizations, because marginalized stakeholders may undermine a conservation plan carried out by others, in which they have no vested interest.

The Process of Local Organization for Watershed Management: A Case Study

This section presents a case study of the process of organizing at the local level for watershed development in order to give a concrete example of how many of the general principles discussed earlier translate into action. The case also highlights the importance of applying these principles in a learning process approach, in which process management constitutes an important role of local organizations.

Organizational and Resource Management Problems in the Río Cabuyal

The Río Ovejas is fed by five main tributaries into the Río Cauca, which is an important source of water and hydroelectric power for the city of Cali downstream in the Cauca Valley in southern Colombia. The water quality of the Río Ovejas was poor as a result of the heavy siltation from upstream clearing and deforestation and pollution by agrochemicals, waste products from artesanal coffee postharvest processing, and the runoff of organic fertilizer, which is used heavily by small farmers in the tributary watersheds. The erosion in the upland watershed is a major cause of annual flooding problems in the lower-lying areas.

The Río Cabuyal catchment, situated in the Andes at an altitude of 1,200–2,200 meters, was selected as the pilot area within the Río Ovejas because of its size and representativity with respect to the soil, agroclimatic, and demographic characteristics of the larger watershed. Small-scale farming and day laboring are the principal sources of income. Coffee, cassava, maize, and beans, and in the upper watershed also fruits, are the principal crops. Livestock production is of minor importance. The lower and mid-altitude parts have relatively good access to markets.

Techniques available to improve the quality and quantity of downstream water included practices such as enclosing areas along riverbanks and around springs where natural vegetation could reestablish itself and planting contour barriers to control soil erosion. However, there was little incentive for upland farmers to adopt any of these practices, because they bore the costs but not the benefits, which accrued to downstream users (Ashby and others 1997). While external agencies concerned with natural resource management saw bad soil management practices as a major problem, farmers were more concerned with the poor availability of credit, which made fertilizer difficult to obtain.

Stakeholders had no common perception of the problems (Ravnborg and Ashby 1996). Inhabitants of the lower and mid-altitude watershed communities were concerned about the seasonal scarcity of piped water, the disappearance of streams, and the decreasing flow of water in the river. They attributed this to deforestation by the inhabitants in the upper-watershed communities. The agencies responsible to downstream users were concerned about water quality in the Cauca Valley, whereas the upper-watershed communities considered their own major problems to be the lack of schools, electricity, all-weather roads, and health services.

Different watershed resources were managed by independent organizations with different objectives. The piped water supply drawn from the headwaters of the tributary watersheds was managed by a local organization, the *junta del acueducto*, while an external agency was responsible for the conservation of forest reserves around the aqueduct intakes. There were no formal channels for any joint decisionmaking or coordination among the local and nonlocal entities. Still worse, several other external agencies active within the catchment area had competing agricultural or conservation objectives. There were no incentives for these agencies to coordinate with one another or with the local communities. Credit, technical assistance, and commercialization outlets were promoted by agricultural development programs, providing upland farmers with incentives to clear and burn secondary bush fallow and forests to plant cassava, notorious locally for its effects on soil degradation.

The Process of Organizational Change

Organization to tackle these problems was initiated by outside agencies interested in watershed development. An interinstitutional coordinating committee was formed with representatives from the public sector, nongovernmental organizations (NGOs), and grower associations to collate all available information and make a joint diagnosis of the problems. The committee, in consultation with local extension agents and community leaders familiar with the Río Ovejas watershed, selected the Río Cabuyal as a pilot catchment area for diagnosis and subsequent intervention. The boundaries of the 7,000-hectare watershed were set to include all persons able to appropriate water from the Río Cabuyal, which involved 1,000 families in 22 communities as local stakeholders.

At the same time, in early 1993, community leaders began to visit different areas of the watershed to motivate local people to collaborate in improving watershed management. They focused their campaign on the decreasing availability of water, which they associated with the clearing and deforestation around the springs and watercourses feeding the reservoirs at the head of the watershed. They rapidly organized a series of *comisiones* or task forces to visit communities in the upper reaches. This was the first time many had gone to these communities. They found much greater poverty than in the lower altitudes: impassable cart tracks, an absence of schools, low-quality housing, poor straggly crops, and an extensive wasteland of bracken-covered fallow left after cropping. Farmers in the upper watershed described their problems and concerns and explained how the clearing and cutting of trees around the reservoirs provided not only cropland but also charcoal, almost their only source of cash income.

The community leaders' strategy to promote conservation of the remaining forest around the reservoirs was to seek support from external organizations, via the coordinating committee, for programs to improve the quality of life in the upper watershed as an explicit reward for the commitment of upland farmers to cease cutting and clearing. They also began to formulate plans for lower-watershed farmers to teach the upland farmers how to improve their crops and to discuss ways in which farmer cooperatives could promote the marketing of produce grown in the upper watershed.

At the same time, external organizations in the coordinating committee began to develop a common set of objectives and activities to achieve them, consolidating their efforts to catch up with the evolving agenda of the community leaders. At a planning workshop in March 1993, participants identified a joint program in which each had a defined role and contribution. It was agreed that conservation activities required complementary agricultural, production, and commercialization activities, mirroring community plans to compensate upland farmers. Each agency began to identify its activities as complementary with those of others. A watershed-based organization, which called itself the Consorcio Interinstitucional para la Agricultura Sostenible en Laderas (CIPASLA), was formed of 12 participating agencies at this workshop and was charged with formalizing agreements among them and raising funds for the joint program.

An important objective of the organizers of CIPASLA was to create a forum in which local inhabitants of the watershed could define, monitor, and enforce their own regulatory rules and determine the benefits going to different stakeholders. As a result, a watershed users association was formally constituted as FEBESURCA (Federación de Beneficiarios de la Subcuenca de Cabuyal). The local leaders, active in promoting the association, defined the basis for membership as representation of all the different local organizations and interest groups, such as the *junta del aqueducto*, schools, cooperatives, women's groups, and village government. Membership reflected the perception of the local leadership that FEBESURCA would be a pressure group composed of people like themselves who could be expected to mobilize around their agenda to protect water sources in the upper watershed, which feed the tanks that store water and supply piped water to the lower areas.

By mid-1993, CIPASLA had developed a substantial budget from outside grants, participating organizations, and participating communities. FEBESURCA had its own share of this budget and provided a forum for local organizations to define an agenda to present to the external organizations in the form of community-based projects. The projects that were executed most rapidly were those that met the demand of mid-altitude farmers to protect their water supply from the upper watershed.

Regulating Resource Use

Local community negotiation to define, monitor, and enforce conservation in the uplands began to replace externally imposed (but generally flouted) regulations. The formation of CIPASLA enabled the local community to initiate new arrangements for conservation zones in the upper reaches of the watershed. Previously, external agencies had been unable to implement a long-standing program to enclose buffer zones to regenerate forest around watercourses. The local *junta del acueducto* and the regional watershed authority, together with community leaders, swiftly began to work with the upper-watershed inhabitants, whose land yielded the springs and streams feeding the water reservoirs. Within a short period, FEBESURCA leaders persuaded the watershed authority to relax its technical recommendations for enclosing and protecting the forest and for regenerating natural forest cover around these water sources. This agency provided technical advice, delegating monitoring and sanctioning to FEBESURCA. This new arrangement enabled community leaders to negotiate with the upper-watershed farmers mutually acceptable limits on use of these areas for agricultural purposes. FEBESURCA leaders organized community labor drawn from upper- and lower-watershed communities to plant trees and enclose areas to be protected by mutual agreement. Over a period of 18 months, 135 hectares were enclosed, and 150,000 trees were planted. The community rapidly mobilized a total of 3,714 person-days for this purpose. They explained their motivation for working together as making an investment in the future of their children, for whom the availability of water will be critical. Also, the *junta del acueducto* announced that it would turn off access to piped water for families who failed to contribute labor.

Subsequently, GIS analysis showed that small tributaries throughout the watershed were as important as the upper-watershed springs and streams for supplying groundwater and stream flow to the lower slopes. This information led to a reevaluation of the rehabilitation strategy, and CIPASLA began to promote conservation in small catchments and downstream tributaries as well.

Conflict Resolution

In 1994 an outbreak of forest fires partly destroyed a buffer zone created by FEBESURCA to protect water sources. The burning was sparked by slash-and-burn land clearance by people colonizing the upper watershed and highlighted the failure to identify the semi-landless as stakeholders in the watershed or as an interest group with representation in the watershed users association (Ravnborg and Ashby 1996). The burning made it clear that securing the buffer zones around watercourses would have to involve the poor and the landless inhabitants of the upper part of the watershed. FEBESURCA had been led by a local interest group concerned with water and reforestation issues, which represented the interests of the mid-altitude farmers. At first, FEBESURCA had not yet evolved as an effective mechanism for channeling demand from a wider constituency, and CIPASLA's organizational structure had not yet encompassed some significant stakeholders whose needs revolved around land use rights and management practices, as opposed to water. By 1994, it was decided that each of the 22 communities should elect a representative to FEBESURCA.

FEBESURCA conducted a stakeholder analysis, facilitated by outsiders, to identify the conflicting interests involved. Participants identified the reasons for burning as well as solutions, including the acceptance of some burning for land preparation where there is a shortage of land and labor. A set of norms, specifying how and when burnings should be conducted, was developed, and some communities formed groups to ensure compliance. Subsequently, farmers generally seemed to be following the recommendations, such as making firebreaks before burning.

These norms were more successful in controlling burning than previous measures because they were developed by farmers themselves and acknowledged burning as the only feasible means of land preparation in some circumstances, rather than simply condemning all burning. Their development illustrates the role of local organizations in providing a forum to analyze and negotiate interests (Ravnborg and Ashby 1996).

Technical Innovation

CIPASLA badly needed technological innovations that would promote conservation while enhancing productivity. In 1993 FEBESURCA formed local agricultural research committees (CIALS) in the upper-watershed communities in return for their commitment to protect the forest and watercourses. CIALS are committees of four or more farmers, elected by their community, who test adaptive technologies in the local environment, combining local knowledge and exotic technologies, on topics chosen by the community (Ashby and others 1995). Based on the CIAL's adaptive testing, new crops for upland farmers were identified, and producer groups formed, funded by CIPASLA. These groups channeled specialist training and market advice from technical agencies to upland farmers. CIPASLA also provided the impetus for the development of small-scale dairying, which began to stimulate changes in land use. Cut-and-carry pastures were planted, replacing annual crops, and a milk producers cooperative was established. FEBESURCA leaders used their political clout to improve roads in the upper watershed to facilitate transport and marketing. Middlemen began to appear in the upper watershed to purchase the product, providing a steady weekly cash income for the farmers. The introduction of commercial production was linked to adoption of the long-rejected contour barriers, live fences, tree plantations, and buffer zones, just as the community leaders had visualized.

References

The word "processed" describes informally reproduced works that may not be commonly available through libraries.

Ashby, Jacqueline, T. Gracia, M. P. Guerrero, C. A. Quirós, and J. I. Roa. 1995. *Institutionalizing Farmer Participation in Adaptive Technology Testing with the "CIAL."* Agricultural Administration (Research and Extension) Network Paper 57. London: Overseas Development Institute.

———. 1997. "Supporting Farmer Experimentation with Local Agricultural Research Committees." In Laurens van Veldhuizen, Ann Waters-Bayer, Ricardo Ramírez, Debra A. Johnson, and John Tompson, eds., *Farmers' Research in Practice,* pp. 245–63. London: Intermediate Technology Publications.

Bardhan, Pranab. 1993. "Analytics of the Institutions of Informal Cooperation in Rural Development." *World Development* 21(4):633–39.

Campbell, A. 1994. "Landcare in Australia: Spawning New Models of Inquiry and Learning for Sustainability." Paper presented to the international symposium for farming systems research and rural development, Montpellier, France, 21–25 November. Processed.

Cernea, Michael M. 1987. "Farmer Organizations and Institution Building for Sustainable Development." *Regional Development Dialogue* 8(2): 1–24.

———. 1993. "Culture and Organization: The Social Sustainability of Induced Development." *Sustainable Development* 1(2):18–29.

Farrington, John, and C. Lobo. 1997. "Scaling up Participatory Watershed Development in India: Lessons from the Indo-German Watershed Development Programme." DFID Natural Resource Perspectives 17. U.K. Department for International Development, London. Processed.

Glance, Natalie S., and Bernardo A. Huberman. 1994. "The Dynamics of Social Dilemmas." *Scientific American* (March):76–81.

Hinchcliffe, Fiona, Irene Guijit, Jules N. Pretty, and Parmesh Shaw. 1995. *New Horizons: The Economic, Social, and Environmental Impacts of Participatory Watershed Development.* Gatekeeper Series 50. London: International Institute for Environment and Development.

Jensen, Jens R. 1996. "Watershed Development: Concept and Issues." In J. R. Jensen, S. L. Seth, T. Sawhney, and R. Kumar, eds., *Watershed Development: Emerging Issues and Framework for Action Plan for Strengthening a Learning Process at All Levels,* pp. 43–63. Proceedings of DANIDA's international workshop on watershed development, DANIDA, Hubli and Bangalore, India, December 1995. WDCU Publication 1. New Delhi: DANIDA.

Jensen, J. R., S. L. Seth, T. Sawhney, and R. Kumar, eds. 1996. *Watershed Development: Emerging Issues and Framework for Action Plan for Strengthening a Learning Process at All Levels.* Proceedings of DANIDA's international workshop on watershed development, DANIDA, Hubli and Bangalore, India, December 1995. WDCU Publication 1. New Delhi: DANIDA.

Knapp, Edwin B., Jacqueline A. Ashby, Helle Munk Ravnborg, and William C. Bell. 1997. "A Landscape That Unites: Community-Led Management of Andean Watershed Resources." Paper presented at Global Challenges in Ecosystem Management in Watershed Contexts, a special conference of the fifty-second annual conference of the International Soil and Water Conservation Society, Toronto, 22–26 July. Processed.

Pretty, Jules N. 1995. "Participatory Learning for Sustainable Agriculture." *World Development* 23(8):1247–63.

Pretty, Jules N., J. Thompson, and J. K. Kiara. 1995. "Agricultural Regeneration in Kenya: The Catchment Approach to Soil and Water Conservation." *Ambio* 24(1):7–15.

Ravnborg, Helle Munk, and Jacqueline Ashby. 1996. *Organizing for Local-Level Watershed Management: Lessons from Río Cabuyal Watershed, Colombia.* AGREN Paper 65. London: Agricultural Research and Extension Network.

Schlager, Edella, W. Blomquist, and S. Yan Tang.

1994. "Mobile Flows, Storage, and Self-Organized Institutions for Governing Common-Pool Resources." *Land Economics* 70(3):294–317.

Shah, Parmesh. 1994. "Participatory Watershed Management in India: The Experience of the Aga Khan Rural Support Programme." In Ian Scoones and John Thompson, eds., *Beyond Farmer First: Rural People's Knowledge, Agricultural Research, and Extension Practice*, pp. 117–24. London: Intermediate Technology Publications.

Sirianni, Carmen, and L. Friedland. 1995. "Social Capital and Civic Innovation: Learning and Capacity Building from the 1960s to the 1990s." Paper prepared for the meetings of the American Sociological Association, Washington, D.C., 20 August. Processed.

Uphoff, Norman Thomas. 1986. *Local Institutional Development: An Analytical Sourcebook with Cases*. Rural Development Committee. Westport, Conn.: Kumarian Press.

———. 1992. *Learning from Gal Oya: Possibilities for Participatory Development and Post-Newtonian Social Science*. Ithaca, N.Y.: Cornell University Press.

White, Thomas A. 1995. "The Emergence and Evolution of Collective Action: Lessons from Watershed Management in Haiti." *World Development* 23(10):1683–98.

11

Innovative Approaches to Technology Generation and Dissemination for Low-Income Farmers

John Farrington and Graham Thiele

Time-series studies of public sector agricultural research have generally demonstrated high rates of return (Evenson and Pray 1991). Methodological refinement has removed some of the earlier upward biases, but estimates remain substantially positive (Farrington, Thirtle, and Henderson 1997). However, many such studies either focus on areas of commercial or semicommercial agriculture alone, where data are good, or use aggregate national-level data and so lump together commercial areas with areas where the returns to agricultural research are likely to be lower.

In areas where commercial agriculture is strong, agroecological conditions tend to be homogeneous (as in irrigated areas), and it is relatively easy to replicate farmers' conditions on research stations. Furthermore, farmers themselves tend to be articulate and self-confident in conveying their requirements to researchers, thus strengthening the client focus of research and extension. By contrast, in many areas of rainfed agriculture, where agroecological conditions are complex, diverse, and risk-prone and where standards of physical (communications, transport) and social (education, health) infrastructure are low, specific efforts will have to be made to elicit farmers' perceptions of needs and opportunities. Our concern in this chapter is with just such areas, where almost 1 billion people live in poverty worldwide.

In many countries of Sub-Saharan Africa and South Asia, population levels continue to rise in these areas. Moreover, the number of rural poor remains so large in relation to the employment opportunities offered in urban areas or in commercial agriculture that migration is unlikely to reduce the overall volume of rural poverty by much in the coming decades.

More and more private commercial companies are penetrating into rainfed areas with, for example, seeds, agrochemicals, and farm machinery. This tends to occur where rainfall is reliable, because fragmented markets and poor infrastructure elsewhere make transaction costs unattractively high. It has been argued that these areas urgently need the dynamism of private sector research and development, but, ironically, they are the least likely to receive it (for research and development on seeds, see Gisselquist and Srivastava 1997). In addition, it is unlikely that the private sector will perform in a socially optimal way in areas where externalities are high, such as soil and water conservation, or in areas where some resources are not privately owned but "common pool."

Attention has therefore turned to ways of improving the effectiveness of the kinds of research institutions likely to remain dominant in these areas in the coming decades, namely those in the public sector. A radical view is that public sector research services in small countries have been so ineffective that they should be shut down and replaced by a low-cost but efficient capacity to "borrow" ideas, for instance through the Internet (Zijp, personal communication). Although we empathize with some of the principles underlying this notion, we see it as impractical because of the substantial knowledge needed of the country's agricultural conditions, the skills needed in interpreting what is often highly condensed documentation of new technology, and, most fundamentally, the capacity needed to conduct adaptive testing of "imported" technologies under local conditions. For better or worse, this requires a continuing research capability. However, as will become clear in the remainder of this chapter, we recognize the limitations of the public sector and see multiagency approaches—not replacement by the private sector or complete abolition of public institutions—as one way of making good some of these limitations.

A number of major international studies have identified these limitations and suggested ways forward. Common threads among these studies are:

- That the public sector has generally not performed well in difficult areas
- That a major reason for poor performance is inadequate orientation toward clients and inadequate participation by them in the design and implementation of research
- That improvements require changes in the performance assessment criteria and reward systems for public sector staff
- That improvements also require additional resources to permit more work in farmers' fields and less work on-station; however, levels of resources available to the public sector are at best static and in many cases declining
- That stronger interaction with local organizations having close contact with farmers (nongovernmental organizations, membership organizations, development projects) can create a "demand pull" on the public sector by helping to articulate farmers' requirements. An extension of this argument is that these demands can be strengthened, and funding problems of the public sector alleviated, where such organizations can either use their own funds or draw on endowment funds, established by donors or governments, to commission research. Such studies include:
- The nine-country study by the International Service for National Agricultural Research (ISNAR) on on-farm client-oriented research (Merrill-Sands and Kaimowitz 1990), whose discussion of mechanisms for linking research and extension could, as Farrington and Bebbington (1993) argue, be adapted to embrace organizations outside the public sector and to strengthen public-private links
- The ISNAR study led by Eponou (1996), which examines the role of formal farmers organizations in the technology development and transfer chain, as does a study by the Overseas Development Institute (Carney forthcoming).
- The 50-country study by ISNAR on agriculture and environmental research in small countries (Eyzaguirre 1996), whose central analytical concept is a portfolio approach to research planning and management in which multiple agencies (universities, government institutes, parastatals, private companies, nongovernmental organizations, and donor projects) work together to address specific problems according to the comparative advantage of each
- The major workshop on future options for agricultural extension held by the World Bank in June 1996 (Zijp 1996), which placed strong emphasis on the desirability of multiagency approaches, because the public sector, especially given its declining level of resources in many countries, is unlikely to be able to meet all the demands made on it.

The proposals for multiagency approaches deriving from this and other work are now being tested in various ways, particularly in externally assisted projects. Some of these projects simply involve contractual relations in which nongovernmental organizations (NGOs), for instance, are contracted to provide extension services in a specific area (see, for instance, the experiment of Chile's Instituto Nacional de Desarrollo Agropecuario described by Berdegue 1994). Others involve the establishment of research grant (endowment) funds that NGOs and others can draw on in order to commission work from the public sector (in some cases also the private sector) that meets clients' needs.

A third category of innovative multiagency approaches, which forms the basis of this chapter, goes beyond narrowly contractual relations and seeks synergistic interactions between different types of organizations. Here, all work together to analyze particular problems, contribute jointly to their solution, review progress in an iterative fashion, and make course corrections by mutual agreement as necessary.

Although some success has been achieved in introducing participatory approaches at the diagnostic stage of the research cycle (for instance, participatory rural appraisal is now widely used, although often badly), much less attention has been paid to ensuring continued interaction between researchers and their clients during the identification, testing, and dissemination of technologies.

Initiatives in the area of agricultural technology need to be set in the context of wider perspectives on scaling up. The major review by Edwards and Hulme (1992) of NGOs' efforts to introduce participatory approaches suggests four types of scaling up:

- Working with government to spread NGO methods and change policy
- Conducting lobbying and advocacy
- Expanding NGOs' own operations
- Strengthening the network of local membership organizations with which NGOs work.

For reasons set out here, the premise underlying this chapter is that public sector resources have to be brought in to permit scaling up in this context. Our discussions therefore relate to the first type of scaling up. The chapter focuses on a specific aspect of the question of scaling up participatory approaches where only limited success has been achieved to date: that is, the organizational and management measures that are necessary if participation is to become part of the mainstream of government research and extension strategy.

Participation

Nongovernmental organizations providing development-related services (as distinct from membership organizations such as farmers associations) tend to work more commonly in health and education than in agriculture and natural resources. Where they do work in agriculture, they often apply the same Freirean principles of participation (Freire 1972) as they do in other sectors. Intensive face-to-face approaches with small, local groups, often over a period of years, facilitate:

- The emergence of a "critical consciousness" of wider social, economic, and political conditions facing rural communities that may bear on agricultural and natural resource constraints
- The identification of how communities can respond to these constraints using their own resources
- The possibilities of drawing on external sources of support where their own resources are inadequate.

This *empowering* approach to participation allows NGOs to take the moral high ground and to dismiss efforts that do not reach these standards. However, from a wider policy perspective, government organizations cannot mandate themselves to concentrate resources in a few villages as do NGOs, and they have to make the best of the difficult job of achieving whatever level of participation is feasible using resources spread thinly over a wide area. This leads government organizations to seek *functional* participation in an effort to generate and disseminate technology more efficiently. Clearly, at times there are gray areas between functional and empowering forms of participation, which represent different ends of a spectrum rather than uniquely distinguishable categories. Nevertheless, as has been argued elsewhere (Farrington 1997), the distinction is important in policy contexts, not least because each approach tends to be associated with very different kinds of organizations. Our concern here is largely with the more functional types of participation.

Much of the literature on participation reports efforts to experiment with different participatory methods, often involving the staff of government organizations, NGOs, universities, or special projects in discussions with individual farmers or, occasionally, village groups, although the basis on which these groups are formed is rarely clear (see, for instance, the International Institute for Environment and Development, London, series notes on the participatory learning approach; Chambers, Pacey, and Thrupp 1989). Although interaction of this kind is clearly feasible and useful, it does not necessarily form a basis for wide-scale implementation. A wide body of evidence now suggests that it is unrealistic to expect low-resource rural people in difficult areas who have long been economically and politically marginalized to have the confidence or skills to engage directly with "outsiders" such as government agencies (Carroll 1992). Many see an important role for NGOs in helping to build such skills and confidence and in helping to support the emergence of diverse kinds of local organizations, so that interaction with government organization staff becomes more cost-effective than if it were conducted solely on an individual basis.

Many of the NGOs involved in agriculture are characterized by:

- Strong interest in low external-input agriculture
- Ability to identify farmers' aspirations and the needs and opportunities to which technologies need to be adapted
- Ability to identify indigenous knowledge and practice and so help in negotiating how they might complement modern technologies
- Awareness of the wider contexts of livelihood in which initiatives toward agricultural change need to be located.

However, many observers raise valid questions over their technical abilities, over possibly unjustified predispositions toward eco-based techniques, and excessive accountability to the rural poor, especially among organizations that rely heavily on external funding.

Farrington and Bebbington (1993) air these potential strengths and weaknesses as a preamble to a major empirical study covering 18 countries and more than 70 case studies of NGOs' work in agricultural and natural resources development. They identify occasional cases in which NGOs and government organizations succeeded in working together in a fully collaborative mode, but many more in which the interaction was tangential, or through service-provision contracts, or in which NGOs were working completely independently of government, having tried and failed to secure agreement or having decided that working with government was practically or philosophically impossible. Their study identified the preconditions for successful interaction among government organizations, NGOs, and farmers groups in implementing participatory approaches but only very few cases in which it was under way and none in which it had been expanded to a substantial scale.

Field-based efforts to establish wide-scale multiagency approaches to the generation and dissemination of technology in agriculture and natural resources have proliferated in the past five years. These have involved varying combinations of government organizations, NGOs, farmers organizations, and, at times, the private commercial sector. This chapter aims to draw out their main features, particularly with respect to the differing modalities for scaling up participatory approaches. It presents three contrasting case studies and then draws lessons across the three.[1] The focus throughout is on organizational and management configurations for participatory approaches that elicit the views of rural people on their needs and their reactions to the technologies on offer.

Essentially, this involves collaborative approaches in which the respective organizations share objectives, make joint decisions on actions and course corrections, and pool some resources to do so. Although one type of organization (for example, a government organization) may help to finance others (NGOs, farmers groups), collaboration toward shared objectives is essentially a higher order of interaction than simple contracting arrangements, which are not considered here.

Case 1: Scaling up Participatory Approaches in an Unstructured Context

Udaipur district in Rajasthan, India, is typical of the approximately 75 districts located in semiarid central India. Its population of some 1.5 million comprises a high proportion of scheduled tribes and castes, annual rainfall is typically 800–1,000 millimeters, but highly variable, irrigation is very limited, soils are poor, and the terrain is undulating (Alsop and others forthcoming). Conflicts among castes at the village level are common. The district has an atypically high number of NGOs, many of whom seek to enhance rural livelihoods through long-term empowerment approaches.

Efforts over three years (1994–97) in Udaipur sought to identify how NGOs and government research and extension services might work more closely together. These efforts were initiated by the Government of Rajasthan and the Ford Foundation. They comprised support to:

- *A central government Farm Science Center.* This support permitted the center (which was located at a prominent local NGO) to set up training courses and experimentation in response to NGOs' requirements, to invite representatives of some 35 NGOs and 10 government departments concerned with aspects of extension or natural resource development plus the State Agricultural University and related research stations to a quarterly forum meeting (which served as a basis for exchanging information, leading to joint field visits and some joint planning of activities), and to create a documentation center for use by government organizations and NGOs in the district and an agricultural research fund to permit NGOs to commission research that was not otherwise being conducted by government organizations.
- *An external organization.* This organization was mandated to assist in developing techniques for monitoring the processes of multiorganizational interaction, supporting individual organizations (primarily NGOs) in adapting and introducing these techniques, and in developing the confidence to deal with senior government organization staff. It was also charged with conducting a number of studies on issues relating to collaboration, with developing a newsletter to document recent developments in NGO–government organization interaction, with generating confidence among organizations to use its correspondence column as a means of drawing the attention of senior officials to inadequacies in collaborative efforts, and with engaging the interest of a local agency (the Farm Science Center) in continuing with its publication.
- *A number of NGOs.* This support allowed NGOs to

collaborate with the extension services of the Department of Agriculture in testing new crop varieties in farmers' fields and funded a course in participatory methods for the Department of Agriculture, NGOs, and others.

A number of successes have been noted. In several cases, NGOs and government organizations have initiated joint projects. Training schedules are responding more closely to NGOs' and farmers' requirements. Government organizations now know more about the programs of other government organizations than NGOs know about those of other NGOs. The flow of information across the NGO–government organization divide has improved. The newsletter on recent developments has grown into a widely read vehicle for information and commentary. A number of organizations see the value of process monitoring techniques, both in making course corrections and in strategic planning. And, after a long struggle, senior government staff have accepted, and are prepared in a limited way to fund, the concept of paraextension workers put forward by an innovative middle-level government organization official and several NGOs for implementation in areas chronically understaffed by government agents. Participatory varietal selection has been adapted to local conditions from a neighboring state and is being led at the village level by farmers' representatives.

However, a number of difficulties remain. From the government organization side, incentive and reward systems in the university are such that researchers (even from the zonal station set up under a World Bank scheme and mandated to service the area's requirements) have rarely participated in meetings, let alone joined in collaborative efforts. Lacking technical support, NGOs initially could not put up clear, researchable proposals under the Agricultural Research Fund (although this has now been addressed to some degree by allocating some of the fund to contracting independent technical support for the preparation of proposals).

The principal constraints, however, are deeply rooted in philosophies and operational procedures on both sides. Government organization programs remain largely centrally determined, with performance targets set in relation to, for instance, the number of demonstrations held. There remains little room for flexibility, and the more enlightened senior government organization staff have difficulty overcoming prejudice, reluctance, and, in some cases, lethargy, lower down. Problems on the NGO side include the reluctance of some to think in terms of strategic development beyond the confines of a few villages in which they are working and failure to adopt a businesslike approach in dealing with well-disposed senior government organization staff. NGO meetings with government organization staff are often characterized more by each NGO's efforts to articulate its philosophy than to agree on concrete plans of action to which government organization staff can respond.

Overall, three years of work have laid some important foundations, generated some collaborative actions (and consolidated others) but led to improvements (as yet unquantified) in agricultural productivity in only a few hundred hectares. While the Government of Rajasthan and others are keen to carry this work forward into other areas, there is a growing consensus that, in order to be more cost-effective, efforts to scale up participatory approaches will, for the future, have to be organized on a more structured basis.

Case 2: Scaling up Participatory Approaches in a Structured Context: An Example from Microwatershed Rehabilitation

Microwatersheds of 500–1,500 hectares containing one or more villages have become the focus of policy attention in several South Asian countries as a natural unit for implementing efforts to intensify agricultural and natural resource productivity (Farrington and Lobo 1997). In India alone, they are attracting some $500 million a year of government investment, with additional funds from numerous donor-supported projects. The biophysical concepts are well proven. Improved management of the common pool forest and pasture lands in the upper slopes:

- Increases the production of biomass potentially useful in livestock production
- Reduces erosion and runoff
- Increases the infiltration of water so that water tables beneath the agricultural land on the lower slopes rise, thereby allowing more water to be lifted to counteract droughts within the season, reducing risk, increasing the productivity of existing crops, and in some cases permitting new crops to be grown.

Improving natural resource management in this way therefore potentially allows a quantum leap in agricultural production technology and in productivity. Pilot experiences typically indicate a doubling in crop productivity, increased employment opportunities, diversification of the village economy, and, where marketing opportunities are good, a several-fold increase in milk production. However, the rehabilitation of resources has proven difficult to achieve sustainably in anything other than small areas: traditional systems of common pool resource management have broken down under population pressure in many areas, and new joint action requires substantial face-to-face effort to create the necessary coherence and confidence at the village level and to define rights and responsibilities in an equitable fashion.

An additional difficulty in India is that several government departments are mandated to deal with aspects of watershed development, and one—the Forest Department—has long had powers of "policing" in order to prevent depletion of the resource. Despite the recognized ineffectiveness of this approach, the department is reluctant in many areas to move to new, joint-management arrangements with NGOs and village groups, in part because these limit rent-seeking on the part of local officials.

Setting the Preconditions for Scaling up at the Local Level

Several factors argue for taking a structured approach to microwatershed rehabilitation:

- The need for a degree of consensus on rights and responsibilities within the village, or among the villages affected, including agreement on measures to protect the resource once recovery is under way (such as restrictions on livestock grazing or the cutting of trees and fodder), on the pumping of water resources to irrigate crops, on the types of crops grown, and so forth
- The need to bring together both villagers and technical specialists in order to identify desirable soil and water conservation and revegetation measures, the levels and division of necessary financial resources and material inputs, and the means by which both sides can monitor progress
- The need to progress along a toposequence from ridge to valley if the physical measures implemented such as contour strips and tree planting are to be effective
- The need for a "learning" approach, in which villages and outside agencies initially tackle a small segment of the watershed, decide on the basis of this experience whether they wish to proceed to the full watershed, and, if they do, incorporate lessons from this experience.

The preparatory phase of the Indo-German Watershed Development Programme (IGWDP) began in 1989, but numerous informal preparations, including experimental watershed sites and contacts with government and other organizations, began in the mid-1980s. Currently with funds of some DM12 million, the program covers 92,000 hectares in 20 districts of Maharashtra, involving 50 NGOs working in 74 watersheds.

The setting of appropriate criteria for the selection of watersheds, villages, and local-level NGO partners and the design of local-level collaborative mechanisms. Criteria for the selection of villages and watersheds include, on the technical side:

- Assured irrigation on no more than 20 percent of net cultivated area
- Notable erosion, land degradation, resource depletion, or water scarcity problems
- Villages located in the upper part of drainage systems
- Watersheds around 1,000 hectares in size and average rainfall around 1,000 millimeters a year
- Village boundaries that correspond closely with those of the watershed
- Cropping systems that do not include long-duration crops with high water requirements, such as sugarcane.

In terms of socioeconomic characteristics,

- Villages should be poorer than average with a high proportion of scheduled tribes and scheduled castes.
- There should be no wide disparities in the size of landholding.
- Villages should preferably have shown a concern for resource conservation and should have a known history of coming together for common causes.

As a condition for selection, villages must commit themselves to:

- Ban the felling of trees
- Ban free grazing and undertake "social fencing" for the protection of vegetation
- Reduce any excess population of nondescript livestock
- Keep water-intensive crops to existing levels, preferably ban them
- Ban deep tubewells
- Contribute voluntary labor to a value of 16 percent of the unskilled labor costs of the project, with landless and poor single-parent households being exempt

- Start a maintenance fund for watershed development
- Take the steps necessary for achieving and maintaining a sustainable production system
- Constitute a village watershed committee to maintain the common assets created.

Criteria for the selection of NGOs to support village organizations include:

- Their reputation and history, the extent to which they have achieved rapport with organizations of the people and of government, their perspective on watershed development, and their technical and managerial capability
- The length of time they have been active in the area
- Demonstrated willingness (in the event of weak familiarity with watershed management) to undertake exposure visits elsewhere, to send village youth and others on specific training programs, and to prepare and implement a demonstration project of at least 100 hectares
- Willingness to accompany village organizations through a capacity-building program and to meet the qualifying criteria before undertaking full implementation.

What is notable about the IGWDP's approach to NGOs is that its early experience of consultation with the larger NGOs already experienced in microwatershed rehabilitation suggested that their commitment to long-term Freirean empowerment-type approaches to group formation and joint action and the inclination of some toward purely indigenous development were likely to predispose them against effective collaboration with government agencies, against improved technologies coming from outside the project, and against rapid scaling up. The IGWDP therefore decided not to work with these organizations.

The design of village-level mechanisms for participatory planning, learning, and implementation. The initial planning approach used by the program was based on gross area planning in which cost norms provided by NABARD—National Bank for Agricultural and Rural Development—for specific types of intervention and specific types of land were routinely applied. A major shortcoming of this approach was that it relied on contour maps, which are inadequate to capture features such as the extent to which individual fields have been leveled. Existing farm-specific improvements, such as leveling and bunding, are a crucial determinant of what further measures are necessary and acceptable. In addition, government maps showing individual landholdings and features such as streams were found to be inconsistent with both contour maps and with reality on the ground.

The net area approach, subsequently developed in the IGWDP relies heavily on consultation with farmers in their own fields. The type and location of interventions agreed with farmers are marked both on the ground (with lime) and on landholding maps. Fields are assessed for slope, soil depth, soil texture, and erosion status, and these data allow the costs of rehabilitation to be derived. The village watershed committees and NGOs are presented with the maps generated in this way and a copy of the principal spreadsheet in the local language, so that they can discuss it prior to submission. The net area planning approach allows villagers to voice their preferences and priorities in the design of watershed rehabilitation and to monitor progress on the ground. However, it also allows technical inputs concerning, for example, tree species, spacing distance, planting arrangements, and bunding from the Forestry Department. Also important, it produces a plan that incorporates agreed work rate and funding norms and is readily accessible (and if accepted, fundable) by the government agency concerned (in this case, NABARD). Using funds allocated to them, the supporting NGO and village watershed committees in each watershed are required to hire a civil engineer for the preparation of full proposals, and engineers are trained in the net planning approach.

A key feature of the IGWDP is that it allows for a capacity-building phase of up to one year in which a small segment of the watershed (typically 100 hectares) is rehabilitated. This allows villagers to develop the requisite skills, to determine whether they feel comfortable with the approach, to suggest amendments as necessary, and to build up group-based capacity for decisionmaking and joint action in the practical context in which these skills will ultimately be applied on a larger scale.

A technical support NGO—the Watershed Organization Trust (WOTR)—created by the IGWDP provides funds to village organizations and local NGOs for the capacity-building phase (currently to a ceiling of Rs500,000—approximately $15,000—per watershed, including administrative costs).

During the capacity-building phase, WOTR provides villagers and the supporting NGO with training in technical skills corresponding to the individual components of watershed development, namely:

- Soil and land management
- Water management
- Crop management
- Afforestation
- Pasture and fodder development
- Livestock management
- Rural energy management
- Other farm and nonfarm activities
- Community development.

These include, for instance, skills in surveying, staking, and nursery raising. Villagers and NGOs are also trained in interpersonal relations, social mobilization, and the management of village-based organizations.

Design of a sustainable mechanism for screening and funding individual proposals submitted for watershed rehabilitation. The IGWDP has established a project sanctioning committee headed by NABARD, which has a strong reputation for probity and technical capability. It is comprised of four representatives of NGOs, the program coordinator, three representatives of the Government of Maharashtra, and one representative of the national government. Proposals have to be structured on agreed financial and technical norms, although there is some flexibility for modifying these to accommodate new technologies. Once a proposal for rehabilitation has been approved, funds are authorized from NABARD for both capacity-building and full-implementation phases, and NABARD is involved in monitoring progress. The central role played by a respected national organization in assessing and channeling finance to donor-supported projects is a cornerstone of replicability: rupees can be channeled through this mechanism once foreign funds have dried up.

Mobilization of administrative and political support from the early stages. The IGWDP focused on obtaining the political support of members of the legislative assembly of Maharashtra, initially by inviting them to see successfully rehabilitated pilot watersheds. It then drew on this support in order to obtain a cabinet resolution implementing joint forest management arrangements in the state. These arrangements, in turn, required the Forestry Department to work with village-based membership organizations and NGOs in designing and implementing rehabilitation measures. Given the high proportion of common property resources falling under Department of Forestry control in the upper slopes of watersheds, their involvement in rehabilitation and agreement to subsequent protection and access arrangements were essential. The approach to obtaining agreement to undertake joint forest management and the experience with its implementation provide valuable lessons for adapting the approach to other areas.

Establishment of channels for drawing on technical expertise following rehabilitation. Watershed rehabilitation increases the amount of fodder available and so may permit the upgrading of livestock and the intensification of production. Similarly, higher water tables allow water to be pumped in order to counteract the threat of drought, extend the season, or cultivate for a second season. The availability of water reduces risk, allows higher-yielding or more valuable crops to be grown, and encourages higher levels of input use. All of this implies a potential demand for new crops and management practices. Linkages with technical agencies in the period following rehabilitation are facilitated by WOTR, which has a central role in the program's philosophy of creating self-sustaining local organizations. It provides NGOs and village organizations with support and training in awareness creation, social mobilization, and the planning, implementation, and monitoring of watershed development projects. WOTR also provides technical and managerial training support and puts NGOs and community-based organizations in contact with line departments of the Government of Maharashtra. WOTR has 29 staff covering the disciplines of social mobilization, women's issues, agronomy, civil engineering, and computer applications. WOTR's training is tailored to specific settings, using a combination of structured workshops and less structured techniques such as village meetings and exposure visits. The WOTR has proposed exploring farm-based and other income-generating opportunities to take advantage of the additional resources created by watershed development and providing extension advice on environmentally sustainable and economically viable dryland farming systems.

Creating a Wider Institutional Structure and Expansion Pathways

For several years prior to implementation of the program, its architects were concerned that participatory watershed development should be replicable over wide areas. Preparatory work was based on the premise that stakeholders need to be engaged at international, national, provincial, and local levels.

At the international level, the program receives funds from two organizations under the German Ministry of Economic Cooperation, both of which have an interest in seeing the program succeed: German Technical Cooperation (Deutsche Gesellschaft für technische Zusammenarbeit, GTZ) provides funds to WOTR for a 12–18-month capacity-building phase, and the German Development Bank (Kreditanstalt für Wiederaufbau, KfW) provides funds to NABARD, which disburses them to local-level agencies for the full-scale implementation phase (four years).

At the national level, the principal stakeholders are the Ministry of Finance (via NABARD) and the Ministry of Agriculture. The Ministry of Finance is ultimately responsible for the disbursement of funds, but the Ministry of Agriculture is keen to see development of watersheds on the ground and is not without influence. Channeling funds through NABARD has several advantages:

- NABARD brings an interest by central government in the performance of the program.
- NABARD has an interest in raising the repayment rates it has achieved historically in rainfed farming areas and so can be expected to commit itself to the success of the program.
- Individual NGOs and village watershed committees can receive foreign funds channeled through NABARD without having to obtain foreign exchange registration.
- Several dozen NABARD staff have technical qualifications in subjects broadly related to agriculture and natural resource management. They feel comfortable discussing technical issues with officials of, for example, the forestry or agriculture departments, and, in turn, command the respect of technical staff in these departments.
- Procedures developed with and through NABARD for the disbursal of foreign funds in this way lend themselves to any subsequent disbursal of Government of India funds.

At the state level, the principal stakeholders are the departments of agriculture, soil and water conservation, and forestry. Ministers overseeing these departments successfully promoted a cabinet resolution in 1992 in support of the program. This has been a key move in facilitating supportive action by line department staff.

At the local level, during the capacity-building phase, the village assembly *(gram sabha)* nominates a village watershed committee, which, in matters relating to forest department lands in the village, works with the forest protection committee. During the capacity-building phase, funds are channeled via WOTR into the NGO's bank account, and the NGO is then responsible for contracting a civil engineer (diploma level) to help in drafting the watershed development plan, together with the villagers themselves, based on net area techniques. The engineer is trained in participatory net-based planning by staff of WOTR. Toward the end of the capacity-building phase, the draft proposal is considered by the village watershed committee and submitted to NABARD.

If approved, funds for the full-implementation phase are channeled into a bank account operated jointly by the NGO and the village watershed committee. WOTR provides ongoing support during the full-implementation phase, and NABARD and the program coordinator are responsible for monitoring and supervision. Management costs go directly to the NGO, whereas project funds go to the joint account of the village watershed committee and the NGO. The expectation is that the role of NGOs will diminish over time as that of local-level membership organizations becomes stronger. Once the rehabilitation works are complete, half of the 16 percent contribution made by the village to the cost of unskilled labor is returned to the village watershed committee to form the core of a maintenance fund.

Two principal expansion pathways are envisaged:

- "Nodes" of approximately 1,000 hectares of watershed are used as a central demonstration that neighboring villages come to see and as a potential training area for new villages forming a village watershed committee; some village watershed committees have already begun to register themselves as NGOs and to obtain the benefit of the funding support available for NGOs while at the same time serving as a vehicle for a type of farmer-to-farmer extension.
- The intention is that the essential features of the Maharashtra model be replicated in other states. These features include a cabinet resolution and various departmental orders analogous to those passed in Maharashtra, the role of NABARD in disbursing funds for agreed proposals, and the role of NGOs in supporting community-based organizations and of WOTR in supporting both the commitments made by villagers and the fusion of local knowledge and technical norms in the net area planning approach. Officials from other states

(such as Gujurat and Andhra Pradesh) have come to observe the approach and are expected to take a franchise on it, allowing it to remain the intellectual property of the IGWDP.

Case 3: Changing the Organization and Management of Government Services to Scale up Participation

This case study reports on the evolution in the eastern lowlands of Bolivia of a model for strengthening links among research, extension, and farmers (Thiele, Wadsworth, and Velez 1998). Central to the model is the highly innovative Research-Extension Liaison Unit (RELU) explicitly charged with linking the state agricultural research center (Centro de Investigaci6n en Agricultura Tropical—CIAT) with a broad range of intermediate institutions, including NGOs, that carry out extension. The dominant function of the unit has evolved over time and with each change new mechanisms for linking with intermediate organizations have been tried and adapted. As a result, although liaison positions or units have been created elsewhere (Ekpere and Idowu 1989), nowhere else, as far as we are aware, has such a diverse yet coherent range of linkage mechanisms been developed.

Origins of the RELU

The Santa Cruz Department in lowland Bolivia is characterized by commercial farming, ranching, and smallholder subsectors. Population density in the smallholder areas is low so that many farmers can only be reached by long journeys across poor roads.

From the mid-1970s to mid-1980s, the government research organization supported a conventional public sector extension service. Plagued by poor communications, long distances, weak infrastructure, and high absenteeism, it had little impact. Subsequently, with advice and support from the then Overseas Development Administration (United Kingdom) and the World Bank, CIAT managers came to see the potential for building on existing links with intermediary organizations.

Links were identified with four main categories of intermediary organizations: NGOs, producers associations (which mainly represented large producers), development projects of the Regional Development Corporation, and suppliers of agricultural inputs. These organizations had a total of 129 full-time extension staff, far more than CIAT's extension offices had ever commanded, and some were already conducting adaptive trials (Thiele, Davies, and Farrington 1988).

An RELU was created in 1988 to link with intermediate organizations and facilitate extension work. The new model fitted well with the then-burgeoning literature on increasing farmer participation (Farrington and Martin 1987), on intermediate organizations pulling down research (Röling 1988), and on multiple centers of innovation (Biggs 1990). It was internally coherent and intellectually attractive but at the time completely untested. RELU staff faced the challenge of making the model work.

Changing Functions

Different core functions have predominated within the RELU without entirely displacing earlier ones. Each function led to the creation of a new set of linkage mechanisms. The dominant function changed partly as a result of external pressures (principally from donors) and partly as difficulties with the previous function became apparent.

When the RELU was created, its job was seen principally as that of transferring technology placed on the shelf by the existing CIAT commodity-based structure. RELU staff were based at the experimental station and spent part of their time working alongside CIAT researchers, whom they were supposed to backstop, to find out about research advances. They advised extensionists on technology available in CIAT, gave talks to farmers, and established demonstration plots.

Verification trials were begun in 1991 and soon became the principal activity for some RELU staff, with a consequent reduction in the importance of the bridging function. This move to adaptive research was due to a number of factors:

- Some of the material presented initially—for example, in agroforestry—was not ready for extension, and further testing was essential.
- Researchers did not always have technologies that were appropriate to farmers' needs.
- A culture of action militated against sifting through research reports to find information on technology.

However, this shift into adaptive research and the strengthening of links with intermediate organizations were accompanied by a weakening of links with CIAT researchers. RELU staff moved from their base in the experimental station to the city office and, in some cases, tested technologies obtained from outside Santa Cruz.

Initially, it had been supposed that the intermediate organizations were relatively competent at extension work. Subsequently, deficiencies were revealed, and a strong demand for training on the part of intermediate organizations was identified. In 1993, following United Kingdom–supported training to its own staff, the RELU began training extensionists. Modules were developed for training in communication skills, including the preparation of talks, extension materials, extension methodology, and participatory rural appraisal.

In 1995 CIAT was reorganized using a matrix management model to promote a stronger farming systems perspective. Farming systems teams, including both researchers and liaison RELU staff, were set up at the local experimental stations. Intermediate organizations with technical backup from the farming systems teams were encouraged to carry out their own verification trials. The RELU synthesized and diffused the results of this work.

RELU Staff

Three types of staff have worked within the RELU, although their roles and duties have shifted over time:

- Subject matter specialists are the largest group and are responsible for groups of commodities. They transform research findings into usable and accessible extension recommendations, transmit this information to intermediate organizations, and feed research back to CIAT researchers.
- Zonal coordinators were established in response to NGO pressure that RELU staff should work on an area and "systems" basis. They help to integrate the extension activities of all institutions working in the area and ensure that extensionists receive the technical information they need from CIAT. In theory, zonal coordinators draw on the skills of subject matter specialists and researchers, when required, but in practice some zonal coordinators have moved into validation, where subject matter specialists have not been able to provide support, or have replaced subject matter specialists on training. Similarly, some subject matter specialists have concentrated their activities in a limited area and become zonalized.
- Communicators, working closely with the two other types of staff, operate an information center for intermediary organizations, prepare and distribute publications targeted at extensionists, and play a lead role in training.

Linkage Mechanisms

Rapid rural appraisals have proved to be a highly effective type of linkage and have helped to raise the quality of interaction and coordination.

Zonal meetings are probably the most powerful of all the linkage mechanisms developed. They bring together extensionists based in the zone, researchers from local experimental stations, and RELU staff. They are organized by the zonal coordinator who refers any problems that cannot be resolved locally to subject matter specialists, commodity researchers, or other experts. During a typical meeting:

- A representative of each group explains the work proposed for the coming month to facilitate the coordination of activities.
- Field problems are discussed and solutions shared.
- A talk or demonstration is given to address problems identified by participants.

Recommendation workshops are organized each year in which researchers, RELU staff, and members of intermediate organizations discuss and update recommendations for a specific commodity, and these are published as a booklet. This process has proved particularly effective at increasing the interactions between commodity researchers and extensionists.

Verification trials are carried out jointly by subject matter specialists, intermediate organizations, and farmers with technical support from researchers. These tools play a useful role, despite inadequate prior research on some technologies, differing degrees of involvement by extensionists, and variable feedback to researchers.

Funds are allocated for collaborative activities with extension organizations. Within the agroforestry program, for example, small amounts of money have been made available to allow these organizations to establish small nurseries. Secondment of researchers into the RELU unit has been used occasionally to promote tighter links with the commodity programs.

A manual explaining the functions of RELU staff, researchers, and extensionists in intermediate organizations and procedures for technology transfer was published and has been widely used. Loose-leaf technical sheets on CIAT technologies organized in a distinctive technical loose-ring binder were produced and have been distributed to 27 intermediate organizations and 250 extensionists. Talks and courses on technologies for extensionists have been given. However, these have tended to respond to immediate needs and have not met the need for longer sustained training courses in more complex technologies. An information center for extensionists was set up. However, because only limited practical information is available, the center is full of research reports and other documents of more value to university students than to extensionists. A database of extensionists and intermediate organizations was established, but high turnover of extensionists has meant that this mechanism has never become fully operational.

Did the RELU Approach Work?

Quantitative evaluation is difficult. The RELU has been involved in so many technological areas that it is hard to assess its impact, there is no clear control for comparison, and it is too early in most cases to conclude much about adoption. But the examples in box 11.1, taken together with other information, allow certain tentative conclusions.

CIAT has made more information available to extension organizations and in some instances (silage), this has led to technological change. However, improved delivery of information about, and follow-up on, promising technologies (granulosis virus) is still needed, and quality needs to be improved. The demand for research has increased (agroforestry), but this has not occurred with all areas of research (silage) and is still not strong enough.

A number of factors contributed to successful implementation of the model.

Box 11.1. Participatory Approaches Scaled up through the RELU

Agroforestry research

Before the RELU was established, agroforestry research was mainly carried out on station with little farmer participation. The agroforestry subject matter specialist chose technologies developed by researchers that seemed most promising for small farmers in frontier areas who were diversifying into livestock and for whom sustaining pasture productivity and fencing were priorities. These included forage alleys, with rows of leguminous forage bushes and timber trees undersown with pasture, living fence posts, mainly for fencing pasture, and windbreaks.

Verification trials were set up to test these options, with the NGOs, subject matter specialist, and farmers working together. Trial design was a single unreplicated block compared with farmer practice.

There was a lot of interest on the part of intermediate organizations in collaborating with the RELU in these validation trials. As of 1995–96 around 60 farmers were participating. However, it later turned out that RELU staff had underestimated the complexity of a difficult technology: what was intended to be verification was in fact the first real testing. Nevertheless, testing of prototype technologies by the RELU cut out research cycles and pulled agroforestry research down to the farm level. As a result of the RELU's work, the agroforestry program itself was modified, and researchers assumed responsibility for testing some technology that had been in verification.

Silage

An important constraint facing small dairy farmers in Santa Cruz is the lack of grazing during the dry winter months. Silage is used widely in highland areas of Bolivia, but CIAT researchers did not regard it as appropriate for Santa Cruz. The subject matter specialist in pastures, without the support of researchers, carried out verification trials with intermediary organizations on the use of silage. Economic analysis showed the technology to be profitable, and farmers were very interested. The subject matter specialist helped local organizations to carry out demonstrations of the technology. By 1995–96, more than 100 dairy farmers had adopted silage use.

Granulosis virus to control soybean pest

In Brazil, a bioinsecticide based on granulosis virus to control the soybean looper (*Anticarsia gemmitalis*) is widely used. CIAT researchers tested this technology and showed it to be effective. Results were written up, but no measures were taken to diffuse the technology. The subject matter specialist produced technical sheets on the granulosis virus for extensionists and arranged for on-farm demonstrations with the producer association for soybeans. Initial results were encouraging, but the subject matter specialist, and several others, left to complete a master's degree with project financing. The unit became overstretched, and the promotion of granulosis virus stopped.

- The model was implemented following a process of consultation with intermediate users.
- CIAT's smallness (it serves only the Santa Cruz Department) allowed flexible implementation and modification of linkage mechanisms, where appropriate.
- CIAT's reputation as a research institution made it a credible partner for intermediate organizations, particularly where it had a "new" technology (such as agroforestry) to recommend.

The following factors limited the success of the model:

- RELU staff had little experience with on-farm testing, systems approaches, and farmer participation focus. Insufficient training was carried out in this area.
- The RELU tended to be more reactive than proactive. Mechanisms exist to prioritize and feed information back to the research agenda, but they are informal and not transparent.
- Operational costs were restricted during periods of financial insecurity.
- Staff had to be trained and the model developed while day-to-day activities went on. This reduced the time and resources available for testing and disseminating technology (see the granulosis virus case).

Conclusions

This chapter has argued that, especially under unreliable rainfed conditions where the private commercial sector is likely to remain weak, there continues to be an important role for public sector research and dissemination. However, in such areas, it has been difficult to maintain an effective public sector capacity on the ground, and research and extension have generally been only weakly oriented toward clients' needs.

Eschewing more radical solutions such as total abolition of public sector research capability and its replacement by electronic media, we argue that the clients of research must place a stronger demand on research if it is to be effective. Participatory approaches are easy enough to implement on a small scale, but specific institutional arrangements have to be made to permit wide-scale implementation. These, we argue, can best be based on partnerships between public and private (especially nongovernmental) organizations and farmers themselves, both to bring additional resources to bear and to help in reorienting the public sector.

The chapter has reviewed partnerships in three contexts: case study 1 reviewed partnerships in a pure, relatively unstructured form; case study 3 reviewed partnerships in a context in which restructuring of the public sector and of public-private relations evolved over time; and case study 2 attempted to capture the major productivity gains offered by microwatershed-based approaches. Microwatershed-based approaches are inherently more challenging for two reasons: first, because of the joint action needed to manage common pool resources and connect them with agriculture and, second, because of the need to integrate and sequence the inputs from several government departments.

These three studies inform policy regarding the institutional preconditions for sustainable scaling up of participatory approaches in several ways. First, it is difficult to reconcile high degrees of participation in decisionmaking on technology options by local organizations (especially NGOs working with low-resource farmers) with the government structure and performance norms designed to deliver services and inputs. Where, as in Udaipur (case 1), an inclusive, participatory, and open-ended strategy involving larger and well-established NGOs is pursued, it has proved difficult to move from talking shop to taking action. The history of relations among NGOs, their differing philosophies, lack of felt need to scale up, continuing (sometimes justified) doubt over what government organizations can offer and whether trying to turn them into client-oriented organizations is worth the effort all have meant that the introduction of practical improvements to technology at the farm level by improving the articulation of farmers' needs has been slow. Although isolated and productive coalitions of interests among certain NGOs and certain individuals or groups within government organizations have emerged, it would take (in addition to a more focused approach by NGOs) major reforms in the structure and performance of government organizations to improve service delivery radically at the farm level. NGOs are predisposed toward group-based approaches, even where agricultural innovation is usually individualistic. There is some justification for this, given the need to strengthen self-confidence among villagers if they are to have any prospect of influencing government services. Nevertheless, as relations with the public sector improve and incomes rise, much group activity is likely to become individualized.

Second, improved management of common pool resources will, however, continue to demand joint action. Although a well-managed connection between common pool and private agricultural resources in semiarid areas offers the prospect of substantial productivity gains and lower risk, it also imposes considerable demands on organizations such as NGOs aiming to support groups and help design joint action. Furthermore, as case study 2 indicates, watersheds cannot be rehabilitated in an unstructured fashion: villagers' views have to be sought and external technology options considered, selected, and built into bankable projects that set out physical and financial norms, the rights and responsibilities of insiders and outsiders, time schedules, sequences of activities, and so on. Microwatershed rehabilitation promises significant and potentially sustainable gains in the productivity of agriculture and natural resources. However, it also poses considerable challenges, both in the structuring and sequencing of interventions and in the requirements of local groups for support in joint action. Certainly, the demands are significantly greater than those made by participatory approaches in agriculture alone, whether individual or group based. In this setting, the project in case 2 decided to work with small, locally focused NGOs that had the necessary value-based philosophy and had (or were prepared to acquire) the necessary skills in management, leadership formation, conflict resolution, and basic technical matters but did not have the entrenched predispositions of some larger NGOs. An additional facet of this case was the scope and quality of arrangements for drawing the technical and financial support of the relevant government organizations into microwatershed rehabilitation and ensuring that such arrangements, where necessary with adaptations, could be transferred to other locations.

Third, case 3 records the substantial changes made in government organization and management in order to facilitate more participatory approaches in an area where public sector research and extension faced chronic obstacles. A new multiagency approach emerged in which research interacted with different kinds of "intermediate" organizations working with farmers on an area-specific basis. In many ways, this represents the type of public sector response that appeared to be desirable in case 1: large farmers already made vocal demands on research services; the intention was that NGOs should do the same for low-resource farmers and that the public sector should be organized and managed in ways designed to allow this. A range of organizational arrangements and linkage mechanisms were used over the several years of experience gained to date in this case.

Overall, there can be little doubt that cases 2 and 3 offer prospects for scaling up participatory approaches to technological change in agriculture and natural resource management. Quantifying the changes on the ground that are attributable to these new institutional arrangements is now of high priority. What is striking is the degree of sophistication of organizational and management arrangements (case 2) and the critical revision of arrangements (case 3) needed to respond to earlier experiences or meet new needs. However, the differing local circumstances, and the detailed local knowledge underpinning changes in organization and management arrangements, are such that it is the principles identified, rather than the actual experiences, that are transferable to other settings.

Note

1. In cases 1 and 2, "structured" implies building on previous experience to enhance efficiency by agreeing in advance to preconditions for successful collaboration, the respective roles of different organizations, shared objectives, and, where possible, intended concrete outputs.

References

The word "processed" describes informally reproduced works that may not be commonly available through libraries.

Alsop, Ruth, John Farrington, Elon Gilbert, and Rajiv Khandelwal. Forthcoming. "Coalitions of Interest: Partnerships for Processes of Agricultural Change." New Delhi. Processed.

Berdegue Julio. 1994. "El sistema privatizado de extensión agrícola en Chile: 17 años de experiencia." Paper presented to the international symposium for farming systems research and rural development, Montpellier, France, 21–25 November. Processed.

Biggs, S. 1990. "A Multiple Source of Innovation Model of Agricultural Research and Technology Promotion." *World Development* 18(11):1481–99.

Carney, Diana. Forthcoming. *Research and Farmers' Organizations: Prospects for Partnership*. London: Overseas Development Institute.

Carroll, T. 1992. *Intermediary NGOs: The Supporting Link in Grassroots Development.* West Hartford, Conn.: Kumarian Press.

Chambers, R., A. Pacey, and L. A. Thrupp, eds. 1989. *Farmer First: Farmer Innovation and Agricultural*

Research. London: Intermediate Technology Publications.

Edwards, M., and D. Hulme, eds. 1992. *Making a Difference: NGOs and Development in a Changing World*. London: Earthscan Publications.

Ekpere, J., and I. Idowu. 1989. *Managing the Links between Research and Technology Transfer: The Case of the Agricultural Extension-Research Liaison Service in Nigeria*. The Hague: International Service for National Agricultural Research.

Eponou, T. 1996. *Partners in Technology Generation and Transfer: Linkages between Research and Farmers' Organizations in Three Selected African Countries*. Research Report 9. The Hague: International Service for National Agricultural Research.

Evenson, R. E., and C. E. Pray. 1991. *Research and Productivity in Asian Agriculture*. Ithaca, N.Y.: Cornell University Press.

Eyzaguirre, P. 1996. *Agriculture and Environmental Research in Small Countries: Innovative Approaches to Strategic Planning*. Chichester, U.K.: John Wiley and Sons.

Farrington, John. 1997. "Farmers' Participation in Agricultural Research and Extension: Lessons from the Last Decade." *Biotechnology and Development Monitor*, 30 March, pp. 12–15.

Farrington, John, and A. J. Bebbington, with D. Lewis and K. Wellard. 1993. *Reluctant Partners? Nongovernmental Organizations, the State, and Sustainable Agricultural Development*. London: Routledge.

Farrington, John, and C. Lobo. 1997. *Scaling up Participatory Watershed Development in India: Lessons from the Indo-German Watershed Development Programme*. Natural Resource Perspectives 17. London: Overseas Development Institute.

Farrington, John, and A. Martin. 1987. *Farmer Participation in Agricultural Research: A Review of Concepts and Practices*. ODI Occasional Paper 9. London: Overseas Development Institute.

Farrington, John, C. Thirtle, and S. Henderson. 1997. "Methodologies for Monitoring and Evaluating Agricultural and Natural Resources Research." *Agricultural Systems* 55(2):273–300.

Freire, P. 1972. *Pedagogy of the Oppressed*. Harmondsworth, U.K.: Penguin.

Gisselquist, David, and Jitendra Srivastava. 1997. *Easing Barriers to Movement of Plant Varieties for Agricultural Development*. World Bank Discussion Paper 367. Washington, D.C.: World Bank.

Merrill-Sands, D., and David Kaimowitz. 1990. *The Technology Triangle: Linking Technology Transfer in Developing Countries*. London: Westview Press.

Röling, Niels. 1988. "Extension Science." In *Wye Studies in Agricultural and Rural Development*. Cambridge, U.K.: Cambridge University Press.

Thiele, Graham, P. Davies, and John Farrington. 1988. *Strength in Diversity: Innovation in Agricultural Technology Development in Eastern Bolivia*. ODI Network Paper 1. London: Overseas Development Institute.

Thiele, Graham, Jonathan Wadsworth, and Roy Velez. 1998. "Creating Linkages: Lessons from Agricultural Research and Extension Liaison in Lowland Bolivia." *European Journal of Agricultural Education and Extension* 4(4):213–24.

Zijp, W. 1996. "Report on a Workshop on Alternative Mechanisms for Funding and Delivering Extension Held at the World Bank, Washington, D.C., 18–19 June 1996." Reported in *Newsletter of the Agricultural Research and Extension Network* 34:8B9. London: Overseas Development Institute.

12

People-Centered Agricultural Development: Principles of Extension for Achieving Long-Term Impact

Roland Bunch

During the past 27 years, a series of principles for making agricultural extension work effective have been developed in Central America and spread to mostly nongovernmental organizations (NGOs) in some 22 nations around the world. In a number of countries, these principles have enabled programs to triple families' basic grain yields at a total program cost of less than $400 per family. Furthermore, many of these families have continued to increase their yields after the program intervention: a recent in-depth study showed that 15 years after program phase-out, yields have once again doubled (Bunch and López 1994).

This chapter describes the inner synergism of people-centered agricultural development, as well as its relationship to the situation of resource-poor farmers in developing nations, the sustainability of agricultural development, and the empowerment of resource-poor people. The chapter also briefly describes the history and spread of the principles of people-centered agricultural development as defined in Bunch (1982), their proven effectiveness (including the results of major evaluations), modifications made in their application, cases in which they have been scaled up, and lessons learned.

The Principles of People-Centered Agricultural Development

Although sometimes called a methodology, people-centered agricultural development is much more flexible than most methodologies. Different organizations in different countries have modified considerably the specifics of the approach, without reducing substantially the effectiveness of the basic ingredients. Therefore, we generally speak of people-centered agricultural development as a series of principles. As long as these basic principles are employed, there is room for a large amount of adaptation to local circumstances, farmer needs, and institutional imperatives, without significantly decreasing the effectiveness of the overall approach.

The basic principles are as follows:

- Motivate and teach farmers to experiment with new technologies on a small scale, thereby reducing the risk of adoption and providing a means for them to continue to develop, adopt, and adapt new technologies in a permanent scientific process of innovation. This principle is frequently referred to as "participatory technology development."
- Use rapid, recognizable success in these experiments to motivate farmers to innovate rather than artificial incentives or subsidies.
- Use technologies that rely primarily on inexpensive, locally available resources.
- Begin the process with a very limited number of technologies, so that the program is focused, achieves the maximum possible success right from the start, and allows even the poorest farmers to become involved in the process.
- Train village leaders as extensionists and support

them as they teach additional farmers, thereby creating and nurturing a community-based multiplier effect. This principle is called "farmer-to-farmer extension" in many nations of Asia.

A good number of organizations already use one or more of these principles. Numerous farmer-to-farmer extension programs in Southeast Asia train villagers as extensionists, while a growing movement in South America uses participatory technology development, a name that emphasizes the development of technology by villager farmers through small-scale experimentation. Nevertheless, long experience in a diversity of cultures has shown that the five principles, when used together, reinforce one another. That is, a synergy exists between the various principles. Training villager extensionists, for instance, becomes much more complicated, and recognizable success much less common, when a program begins with multiple technologies. It also becomes virtually impossible for farmers to experiment when the number of technologies is unlimited, especially when several of them are quite expensive. Thus the principles achieve considerably more impact when they are applied as a group (see figure 12.1).

People-Centered Agricultural Development and the Situation of Resource-Poor Farmers

All the major systems of agricultural extension used in developing nations were transplanted from the industrial countries. These systems were not used in industrial nations when they were going through their own processes of agricultural development but rather evolved well after they were economically developed. Therefore, in many ways these systems are not appropriate to the situation of resource-poor farmers in the developing world.

In contrast, people-centered agricultural development principles grew out of the experience of a group of NGOs in Central America. They take into account the social, economic, and ecological conditions that exist in developing nations.

The appropriateness of people-centered agricultural development for developing nations is further evidenced by the fact that these principles have been discovered independently by a whole series of people and organizations around the world, from Polan Lacki (1993) in Chile to PATECORE in Burkina Faso (Atampugre 1993) and the Food and Agriculture

Figure 12.1. The Synergy between the Principles of People-Centered Agricultural Development

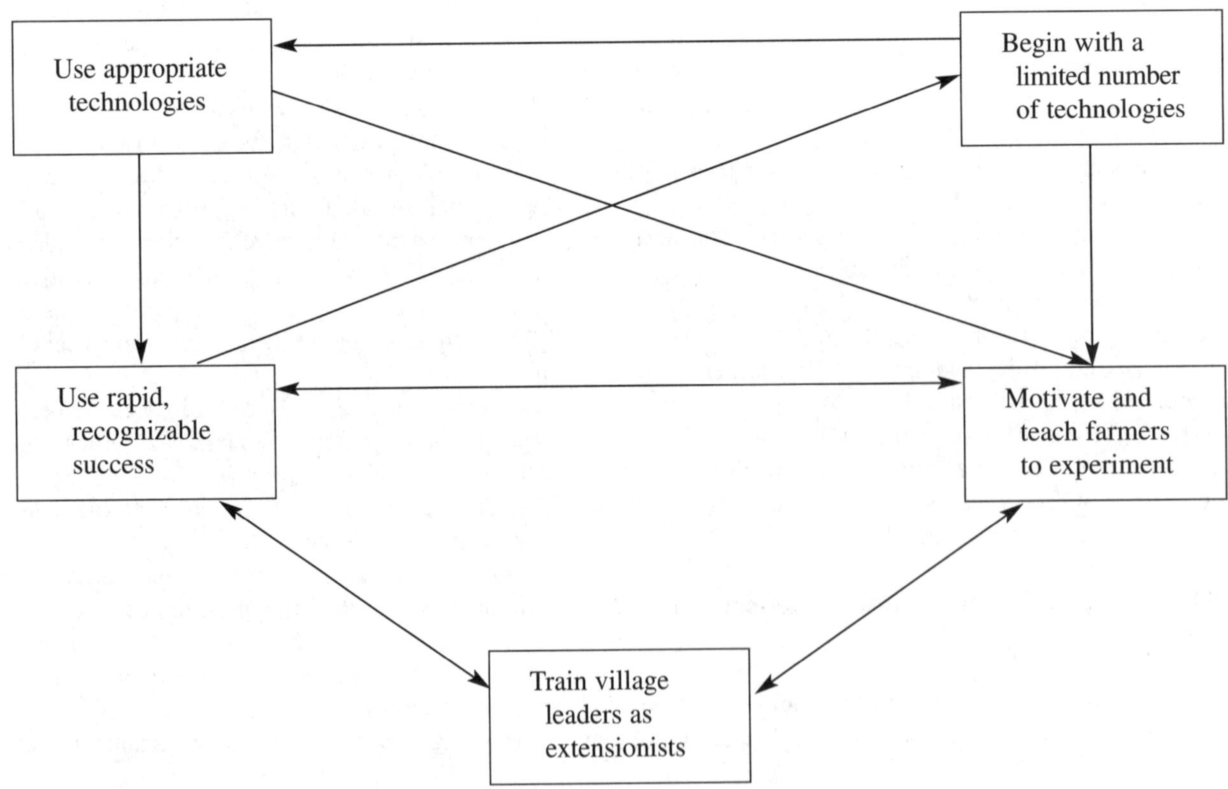

Organization's farmer field schools in Southeast Asia.

Figure 12.2 illustrates this relationship between the principles of people-centered agricultural development and the situation of resource-poor farmers. On the left is the list of principles and on the right are the characteristics most often found in small rural communities of the developing world. In such situations, we find not only obstacles and problems that a system of agricultural development must overcome, such as lack of formal education and economic infrastructure, but also very important resources that a system of agricultural development must take advantage of if it is to be as effective and as efficient as possible. These include an abundance of well-motivated human resources and an ecology that in many cases is highly favorable to very high rates of agricultural productivity.

The arrows indicate causal relationships. For example, the first principle—to motivate and teach farmers to experiment—has a causal relationship with four characteristics of rural areas in developing nations. In the first case, people who have very little or no formal education are not accustomed to learning through the written word or by listening to presentations. They learn much more efficiently through their own experience. Thus, when they themselves experiment with new technologies, they learn in the best way they can.

In the case of the second arrow, villager experimentation has a special value because many of the technologies "transferred" to developing nations come from industrial nations. Thus it is extremely important for villager farmers not only to validate these technologies under their own particular conditions but also to experiment with how they could be modified, where necessary, to become more appropriate to the different circumstances in which they are being applied.

In the third case, being able to validate and modify technologies for their own use, as well as to create entirely new technologies, makes the farmer much less dependent on outside sources of information. This empowers the farmer. And in situations where infrastructure and information services are lacking, the ability to adapt technologies and invent new ones becomes extremely important.

The last of these causal relationships is perhaps the most important of the four. If the world is to feed its rapidly growing population without destroying the

Figure 12.2. The Relationship between People-Centered Development and the Situation of Resource-Poor Farmers in the Developing World

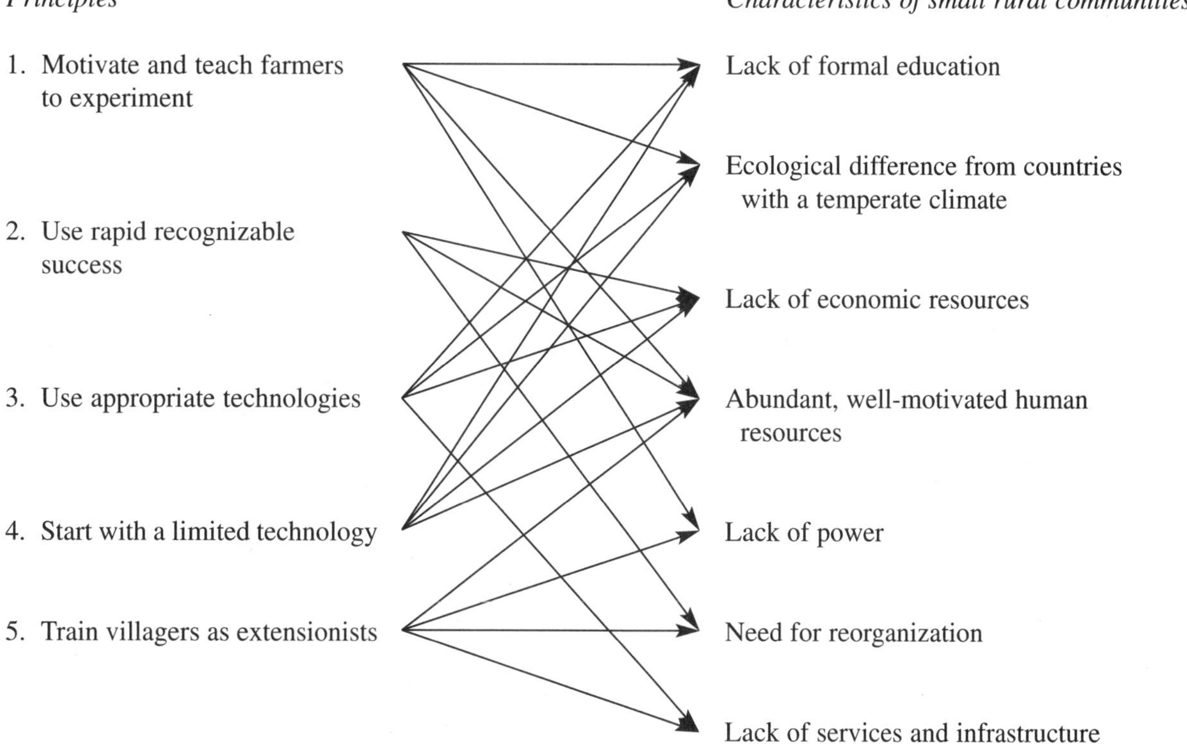

planet's environment, we have to find a large number of agroecological technologies that are appropriate to the social, economic, and ecological circumstances of developing nations. If we fail to involve in this monumental effort the millions of well-motivated villager leaders in the hundreds of thousands of villages in the developing world, the threats of poverty and famine may well become acute.

The Sustainability of Agricultural Productivity

In many developing countries extension services are overstaffed and underpaid. They lack means of transport and often lack cutting-edge messages. Even if they were clearly ahead of innovative farmers, extensionists are seldom working with more than one out of every four farmers, and even then farm visits may be so infrequent as to be largely ineffective (Saravia 1983). The result of this lack of resources is that governments often target areas of high agricultural potential, with the result that problems of poverty and inequality become even more severe (information presented at an International Food Policy Research Institute workshop on land degradation in the developing world, Annapolis, Md., 4–6 April 1995).

A way of meeting this challenge is for governments to enlist farmers in solving their problems of low productivity. Widespread experience on three continents shows that farmers can fulfill three major roles (and many minor ones) in furthering their own agricultural development:

- Establish and manage experiments in order to modify those technologies already known and develop new ones
- Spread knowledge of useful technologies from one farmer to another
- Carry on, by themselves if necessary, the processes of agricultural investigation and extension, once they have learned them, thereby continuing to increase their yields.

If governments are to feed their growing populations, at a cost within reason, they must involve farmers in these three activities.

The people-centered agricultural development process teaches villagers how to experiment and teach one another through the most efficient teaching process we know: learning by doing. Repeatedly, year after year, farmers manage experiments. Some of them also become extensionists, teaching farmers in other villages what they themselves have learned. And the success of the innovations they learn through the people-centered agricultural development process motivates them to continue experimenting and to share what they know.

The third activity, that of villagers' sustaining the growth in agricultural productivity wholly or largely by themselves, is a much more complicated issue. First we must ask ourselves what factors must exist in a community for the farmers to, at the very least, maintain high levels of productivity. Personnel from COSECHA (Asociación de Consejeros para una Agricultura Sostenible, Ecológica y Humana) have asked this question of groups of agronomists in some two dozen nations. The resulting list is virtually the same anywhere the question is asked and almost always includes the following:

- The motivation to continue the development process.
- Self-confidence and a respect for their own knowledge and culture. People who are convinced they are ignorant or incompetent all too often become incompetent (Schumacher 1973: 192).
- The ability to organize and manage experiments. New pests attack crops, seeds degenerate, input prices rise, old markets dry up, and new ones appear. The only way farmers can maintain productivity and profitability in a modern, rapidly changing environment is constantly to be trying out new technologies.
- Medium- to long-term use rights over a certain minimum of natural resources that are in satisfactory condition. Without a minimum of certain resources—land and water—no one can produce enough food to live well.
- Access to or ownership of adequate financial resources. This need not be very much. Most current loan programs handle much more money than small farmers really need. But farmers do need at least some extra capital to risk in their experimentation and invest in improvements. Most of this capital will usually result from their own higher productivity.
- A certain basic knowledge of biological and agronomic processes. This knowledge is necessary in order to understand experimental results and decide what possibilities of improvement will be most promising for future experimentation.
- A diversified agriculture. Knowledge of a series of crops, animals, and trees reduces risk and provides a basis for future innovations.
- The ability and motivation to share information about agricultural technologies with other farmers.

No one farmer can experiment enough to continue improving his or her productivity. The only way whole villages or areas can solve their problems and move ahead is for each farmer to be learning from the experiments of dozens of other farmers.

- Organization-building capacity. With constant innovation, new needs and new opportunities will present themselves. These will often best be seized or solved not through some preexisting structure, but by new organizations, either permanent or temporary, that people will create if and when they are needed.

Catalysts in this process will be:

- Contacts with outside sources of information and support
- Administrative capabilities, meaning the ability to plan strategically, to handle money and accounting procedures, and to manage group dynamics
- Minimal rural infrastructure and access to markets
- A high rate of literacy among farmers. Although the process has worked in areas where functional adult literacy is as low as 20 percent, higher literacy rates make the process more efficient

The list appears to be long and difficult to achieve. However, in a people-centered program, very few of these attributes require any special effort; *most of them are reinforced each growing season by the principles used to design the program.* Figure 12.3 compares the principles of people-centered agricultural development with the factors of sustainability. Each arrow indicates a causal relationship between competent use of the principle and the strengthening of one of the factors needed for sustainability.

To take the first principle, for example, by experimenting, farmers gain the ability to manage experiments through the time-honored method of learning by doing. When farmers experiment, they gain a good deal of basic agricultural knowledge. When farmers know how to experiment, and are motivated to do so, they gain the ability to continue to diversify their agriculture. And when they are capable of constantly acquiring information in this manner, they will, on a sustainable basis, have something valuable to share with others.

In conclusion, people-centered agricultural development principles go a long way toward strengthen-

Figure 12.3. The Relationship between People-Centered Agricultural Development and the Factors That Make the Agricultural Development Process Sustainable

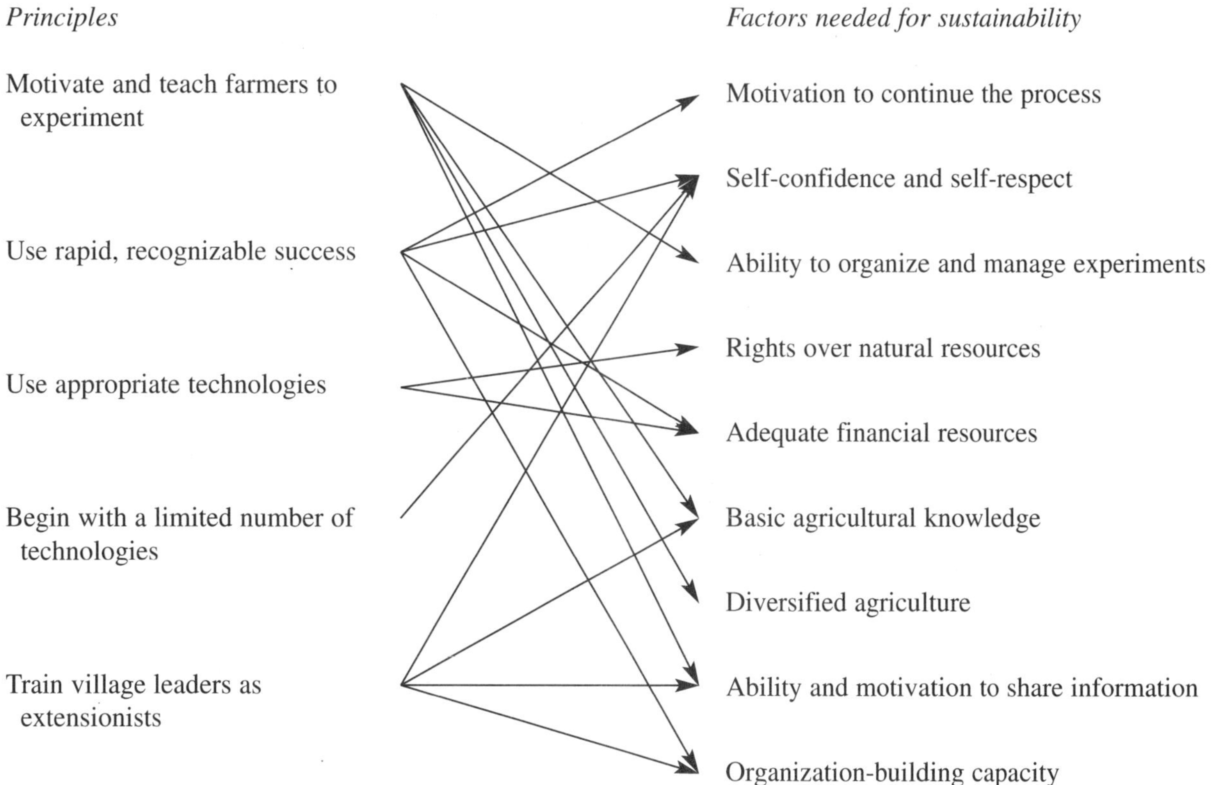

ing precisely those factors that can make the development process self-sustaining at the village level. Thus, it is the extension methodology—*how* we teach—that helps a program attain sustainability, much more than the technology—*what* we teach.

People-Centered Agricultural Development, Sustainability, and Empowerment

It is also worthwhile to compare the factors needed for sustainability with the factors identified by social scientists as sources of power in any society (figure 12.4). One can see immediately that virtually all the sources of power are inherent to one or several of the factors of sustainability. That is, empowerment is inherent to any process that permits villagers to sustain the process of agricultural development. If we wish agricultural development to be an ongoing process—to be sustainable—it must include empowerment. Unempowered farmers are simply incapable of carrying on the process.

But villagers do not merely gain some sort of agricultural self-confidence. They gain self-confidence in general, which gives them more competence in many areas of life. Villagers' increased economic resources do not have to be used in agricultural activities; they can be used in any walk of life. Nor will their organizational skills be used to build agricultural organizations exclusively; they can be used to establish organizations to further any ends the people choose.

Thus, while the process creates a sustainable agricultural development process, it also, by its very nature, empowers people in ways useful for all sorts of development efforts. That is, we are not talking of an effort that only makes agricultural development sustainable. Rather, we are talking about a process that provides people with a series of skills and attitudes that will enable them to improve their own well-being in many aspects of life, from organizational management to health improvement, and from employment-creation activities to home betterment.

In a way, the people-centered agricultural development process uses agricultural improvement, a very widely present "felt need," to get people involved in a process that can start them on the road toward, and provide them with many of the skills and attitudes necessary for them to continue, an overall process of integrated rural development.

Figure 12.4. The Relationship between the Factors of Sustainability of the Agricultural Development Process and Villager Empowerment

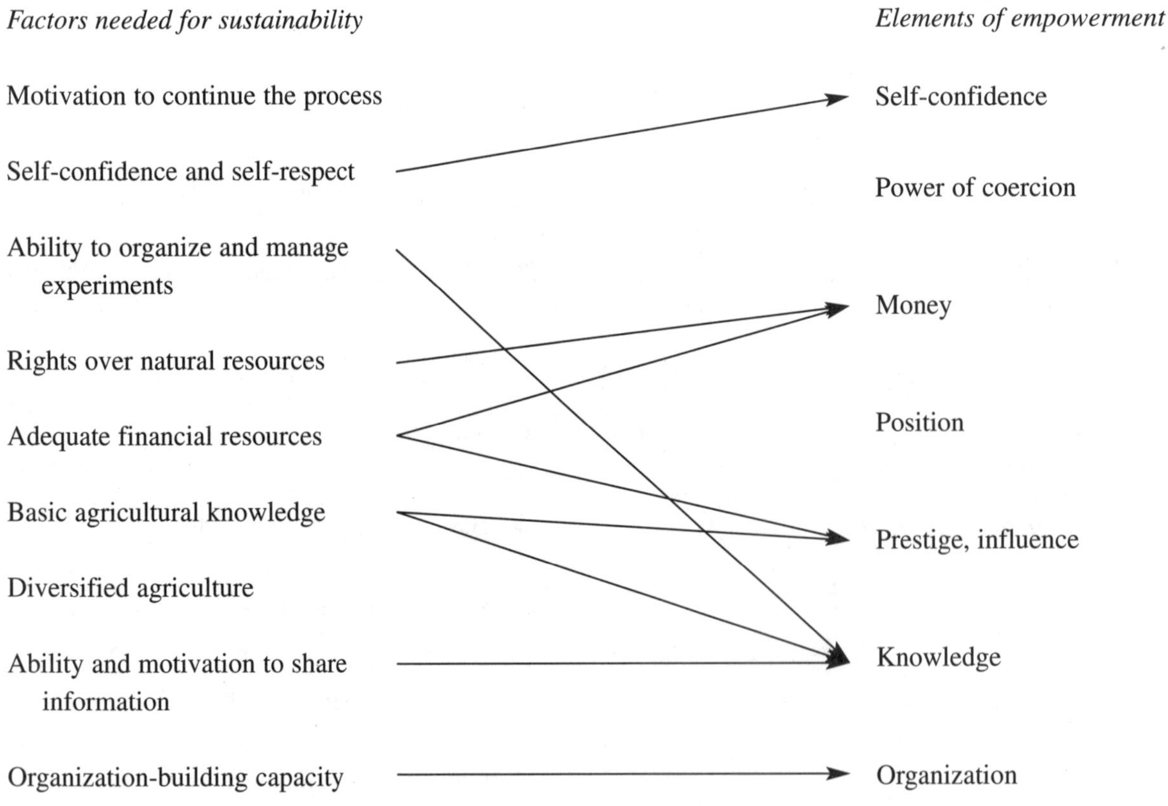

People-Centered Agricultural Development History and Present Dissemination

Most of the principles now included in people-centered agricultural development were being tried out by one or another institution in the highlands of Guatemala during the 1960s. In 1972, World Neighbors, with OXFAM /U.K.'s support, established the San Martín Jilotepeque Integrated Development Program. This was the first program to employ all the principles of people-centered agricultural development in a single program. The results, partially described here, greatly exceeded expectations. External evaluations confirmed the success, in agricultural productivity, health improvement, organizational quality, and so forth. Development Alternatives, Inc., after studying 41 exemplary programs in seven nations around the world, gave its number one rating to the San Martín Program.

Impressed by the results of this and a later sister program in San José Poaquil, Guatemala, World Neighbors decided to apply people-centered development principles to programs in Bolivia, Haiti, Honduras, Indonesia, Kenya, Mexico, Peru, the Philippines, and Togo. In every case except Kenya, these programs produced such significant results that they were heavily visited and considered worthy of emulation.

Little by little, a series of other institutions have adopted people-centered agricultural development principles for their own programs. Organizations as diverse as the World Wildlife Fund, Trees for People/Germany, the Mennonite Central Committee, World Vision, Catholic Relief Services, the Aga Khan Foundation, Cooperative for American Relief Everywhere (CARE), Campesino a Campesino (Farmer to Farmer), German Technical Cooperation (GTZ), Peace Corps, and Rodale International have applied the principles to programs they manage or fund in one or more nations around the world.

The Effectiveness of People-Centered Agricultural Development Principles

The people-centered agricultural development approach not only has proven itself capable of achieving high rates of farmer adoption and increased productivity but, even more important, has done so at much less cost than other extension systems. Inasmuch as farmers become a major factor in the spread of both agricultural technologies and the development process, costs are considerably reduced. Instead of having to work with each individual group of farmers, the more expensive professional personnel become managers of farmer extensionists who do most of the teaching. Therefore, each professional agronomist trains, supports, and backs up a group of 6 to 10 farmer extensionists. In this manner, each agronomist can reach, intensively but indirectly, some 150 to 250 farmers. Even though the farmer extensionists are paid a small stipend (after the first year or two), the reduction in costs per farmer reached is dramatic.

World Neighbors and COSECHA in Central America have a standard of being able sustainably to triple the productivity of farmers' basic grain production (from levels of between 400 and 800 kilograms per hectare) for a total program cost of less than $400 (in 1996 dollars) per farm family, including administration, salaries, transportation, and so forth. That is, if a program costs $50,000 each year for eight years, for a total of $400,000, then about 1,000 farmers should have sustainably tripled their productivity of basic grains.

This $400 per family cost includes leadership training, organizational strengthening, training in the basic principles of biology and agriculture, and, in most cases, agricultural diversification. Some of World Neighbors' programs have not attained this goal. Their first programs in Mexico (in Oaxaca and Tlaxcala states) were more costly. However, the San Martín Program in Guatemala was a good deal more efficient than this, and the Guinope Program in Honduras cost almost exactly what the goal would indicate. The World Neighbors' Cantarranas Program cost about $450 per family and tripled productivity, while a program in Honduras, run by students from a college in Oregon, United States, spent less than $300 per family and tripled productivity.

In any case, these figures are a good deal lower than those of virtually any other methodology. Many program proposals describe efforts that cost several thousand dollars per family benefited, while some cost well over $10,000 per family (25 times less efficient than the $400 goal). And most of the programs resulting from these proposals never reach all the people they hoped to, nor do many of those people ever manage to triple their productivity sustainably. A fair number of agricultural development programs spend more than the $400 per family on artificial incentives alone.

Nature and Sustainability of the Impact

A key question is how much of the increases in productivity are sustained after the program's termination. Are yields maintained after the outside intervention ends? A recent study of three of the earlier programs (in San Martín Jilotepeque, Guatemala, and in Guinope and Cantarranas, Honduras) showed that yields not only maintained themselves, but continued, in general, to increase significantly (Bunch and López 1994). In San Martín, for instance, average maize yields in the four villages studied were 400 kilograms per hectare before the program (1972) and approximately 2,000 kilograms per hectare at program termination (1979). But 15 years later (1994), with virtually no additional intervention in agricultural development by any institution, maize yields averaged close to 4,500 kilograms per hectare. That is, yields once again doubled. In the same 15 years, bean yields (the second most important staple food in the area) increased 75 percent. The villagers themselves had successfully carried on the process of agricultural development.

The picture was not, however, uniform. In the 25 to 30 percent of villages where the response was best, yields continued to increase after program termination. In the 40 to 50 percent of the average villages, yields more or less maintained or increased only marginally, whereas in the 20 to 35 percent of the villages where impact was poorest, yields decreased, although they continued to be better than the yields at program initiation.

Other signs of innovation were roughly proportional. In the villages of poor impact, there was virtually no ongoing innovation, whereas in the best villages, literally hundreds of experiments were made after program termination, and often more than 10 successful innovations per community were being used.

In San Martín, whole new systems of agricultural production appeared. One of the four villages studied developed an intensive cattle-raising system in which they were raising some seven head per hectare. Another village became a major producer of fruit, while two others developed an intensive coffee-producing system. And one of the villages even developed a small-scale (less than 1 hectare per family) sustainable forest management system, earning good incomes by planting trees in open spaces (without using a nursery) and cutting (on average) the number of trees they could harvest on a sustainable basis each year.

Other impacts were also studied. In the high-impact communities, the number of organizations increased, land prices increased dramatically, incomes increased, and outmigration either diminished drastically (in San Martín) or was reversed. In the Guinope program villages, net inmigration occurred between 1990 and 1994, in spite of heavy net outmigration during the 1970s—a phenomenon without precedent in Honduras, as far as we know.

One of the major conclusions of the study was that sustainability does not reside in technologies. Whereas the technologies taught in these programs had a half-life of perhaps five years (after five years, half the technologies had been abandoned), the agricultural development process and increases in yields continued. Thus sustainability is not to be found in the correct choice of technologies. Markets change, input prices increase, new technological opportunities appear, varieties degenerate, pests spread, and competition becomes stiffer. Most technologies disappear, sooner or later, in this fast-changing modern world. The hope for sustainability of agricultural development is not in the nature of the technology, but in the nature of the process. The study confirms that the social process—one of widespread experimentation and sharing of information, innovation, and group problem-solving—is sustainable, year after year. Although formal studies have not yet been done, preliminary empirical evidence indicates that similar phenomena have occurred in many other programs using people-centered agricultural development principles around the world.

Obstacles to Achieving Such Impact

People-centered agricultural development principles are not capable of overcoming all obstacles. They do require well-motivated extension personnel and a good understanding of how the principles can be applied in each particular situation. Although some World Wildlife Fund–supported programs in Mesoamerica have been highly successful, such as the Defensores de la Naturaleza Program in Guatemala and the Calakmul and Línea Biósfera programs in southeastern Mexico, other programs using people-centered principles, because of a lack of motivation on the part of personnel or extremely poor, top-down administration, have failed.

We are often asked under what farming conditions the people-centered agricultural development principles would *not* work. We suspect they would not work with mechanized, highly capitalized farmers, because these farmers would probably not be willing to teach one another. They also might not work in areas for which no known group of two or three technologies would bring significant increases in yields, such as the semiarid areas of the African Sudan and the higher reaches of the Peruvian Andes.

The people-centered agricultural development process is also very difficult, or even impossible, in an institutional framework that discourages decision-making at a level close to the village extensionist (Pretty and Chambers 1994). Both the principles and the technologies used must be adapted according to differences in farming systems, cultures, land tenure, environmental factors, accessibility of villages and markets, and so forth. Institutions that fail to provide the space for such adaptations will rarely be successful in using the people-centered agricultural development approach.

Successful application of the methodology is also made much more difficult where previous development programs were highly paternalistic, giving away all sorts of inputs, food, and other artificial incentives. Nevertheless, even in these sorts of situations, significant successes have been achieved.

Important Modifications in the Application of People-Centered Agricultural Development Principles

As people-centered development principles have been applied around the world, they have been modified considerably. This modification, by and large, has been a very healthy one, not only allowing the process to take into account local variations and differing cultural or economic factors but also helping to understand the limits to modifying their application without reducing their effectiveness.

What are the major modifications and what have we learned from these experiences? First of all, we have learned that limiting the initial number of innovations taught is easily the most important of the principles, while at the same time being the one about which program personnel, at first, are usually the least enthusiastic. In fact, there seems to be a remarkable inverse correlation between a program's cost-efficiency (measured in the number of families whose productivity in basic grains can be tripled for each $10,000 spent) and the number of innovations introduced in a village in the first year of the program. Even programs that did not consciously use people-centered agricultural development principles, but whose early innovations were limited by other factors (for example, in the PATECORE Program in Burkina Faso, technological offerings were limited by the severity of the environment and the paucity of innovations known to be successful there), have shown remarkable levels of impact.

A second major modification has been that of the Campesino a Campesino Program in Nicaragua. In this case, taking advantage of the villager dynamism that coursed through the country in the 1980s, the program initiated a process of villagers teaching villagers that spread a small number of beneficial agricultural innovations across a much wider area much more rapidly than any other program using people-centered agricultural development principles. They did this almost totally with volunteer villager labor, with an absolute minimum of organizational and financial support, and with the involvement of a minimum of professionals. Thus a program working in some 35 *municipios*—probably in excess of 300 villages—did not have, until recently, more than four professional agronomists on its entire agricultural extension staff nationwide.

Some observers claim that this incredibly rapid expansion with apparent far-reaching success has resulted in low percentages of villagers actually innovating within most villages. Other observers feel that such rapid expansion with minimal supervision could only occur in a nation in which major change has occurred and villagers are fairly well-organized and highly motivated, as is the case in Nicaragua presently. Unfortunately, we in COSECHA have not been able to follow the progress of Campesino a Campesino closely enough to know whether these observations are justified or not.

Some organizations have tried using criteria for selecting villager extensionists. These criteria often include having a grade-school education, having established leadership or prestige in the community, and having practiced the suggested innovations successfully on their own farm. The first of these criteria has proven counterproductive in most cases, and the second is largely irrelevant. The third criterion, however, has proven to be of tremendous value, because anyone who produces three times what his or her neighbors produce will have all the credibility he or she needs.

A fourth innovation has been that of trying to arrive at scientifically valid conclusions at a 1 or 5 percent level of significance from the farmers' numerous and varied experiments. One methodology, presently being tried in World Neighbors' South American programs, has been to convince each farmer to repeat each experiment several times (Beingolea Ochoa n.d.). Statistical analysis, using either hand calculations or modified stability analysis, has produced statistically valid results (Hildebrand 1993). However, some observers suspect that farmers will not continue to repeat their experiments, because they have little personal motivation for doing so.

Another approach has been to use the data from a large number of farmers who are doing a similar experiment without repetitions on any one farm. Again, the calculations can be done by hand or computer, and results have been found to be more reliable and predictive than those from experiment stations (Rzewnicki and others 1988). This procedure could be extremely important as more and more farmers begin to experiment. It will help farmers and scientists to enter into a mutually profitable dialogue and will make it possible for the tremendous amounts of valuable information that farmers produce to enter into the more formal repositories of agricultural knowledge.

An additional variant is to incorporate people-centered agricultural development principles into other kinds of rural development programs. Many of these principles have been used in health programs, family planning programs, vocational education programs, and even a participatory post-earthquake housing program. Indications are that, where applicable, they have increased the effectiveness of these programs, also.

The Experience of Large-Scale People-Centered Agricultural Development Programs

In two cases, people-centered agricultural development principles have been tried out on a large scale. The excellent work in Nicaragua of Campesino a Campesino, assisted by the collaboration of a good number of NGOs, whose collaboration Campesino a Campesino encourages, has already been mentioned.

The second case is that of EPAGRI and the Souza Tobacco Company, which collaborate with each other in the State of Santa Catarina in southern Brazil. Well over 100,000 farmers have made major changes in their agricultural practices (including use of cover crops, zero tillage, and improved pig raising) that have at least tripled yields. Many farmers are now harvesting maize at a rate of 7 tons per hectare and more.

In this case, EPAGRI uses people-centered agricultural development principles, except that it has only recently begun using farmer extensionists, and on an experimental basis. In contrast, the Souza Tobacco Company—which limits its work to tobacco production, including the introduction of cover crops—employs a large corps of farmer extensionists. Both programs, although very large, have had excellent levels of impact. Of the large-scale programs working with small- to medium-scale farmers, EPAGRI's is perhaps the most successful in Latin America.

Perhaps the major innovation that EPAGRI has used to make people-centered agricultural development principles work in a large-scale program is to organize agronomists into regional teams. This procedure has been used precisely to reduce the verticality or bureaucratic nature of the organization. Although there is a team leader, personnel feel a sense of teamwork, of everyone working together to achieve a common goal. This allows closer personal ties among extensionists and administrators, more openness to discuss issues openly and learn from one another, more freedom for decisionmaking at the village level, more upward flow of information from the field, and less organizational distance between farmers and statewide administrators. Each one of these factors tends to increase the effectiveness of large-scale agricultural extension programs (Pretty 1995).

Lessons Learned

Through the years, and having visited more than 250 agricultural development projects around the world, we in COSECHA have learned the following lessons:

- The people-centered agricultural development approach can be very effective in a wide number of cultures, environments, and political settings.
- Limiting the initial number of technological innovations to one or two simple technologies that will provide a dramatic impact and get the social process moving is probably the most important principle.
- Sustainability of the process of agricultural improvement is not to be found in the technologies introduced, but in the social process of active, farmer-run innovation and dissemination of ideas.
- The achievement of this sustainable social process requires a series of conditions that include villager empowerment as a necessary ingredient.
- The people-centered agricultural development process requires well-motivated personnel, a good deal of decisionmaking freedom at the grassroots, and an ample understanding of the principles and how they can and should be applied in each partic-

ular situation.

- The self-confidence, organizational abilities, higher incomes, and overall empowerment achieved are applicable to all participatory development efforts. This style of agricultural development work can help to stimulate a much broader process of integrated rural development.

References

The word "processed" describes informally reproduced works that may not be commonly available through libraries.

Atampugre, Nicholas. 1993. *Behind the Lines of Stone: The Social Impact of a Soil and Water Conservation Project in the Sahel.* Oxford: OXFAM.

Beingolea Ochoa, Julio. n.d. "Planning and Conducting Experimental Field Trials in Peasant Communities." Pamphlet published by Office for the Andean Area, World Neighbors, Santiago, Chile. Processed.

Bunch, Roland. 1982. *Two Ears of Corn: A Guide to People-Centered Agricultural Improvement.* Oklahoma City: World Neighbors.

Bunch, Roland, and Gabino López. 1994. "Soil Recuperation in Central America, Measuring the Impact Three to Forty Years after Intervention." Paper presented at the international policy workshop of the International Institute for Environment and Development, Bangalore, India, 28 November–2 December. Processed.

Hildebrand, Peter E. 1993. "Steps in the Analysis and Interpretation of On-Farm Research-Extension Data Based on Modified Stability Analysis: A Training Guide." Staff Paper Series. Food and Resource Economics Department, University of Florida, Gainesville. Processed.

Lacki, Polan. 1993. *Development of the Small Farm: From Dependency to Self-Reliance.* 2d ed. Santiago: Food and Agriculture Organization of the United Nations.

Pretty, Jules N. 1995. *Regenerating Agriculture: Policies and Practice for Sustainability and Self-Reliance.* London: Earthscan Publications.

Pretty, Jules N., and Robert Chambers. 1994. "Towards a Learning Paradigm, New Professionalism and Institutions for a Sustainable Agriculture." In Ian Scoones and John Thompson, eds., *Beyond Farmer First: Rural People's Knowledge, Agricultural Research, and Extension Practice.* London: International Institute for Environment and Development.

Rzewnicki, Phil E., and others. 1988. "On-Farm Experiment Designs and Implications for Locating Research Sites." *American Journal of Alternative Agriculture* 3(4):168–73.

Saravia, Antonio. 1983. *Un enfoque de sistemas para el desarrollo agrícola.* San José: Instituto Interamericano de Cooperación para la Agricultura.

Schumacher, E. F. 1973. *Small Is Beautiful: Economics as If People Mattered.* New York. Harper and Row.

13

Making Market-Assisted Land Reform Work: Initial Experience from Colombia, Brazil, and South Africa

Klaus Deininger

Theoretical reasons and empirical evidence suggest that land reform may provide benefits in terms of both equity and efficiency. First, a large body of research has demonstrated the existence of a robustly negative relationship between farm size and productivity due to the supervision costs associated with employing hired labor. This implies that redistributing land from large to small farms can increase productivity (see Binswanger, Deininger, and Feder 1995 for references). Second, in many situations, landownership is associated with improved access to credit markets, providing benefits as an insurance substitute to smooth consumption intertemporally. By enabling the poor to undertake indivisible productive investments (or preventing them from depleting their base of assets), this could lead to higher aggregate growth (see Bardhan, Bowles, and Gintis forthcoming for references). Finally, even in aggregate cross-country regression, the distribution of productive assets—more than the distribution of income—seems to have an impact on aggregate growth (Birdsall and Londoño 1997 and Deininger and Squire forthcoming).

Notwithstanding this apparent potential, actual experience with land reform has been disappointing in all but a few exceptional cases such as Korea, Japan, Taiwan (China), and to some extent Kenya. Despite—or because of—this, land reform remains a hotly debated issue in a number of countries (Brazil, Colombia, El Salvador, Guatemala, South Africa, Zimbabwe) some of which are spending considerable amounts of resources for this purpose. A mechanism for land reform that allows small farmers to realize their productive potential at a cost that is comparable to other types of government interventions (education) would be very desirable.

This chapter describes a new type of negotiated land reform in which land transfers are based on voluntary negotiation and agreement between buyers and sellers and in which the government's role is restricted to providing a land purchase grant to eligible beneficiaries. The main focus is on Colombia, where legislation mandating negotiated land reform was passed in 1994. The chapter describes the difficulties encountered in implementing the new model, reviews solutions that have emerged, and, by comparing the initial experience to that of Brazil and South Africa—two other countries that recently began implementing a negotiated approach to land reform—attempts to draw some lessons that might be of use to other countries where the distribution of land is very unequal and agrarian reform is rapidly moving up on the political agenda.

The chapter is structured as follows. We begin by discussing some of the reasons underlying the change of the government's land reform policy and then elaborate on the principles of market-assisted land reform and some of the difficulties experienced in making it operational. We then discuss the mechanisms developed in five pilot *municipios* and, by comparing them with prevailing practices, describe some of the details associated with implementation as well as farm models that have been developed in these *municipios*. This is followed by a section comparing the mechanisms used in Colombia to those adopted in Brazil and South Africa and briefly discussing some of the implications for monitoring the new approach.

The Rationale for Market-Assisted Land Reform

In Colombia, land reform has been a long-standing concern to correct the inequitable distribution of land, to increase the productivity and environmental sustainability of agricultural production, and to reduce widespread rural violence. Maldistribution of land in rural areas, dating back to the *encomiendas* given out following the Spanish conquest, has been reinforced and exacerbated in more recent times by a number of factors:

- Tax incentives for agriculture that imply that rich individuals acquired land in order to offset taxes on nonagricultural enterprises
- Legal impediments to the smooth functioning of the land rental and sales markets. Share tenancy has been either directly outlawed or, when this was lifted, discouraged by the fact that tenants receive property rights to whatever land improvements they have made, making it in principle impossible to terminate their leases
- Credit and interest rate subsidies plus disproportionate protection of the livestock subsector that provide incentives for agricultural cultivation with very low labor intensity (World Bank 1996).
- The use of land to launder money acquired by drug lords.

These factors have implications for factor use, employment generation, and welfare in rural areas. First, while small farmers were often driven off their traditional lands and forced to eke out a living in marginal and environmentally fragile areas, much of the best agricultural land continued to be devoted to extensive livestock grazing or was not farmed at all due to violence (Heath and Binswanger 1996). Only 25 percent of the land suitable for crop production was devoted to this use, while the rest was left to pasture. This suggests that there are indeed large tracts of unused or underused land that could be subjected to land reform in order to increase agricultural productivity—a notion in line with available empirical evidence.[1]

Second, rural employment growth since the 1950s has been dismally low, significantly below the standard even of Latin American countries (Misión Social 1990). This appears to have increased peasants' inclination to support, or at least live with, exceptionally high levels of rural violence that increasingly constitute a drag on the whole economy (estimates in the Colombian press put the losses associated with rural violence at about 15 percent of gross domestic product). The government sees the reduction in rural violence as an important goal of land reform.

Third, structural adjustment made the lack of adaptability in the large-farm sector particularly blatant. Large mechanized farms that cultivated mainly traditional crops with minimal inputs of labor were highly indebted and therefore unable to adjust to the new environment and take advantage of the opportunities for exports of nontraditional crops. Unable to respond to the loss of agricultural protection in a productive way, the large-farm sector resorted to large-scale lobbying. Establishment of a dynamic small-farm sector would, it was hoped, enable Colombia to capitalize on its agroecological diversity and significantly increase its exports of traditional and nontraditional crops.

None of these concerns is new. In the 1950s a World Bank mission identified the maldistribution of productive resources, especially land, as one of the root causes of economic stagnation. In 1961 the government established the National Land Reform Institute (Instituto Nacional Colombiano de Reforma Agraria—INCORA) to bring about a more equitable distribution of assets in the rural economy. However, for most of its history, this institute followed a centralist and paternalistic approach to land reform, until 1994, when a law was passed to create the basis for a market-oriented and more beneficiary-driven approach. We will return to the details of this law after briefly examining the rationale that motivated the government to shift to the market-oriented or negotiated approach.

Why through the Market?

From an institutional point of view, the main reasons for the shift from an interventionist to a market-assisted model of land reform were the limited success of centralized land reform and the elimination of the traditional source of finance for INCORA. Even though considerable amounts of resources were spent on land reform (INCORA's average annual budget in the late 1980s was about $140 million), almost 35 years of operations produced little visible effect. In the early 1990s the administrative costs of transferring land were very high, amounting to about 50 percent of the total land reform budget or about $15,000 per beneficiary. INCORA appeared to be more effective in regularizing spontaneous settlement on the frontier than in converting the landless into successful agricultural entrepreneurs in areas transferred from large owners. Many beneficiaries of the INCORA program abandoned full-time agriculture and rented out most or all of their land, in many cases to the old landlord. In the aggregate, almost 35 years of state-led land reform hardly affected overall land distribution; between the 1960s and 1990, the Gini coefficient of the operational land distribution fell only 3 percentage points, from 0.87 to 0.84.

Traditionally, INCORA was financed by a share of duties on agricultural imports. The elimination of these duties in the course of structural adjustment meant that other sources of funding—ones subject to greater public scrutiny—had to be found. INCORA's agreement to adopt a less paternalistic approach freed it from the burden of many of its erstwhile functions, including selection of beneficiaries, adjudication of land, provision of technical assistance, and accounting, enabling it to focus more effectively on regulatory oversight. In fact, this reorientation was a condition of the continued existence of the institution.[2]

In addition to these institutional factors, structural adjustment enhanced the scope for market-assisted land reform by increasing both the supply of and the demand for land. On the supply side, adjustment reduced the profitability of large-scale farming by eliminating agricultural protection while increasing the opportunities for investment in other parts of the economy. This made selling a more attractive proposition for landowners. However, the dynamic response to structural adjustment and the simultaneous decline in annual crop production widened the inequality between regions and worsened rural poverty in a number of areas. This was accompanied by higher levels of rural violence and demand for land. All of this suggests that achieving a sustainable reduction of rural poverty will require more far-reaching measures, possibly including a redistribution of assets.

Principles of Negotiated Land Reform

The legal framework for negotiated land reform is characterized by three main principles:

- Potential beneficiaries (with assets below a certain minimum level) can negotiate independently with landowners and, once a deal has been struck, are eligible for a grant of up to 70 percent of the land's purchase price (subject to an upper limit).[3]
- INCORA's role is limited to that of regulatory oversight and grant disbursement.
- A decentralized institutional structure ensures that land reform is driven by demand and coordinated with other government programs. This structure revolves around the Consejo Municipal de Desarrollo Rural (CMDR), which, through a subcommittee for land reform issues (Comité Municipal de Reforma Agraria, CMDR), is intended to create the institutional structure to integrate various entities and make land reform an integral part of the municipal development strategy.

Despite favorable preconditions, and the government's expressed determination to distribute 1 million hectares within four years, the land reform program had a disappointingly slow start.[4] In the remainder of this section, we consider some of the obstacles that have contributed to this lack of progress and review regulatory measures that have been taken in the meantime to address and, if possible, eliminate some of them.

What Is to Be Financed?

It is well accepted that enabling the poor to establish viable agricultural enterprises without being burdened by excessive debt requires a grant (Binswanger and Elgin 1988 and Carter and Mesbah 1993). Based on historical experience, the Colombian land reform law establishes that the maximum possible grant is 70 percent of the price of land for a viable family farm (defined as one that generates at least three minimum salaries).[5] The remaining 30 percent of the purchase price for land, plus any additional start-up investment, has to be obtained from other sources—either the farm household's own resources or a regular loan from a financial institution. It was hoped that by sharing the risk, financial intermediaries would provide additional assurance of the economic viability of land reform enterprises.

A shortcoming of this arrangement is that grant resources are restricted to the purchase of land, even though it is recognized that establishment of a farm enterprise as envisaged in the law will require funds to establish on-farm infrastructure and improvements like drainage, sheds, and perennials as well as working capital like animals, tools, and machinery. This restriction has two undesirable consequences.

First, the explicit prohibition of using the grant for anything except land purchases may drive up the price of land. Instead of providing an incentive for beneficiaries to bargain and obtain land at the lowest possible price, it creates incentives for collusion between sellers and buyers to overstate land prices, divide the surplus between them, and let the government foot the bill. This has indeed happened; landlords in many instances have overstated the price of land, and—by covering the complete land value with the 70 percent grant—have obtained a subsidy element of 100 percent.[6] Consequently, in 1996 the price of land acquired through direct intervention by INCORA was lower than the price of land acquired by beneficiaries through negotiated land reform in the open market, leading to widespread dissatisfaction and calls for a return to the interventionist paradigm.[7]

Second, even in the absence of such creative solutions, the law created an incentive structure strongly biased toward the transfer of well-developed agricultural land that was already located close to infrastructure and endowed with the necessary complementary investment. This makes it more difficult to target the program to the truly poor in outlying areas, an issue that is of particular importance because land reform has considerable scope to target poor individuals who are usually beyond the reach of more conventional government programs. The inability to use the grant for investments such as drainage, irrigation, or trees that are attached to the land is clearly contrary to the goal of developing underused lands in inaccessible areas.[8] It reduces land reform to a mere redistribution of existing assets rather than the creation of new ones, even if the aggregate cost per beneficiary were the same as under a program targeting underused areas. In the more likely event that undeveloped land could be acquired more cheaply and beneficiaries' own contribution would reduce the cost of establishing on- and off-farm infrastructure, such an approach would be associated with direct inefficiencies and a reduction in the number of beneficiaries who could be served by any given budget.

What Is the Goal of Land Reform?

To have land reform contribute to the creation of viable agricultural enterprises, rather than a rural proletariat with plots too small to generate a sustainable livelihood, the government established that the land transferred under the land reform program would have to provide beneficiaries with an income of at least three minimum salaries. In practice (assuming average land quality), this works out to a farm size of about 15 hectares. Although the government's desire to prevent the land reform program from contributing to *minifundización* (the creation of commercially nonviable garden plots or periurban squatter schemes) is understandable and legitimate, the imposition of a minimum farm size is unlikely to achieve this goal.

First, such a requirement concentrates a limited amount of resources among very few beneficiaries, creating an agrarian bourgeoisie much better-off than the average farmer in any given *municipio*. It also neglects the considerable human capital and other assets as well as experience with financial and marketing institutions that are necessary to operate a successful 15-hectare farm.[9] This tendency is exacerbated by the provision of a land purchase grant only for the transfer of land without complementary investments. Conversations with farmers and land reform beneficiaries suggest that they prefer the provision of smaller plots together with funds for complementary investments over the provision of large plots without such investments. This preference is supported by the experience of many past reform beneficiaries who were forced to rent out the larger part of their holdings owing to a lack of complementary resources.[10]

Second, a centrally imposed minimum size requirement neglects the potential of the rural poor—especially those in proximity to urban areas—to derive income from a variety of sources. It leaves little room for beneficiaries to expand their holdings gradually by renting or purchasing additional land and fails to account for the considerable heterogeneity of conditions prevailing in different parts of the country. Decentralized administration of land reform is more likely to accommodate this heterogeneity by permitting local institutions to determine matters such as the optimum farm size given local conditions. Local institutions should likewise decide how grant elements should be distributed, striking a balance between providing them to comparatively few beneficiaries or to so many beneficiaries that available resources are spread too thinly for farms to be viable.

Who Are the Main Actors?

The Colombian law clearly recognizes that negotiated land reform can work only if it is built on local initiative and if beneficiaries and their representatives coordinate the functions previously performed by a centralized institution. The institutional structure established is exemplary from a conceptual point of view, envisaging (a) decentralized decisionmaking characterized by maximum local participation expressed through the preeminent role of the local councils (CMDR and CMRA); (b) private sector involvement in bringing together potential buyers, provision of complementary credit, and technical assistance to continue during the first two years of production on the land received; and (c) the limitation of INCORA to preventing the misuse of funds, ensuring that regulatory requirements are met, and coordinating the different government agencies involved.

The structure, however, functioned poorly because poor dissemination of the law—among both beneficiaries and INCORA staff—prevented the effective participation of beneficiaries. By default then, INCORA interpreted its mandate so broadly as to run the program without private sector participation and in many cases without the participation of the institutions that were expected to represent beneficiary interests at the local level. This provided little incentive for the private sector to assume the functions envisaged in the law. Private real estate agents, who were supposed to act as information brokers in making potential buyers and sellers aware of market demands, found it difficult to compete with comparable INCORA services that were offered free of charge. In addition, INCORA's continued involvement in the decisionmaking process—including its ability to manipulate political levers—put the institution in a position to virtually guarantee a "successful" outcome.

"Success" was defined more in terms of transferring land and exhausting available budgets than in establishing viable rural enterprises. Consequently, even approved projects were deprived of economic analysis and reliable sources of technical assistance. It is thus no surprise that the financial sector, still conscious of a long history of forced lending to land reform, remained unconvinced and cautious. Financing of the land as well as any complementary credit therefore remained a near-monopoly of the government-owned Caja Agraria where political demands to extend credit to land reform beneficiaries clashed with the desire to restructure this enterprise, leading to prolonged indecision and deadlock. "Beneficiaries" meanwhile were sitting on the land without complementary credit, without a clear idea of how to organize production or marketing, and with the typical clientelist mentality ("INCORA will provide") that market-assisted land reform was meant to eliminate.

Measures Taken to Overcome Initial Constraints

To overcome these constraints, the following changes were made in the legal framework.[11] It was clarified that the goal of market-assisted land reform is the establishment of viable productive projects (*proyectos productivos*) rather than the mere transfer of land. A mechanism was devised to facilitate the use of grant funds to finance nonland investments, thus overcoming the bias inherent in previous legal provisions.

Although the concept of a minimum target income for land reform beneficiaries was maintained, this target income was reduced to two, rather than three, minimum salaries. The income-generating capacity of the productive project no longer is based on *municipio*-level averages but has to be demonstrated by the beneficiary in an economically viable farm plan, which can include income from nonagricultural sources.

Responsibility for the approval of land transfers was decentralized from INCORA's national headquarters to the regional offices, while mechanisms for increased accountability to prevent abuse of this authority were introduced. It was clarified that a functioning municipal council and a municipal land reform plan were preconditions for the introduction of land reform in any given municipality.

The Practice of Market-Assisted Land Reform

The experience gathered during the first two years of the law's implementation tells us little about the mechanisms required for the success of a program of this nature. To generate such insight in a rapid and cost-effective way, the government has identified five pilot *municipios*, selected to reflect the heterogeneity of the country, where new methods of truly market-assisted land reform will be introduced.[12] Comparing the experiences of these pilots with INCORA land reform practices illustrates a number of important issues (see table 13.1).

Two key insights have emerged from these comparisons: the central importance of productive projects and the great benefit that can be gained from a structured and well-defined process at the local level that leads to the elaboration of a municipal land reform plan, especially in situations where preexisting capacity is limited. The next section describes the main elements of such locally devised plans and briefly summarizes the experience of their implementation. We then touch on the arrangements for monitoring and evaluating the pilot.

Table 13.1. Mechanisms to Implement Market-Assisted Land Reform in Colombia, Brazil, and South Africa

	Colombia		*Brazil*		
Task	*INCORA*	*Market assisted*	*INCRA*	*Market assisted*	*South Africa, pilots*
Land selection	Selection by INCORA based on political pressure; cost of $18,000–$22,000 per family	By beneficiaries; with means to increase transparency and provide technical assistance	Purchase or expropriation; average cost of $11,600 per family; mainly legalization of occupied lands	Negotiated by community; willing seller (including banks)–willing buyer; expected cost of $3,000 per beneficiary	Community initiative
Land financing	70 percent of land value (up to $22,000) as grant (20 percent cash; 50 percent bonds); 30 percent through Caja Agraria credit (lengthy delays)	Administration of grant resources and provision of additional credit for land purchase and working capital by commercial bank	Agrarian reform bonds for unimproved land and cash for improvements and crops; beneficiaries in theory expected to pay back, but not enforced	Loan to approved beneficiaries from a commercial bank (considerable subsidy element)	Maximum grant of R15,000 for planning and land purchase
Beneficiary selection	Point scheme for social need and agricultural experience; in practice ad hoc selection based on individual farms	Comprehensive registration; preselection based on social criteria, and final selection based on productive projects	Through INCRA based on examination of agricultural knowledge; in practice almost all are regularized squatters	Self-selection of beneficiaries; clearance of price and title by State Land Institute; decentralized approval; occupied lands ineligible	Self-selection of beneficiaries subject to maximum income criterion (less than R1,500 per month)
Farming project definition	Perceived to be necessary only to obtain a bank loan	Key issue for selection; different forms of technical assistance available; farm models available at municipal level	No specific arrangements	Up to 8 percent of project value is available for technical assistance in project preparation and implementation; farm models elaborated at state level	Provincial plan as a precondition, but few specific guidelines provided and no farm models elaborated
Other financing	Credit for land and working capital, the big bottleneck causing implementation delays	Independent financial institutions to provide integrated credit for the whole project	Credit of up to $1,150 (average $610) for food and housing and $7,500 (average $4,500) for working capital; 70 percent subsidy element; minimal cost recovery	Access to PROCERA credit like other land reform beneficiaries	Responsibility of beneficiaries
Off-farm investments	Complex set of interinstitutional coordination with little results up to now	Identified and costed in municipal land reform plan	Provided by INCRA ($3,200 in 1994, now up to $8,000), almost all for roads	Grant of $4,000 per beneficiary, disbursed directly to the community	Services and to some extent infrastructure provincial responsibility; coordination with the center still weak

The Municipal Land Reform Plan

Three key functions of the municipal land reform plan are to provide a more systematic way of identifying potential beneficiaries and land, to provide farm models that can be used to elaborate productive projects by beneficiaries, and to clarify and establish linkages to other institutions. The preparation of this plan, its final approval, and the monitoring of its implementation are the responsibilities of the municipal land reform committee. The existence of a land reform plan of this type is emerging as a precondition for going ahead with land reform in a given *municipio.*

Identification of Potential Beneficiaries

Even though a needs-based qualification system—which assigned points to potential beneficiaries based on their socioeconomic needs—existed under INCORA, in practice the resulting decisions were almost always overridden by considerations other than maximizing equity or productivity. In order to prevent what it saw as the unrealistic expectations and waste of resources that would result from potential beneficiaries negotiating directly with landowners interested in selling, INCORA continued to select beneficiaries on a case-by-case basis once a given farm had been put up for sale and central approval for the release of the necessary land purchase funds had been obtained. In these cases, to be able to disburse funds quickly, sales of farms were often quite secretive. The selection committees that were established included workers of the existing farm who were generally careful not to admit too many contenders from outside.[13]

By contrast, the pilot approach is based on the principle that establishing clear rules of the game, ensuring the participation of beneficiaries in all stages of the process, and increasing the flow of information are the best ways to generate "realistic expectations" and create a transparent and competitive market for land. The process starts with a systematic information campaign throughout the *municipio*, which should result in a fairly complete inscription of all potential land reform beneficiaries (*aspirantes*) in a registry to be maintained by INCORA. A brief questionnaire provides basic information on beneficiaries' educational level, their agricultural experience (if any), their sources of income, and their access to other types of government services such as education or health. This information is then used to conduct a prequalification, which is essentially a means test based on assets. The whole process provides an opportunity to integrate land reform into a broader program of capacity building and social assistance at the municipal level. In a demand-driven and decentralized process, this could resolve at least some aspects of the potential conflict between the dual objectives of equity and efficiency, objectives that are to some extent unavoidable if land reform is to make a long-term sustainable contribution to poverty reduction.

Enabling potential beneficiaries to register at public offices and police stations in outlying villages has had a considerable impact on the number of registrations. Contrary to the procedures followed earlier, the information supplied is checked for consistency. The names of rejected and accepted *aspirantes* (with reasons for rejection) are posted publicly. The publicity of the selection process seems not only to have increased accountability but also to have led local authorities to understand the tradeoffs between different programs, as well as the scope and limitations of land reform. In all of the *municipios* involved in the pilot, alternative programs have been initiated to (temporarily or permanently) take care of the specific needs of groups who will not benefit from land reform in the immediate future. These include chicken hatcheries and other microenterprises for female household heads, construction of rural roads under seasonal food-for-work schemes, as well as relocation of farmers from environmentally fragile areas and reforestation of critical zones.

The information obtained through the beneficiary questionnaire is then aggregated at the *municipio* level and used to provide a profile of potential beneficiaries that contains, among other things, a description of their capacity and human capital endowments, their current employment status, agricultural experience, asset position, and access to land (as sharecroppers or workers), their organization in groups (with some form of organizational structure), their current and past involvement in credit and financial markets, and access to other forms of government assistance.

Based on its profile of *aspirantes*, the local council, in close collaboration with the responsible institutions, develops a training program for potential beneficiaries who passed the first stage of prequalification. The rationale is that, without a basic familiarity with the economic requirements of specific types of production and with their own capacity to meet these requirements, beneficiaries will be unable to judge the productive capacity of a given farm or meaningfully negotiate with landlords. Most *municipios* seem to have decided to offer training to about twice the number of those tentatively qualified in order to allow for attrition and to ensure that potential beneficiaries do indeed compete in trying to put together the most viable farm projects.

Selection of Land

The availability of large amounts of unused or underused land in large holdings implies that in a reasonably fluid land market there would be plenty of supply to enable potential beneficiaries to choose the most suitable lands and negotiate a competitive price.[14] In practice, rationing was largely by nonprice means, including outright corruption, and the main challenge for landowners willing to sell was to obtain clearance from the regional and central INCORA offices regarding the availability of funds to purchase their land, which was seen as the precondition for proceeding with negotiations.

In the pilot *municipios*, this has been replaced by a more systematic procedure. The first step is to determine ecologically suitable zones and, based on cadastral information, to establish an inventory of the land according to size classification that could be used to identify target areas for agrarian reform. Areas where land reform would result in environmental hazards, where soil fertility is insufficient, or where the existing ownership structure is already characterized by small- to medium-size holdings are thus eliminated a priori. This gives beneficiaries a better idea of the potential supply of land for land reform, helps to set realistic goals, and puts into perspective the potential contribution of land reform for solving the social problems of a given *municipio*. By relating effective supply and demand, the systematic approach to land selection that is required to establish a municipal land reform plan makes evident the potential of land reform to push up land prices within the *municipio*. This in turn identifies *municipios* in which outmigration (or use of very small plots of land) is the only viable option for executing land reform in a cost-effective manner.

Experience from the pilots indicates that land that had traditionally been registered with INCORA was often of marginal quality and hardly suitable for land reform, while some of the best land continued to lie idle or underused. Specific measures to resolve this issue have included:

- More effective and systematic dissemination of information not only among potential buyers but also among any sellers of land, specifically information on the mechanisms of market-assisted land reform and the modalities of payment under this program
- Increasing sellers' awareness of the scope and potential for alternative forms, such as land rental, that could temporarily or permanently provide potential beneficiaries with access to land, immediately increasing efficiency and perhaps serving as a springboard for the landless to acquire the information and agricultural experience necessary to put together a productive project
- Encouraging more effective collection of existing municipal land taxes—a strategy in line with the central government's desire to enlarge the revenue base of local governments and to reduce gradually the need for central transfers.

Because there is little justification for a land reform program that would push up prices for land, thus transferring resources to rich landlords rather than to poor beneficiaries, *municipios* are eligible for land reform grants only if they can demonstrate that existing supply is between double and triple the amount of land to be transacted so as to ensure a truly competitive land market.

Farm Project Development

The most fundamental challenge in the transition from a centrally administered and paternalistic land reform into a market-driven one is to change participants' mentality. The outcome of land reform will rely heavily on the initiatives and entrepreneurship of beneficiaries who must themselves identify opportunities for profitable economic development. This represents a profound reorientation on the part of farmers who are accustomed to waiting passively to be settled on plots preselected by government officials. Without the change in mentality that this reorientation implies, the potential of market-assisted land reform to alleviate poverty and increase productivity is vastly diminished.

Thus, in contrast to the approach followed by INCORA, in which agricultural production and productivity were given little if any consideration, the articulation of viable and economically feasible farm plans takes a pivotal role in the pilot. Under the traditional approach, beneficiaries generally elaborated their "productive projects" *after* obtaining access to the land, with little systematic guidance and no discretion in the use of the technical assistance funds administered by INCORA. Farm plans were regarded as a tedious necessity for obtaining access to complementary credit rather than as a legitimate justification for receiving the award of public funds.[15] Without a clear understanding of the economic potential of the farms to be established, expected returns, and alternative offers of comparable land, beneficiaries' ability to engage in substantive bargaining was greatly diminished. It was only natural that INCORA generally took the lead in "negotiating" with the landlord. This negotiation ordinarily amounted to a mere formality revolving around the price set by an official appraiser. Without any training or capacity building before accessing the land, it is not surprising that beneficiaries showed little awareness of the importance of support services or marketing. Even at its best, land reform remained a largely closed affair that would not generate significant externalities beyond the properties concerned.

Compared with this, the process introduced in the pilot *municipios* attempts to provide beneficiaries with a clear idea of productive opportunities consistent with their abilities *before* they formulate productive projects that form the basis for "shopping" for land. To this end, agricultural professionals, either from government agencies or from local NGOs, are employed to establish crop budgets for a range of options actually practiced in the *municipio* and to conduct training courses and hold meetings to disseminate them. Aggregation of these into farm plans involving more intensive land use and sustainable income for beneficiaries provides the basis for the formulation of "productive projects." Once potential beneficiaries understand them, they can adapt the model budgets to specific farms that are available on the market and use them in subsequent price negotiations.

The shift in emphasis away from maximizing the amount of land and minimizing additional capital investment and toward fully using available family labor has resulted in productive projects that are quite different from those promoted before.[16] Three differences are particularly noteworthy. All of the projects include a "garden plot" component, setting aside about 1 hectare for domestic consumption needs (including chickens, one pig, and a cow) and intensive cultivation of vegetables or fruits, with the surplus to be sold in the market. Emphasis is on providing full employment of the family's labor force throughout the year, meaning that farm plans focus on high-value crops rather than traditional bulk commodities, more diversified crop production, and an important livestock component in virtually each of the cases.

The requirements of land for the average land reform beneficiary have more or less halved, from about 15 hectares under the INCORA-based market-assisted model to about 8 hectares on average. The average land purchase grant has shrunk from Col$22.5 million to Col$8.3 million (with an exchange rate of slightly less than 1,000 pesos to the dollar). If projections about employment generation are correct, the land purchase grant needed per additional employment created is about $3,000, a figure that compares favorably with other types of government interventions.

A brief summary illustrates not only the diversity of models that are suggested for different agroecological conditions but also the high labor intensity of the farm plans adopted, the importance of complementary investments, and the high working capital requirements in most of the models (see table 13.2).[17] The models are characterized by high financial and economic rates of return (between 60 and 90 percent), suggesting that even if some difficulties are encountered during implementation, they are likely to be economically worthwhile. Beyond these figures, the farm models have proven to be of great importance in the debate about land reform, especially at the municipal level.

Information concerning productive projects, demand for land by different groups in the population, current land values, available supply of land at various prices, the social and productive infrastructure required to make these productive, and estimated costs of different components of land reform (land, complementary on-farm investments, and off-farm investments) are all incorporated in the municipal land reform plan. This, together with information on the contributions expected from different participants (beneficiaries, central government, local institutions), allows local authorities to put together a more comprehensive and coordinated program of land reform that is in line with the specific needs and opportunities (including the fiscal capacity) of the *municipio*. This information identifies the need for governmental and nongovernmental institutions (NGOs, banks, local governments, providers of technical assistance, universities) that are supposed to play key roles in the dissemination, capacity building, execution, and follow-up of the land reform process. This serves to alert municipal councils to cases in which, due to very low levels of local capacity and beneficiary organization, broad participation of beneficiaries cannot be taken for granted.

In such circumstances, municipal councils may consider specific measures to improve dissemination and capacity and to increase competition among the institutions involved in land reform execution.

Financing and Linkages to Other Institutions

Given the scant attention traditionally devoted to beneficiaries' economic viability and the high transaction

Table 13.2. Main Economic Parameters of Productive Projects Developed for Four Pilot *Municipios*, by Model Number, Colombia

	Rivera			*Montelibano*		*Puerto Wilches*				*Fuente de Oro*		
Indicator	*(1)*	*(2)*	*(3)*	*(1)*	*(2)*	*(1)*	*(2)*	*(3)*	*(4)*	*(1)*	*(2)*	*(3)*
Farm												
Farm size (hectares)	6.0	5.0	4.0	10.0	7.0	12.5	8.5	6.5	6.5	6.0	8.0	6.0
Labor days used	917	660	537	224	380	1,543	492	1,170	1,385	1,187	1,064	4
Total investment	15.2	14.2	16.7	10.9	13.3	11.3	12.4	15.4	9.2	13.3	13.0	13.3
Percentage for land	9.2	8.2	9.7	5.6	7.0	4.8	6.4	10.4	4.2	9.0	8.8	9.0
Percentage for company	6.0	6.0	7.0	5.3	6.3	6.5	6.0	5.0	5.0	4.3	4.2	4.3
Grant	9.4	8.7	10.5	5.7	8.0	6.4	7.6	10.8	6.4	8.8	8.6	8.9
Own contribution	1.8	1.8	0.9	1.6	1.9	5.1	3.0	2.3	3.3	1.3	1.4	1.3
Bank credit	4.0	3.7	5.3	3.6	3.4	–0.2	1.8	2.3	–0.5	3.2	3.0	3.1
Grant/ employment	2.3	2.9	4.3	5.6	4.6	0.9	3.4	2.0	1.0	1.6	1.8	1.2
Working capital	11.6	4.4	8.0	13.0	13.3	7.6	6.6	4.0	6.6	10.6	12.2	16.0
Output												
Total	28.6	19.6	30.4	18.3	14.7	14.7	14.8	13.5	13.0	22.5	20.3	32.3
Per person	6.9	6.5	12.5	18.0	8.5	2.1	6.6	2.5	2.1	4.2	4.2	4.4
Land allocation (hectares)												
House plot	1	1	0.5	0.5	0.5	1.5	1.5	1.5	1.5	1	1	1
Tomatoes, beans	2											
Tobacco, maize		2	1.5									
Maize, rice, sorghum				3							6	
Plantain				0.5						4		2
Rice, vegetables, maize						5						
Manioc						2						
Oil palm						3		5				
Citrus							1					
Papaya												1
Rubber									5			
Improved pasture	3	2	2	6	1.5	6	6			1	1	2

Note: Costs are in millions of 1997 pesos (with 980 pesos to the dollar).
Source: Municipal land reform plans for the four *municipios.*

costs associated with repossession of the land that served as collateral for land reform loans, it is not surprising that commercial banks are extremely hesitant to advance credit to land reform beneficiaries. The resulting de facto monopoly of Caja Agraria in providing credit opened the door to political maneuvering and—because this institution is itself undergoing restructuring—lengthy delays and bureaucratic requirements in the award of credit. As a result, many beneficiaries who have already accessed land under the INCORA mechanism continue to wait for the sale to be finalized. In the meantime, the inability to obtain even working capital credit severely impedes their productive performance (and motivation).

Under the pilot, a number of cooperative banks already active in rural areas have agreed to lend to land reform beneficiaries and thus create competition with Caja Agraria. The preferred arrangement bears similarity to contract farming. The banks work closely with the providers of technical assistance (ensuring that the farm business established by beneficiaries will indeed generate the desired revenues) and help farmers market produce. This enables them to supervise the use of credit more closely, to ensure that enterprises are indeed developing their productive potential, and to deduct loan repayments at the source, rather than relying on unrealistic expectations of foreclosure.

Monitoring and Evaluation

Although use of a grant-based mechanism that relies on market transactions to redistribute productive assets is an innovative approach, it involves a number of largely unproven procedures and, without close monitoring and supervision, may produce unexpected and unintended outcomes. It is therefore essential to have an information system that continuously monitors each successive stage of the land reform's implementation in addition to assessing its long-term impacts. This makes it possible to quickly identify, evaluate, and rectify unforeseen deviations from the program's overall objectives, as well as determinants of successful implementation, lending the process a responsiveness so sorely lacking in INCORA practices. This is particularly important in view of the focus on decentralized implementation. Table 13.3 relates key components of monitoring and impact evaluation to the issues discussed in the

Table 13.3. Key Variables for Land Reform Planning, Monitoring, and Impact Assessment

Unit	*Municipal land reform plan*	*Monitoring*	*Impact assessment*
Beneficiaries	Beneficiary profile (capacity, welfare) Training requirements	Grant per beneficiary and employment Group formation Additional employment generated Targeting efficiency Improvements in access to land	Increase in income Consumption smoothing (assets) Credit market access Social services Reduction of violence
Projects	Demand and supply of land Characteristics of productive projects Complementary investments needed Cost by component (who pays?)	Characteristics of farms transferred Implementation of projects Repayment performance (planned and actual)	Agricultural productivity Environmental sustainability
Institutions	Institutional capacity (technical assistance, finance, marketing)	Effectiveness in dissemination and capacity building Efficiency of land transfer process Private sector participation	Strengthened local government Fiscal sustainability

municipal land reform plan. Key questions to be answered in this context are:

- What instruments are best suited to reach particular target groups, maximize the net benefits of land reform (or minimize the cost of its execution), and are compatible with rapid attainment of the government's quantitative goals? How do they perform in terms of beneficiaries' adherence to the project plans they have established (and the associated economic benefits)?
- What are the direct and indirect impacts on agricultural productivity and poverty? How does this compare with other instruments at the government's disposal, such as construction of infrastructure or support to education? And under what conditions can land reform constitute a sustainable and worthwhile use of public funds?

Monitoring describes what is happening on the ground, compares it to what had been set out in the municipal land reform plan, and uses the information to update the parameters contained in the municipal land reform plan. In addition, the information provided can be used to (a) assess whether the mechanisms used reach target populations and areas (poor rural dwellers and underused lands), (b) provide an ex ante estimate of the expected benefits of land reform, both for productivity and for poverty reduction, and (c) provide information on whether implementation is progressing as expected. *Impact assessment*, in turn, is concerned with the ultimate impact, both direct and indirect, of the program on household well-being, agricultural productivity, environmental sustainability, and institutional strengthening.

Comparison with Market-Assisted Land Reform in Other Countries

Brazil and South Africa have also, under different conditions, recently initiated programs of negotiated land reform. With an institutional background very similar to that of Colombia (presence of land reform legislation and a central land reform institute dating from the early 1960s), negotiated land reform in Brazil has been initiated by individual states. The purpose of the Brazilian interventions is to establish cheaper, more agile policy alternatives to centralized land reform in an environment where the issue of land reform is high on the political agenda and potential beneficiaries have at least some idea of what to do with the land. By contrast, negotiated land reform in South Africa has been adopted in the context of the national reconstruction program, in an environment in which productive small-scale agriculture was eradicated almost a century ago. Here, much greater effort is required to establish the decentralized infrastructure necessary to implement land reform, to provide complementary services such as marketing and technical assistance, and to increase beneficiaries' agricultural and entrepreneurial capacity.

Brazil

With a land distribution among the most unequal in the world, Brazil's situation is similar to Colombia's in a number of respects. There is a very large and vocal political demand for land reform; a recent study by the Food and Agriculture Organization estimates that 2.5 million families are potential candidates for land reform. A land reform institute (INCRA) was established in 1969, distributing 10 million hectares to 200,000 families and colonizing about 14 million hectares for about 75,000 beneficiary families since then. Land reform has recently acquired considerable political importance; a federal minister for agrarian reform was appointed in 1996, and the land reform budget tripled from $0.4 billion in 1994 to $1.3 billion in 1995 with a further increase to $2.6 billion proposed in 1997.

Although the large majority of these federal funds will be spent according to the old process, which—at a cost of about $30,000 per beneficiary—is not only costly but also lengthy and bureaucratically cumbersome, state governments in the Northeast have been moving ahead and setting up a decentralized market-assisted pilot scheme.[18] This scheme is intended to speed up the land reform process, reduce costs by more than half, and provide the basis for a model that could eventually be adopted nationwide.

The main differences from the old mechanism are the following (compare table 13.1):

- Rather than relying on a lengthy process of expropriation, land is selected by community groups on a willing seller–willing buyer basis. This is expected to reduce the price of land from $11,000 to $3,000, mainly by avoiding the need to pay for expensive land improvements that are of little use for small-scale agriculture.
- Landlords are compensated with cash instead of highly discounted government bonds. This provides a strong incentive for landowners—including many banks who hold title to large tracts of land as collateral for nonperforming loans—to sell

land to land reform beneficiaries.

- Government's role is reduced to providing assurance that there are no problems with the land titles and ensuring that the price negotiated between community groups and landlords is within acceptable boundaries. Projects are approved at the state level.
- Technical assistance is provided on a strictly demand-driven basis; beneficiaries can use part of the community grant made available under a World Bank loan to contract private providers; CONTAG (Federation of Rural Workers) participates in the state councils and assists with information dissemination and land purchase negotiations.

The only commonality between pilots and the nationwide land reform process is the fact that beneficiaries under the new process have access to a subsidized loan under a special program (PROCERA) for land reform beneficiaries.

Although the broad principles are similar to those in Colombia, the process is considerably more flexible and agile. The main points of difference are the following:

- Because grant financing is provided for complementary and community-level infrastructure rather than land itself, beneficiaries who expect to repay their land purchase loan have an incentive to reduce the price of land as much as possible. This creates an incentive to focus on lands that are currently underused, thus reducing the expected purchase price and directing land reform to areas where the social gains from the intervention are maximized.
- The process of beneficiary selection is less formal and bureaucratic than in Colombia, relying on an infrastructure of existing community associations to conduct information campaigns and ensure broad representativeness. Community control facilitates greater flexibility in project execution and should be particularly adequate if, through close supervision and ex post accountability, a minority of politically vocal and well-represented farmers can be prevented from appropriating most of the benefits.
- As long as potential beneficiaries already have experience with the technology to be used on the new farms, as tenants for instance, repayment incentives are strong.[19] And as long as beneficiary organizations exist that can assist farmers in preparing projects and in the initial stages of establishment, the solution chosen—where technical assistance can be financed, on a voluntary basis, through the community grant—is appropriate. However, evaluation of the initial projects should examine whether this excludes certain target groups or whether more stringent requirements may improve the quality of the projects established, as well as beneficiaries' repayment capacity and the overall benefits of land reform.
- Since the working capital credit provided to beneficiaries is subsidized (with a subsidy element of 70 percent), other sources of working capital, as well as marketing channels, need to be available to ensure the longer-term economic viability of land reform beneficiaries. This may well be one of the most critical aspects of land reform.

Given the political importance of land reform and the limited knowledge of both the most appropriate mechanisms to implement this reform as well as the magnitude of the productivity and poverty impact, the government has established the Central Institute for Agrarian Studies to (a) encourage discussion between all parts of civil society, academics, and politicians on land reform issues, (b) carry out a thorough and careful monitoring and evaluation of the land reform process either directly or through local institutions, and (c) make the data collected in this process available to national and international researchers, thus acting as a hub in a broader network of countries and researchers interested in negotiated land reform.

South Africa

While South Africa shares with Brazil and Colombia a highly unequal land distribution, policies that neglected the land rights of the nonwhite population have systematically exacerbated these inequalities. The Native Lands Act of 1912 prohibited blacks from establishing new farming operations, sharecropping, or renting land outside the reserves, which made up only 7.7 percent of the country's area. Inside the reserves, an artificial form of "traditional" tenure with maximum holding sizes and restrictions on land transactions was imposed. Subsequent policies of "black spot removal" transferred the large majority of black farmers who legitimately owned land outside the reserves into the homelands where tenure restrictions, high population density, and lack of capital and market access made commercial agriculture virtually impossible. Labor laws that discriminated against blacks in favor of whites and generous capital subsidies contributed to successive evictions of large parts of the black population from white farms, where they had been employed as labor tenants and farm workers.

Although the Native Lands Act was repealed in 1993, the momentous task of reversing these policies and their consequences was left to the government that entered power following the 1994 elections. In attempting to do so, this government had to contend not only with the extremely unequal land distribution (the average amount held per person was 1.3 hectares by blacks compared with 1,570 hectares by whites) but also with the lack of any local government structure, widespread absence of administrative capacity, a highly indebted large-farm sector, and fear that redistribution would wreak havoc on agricultural productivity and jeopardize national food security. The government decided to adopt a land reform policy that would redress the injustices of apartheid, foster national reconciliation and stability, underpin economic growth, improve household welfare, and alleviate poverty (Government of South Africa 1996). The three central components of this policy are as follows:

- *Restitution.* Legal processes have been put in place to compensate (in cash or in kind) individuals who were victims of forced removals after 1913. All restitution cases are dealt with through the Land Claims Court and Commission, established in 1994, to which claims have to be submitted within three years (that is, by the end of 1997). Even if the legal process can be completed in a speedy manner, the inability of the vast majority of the population to furnish written evidence makes this option feasible for only a small part of the population.
- *Land tenure reform.* This component seeks to improve tenure security of all South Africans by recognizing individual as well as communal ownership rights to land, giving people the right to make decisions about their own tenure system, adjudicating disputes, reforming tenancy laws, and attempting to end discrimination against women in land allocation and holding. It is intended to create the administrative infrastructure that will provide hitherto disadvantaged groups with access to land under a wide array of arrangements that are in line with agroecological endowments and community characteristics. It is hoped that this will provide the regulatory environment for a land rental market by transferring land to more productive users and redressing the inefficiencies of the apartheid system.
- *Redistribution.* As the main component of the government's land reform policy, redistribution aims to complement the market by providing land for productive and residential purposes to a large number of rural blacks who were dispossessed during apartheid and who are interested in obtaining land. It aims to do so by providing a one-time settlement/land acquisition grant of R15,000 ($3,300), equivalent to the national housing subsidy available in urban areas to eligible beneficiaries, defined as anyone with a monthly salary below R1,500. The choice of land redistribution rather than expropriation (which, as in Colombia, can still be used as an instrument of last resort) was based on the need to maintain public confidence in the land market and more generally to affirm the government's respect for individual property rights. It also reflects the recognition that expropriation in other countries has failed to provide rapid access to land for a large number of people, instead degenerating into lengthy political maneuvering and rent-seeking. The number of potential land reform beneficiaries is considerable; estimates indicate that there are about 200,000 labor tenants, 1 million farm workers, and as many as 7 million–8 million blacks in the reserves.

Capacity building is particularly important because the large majority of the rural population has never seen a successful and productive small farm, and many land reform beneficiaries themselves seem to believe that efficient agricultural production is possible only on large farms. The South African government is well aware of these issues and has set up a number of pilots in different provinces to accumulate experience and improve the execution of land reform. These projects are gathering momentum and have provided a number of valuable lessons.

The almost complete fungibility of the land purchase grant represents an important advantage over the Brazilian and Colombian models. It prevents individuals without comparative advantage in farming from becoming land reform beneficiaries just to secure the government subsidy, while eliminating the possibility that land reform will inflate land prices.

Although the government wishes to execute land reform with maximum local participation, the lack of an institutional structure has up to now made effective decentralization difficult.[20] This has, at times, resulted in confusion regarding the criteria and responsibility for making decisions across alternative projects.[21]

Although land reform in South Africa is the

responsibility of the Department of Land Affairs, provincial governments determine how and where complementary services and infrastructure are provided. This would be a difficult arrangement in the best of circumstances and threatens to fragment the land reform program in an environment in which administrative capacity is limited, few well-established procedures and mechanisms exist, and ethnic and political differences between provincial and central authorities create a feeling of animosity and distrust.[22] These shortcomings can be addressed by making the process more demand driven and incorporating current landowners more systematically in the selection of beneficiaries, elaboration of farm plans, and use of technical assistance.

The applicable regulations envisage that a land use plan should lay out the farm model to be adopted by beneficiaries. Most of these plans seem to have been elaborated by authorities without the significant participation of beneficiaries, limiting their sense of ownership and of responsibility for the formulation and outcomes of projects. Although the lack of beneficiary capacity makes this a difficult process, a better understanding of the target group's specific needs and aspirations, and of the productive opportunities in given local areas, could have very beneficial effects. The experience from land reform in Latin America illustrates that, in the absence of such a plan and associated productive projects, there is a danger that beneficiaries will revert to survival farming and that land reform will adopt a "populist-welfarist" paradigm that, by negating the impact of land reform on productivity, will threaten the poverty-reducing potential of the intervention (Cross and others 1996).[23]

Conclusion: Comparing Different Approaches

Although it is too early to judge whether negotiated land reform can rise to the challenges that administrative land reform has failed to solve, the unfolding experience in the three countries described can at least provide the basis for a first assessment. Comparison of the approaches taken in different countries suggests that (a) land reform through negotiation can only succeed if measures are taken to make the market for land sales and rental more transparent and fluid; (b) productive projects are a core element of market-assisted land reform, which is designed to establish economically viable and productive projects at a socially justifiable cost rather than to transfer assets; (c) the only way to achieve effective coordination of the various entities involved in this process is through demand-driven and decentralized implementation; and (d) the long-run success of land reform is likely to depend on getting the private sector involved in implementation and the ability to use the land purchase grant to "crowd in" private money.

Making land markets more transparent and fluid. Selection of land should be demand driven and effected through beneficiaries. However, information on land prices as well as beneficiaries' ability to assess realistically the value of a piece of land or the potential productive returns to its more intensive cultivation are often limited. Three mutually reinforcing strategies to deal with this constraint are (a) the provision of technical assistance at the community level, including assessment of the adequacy of the land price at the point of transaction; (b) the cofinancing of the land purchase through a private bank that, because it shares in the risk of default, will have an incentive to assess the economic feasibility of the proposed farming project; (c) the creation of a "market information system" in the form of a public announcement of land prices for plots transacted in the market, both with and without use of a land purchase grant. It should be clear that negotiated land reform is intended to be a complement, rather than a substitute, for other forms of gaining access to land. This understanding is fostered when it becomes clear that under a regime of negotiated land reform, land rental is a means for beneficiaries to accumulate experience and start-up capital, not something that augurs subsequent government expropriation. Moreover, whatever restrictions on land rental persist should be eliminated. Up to now none of the models considered here has contemplated in depth the potential benefits of interregional migration and ways to encourage such migration to reduce the cost of a land reform program and at the same time enhance its impact on productivity.

Focusing on productive projects. Productive projects are likely to be the key of market-assisted land reform (a) because productivity is the only objective criterion to put an upper bound on the price that can be expected to be paid, (b) because it provides a necessary basis for financial intermediaries to evaluate and eventually support such projects, and (c) because it requires beneficiaries to familiarize themselves with the realities they are likely to confront as independent farmers, and in the process provides them with greater clarity on their own aspirations as well as the potential—and the limitations—of land reform to contribute to the attainment of these goals.

Decentralizing implementation. Experience with centralized land reform has revealed it as costly and inefficient, which is incompatible with the very principle of decentralized negotiation. In Colombia, a two-year attempt to execute negotiated land reform in a centralized manner has convinced all but the most hard-headed INCORA staff of the virtues of a decentralized participatory approach. "Absence of capacity" is unlikely to be an excuse for choosing a more centralized approach: in most cases peasants were much faster to figure out how to work with the system than was INCORA's bureaucracy. Avoidance of such a costly learning process in Brazil owes much to the presence of community organizations that were ready to apply to land reform the experience gained in decentralized poverty alleviation projects. Even in South Africa, where the absence of local institutions at the start of the program made a relatively centralized mode of implementation necessary, the rapid emergence of local governments has already facilitated significant steps toward greater decentralization.

Maximizing private sector involvement. Two lessons have become evident about the financing of land reform. First, selective subsidization of a specific part of the land reform package to the exclusion of others (as was originally the case in Colombia) is likely to be counterproductive, and a flat grant that can be used for all types of expenditures (as in South Africa) is clearly preferable. Second, ensuring access to credit for working capital and other requirements as a part of the land reform package seems to be absolutely essential for beneficiaries to develop financially sustainable operations. Given the high transaction costs of providing credit in rural areas and the greater need for monitoring when most clients lack previous exposure to credit, a unified source of financing that caters to beneficiaries' needs for land purchase as well as working capital is most desirable. Full support by the commercial private sector, private and cooperative banks in particular, on whose continued participation the success of market-assisted land reform is likely to depend most, relies heavily on the formulation of economically viable and technically feasible productive projects.[24] Getting participants—government bureaucrats and NGOs as well as potential beneficiaries—to face the realities of the market is essential, not only to convince doubtful private financiers but also to develop a better understanding of the requirements and the alternatives available for individuals who for various reasons are not the most promising candidates for land reform. If, by doing so, the process of negotiated land reform contributes to changing beneficiaries' attitudes and manages to transform them from passive objects who expect government to deliver a turnkey solution into subjects of the process who are able to convert a one-time subsidy into a permanent improvement in their livelihood, it will have more than achieved its purpose.

Notes

1. Balcazar (1990) summarizes the existing literature on three points, namely (a) small farms farm more intensively than large farms, as measured by value of output per unit of area, (b) between 1973–76 and 1988 average physical yields on small farms increased about 82 percent (this seems to be an unweighted average across different types of crops) whereas those on large farms remained stagnant, (c) regional as well as commodity-specific studies do not find any systematic relationship between farm size and adoption of new technology or improvements in productivity.

2. Although one of the main arguments for keeping INCORA was that its field presence would speed up implementation of the program, this has not materialized. Three years after passage of the law that mandated market-assisted land reform, the level of execution is far behind schedule and INCORA still has a staff above the level that many outside observers consider necessary to execute effectively a truly decentralized and demand-driven market-assisted land reform program. Restructuring (or complete closure) of INCORA without a proven alternative would clearly have been difficult from both an operational and a political point of view. However, land reform would probably have been implemented faster if the government had immediately initiated decentralized pilots of the type described in this chapter. Instead, the government waited for the regulatory framework of the new law (which ironically was entrusted to INCORA) to be almost completed, only to discover that considerable redrafting was required to make the regulations compatible with executing land reform through the market.

3. Modalities of financing are quite flexible—sellers are supposed to obtain 50 percent of the land's

purchase price in cash (a 30 percent contribution by the beneficiary and 20 percent from the government grant) and the remainder in government bonds, which are traded at a considerable discount in the secondary market. A main boon for potential sellers is that any sales of land under the land reform are exempt from capital gains tax.

4. Following passage of the law in 1994, it was envisaged that INCORA would spend about 50 percent of available funds for market-assisted land reform and 50 percent according to the old model (direct intervention) in 1995. Because the regulations to complement the law (for which INCORA had the main responsibility) proceeded very slowly, INCORA first attempted to shift resources from the market to the administered component and, after this proved impossible, had to disburse almost $20 million during a very short period at the end of the year to avoid losing the funds. Given the time pressure, it is not surprising that the processes used were not necessarily in line with the principles of market-assisted land reform.

5. To arrive at the maximum grant size per beneficiary, the total land reform budget was divided by the average number of past beneficiaries.

6. Although these practices were encouraged by the cumbersome mechanisms involved with obtaining credit to finance the 30 percent of land value not covered by the government's purchase grant, this is hardly an excuse for them.

7. Even though all land reform transactions are to be conducted through the mechanism described, INCORA maintains a budget for direct intervention in cases of social or other emergency.

8. As a result, it is possible to use the grant to finance coffee trees that are already planted but not that have yet to be established.

9. The concentration of benefits may be expedient from a political point of view, especially if the benefits can be appropriated by the about 10–15 percent of the peasantry that is politically well organized and collaborates closely with INCORA. With respect to farm size, even in the European Union, the average farm size in 1990 (including the former German Democratic Republic) was, with 14.8 hectares, slightly below the Colombian "minimum size" of 15 hectares. While average farm sizes in Greece and Italy were only 4 and 5.6 hectares, respectively, even in the Netherlands and Germany, average farm size was 16 hectares (Eurostat 1995).

10. The categorical prohibition on rental of land reform land (included in the 1994 law) is unlikely to be enforceable and may even be counterproductive in that it discourages successful reform beneficiaries from intensifying production on part of their land and renting out the other part, while preventing unsuccessful beneficiaries from exiting.

11. Because changing the law itself was considered to be too cumbersome and lengthy, the changes were introduced through specific regulations governing implementation of the law.

12. These *municipios* are San Benito Abad in Sucre, Rivera in Huila, Fuente de Oro in Meta, Montelibano in Córdoba, and Puerto Wilches in Santander.

13. Although one would expect that the transfer of large and relatively extensively cultivated farms would provide an opportunity for accommodating additional workers, the opposite often happened in practice, due to the unwillingness of existing workers to reduce their share and the associated incentives to exaggerate the amount of land needed to establish a productive unit. This tendency to expel laborers is well known from the theory of cooperatives.

14. This is supported by anecdotal evidence—in all of the places visited, potential sellers of land were the most ardent. There is a need to quantify the magnitudes involved—that is, both the degree of underuse and the amount of land that hypothetically would be available for a productivity-enhancing land transfer.

15. This perception is illustrated by the fact that official records for the land transferred in 1996 indicate that in all cases where no credit from Caja Agraria was needed (because the seller advanced the 30 percent complementary credit to the buyer), a farm plan or productive project was "not required."

16. Indeed, examination of the farm models recommended by INCORA for some of the land reform settlements indicated that—despite their considerable physical extent—these models failed to provide year-round full employment for the beneficiary and his or her family.

17. Even model 1 in Montelibano—a model of very low labor intensity—is considerably more labor intensive than the average INCORA project in this *municipio*.

18. The steps involved in the process of expropriation, which is applicable to land that is used to less than 80 percent, are as follows. First, there is a visit by an INCRA mission to assess the value of land and improvements, followed by expropriation through presidential decree and confirmation of expropriation through a federal court (*emissão de posse*) in a

process that takes about a year. Once this is accomplished, landowners are compensated with *titulos da reforma agraria* (with a real interest rate of 6 percent; bearing a discount of 25–40 percent in the market); anecdotal evidence of excessive compensation abounds. The necessary infrastructure investment is then included in the subsequent year's INCRA budget (it takes one to two years for the infrastructure to be established), and INCRA announces that the land is available, selecting beneficiaries based on agricultural skills. In practice, all cases are limited to regularization of existing squatter settlements. Once selected, beneficiaries are eligible for credit from PROCERA (maximum, $7,500; average, $4,500), a special program for land reform beneficiaries that is administered through INCRA (with an effective subsidy of about 70 percent).

19. Because it is likely that this will require some monitoring (and thus increase costs), repayment will depend on the structure of the bank's incentives, in particular who will bear the cost of a possible default on the loan and the nature of the compensation the bank receives.

20. Considerable progress along these lines has been made, for example by contracting out monitoring and evaluation of the land reform pilots to local (traditionally black) universities, a step that, it is hoped, will not only increase the quality of evaluation but also enhance the capacity of these institutions, which has traditionally been quite weak.

21. The green paper states that (a) most critical and desperate needs will be attended to first, (b) priority will be given to projects where institutional capacity is sufficient to allow quick and effective implementation, (c) sustainability and viability of the project need to be ensured, and (d) government will spread its efforts geographically and across different types of beneficiaries.

22. Based on a case study of a provincial land reform pilot in Kwazulu Natal, Cross and others (1996) report four main shortcomings, namely (a) a complete lack of interinstitutional coordination (concerning access to water resources), (b) underestimation of the time and energy required for legal issues such as adjudication between conflicting land claims, (c) political tensions between the African National Congress and Inkafha Freedom Party (at the national and provincial level, respectively) which prevent the effective delivery of services other than land to the beneficiaries, and (d) a worrying tendency of local chiefs to interfere because they perceive land reform mainly as a means to enhance their own power, which depends on the number of people they are able to accommodate rather than the productive success of the programs that are established.

23. "NGOs and pilot structures are ... beginning to be afraid that they would not be able to do more than provide land redistribution beneficiaries with the minimum of land and secure tenure and that this alone would amount to just dumping them—equivalent to the apartheid practice of dumping removals victims in the middle of the veld with no shelter and no way to make a living" (Cross and others 1996: 166).

24. There seems to be broad consensus in the literature on land issues that the full benefits from landownership materialize only if landowners have access to financial markets (Carter and Zegarra 1995).

References

The word "processed" describes informally reproduced works that may not be commonly available through libraries.

Balcazar, Alvaro. 1990. "Tanao de finca, dinamica tecnologia y rendimientos agricolas." *Coyuntura Agropecuaria* 7(3):107–25.

Bardhan, Pranab, Samuel Bowles, and Herbert Gintis. Forthcoming. "Wealth Inequality, Wealth Constraints, and Economic Performance." In Anthony Atkinson and François Bourguignon, eds., *Handbook on Income Distribution*. Amsterdam: Elsevier North-Holland.

Binswanger, Hans P., Klaus Deininger, and Gershon Feder. 1995. "Power, Distortions, Revolt, and Reform in Agricultural Land Relations." In J. Behrman and T. N. Srinivasan, eds., *Handbook of Development Economics*. Vol. 3. Amsterdam: Elsevier Science B.V.

Binswanger, Hans P., and M. Elgin. 1988. "What Are the Prospects for Land Reform?" In A. Maunder and A. Valdes, eds., *Agriculture and Governments in an Interdependent World*. Proceedings of the twentieth International Conference of Agricultural Economists, Buenos Aires. Aldershot: Dartmouth.

Birdsall, Nancy, and Juan Luis Londoño. 1997. "Asset Inequality Matters: An Assessment of the World Bank's Approach to Poverty Reduction." *American Economic Review* 87(2):32–37.

Carter, Michael R., and Dina Mesbah. 1993. "Can Land Market Reform Mitigate the Exclusionary Aspects of Rapid Agro-Export Growth?" *World Development* 21(7):1085–100.

Carter, Michael R., and Eduardo Zegarra. 1995. "Reshaping Class Competitiveness and the Trajectory of Agrarian Growth with Well-Sequenced Policy Reform." Agricultural Economics Staff Paper Series 379. University of Wisconsin-Madison. Processed.

Cross, Catherine, T. Mngadi, S. Sibanda, and V. Jama. 1996. "Making a Living under Land Reform: Weighing up the Chance in KwaZulu-Natal." In M. Lipton, F. Ellis, and M. Lipton, eds., *Land, Labour, and Livelihoods in Rural South Africa.* Vol. 2: *KwaZulu-Natal and Northern Province,* chap. 6. Natal: Indicator Press, University of Natal.

Deininger, Klaus, and Lyn Squire. Forthcoming. "New Ways of Looking at Old Issues: Inequality and Growth." *Journal of Development Economics.*

Eurostat. 1995. *SPEL—Daten für die Landwirtschaft der EU = SPEL—Data for EU Agriculture = Dounées SPEL pour l'Agriculture de l'EU.* Luxembourg: Office des Publications Officielles des Communautés Européenne.

Heath, John, and Hans P. Binswanger. 1996. "Natural Resource Degradation Effects of Poverty and Population Growth Are Largely Policy-Induced." *Environment and Development Economics* 1(1):65–83.

Misión Social. 1990. *El desarrollo agropecuario en Colombia.* Vol. 1. Bogotá: Ministerio de Agricultura, Departamento Nacional de Planeación.

South Africa, Government of. 1996. "Our Land. Green Paper on South African Land Policy." Head Office, Department of Land Affairs, Pretoria. Processed.

World Bank. 1996. *Review of Colombia's Agricultural and Rural Development Strategy.* Washington, D.C.

14

Decentralization and Biodiversity Conservation

Julian Caldecott and Ernst Lutz

This chapter adopts a particular perspective within a much broader study of decentralization (Binswanger, Shah, and Parker 1994), which is itself only one aspect of the greater issue of good governance. Here, we focus on one aspect of decentralization: that is, whether or not it promotes the conservation of biological diversity and, if so, under what conditions? We know that conservation is part of sustainable development and good governance (WCED 1987, IUCN, UNEP, and WWF 1991, UNCED 1992, WRI, IUCN, and UNEP 1992, and UNEP 1995), that participation by local communities is needed for effective conservation, and that this may have to be supported by a significant degree of decentralization (Bhatnagar and Williams 1992, Wells, Brandon, and Hannah 1992, and World Bank 1994). But questions remain concerning the kind and scale of decentralization, its linkage with participation, and the role of other elements such as incentives, enforcement, and institutional capacity in achieving a socially optimal degree of conservation. We now need to discover how exactly and by what mechanisms these various factors are linked.

To seek answers to these questions, case studies were undertaken in 10 countries: Colombia by Eduardo Uribe, Costa Rica by Julian Caldecott and Annie Lovejoy, India by Shekhar Singh, Indonesia by Julian Caldecott, Kenya by Joyce Poole and Richard Leakey, Nepal by Uday Sharma and Michael Wells, Nigeria by Julian Caldecott and Andrew Babatunde Morakinyo, the Philippines by Maria Dulce Cacha and Julian Caldecott, Russia by Margaret Williams and Michael Wells, and Zimbabwe by Brian Child (see Lutz and Caldecott 1996). In addition, Ajit Banerjee and Ernst Lutz (1996) reviewed 32 World Bank and Global Environment Facility projects with conservation goals. They focused on project-related variables and their contribution to the effectiveness of habitat conservation.

Part of the background to all this work is the knowledge that many past attempts to conserve biodiversity failed. One reason was an overly centralized approach, often involving top-down planning by technicians and bureaucrats without concern for the opinions or well-being of the people affected by their decisions. Examples are common and include efforts of the former Soviet Union, whose system of parks and reserves helped conserve much of the country's biodiversity, but whose centralized decisionmaking was seriously distorted by inefficiency and corruption. For the system to function, government needed to have enough power to be able to ignore local aspirations and local conditions, and when that power failed, many nature reserves were immediately exploited by local groups that could conceive of no reason not to do so. In many other countries too, the viability of centrally planned "fortress reserves" has been undermined by their cost and also by the democratic deficit built into them (Bonner 1993). Complete decentralization, however, can also be counterproductive to conservation. Under those conditions, local people, perhaps in cooperation with outside entrepreneurs, may simply degrade and deplete the resources faster and more efficiently. *So there clearly remains a continuing role for central governments. The challenge is to find the degree of decentralization that is most appropriate.*

These issues will perhaps become clearer if conservation is taken to mean improving the allocation and use of environmental resources, including the various components of biodiversity (distinct populations, species, ecosystems, and so forth) and the range of ecological services and goods they provide. These resources are consumed at various levels of society, which can be classified broadly as local (comprising the community and bioregional levels), national, and global. Many different costs and benefits are associated with managing resources in a particular way at each level of society, and managing them to maximize benefits at one level may impose costs on other levels. Thus what is internal to one level may be external to another.

To illustrate the effect of this, assume that an action at the local level creates 10 units of benefit at that level, while causing negative impacts (or externalities) worth eight units at the national level and two at the global level. These impacts can take many forms, for example the loss of species, watersheds, fisheries, or carbon stores. One way for nonlocal society to avoid such impacts is to prevent local people from acting in their own interests, such as by imposing a fortress reserve on them. But a fairer alternative would be to pay the local level 10 or more units of resources to refrain from undertaking the action concerned. This transfer payment may take the form of a rural development project offering alternative benefits and is the rationale behind many of the integrated conservation-development projects that have been undertaken in recent years (McNeely 1988, Ruitenbeek 1990, 1992, Brandon and Wells 1992, Wells 1992, and Wells and Brandon 1993), albeit with mixed success.

Thus each level of society has its own interests in how biodiversity resources are used, and payments from one to another can in principle be used to compensate losing parties so that all sides are satisfied. Decentralization can be seen as a process by which property rights and bargaining powers are redistributed among the levels of society, and there are at least three distinct aspects to this. First, there is the educational (or self-discovery) process by which each group identifies itself, its priorities, and hence its interests.[1] Second, there is the empowerment process by which each group obtains the bargaining power needed to advance those interests. Finally, there is the process of communication and negotiation by which the various groups agree to the terms of transactions among them. These transactions represent the exchange of rights and obligations or the payment of compensation for impacts caused by one group fulfilling its own set of interests on the opportunity for others to fulfill theirs.

It seems reasonable to facilitate these processes, thus reducing the average cost of transactions for the benefit of society as a whole. One way to do so is to assist in the peaceful redistribution of power, while also establishing mechanisms for bargaining based on the new power structure. In overly centralized societies, this may mean promoting decentralization as one component of a broader reform process. This could involve, for example, giving local groups the authority to manage environments and to tax and spend in order to do so. But not all functions of government can usefully be decentralized (Prud'homme 1995), and nonlocal groups may be in a better position than local ones to appreciate long-term or large-scale issues and to act as disinterested arbiters of local disputes that cannot be solved locally. This nonlocal perspective is vital in conservation, which fundamentally concerns avoiding or managing interspecies, intergenerational, interregional, and international conflicts of interest. *Empowerment of local groups should therefore be balanced by a continuing role for central government* to deal with market failures and to ensure both social equity and environmental protection.

Local activities occur in the context of national policies, laws, and institutions, but often out of sight of the agents of government (Caldecott and Fameso 1991). They can directly affect the components of biodiversity, in which the national and global levels of society are now increasingly interested. These levels wish to protect viable and representative natural ecosystems, particularly terrestrial ones in the humid tropics, where most of the world's estimated 10 million–100 million species exist (UNEP 1995). Tropical species are especially vulnerable to major and widespread habitat change, such as that brought about by clear-felling, fire, or agriculture. Thus the single best way to maintain richness of global species overall is to create and manage an adequate system of protected areas in the tropics. Other approaches complement this aim or can serve others, including the use of off-site techniques such as captive breeding and the conservation of wild populations under logging, hunting, fishing, and other kinds of harvesting to the extent this can be done sustainably.

Given the externalities involved, certain resource management options may exist for which the global social net benefit is maximized and where, with appropriate compensation, all actors at various levels can be equally or better off. A complication emerges, however, if people at the local level are unaware of the value of wild species, for example, or of the consequences of changing land use in water catchments, either for themselves or for society as a whole. A rational response to this at the national and global levels, in addition to or instead of compensation or other incentives, would be to invest in education at the local level. This should reduce the cost of any transfer payments that may still be required, while taking advantage of the fact that education, once paid for, continues to yield benefits without further cost (Tilak 1989, Haddad and others 1990, and Herz, Subbarao, and Raney 1991). The high long-term cost-effectiveness of education helps to explain why education activities are now so often linked to conservation projects (for example, GEF 1995).

Once conservation is adopted as a policy priority, a special challenge is created for good governance, which is uniquely concerned with managing actual or potential conflicts between the needs of people and those plants and animals, between the needs of different generations of people, and between the needs of widely separated populations of people. This chapter aims to explore some of the implications of entering this new territory. In doing so, it draws insights from field experience that can help to make sense of the complex interaction among all the factors involved, thereby helping to clarify the relationship between decentralization and biodiversity conservation.

Processes of Decentralization

The country studies reveal certain common features of decentralization processes. Until the mid-1970s in Kenya, Nepal, and Zimbabwe or the late 1980s and early 1990s in Costa Rica and Colombia, countries had a centralized but fragmented approach to environmental management and rural development (table

Table 14.1. Changes in National Conservation Institutions in Select Countries

Country	*Fragmented and centralized*	*Integrated and centralized*	*Mediating body*
Colombia	1968–93: National Institute of Natural Renewable Resources	1993: Ministry of the Environment	1993: reformed autonomous regional corporations
Costa Rica	Pre-1989: National Parks Service, Forestry Directorate, Wildlife Service	1989–95: National Parks Service, National System of Conservation Areas (SINAC)	1987: National Parks Foundation 1989: National Biodiversity Institute (INBio)
Kenya	Pre-1976: National Parks Board, Game Department, Forestry Department	1976–90: Wildlife Conservation and Management Department	1990: Kenya Wildlife Service (parastatal)
Nepal	Pre-1973: royal hunting preserves	1973: Department of National Parks and Wildlife Conservation	1986: King Mahendra Trust for Nature Conservation
Philippines	Pre-1992: Protected Areas and Wildlife Bureau, Tourism Authority, National Power Corporation, Department of Agriculture	1992: National Integrated Protected Areas System (NIPAS), Department of Environment and Natural Resources	1992: Foundation for the Philippine Environment
Zimbabwe	1960–75: Department of National Parks and Wildlife Management	1975: Department of National Parks and Wildlife Management and intensive conservation areas	1989: CAMPFIRE Association

14.1). Control over resources was divided horizontally and vertically among line ministries and other agencies, which competed among themselves for power and funds. It is unclear what prompted changes to this system in each country, but we know that attempts were made to rationalize and consolidate authority over protected area systems, creating more integrated but still centralized arrangements.

As these new institutions were created, it became possible to reform them by amending the laws under which they were established. Thus Nepal's National Parks and Wildlife Conservation Act of 1973 was amended in 1989 to authorize multiple-use conservation areas and the involvement of nongovernmental organizations (NGOs) and again in 1993 to authorize buffer zones and local revenue-sharing arrangements. Meanwhile, Zimbabwe's Parks and Wild Life Act of 1975 was amended in 1982 to empower the district councils over wildlife. In some cases, reform was driven by events in the field, rather than by the national legislature. Thus in the early 1990s, the Guanacaste conservation area in Costa Rica and the first CAMPFIRE districts in Zimbabwe went beyond what was then current law to develop local arrangements for conservation.[2] In these cases, leadership by inspired individuals encouraged local groups to take risks that were rewarded later when laws were changed to authorize what had already been implemented on the ground. But in other cases, local environmental activism provoked reactionary efforts to restore central control. An example is Ogoniland in the Niger delta region of Nigeria, where local action against environmental damage caused by the oil industry was harshly suppressed by the Nigerian military.

In Indonesia, those field projects that have been effective in influencing central government seem to have done so by cautiously going beyond what was the previous official interpretation of certain laws. Examples include projects at Arfak in Irian Jaya and Kayan-Mentarang in East Kalimantan, where community mapping of resource boundaries and local participation in resource use helped to create subtle but important changes in official attitudes to local empowerment. In Costa Rica and Zimbabwe, reforms that gave local people far greater control over and opportunity to profit from the management of wild species and natural ecosystems prompted political reaction. Aggressive questioning of CAMPFIRE began in Zimbabwe in about 1992, and SINAC reforms were suspended by a conservative regime in Costa Rica from 1990 to 1994.[3]

A Role for Mediating Organizations

These experiences confirm that decentralization is inevitably and fundamentally a political process because it redistributes power. Because of this, an important factor in decentralizing a country is the presence of a body able to mediate among various interest groups and to promote smooth and effective transfers of power. These are often officially sponsored NGOs, such as the CAMPFIRE Association in Zimbabwe, the National Biodiversity Institute (INBio) in Costa Rica, the King Mahendra Trust for Nature Conservation in Nepal, and the autonomous regional corporations in Colombia. All were created by government and have close formal and especially informal links with government, but they also have a clear, independent role and much operational and financial autonomy in fulfilling it. Other attempts at making such arrangements are represented by the parastatal Kenya Wildlife Service and the national parks boards in Nigeria, but these have tended to be vulnerable to political interference.

In the absence of an effective mediating body, other NGOs may adopt a less formal but analogous role. This may involve a local NGO developing a partnership with official agencies in its area of operations. An example is that in Nizhny Novgorod, Russia, between Dront Eco-Center (the NGO), the regional unit of the Ministry of Environmental Protection, and the managers of the federally funded Kerzhenski reserve. Here, the three groups collaborate to devise and implement regional conservation initiatives. International conservation NGOs such as the World Wide Fund for Nature also often position themselves as intermediaries between official agencies and the public in the countries where they work. Some specialize in integrating themselves as closely as possible with governments and can be effective in providing technical advice and policy guidance to them. Conservation NGOs with close links to government are evident, for example, in Indonesia and Nigeria, but before 1993 were kept more remote from government in Colombia and still remain so in Kenya. Establishing a new and effective mediating body in a country means that existing NGOs have to reposition themselves accordingly, and adjustments of this kind (which are seldom smooth) have been seen in Costa Rica, Zimbabwe, and elsewhere.

Role of Local and International NGOs

Local and international NGOs have several important

roles to play in conservation and rural development projects (Wells, Brandon, and Hannah 1992 and GEF 1993). They tend to be the seed bed of new concepts that are later adapted and used by official donors and governments. This role is mainly due to their ability to attract innovative individuals, to promote their ideas through advocacy, and to try out their concepts through small and highly visible pilot projects. Because their approach tends to be flexible, consultative, and sensitive to local needs, NGOs can also act as buffers between large projects and small communities, helping to scale down project activities to a more appropriate level. Local NGOs are therefore often used as contractors to deliver community services on behalf of large projects. Finally, NGOs can reduce the shortfall between the local empowerment planned by a project and that actually achieved, as evidenced by Banerjee and Lutz (1996). This is because they are sensitive to the betrayal of ideals, will campaign in public if they see this happening, and therefore have an important role to play in keeping elite institutions honest.

Fiscal Decentralization

The most potent form of decentralization is the transfer of authority over spending decisions, and elite groups retain these particular powers far more diligently than those over other administrative functions. Thus the most complete cases of decentralization are those where local people collect revenues and decide locally how they will be spent. This was the key to the strength of the CAMPFIRE districts in Zimbabwe and the Guanacaste conservation area in Costa Rica, and the same kind of autonomy is seen to some extent in the joint forest management areas in India, buffer zone areas in Nepal, autonomous regions in Colombia, newly empowered municipalities in the Philippines, and certain project areas in Indonesia.

Arrangements for making decisions on how to spend local funds vary, and some are more transparent than others. Zimbabwe's CAMPFIRE communities give an example of full participation in a process of debate and decision, with allocations to public works and private pockets being made openly and in cash. At the other extreme is the Masai Mara reserve in Kenya, which is under the control of the district council but is affected by a serious lack of accountability and transparency in decisionmaking. There, the large sums of money collected from entrance fees contribute little either to maintaining the wildlife and biodiversity assets on which nature tourism is based or to creating incentives to encourage conservation among the local population.

Ecosystems within nature reserves contain resources whose use can contribute to local revenues under certain circumstances. Local revenues vary according to the nature of the resource concerned and other factors such as the presence of markets. Selling trophies to hunters (as in Zimbabwe), or biodiversity prospecting permits to pharmaceutical companies (as in Costa Rica), or lodging and food to tourists (as in Nepal or Indonesia) all require technologies and attitudes that may not have existed previously in each country. If arrangements allow revenues to be captured by local people, then this will tend to promote decentralization and, if other conditions are favorable, conservation. But the opposite effects are seen if most benefits are obtained by nonlocal groups, such as corporations exploiting timber, fish, minerals, or tourism without the consent of local groups. There are many such cases in the country studies, including hotels in the Royal Chitwan national park in Nepal and the Amboseli national park in Kenya and plantation developments and logging concessions in and around the Okomu reserve and Cross River national park in Nigeria.

Nature reserves have potential value, which can be turned into revenue if the technology and markets to do so exist. But if those revenues are large, they may attract the interest of powerful groups or individuals. Thus as wildlife resources in Zimbabwe came to generate local revenues for the CAMPFIRE districts, pressures grew to use them in ways that relieved central government of its spending responsibilities. This had an effect like one of expropriating local funds, and such a response can be expected when central governments with limited resources see some groups of citizens becoming wealthier than others. Unless the central government has a special reason not to intervene, or the newly enriched groups have enough political influence to defend their wealth, at least part of it will be taken away from them. Thus at some stage, a political response will be needed to ensure that enough benefits are retained by local people to sustain the decentralized decisionmaking unit.

Some Reasons for Decentralization

The fact that many countries are tending to decentralize might be explained in two ways. First, rural people may be becoming better organized, better educated, and more aware of economic and ecological realities and hence more assertive in protecting their interests. Alternatively, the rapid degradation of ecosystems in most countries may simply be making those ecosystems less valuable and less interesting as objects of control by powerful groups and individuals. Virgin rain forests and unfished seas can be exploited profitably by corporate investors, but hacked forests and exhausted fisheries are not worth such investment and may be passed by the elite to local control until they recover or until new technologies or markets change their value. Both processes are probably at work and will interact in a complex and patchy way depending on many factors in each country. The balance emerging in each place and time reflects the level of government commitment to decentralize, but this is not enough on its own, and appropriate laws, policies, human capacity and institution building, and time are all needed as well. These other requirements flow from the many ways in which local autonomy can be threatened and the many kinds of responses needed to overcome those threats.

Links to Conservation Effectiveness

Some threats to communities and their environments can be solved locally, an example being in the Masoka area of Zimbabwe where secure land tenure allowed local people to evict nonlocal squatters. But well-armed poaching gangs provide a challenge that local people alone may be unable to meet, as happened in several of Kenya's national parks and reserves. This also applies to planning failures and cases where impacts on the environment were not anticipated, such as impacts caused by road projects in Colombia's Salamanca national park and by water diversions in Kenya's Amboseli and Tsavo West national parks. A condition of local autonomy is therefore the protection of local environments through proper central planning and properly enforced laws on assessing and managing environmental impacts. This is one reason why *conservation can be seen as a joint venture between local and nonlocal levels of society*. It suggests that as local actors become stronger, there is a need to strengthen the willingness and capacity of nonlocal institutions to support decentralization. This highlights the role of central government in promoting actions, such as conserving biodiversity, that local people may not be able or willing to perform unaided.

Incentives and Enforcement

Willingness of local people to respect boundaries and rules protecting nature reserves should be strengthened by obtaining their prior agreement to them. The incentive for government to seek this agreement lies in the need to spend fewer resources on policing where there is strong local support for conservation.[4] As shown in the Arfak mountains in Irian Jaya, Indonesia, for example, setting reserve boundaries by local consultation can help people to accept them as permanent and inviolable and are therefore much cheaper to maintain than boundaries imposed by central authority alone. Similarly, from Zimbabwe we know that local participation in wildlife management decisions and associated revenues can reverse long-standing hostility to wildlife (see also Pye-Smith and Feyerabend 1994).

An apparent paradox here is that local threats to nature reserves can also arise from private motives to exploit resources. Examples can be drawn from the country study on Russia, which documents the effects of a sudden collapse of central power after a long period of highly centralized governance. Violations of reserves by or with the connivance of local people and local authorities are now common and include poaching endangered species (Siberian tigers in Lazovsky and Sikhote Alinski), constructing roads (in Samarskaya Luka) and buildings (in Pri-Oksky Terrasny), grazing livestock (in Daurski and Altaiski), fishing (in Magadanski and Kostomuksha), and clear-cut logging in several reserves in European Russia. In these cases, local groups have taken power without being able to limit resource exploitation, either because local people are not aware of the value of nature reserves to themselves or because they have no tradition of holding those in power accountable, or both. Similar problems have arisen elsewhere, where local elites have received new opportunities to profit by the arrival of mining or logging companies. In these cases, much damage can be done before local awareness of environmental consequences gives birth to local opposition to the elite's decisions.

To offset such risks requires appropriate incentives and policing to complement community-level motivations and action. Enforcement may be welcomed by local people if it is carried out justly and impartially, if they have a role in defining its scope and approach, or if it clearly benefits them directly or indirectly. An example of the latter can be found in northeastern Kenya, where antipoaching work by the Kenya

Wildlife Service is welcomed locally because it tends to protect local people against armed Somali intruders. For government, a key issue guiding investment in policing is the likely cost-effectiveness of different kinds of effort, and two factors contribute strongly to this (Caldecott 1996). First, there is its style, which is influenced by factors such as vigilance, consistency, public relations effort, staff quality, and legal process. Second, there is the level of local compliance, which is influenced strongly by the style of policing, the extent of local involvement in setting rules and awareness of why they are needed, and the economic incentive structure applied to local people's use of nature reserves and wild species. These factors all interact with one another, and proper policing has a positive effect on compliance, while improper policing is ineffective or worse.

Proper policing requires that those who undertake it are seen by local people to act on behalf of an authority that has a legitimate claim on their loyalty. This can occur at the community level of society, and such social pressure is the main means by which endogenous cultural rules are enforced in any society. Examples include hunting taboos and sacred forests, both of which are common among tribal peoples (Caldecott 1988 and Kiss 1990). Exogenous rules, however, such as those protecting biodiversity, may have no equivalent basis in people's upbringing and may be hard to enforce at the community level because of kinship ties and friendships among the people responsible. Finally, most rural communities lack an adequate tax base with which to finance more than the most rudimentary volunteer protection efforts, which will only be effective in exceptional cases.

Conservation enforcement services have the best morale and are best accepted locally when they work on behalf of a level of society beyond that of the community but not as remote as that of the nation, which we call the bioregional level. This is the case in the Guanacaste conservation area in Costa Rica, where wardens work on behalf of a regional committee of local stakeholders. The main drawback here is likely to be one of resources, which are limited in most rural bioregions.[5] Without adequate funds held locally, and considering that threats can occur through external planning failures, the national level of society will need to accept considerable responsibility for conservation policing. The latter is therefore often best done on behalf of local authorities with the assistance and supervision of the national level, which should provide block grants and other forms of assistance such as help in monitoring and applying environmental impact assessment and spatial planning laws.

Scientific Expertise

A final issue is whether policies and decisions are scientifically sound. Conservation biologists, for example, are trained to make judgments on the minimum size of viable reserves, the location of their management zones, and the likely consequences for wildlife of the hunting laws applied in their buffer zone. Unless they themselves possess such skills, local groups will need support from specialists if they are to make such judgments or if they are to set up the monitoring systems for ecosystems and populations without which no management system that is more than moderately intensive can succeed for long. Conservation requires paying just as much attention to ecology as to the social, managerial, and financial aspects of proper policing and to the economic and psychological factors that operate through incentive structures to regulate the use of resources and thereby increase the likelihood of conservation success.

Concluding Observations

The diverse experiences reported in the country studies and project analyses show that decentralization and conservation are complex processes that interact with one another in many ways. From the historical reviews, we can conclude that centralized, top-down conservation has seldom been effective, except where large budgets are available for enforcement and the society concerned is willing to accept a rather undemocratic conservation process. Looking at the more recent experience of countries in giving new responsibilities to local government units and NGOs, we can see that this creates both opportunities and potential problems. To take advantage of the former while avoiding the latter, a cluster of arrangements must be made *as a whole* if conservation is to work well in a decentralized setting. Of these, seven merit special attention:

- Local participation, especially in a way that allows local people to understand and endorse the boundaries and management plans of nature reserves and that promotes clear tenure over land and other resources in and around the reserves
- Capacity building, especially to increase skills and accountability among local government units and

NGOs, so they can work together to promote conservation and rural development

- Incentive structures, especially those that allow local communities to keep income from the sustainable use of nature reserves and other biodiversity assets
- Conditional subsidies, especially where divergent costs and benefits of conservation are experienced by local and nonlocal groups, making it necessary for global and national society to bridge the gap with livelihood investments or grants
- Appropriate enforcement, especially against powerful local or central interests, and always in the context of education and public relations activities
- Stakeholder forums and ecoregional executives, which need decisionmaking and fiscal authority to fulfill their three main roles of avoiding conflict through dialogue, authorizing conservation action, and requesting help from nonlocal society to meet local development priorities
- Enabling policies, laws, and institutions to provide a clear and supportive framework for conservation on behalf of national government, creating incentives at the local level to harmonize development with conservation and thereby reducing the need for enforcement.

The country studies also illustrate many of the key issues in sustainable development, even though they report only a small sample of recent global experience. A review of this material and other evidence in this volume and elsewhere (Barzetti 1993, IIED and ODA 1994, Western, Wright, and Strum 1994, White and others 1994, UNEP 1995, and Caldecott 1996) leads us to conclude that conservation problems can usually be traced to one or more of the following underlying causes. First, people who depend on an ecosystem may not be aware of the connection between the ecosystem's well-being and their own. Second, they may be incautious in changing the ecosystem to make it more productive, without realizing that this can do more harm than good to the interests of the majority. Third, the presence of external effects may be connected with local actions. Fourth, local people may have no accepted rules governing the use of the ecosystem, often because a traditional management system has broken down or has been replaced by a central and ineffective one. Fifth, they may not have the authority or ability to manage the ecosystem exclusively in their own interests. Finally, key decisions changing the use of the ecosystem may be made without involving all the people affected by those decisions.

Weak environmental awareness, caution, self-regulation, tenure, and accountability are thus typical of conservation problems. They can usually be found—alone or in combination, obviously or disguised—wherever the components of biodiversity are being eroded by human action. Decentralization can directly help to solve problems of self-regulation, tenure, and accountability but has a more indirect role in improving environmental awareness and caution. This awareness depends on public understanding of how ecosystems work, and decentralization can help to increase local responsibility for environmental management, thus making this process more relevant and interesting to local people. Greater interest then tends to promote willingness to learn about ecology.

Self-regulation, tenure, and accountability are strongly interconnected. Some reforms that a country can make will help to ensure that people who decide how to use environments are directly affected by the consequences of those decisions. By shortening the feedback loops between decisions and their effects, such reforms reward a cautious approach to decisionmaking. Moreover, changes that give authority specifically to people living within the managed environment encourage decisions that are responsive to local conditions. If, in addition, other local stakeholders are encouraged and enabled to question those decisions, then responsibility is promoted and a strong force for good governance created.

All such changes tend to improve environmental management and are often consistent with key elements in the process of conservation (see Janzen 1991, 1992 and WRI, IUCN, and UNEP 1992). The first of these is that the components of biodiversity must be saved in order to preserve the option to use them. The second is that people must study and learn about what those components are and how they might be used. The third is that they must be used sustainably in order to meet the economic, intellectual, and other needs of society. The final element is that they and their uses must be taught about, so they will be valued by people. From this point of view, then, decentralization can be important in allowing biodiversity to be perceived as a local resource for local husbandry, which in turn can motivate local people to preserve the biodiversity in their own environment.

Reforms of this kind yield benefits that are not limited to local people, because they can also reduce the cost to a government of meeting its own conservation aims (Caldecott, Bashir, and Mohamed 1995).

They do so partly by encouraging local understanding of how to make better use of ecosystems (thus reducing the need for inducements to do so) and by prompting people both to insist on and assist in proper environmental policing (thus reducing its recurrent cost). They also promote dialogue and trust and may reduce the cost of negotiating settlement of claims arising from impacts within and between local and other levels of society. If combined with strengthened resource tenure, participation can help local people to capture some of the economic benefits of conservation, further rewarding local policing effort at little cost to government. Finally, better environmental management helps governments to avoid the cost of rehabilitating societies that have been blighted by environmental damage. By promoting local participation, decentralization can have an important role in all this, but it is not a panacea and must always be seen, judged, and planned in context. This leads us to five main observations:

- Precipitate and unplanned decentralization can neutralize national and global influence, while giving powers to local societies that may lack adequate skills and accountability to use those powers properly. This means that the social context should be analyzed carefully before changes are attempted.
- Redistributing power may affect those who were formerly doing well, prompting them to resist change. Thus mediating bodies will be needed to smooth the transfer of power, and these must be trusted enough by all sides to be able to reassure them. Such bodies may have to slow down the process while studies and consultations occur and while people seek alternative livelihoods and adjust their expectations of the society in which they live. These bodies can take many forms, but they must be genuinely independent of the main parties and competent to act, as well as be trusted and trustworthy.
- A locality or bioregion can empower itself by unilateral action, but this can provoke efforts to reassert control by groups that feel threatened. These groups might seek to reclaim bioregional revenues or to reverse events through political, administrative, or military intervention in the name of national unity. To prevent this, other changes in law and policy may be needed to protect the newly empowered bioregion and to permit it to sustain itself.
- There is the risk that a bioregion or community that is no longer sheltered by a national government may become vulnerable to groups wishing to exploit it, for example as a source of raw materials or as a dumping ground for wastes. Where national governments are no longer able to control such threats, localities must be helped to communicate and collaborate to prevent them from being singled out and overwhelmed one by one.
- Finally, uncertainties abound in any political process, including decentralization. Thus there is always a risk that the need to protect nature reserves may be forgotten for a time.

The last observation draws attention to the fact that the process of decentralization in many countries is not driven by public interest in biodiversity conservation, but rather by a desire for better access to the fruits of economic development through democratic participation. Conservation will benefit from this only to the extent that ecosystems and the biodiversity they contain are seen as resources to sustain development, in other words as valuable resources that people (or different levels of society) may wish to control for their own particular benefit. If there is no such perception, then conservation benefits will accrue from decentralization only accidentally and, if biodiversity continues to be perceived as valueless by newly empowered local groups, only temporarily.

Because conservation requires *permanent* solutions to problems of species extinction and environmental degradation, it must involve changing perceptions and values among people who control the fate of ecosystems. In decentralized circumstances, this means local people. Because their main motive for seeking more power is likely to be to enhance their economic position, the link between conservation and local enrichment (or poverty avoidance) must be made very explicit. This is a major theme in modern conservation projects, which have to show that real benefits can come from conservation. To do this, the definition of wealth may have to be expanded to include its biological and cultural dimensions as well as its more conventional economic aspect. People may also have to be helped to perceive the value of long-term and future benefits as higher than that of temporary get-rich-quick schemes.

These new and demanding conditions for successful conservation projects imply the need to identify and sustain those features of the project area that allow wealth creation and ensure poverty avoidance. An aim is to neutralize forces that may damage local ecosystems and erode biodiversity, while promoting the local capture of economic benefits and the more productive use of natural and artificial ecosystems. These all depend on adequate levels of environmental awareness and security of resource tenure. If these are achieved in a project area, they will create more favorable conditions for other forms of investment, both by strengthening local institutions and by improving local knowledge and thinking skills. The interplay among all of these factors means that appropriate decentralization of relevant functions is as much a necessary (albeit insufficient) condition for conservation as conservation is for sustainable development.

Notes

1. Or *conscientização* (Freire 1984), that is, learning to perceive and act against oppressive elements of reality (which in our view include harmful environmental externalities).

2. Under the Communal Areas Management Programme for Indigenous Resources (CAMPFIRE), local people own the rights to manage and profit from wildlife resources in their communal lands. This gives them a strong incentive to keep wildlife populations as productive assets, rather than allow them to be overharvested or lost through habitat destruction.

3. Sistema Nacional de Areas de Conservación (SINAC), Costa Rica's National System of Conservation Areas, incorporates a number of reserves, each of which is managed by a locally appointed and locally accountable committee with a very high degree of autonomy in all areas of decisionmaking.

4. The term policing refers here to all forms of publicly authorized surveillance and intervention that aim to prevent harm.

5. Although not to the usual extent at Guanacaste, which by 1992 had access to an externally financed $11 million endowment fund.

References

The word "processed" describes informally reproduced works that may not be commonly available through libraries.

Banerjee, Ajit, and Ernst Lutz. 1996. "Analysis of World Bank and GEF Projects." In Ernst Lutz and Julian O. Caldecott, eds., *Decentralization and Biodiversity Conservation.* Washington, D.C.: World Bank.

Barzetti, V., ed. 1993. "Parks and Progress: Protected Areas and Economic Development in Latin America and the Caribbean." World Conservation Union, Gland, and Inter-American Development Bank, Washington, D.C. Processed.

Bhatnagar, Bhuvan, and Aubrey C. Williams. 1992. *Participatory Development and the World Bank: Potential Directions for Change.* Discussion Paper 183. Washington, D.C.: World Bank.

Binswanger, Hans, Anwar Shah, and A. N. Parker, comps. 1994. "Decentralization, Fiscal Systems, and Rural Development, Revised Proposal/ Request for Research Support Budget Funding." Rural Development Department, World Bank, Washington, D.C. Processed.

Bonner, R. 1993. *At the Hand of Man: Peril and Hope for Africa's Wildlife.* London: Simon and Schuster.

Brandon, Katrina E., and Michael Wells. 1992. "Planning for People and Parks: Design Dilemmas." *World Development* 20(April):557–70.

Caldecott, Julian O. 1988. *Hunting and Wildlife Management in Sarawak.* Gland: World Conservation Union.

———. 1996. *Designing Conservation Projects.* Cambridge, U.K.: Cambridge University Press.

Caldecott, Julian O., S. Bashir, and S. Mohamed. 1995. *Issues in Sustainably Financing Coastal Conservation.* Proceedings of the International Coral Reef Initiative South Asia regional workshop, the Maldives, 29 November–3 December. Chatham, U.K.: Natural Resources Institute; London: Overseas Development Administration.

Caldecott, Julian O., and T. F. Fameso. 1991. "TFAP Nigeria: Findings of the Preliminary Conservation and Environment Study Mission." Commission of the European Communities, Brussels. Processed.

Freire, P. 1984. *Pedagogy of the Oppressed.* New York: Continuum.

GEF (Global Environment Facility). 1993. "The Social Challenge of Biodiversity Conservation," ed. by S. H. Shelton and written by G. Castilleja, P. J. Poole, and C. C. Geister. Working Paper 1. Washington, D.C. Processed.

———. 1995. "Quarterly Operational Report, August 1995." Washington, D.C. Processed.

Haddad, Wadi D., Martin Carnoy, Rosemary Rinaldi, and Omporn Regel. 1990. *Education and Development: Evidence for New Priorities.* Discussion Paper 95. Washington, D.C.: World Bank.

Herz, Barbara, K. Subbarao, and Laura Raney. 1991. *Letting Girls Learn: Promising Approaches in*

Primary and Secondary Education. Discussion Paper 133. Washington, D.C.: World Bank.

IIED and ODA (International Institute for Environment and Development and Overseas Development Administration). 1994. *Whose Eden? An Overview of Community Approaches to Wildlife Management.* London.

IUCN, UNEP, and WWF (World Conservation Union, United Nations Environment Programme, and World Wide Fund for Nature). 1991. *Caring for the Earth: A Strategy for Sustainable Living.* Gland.

Janzen, D. H. 1991. "How to Save Tropical Biodiversity." *American Entomologist* 37(3, fall):159–71.

———. 1992. "A South-North Perspective on Science in the Management, Use, and Economic Development of Biodiversity." In O. T. Sandlund, K. Hindar, and A. H. D. Brown, eds., *Conservation of Biodiversity for Sustainable Development.* Oslo: Scandinavian University Press.

Kiss, Agnes, ed. 1990. *Living with Wildlife: Wildlife Resource Management with Local Participation in Africa.* Technical Paper 130. Washington, D.C.: World Bank.

Lutz, Ernst, and Julian O. Caldecott, eds. 1996. *Decentralization and Biodiversity Conservation.* Washington, D.C.: World Bank.

McNeely, Jeffrey A. 1988. *Economics and Biological Diversity.* Gland: World Conservation Union.

Prud'homme, R. 1995. "The Dangers of Decentralization." *The World Bank Research Observer* 10(2):201–20.

Pye-Smith, C., and G. B. Feyerabend. 1994. *The Wealth of Communities: Stories of Success in Local Environmental Management.* London: Earthscan.

Ruitenbeek, H. J. 1990. *Economic Analysis of Conservation Initiatives: Examples from West Africa.* Godalming, U.K.: World Wide Fund for Nature.

———. 1992. "The Rainforest Supply Price: A Tool for Evaluating Rainforest Conservation Expenditures." *Ecological Economics* 6(July):57–78.

Tilak, Jandhyala B. G. 1989. *Education and Its Relation to Economic Growth, Poverty, and Income Distribution: Past Evidence and Further Analysis.* Discussion Paper 46. Washington, D.C.: World Bank.

UNCED (United Nations Conference on Environment and Development). 1992. *Agenda 21.* Rio de Janeiro: United Nations Conference on Environment and Development; Nairobi: United Nations Environment Programme; New York: United Nations Secretariat.

UNEP (United Nations Environment Programme). 1995. *Global Biodiversity Assessment.* Cambridge, U.K.: Cambridge University Press.

WCED (World Commission on Environment and Development). 1987. *Our Common Future.* Oxford: Oxford University Press.

Wells, Michael. 1992. "Biodiversity Conservation, Affluence, and Poverty: Mismatched Costs and Benefits and Efforts to Remedy Them." *Ambio* 21(3):237–43.

Wells, Michael P., and Katrina E. Brandon. 1993. "The Principles and Practice of Buffer Zones and Local Participation in Biodiversity Conservation." *Ambio* 22(2-3):157–62.

Wells, Michael, Katrina E. Brandon, and Lee Hannah. 1992. *People and Parks: An Analysis of Projects Linking Protected Area Management with Local Communities.* Washington, D.C.: World Bank.

Western, D., M. Wright, and S. Strum. 1994. *Natural Connections: Perspectives in Community-Based Conservation.* Washington, D.C.: Island Press.

White, A. T., L. Z. Hale, Y. Renard, and L. Cortesi, eds. 1994. *Collaborative and Community-Based Management of Coral Reefs: Lessons from Experience.* West Hartford, Conn.: Kumarian Press.

World Bank. 1994. "The World Bank and Participation." Operations Policy Department, World Bank, Washington, D.C. Processed.

WRI, IUCN, and UNEP (World Resources Institute, World Conservation Union, and United Nations Environment Programme). 1992. *Global Biodiversity Strategy: Guidelines for Action to Save, Study, and Use Earth's Biotic Wealth Sustainably and Equitably.* Washington, D.C.: World Resources Institute; Gland: World Conservation Union; Nairobi: United Nations Environment Programme.

15

The Importance of Gender Issues for Environmentally and Socially Sustainable Rural Development

Agnes R. Quisumbing, Lynn R. Brown, Lawrence Haddad, and Ruth Meinzen-Dick

For several decades, food, agricultural, and natural resource management policies have been designed without acknowledging that rural men and women may have different preferences, face different constraints, and respond differently to incentives. Indeed, until Boserup's (1970) path-breaking work, women were largely "invisible" within agriculture, and the predominant assumption was that the male heads of household made most, if not all, farm allocation and production decisions. Subsequent work (for example, Dixon 1982 and Safilios-Rothschild 1985) has shed light on women's contribution within different farming systems, whether as farm managers, laborers, or managers of natural resources. More recently, a growing literature has questioned the assumption that households make decisions as one (see Alderman and others 1994 and Haddad, Hoddinott, and Alderman 1997 for reviews). This body of work shows how policymakers may not be able to predict accurately individuals' responses to policies due to different preferences as well as asymmetries in the distribution of rights, resources, and responsibilities within households.

In many societies, gender is an important determinant of the distribution of rights, resources, and responsibilities among individuals, households, and communities (Moser 1989). Because this distribution is socially determined, rather than innate or biological, it is also subject to policy intervention. Project experience shows how neglecting the gender dimension in development policy design has led to failures in project implementation, failures in the adoption of new agricultural technologies, or the adoption of new technology but with negative unanticipated impacts (see Alderman and others 1994 for a review).[1] At the same time, gender roles in agriculture are not immutable and may be affected by economic incentives: the gender division of labor by task may change, as expedient (see Tisch and Paris 1993 on the Philippines), and women farmers in Africa may increase their production of cash crops, traditionally a male domain, as these become more profitable (Saito, Mekonnen, and Spurling 1994). Project experience suggests that knowledge of the relative bargaining power of men and women could help to predict the outcome of development projects and promote a design that would not lead to a deterioration of women's access to and control over resources (von Braun and Webb 1989).

This chapter argues that policymakers would do well to take gender issues into account when formulating food, agricultural, and natural resource management policies. Based on a review of empirical evidence, it shows that reducing inequalities between men and women in human and physical capital can lead to major efficiency and productivity increases in agriculture. Moreover, addressing inequalities in the underlying distribution of property rights between men and women may be essential not only for increased agricultural productivity but also for environmental and social sustainability. The second-round effects are also important: this chapter shows that increases in income, when controlled by women, have a greater impact both on household food security and on investments in child health and schooling. Finally, paying attention to gender issues in the design of policies and projects is crucial to designing projects that are both successful and sustainable.

Gender-Specific Constraints in Agriculture

Despite their importance in agriculture, women usually have lower levels of physical and human capital than men. In many developing countries, women have lower school enrollment rates, literacy rates, and schooling attainment. In the early 1980s, average literacy rates for men in developing countries were over 50 percent, while more than two-thirds of women were still illiterate (Seager and Olson 1986). This disparity continues to be larger in rural areas, where educational attainment is lower, and persists despite high private rates of return to women's schooling (Quisumbing 1995) and high social returns to women's education (Subbarao and Raney 1996).

Laws governing women's rights to land and other property—whether legal or customary—typically provide them weaker rights to land. Many religious laws also discriminate against women in land rights (see Agarwal 1994 for examples for South Asia).[2] Women's use rights under customary law in many African countries are allocated from their husbands and natal families based on their position within a kinship group and, in particular, on their relationship to a male relative (father, brother, husband). These rights entitle women to farm the land, often in exchange for labor on their husband's and other family plots. However, these indigenous customs have been modified by Western colonization, which introduced private ownership by individual registration of land and often discriminated against women in titling. Furthermore, because women obtain land rights usually through a male relative, there is no guarantee that they will retain these rights in the case of death or divorce. Moreover, a farm household survey in Kenya and Nigeria (Saito, Mekonnen, and Spurling 1994) finds that more male than female heads of households, and more male than female farmers, are able to exercise their land rights fully. Absence of formal land rights and smaller land sizes cultivated by women may be critical because land is usually needed as collateral in credit markets.

Access to credit, both formal and informal, has important implications for the ability to attain a stable standard of living and undertake productive activities. Credit is particularly important during the planting season, when seeds and cash inputs have to be purchased, and in the lean season before the harvest, as stocks for consumption are depleted. Credit may also be essential to smoothing consumption in case of crop failure or drought.

Collateral requirements, high transaction costs, limited education and mobility, social and cultural barriers, and the nature of women's business limit women's ability to obtain credit. Property that is acceptable as collateral, especially land, is usually registered in men's names, and the types of valuables women have, such as jewelry, are often deemed unacceptable by formal financial institutions. The transaction costs involved in obtaining credit—transportation costs, paperwork, time spent waiting—may be higher for women due to higher opportunity costs from forgone activities. Indeed, in rural Kenya, the distance to a bank is a significant determinant of women's probability of obtaining credit but does not affect men's borrowing behavior (Saito, Mekonnen, and Spurling 1994). Women's lower educational levels relative to men, their lack of familiarity with loan procedures, and social and cultural barriers may constrain their mobility and interaction with predominantly male credit officers or moneylenders. Moreover, women may be credit constrained because their role as primary caregivers and the health risks associated with childbearing lead to intermittent employment, which makes them risky clients for banks. Women also tend to be involved in the production of relatively low-return crops that are not included in formal sector lending programs. Exclusion from local groups, such as farmers groups, may prevent women from receiving both credit and extension advice, particularly if the extension worker plays an important role in credit delivery. Moreover, extension workers tend to favor landowners (Agarwal 1994), who are rarely women, thereby giving men preferential access to information.

In spite of women's important role as farm managers and agricultural workers, whether as family or hired laborers, they have seldom been beneficiaries of extension. Traditional extension systems based on single-commodity extension often fail to consider the crops and activities in which women are involved. Community or rural extension, in contrast, may cover the broad spectrum of women's activities but runs the risk of treating specific tasks superficially. Furthermore, extension systems in many developing countries are overloaded; in the late 1980s, ratios in

Africa, Asia, the Near East, and Latin America range from one agent to 2,000 or 3,000 farmers (Saito and Spurling 1992). In contrast, in Europe and North America, one extension agent serves 300–400 farmers. In addition, women are underrepresented among extension agents. Even in regions with a long tradition of female farming, such as Africa, in the early 1990s, only 11 percent of extension staff and 7 percent of field extension staff were women. Moreover, although female extension workers may be trained in agriculture, they are often mandated to give advice only on home economics subjects. This may constrain the delivery of agricultural extension messages to female farmers, who may be restricted from interacting with male extension agents and who prefer to interact with female agents.

Lastly, time is a scarce resource for rural women. Many studies have documented the long hours that women spend in productive activities compared with men (Brown and Haddad 1995 and McGuire and Popkin 1990). In Africa, women spend as much as two hours a day caring for children, three hours preparing food, and two hours fetching fuel and water. In rural Asia, food-processing activities take up two to three hours daily, and in rural Bangladesh, women may spend as long as six hours a day fetching water. Where women's nutritional status is already compromised or at risk, demanding physical labor may have significant negative effects on their nutritional status, as suggested by Higgins and Alderman (1997), based on a Ghanaian study. The extent of rural women's multiple responsibilities often imposes time and energy constraints on their participation in programs designed to increase their incomes and on their willingness to adopt new technologies.

Productivity and Efficiency

Women's limited control over productive resources has important consequences for agricultural productivity. Econometric evidence on gender differences in agricultural productivity points to the importance of investing in women by increasing their human capital, through education and extension, and by increasing their access to physical and financial inputs (Quisumbing 1995, 1996).

Technical Efficiency of Male and Female Farmers

Numerous studies have found that, where male and female farmers manage separate plots, as in many African farming systems, plots controlled by women have lower yields than plots farmed by men. Studies analyzing the technical efficiency of male and female farmers have suggested that this is due to lower levels of input use on women's plots and not to inherent managerial differences between male and female farmers. For example, studies from Kenya, Korea, Nigeria, and Thailand, summarized in table 15.1, show that female farmers are equally efficient as male farmers, once individual characteristics and input levels are controlled for (Moock 1976, Saito, Mekonnen, and Spurling 1994, and Jamison and Lau 1982).

There may be significant gains from increasing women's level of physical and human capital, as suggested by simulations using coefficients from the production functions estimated in the studies cited. Simulations performed using Moock's (1976) coefficients for female farmers suggest that if female maize farmers were given sample mean characteristics and input levels, their yield would increase 7 percent. If they were given men's mean input levels and other characteristics, yields would increase 9 percent. Giving all women at least a year of primary education would raise yields 24 percent, which reflects the gains to providing primary education in a setting where women have very low educational levels. Simulations with the Saito, Mekonnen, and Spurling (1994) coefficients suggest a 22 percent increase in women's yields on plots of maize, beans, and cowpeas if women farmers were given the characteristics and input levels of male farmers. Increasing land area and fertilizer use to the levels of male farmers would increase women's yields 10.5 and 1.6 percent, respectively. However, these studies do not take into account the relationship between input application and farmer characteristics. Since more-educated farmers are more likely to adopt new technologies and apply modern inputs, the contribution of such inputs may be overstated and that of education underemphasized. This may underestimate the consequences of underinvesting in women's education in rural areas.

More recent literature suggests that the asymmetry of roles and obligations within the household, particularly in African farming systems, may underlie women's lower yields. A study in Burkina Faso (Udry 1996) finds that plots controlled by women have significantly lower yields than similar plots within the household that are planted with the same crop in the same year, but controlled by men. The yield differentials are due to lower input intensity on female-managed plots: much less male labor per hectare is devoted to plots controlled by women than to similar plots controlled by men. Child labor and unpaid exchange labor are also applied more intensively to plots controlled by men. Lastly, virtually all fertilizer is concentrated in plots controlled by men, even though the marginal product of fertilizer diminishes. These differences in input intensity between male- and female-managed plots persist even after land quality, measurement error, or risk management behavior are taken into account.

If the household had common objectives and preferences, factors of production would be allocated efficiently to the various productive activities of the household. The consistently higher input intensities on men's plots suggest misallocation of household resources, which is traced to stronger incentives for individuals to achieve high output on their own plots, as well as imperfect labor allocation processes within the household. Estimates show that reallocating currently used factors of production across plots could

Table 15.1. Production Function Studies with Estimates of Male-Female Differences in Technical Efficiency

Study	*Sample*	*Sex of farmer variable*	*Coefficient*	*Dependent variable definition*
Kenya				
Moock (1976)	All farmers: males = 101; females = 51	Female dummy x log of area planted in maize	0.090	Log of maize output per acre
		Female dummy x log of plant population per acre	–0.280*	
		Female dummy x log of labor input per acre	0.108***	
		Female dummy x primary schooling dummy	0.167***	
		Female dummy x log of extension contact index	–0.028	
Bindlish and Evenson (1993)	All farmers: male heads = 434; female heads = 241	Female head dummy	–0.022	Log (ln) of crop production in 1990
Saito, Mekonnen, and Spurling (1994)	All farmers: males = 306; females = 147	Male farmer dummy	–0.017	Log of total value of crop production at farmer level
Nigeria				
Saito, Mekonnen, and Spurling (1994)	All farmers: number of farmers = 226; males = 210; females = 15	Male farmer dummy	–0.130	Log of total value of production at farmer level
Korea, Rep. of				
Jamison and Lau (1982)	Mechanical farms = 1,363 (90.2 percent male heads)	Male head dummy	0.95***	Log of value of agricultural crop output
	Nonmechanical farms = 541 (87.6 percent male heads)	Male head dummy	0.059	Log of value of agricultural crop output
Thailand				
Jamison and Lau (1982)	Chemical farms = 91 (97.7 percent male heads)	Male head dummy	0.076	Log of output (kilograms)
	Nonchemical farms = 184 (99.5 male heads)	Male head dummy	0.269	Log of output (kilograms)

* Significant at 10 percent.
** Significant at 5 percent
*** Significant at 1 percent.

increase the value of household output by 10 to 20 percent (Alderman and others 1995).

Technological Adoption by Male and Female Farmers

In some countries, especially where new technologies are associated with "male" crops or activities, women may be less likely to adopt new crops or technologies. Female farm managers are less likely to grow coffee in Kenya and rear livestock in Tanzania but are equally likely to have cattle in Kenya and to grow coffee or cocoa in Côte d'Ivoire (Appleton and others 1991). Female farmers in Zambia are less likely to use oxen in cultivation (Jha, Hojjati, and Vosti 1991), because plowing is traditionally a male activity. However, women may more readily adopt technologies related to the tasks they perform, particularly if the extension agent is female. Evidence from Kenya suggests that female farmers are equally likely to apply technical advice from extension agents and are even more likely to adopt relatively complex practices such as top dressing, chemical use, and stalk borer control (Bindlish and Evenson 1993). Another study from Kenya (Saito, Mekonnen, and Spurling 1994) indicates that female farmers are more likely to adopt improved seed and use agrochemicals. In Nigeria, although male farmers are more likely to use insecticide, male and female farmers are equally likely to use fertilizer.

Despite the mixed evidence on technological adoption by gender, most of the studies find that more-educated farmers are more likely to adopt new technologies. In Kenya, more-educated females, in particular, are more likely to adopt coffee. Raising the educational level of female farmers by giving them universal primary education has higher marginal effects on the probabilities of adopting coffee than raising the educational level of all farmers, due to the generally lower levels of female education in Africa (Burger and Gunning, personal communication). Simulations suggest that a 10 percent increase in the percentage of women having completed primary schooling leads to a 6 percent increase in early adoption and a 14 percent increase in late adoption. Increasing female education also has a higher impact on adoption than increasing land sizes—a 10 percent increase in arable land increases early adoption only by 2 percent and late adoption by 6 percent.

Providing universal primary education may also stimulate early adoption by female farmers, whom other women are more likely to imitate. Other simulations by Burger and Gunning suggest that female decisionmakers are unresponsive to increases in early adoption by male farmers. This is consistent with other findings that previous awareness and adoption of the technology, particularly by farmers of the same sex, also increase the probability of adoption. The significance of gender-specific copying effects highlights the need not only for female extension agents to work with female farmers, but also for female contact farmers to be chosen. The importance of cooperatives and extension in many of these studies emphasizes the need to provide support infrastructure to rural areas.

Most of the studies reviewed also suggest that farmers with larger areas cultivated and higher values of farm tools are more likely to adopt new technology. To the extent that women farmers may have less education, have less access to land, and own fewer tools, they may be less likely to adopt new technologies. A relatively unexplored area is the extent to which weaker property rights to land and trees diminish the incentives to adopt higher-valued, permanent crops and environmentally sustainable farming practices.

Property Rights and Natural Resources

The distribution of property rights to land and natural resources underlies men's and women's differential control over productive resources. Not only does addressing gender asymmetries in control over resources improve productivity, equity, and human development, it also is critical to environmental sustainability. One of the basic arguments for paying attention to property rights is that secure tenure encourages investment, which leads to more productive and sustainable use of the resource (see Besley 1995 and Place and Hazell 1993). But if women are blocked from certain avenues of investment (for example, tree planting), or if they know that particular investments and increases in productivity will lead to the loss of their access to land, their insecurity of tenure can be a barrier to productivity and their incentives to sustain the resource over time. This is often the case when agriculture becomes commercialized, especially where food production is under women's control, but the income from cash crops goes to men (for example, see Chambers and Morris 1973 and Dey 1981).

Official policies toward granting rights from the state range from favoring men (as in many irrigation systems; see Zwarteveen 1997) to giving preference to women or women's groups. Even where official policies are gender-neutral in government-allocated rights, women may have difficulty acquiring those rights (see Lastarria-Cornhiel 1997). Limitations to access include money, where legal or illegal administrative payments are made, little knowledge of the public institutions, and distance, which may constrain women more due to their lower educational attainment and limited mobility. Limited access also results from traditional expectations of women's place and behavior that keep women in the private domain (see Freudenberger 1994).

The reasons for divergence between de jure and de facto equity are particularly important for program design. For example, acquiring resource rights through labor contributions ("sweat equity") appears to be a more equitable route than cash purchase or inheritance for resource-poor households and individuals. But this does not always hold in practice. Women may also be constrained by time, owing to their additional domestic responsibilities, from contributing sweat equity to irrigation, trees, and land development projects.

There is a tendency to idealize community resource allocation as being very equitable. Indeed, communal tenure systems often provide for all households to have some land (although women's rights may be subordinate to men's; see Lastarria-Cornhiel 1997). The poor depend heavily on common property resources such as pasture or forests that are available to members of a community (Jodha 1992), and many communities have norms that no one should be denied access to basic drinking water. However, community norms regarding the appropriate status for women may be the greatest barriers to women's control over resources, especially independent rights to the resource. Clearing trees is often a way to establish land rights under communal land tenure systems in Africa, but tree clearing is almost exclusively a male task in most societies and thus precludes women from establishing a land right. In other cases, planting trees to prevent soil erosion or improve output requires clear land rights. Even use of the commons may require access to other complementary means of production, such as animals to use grazing lands, favoring those with more resources (see Brouwer 1995).

Looking at the complementary inputs required to obtain rights sheds light on the barriers to equitable access. Land titling requires political connections and know-how, as much as cash. Even common property or open-access resources may require some private resources to exploit (for example, grazing lands require cattle, marine fisheries require boats, groundwater requires wells and pumps). Creating a level playing field for women may require addressing these hurdles as much as the formal rules and laws relating to resource tenure.

These issues speak principally to formal rights of control, such as exclusion, sale, and availability of land to others. Women's *access* to resources for their own productive and reproductive activities is more prevalent than their control. The patterns of access are more complex and more nuanced. It is in the informal use rights where we are more likely to see the flexibility and subtleties that characterize actual practice. But the process of formalizing property rights though privatization and titling of resources reduces the complex bundle of rights into a single unitary right. When customary use rights are not recognized, many women and marginal users lose out. The effect of gender relations on formal and informal property rights suggests that policymakers need to look beyond legal rights and to consider removing gender-based constraints to other services and rights, which combine to limit women's access to property (Meinzen-Dick and others 1997).

There is also a need to explore alternatives to freehold tenure that allow more flexible use patterns, which can benefit women as well as men. To do this requires good examples of tenure arrangements that accommodate multiple users. Because many of these are customary rather than statutory arrangements, there is a need for more written documentation that could be disseminated to policymakers and others involved in shaping tenure arrangements. Especially important in this regard are examples where the resource or the products of that resource have entered the market, rather than only subsistence products with more limited demand.

Looking specifically at women's uses of natural resources allows us to see a wider variety of uses and hence gives a more accurate accounting of productivity. Hedgerows, wetlands, forests, and other niches are sources of food, fuel, thatching, mats, or medicinal plants that are especially important to women (Rocheleau and Edmunds 1997). Property regimes and resource management systems that maximize output of a single commodity may appear to give the highest returns under conventional analysis, but when we look at the full spectrum of uses, other property regimes may have a higher value of output.[3] For example, looking only at marketed logs as the output of a forest misses the total value of fodder, fruits, "minor forest products," and kindling. Measuring only the grain output of an irrigation system misses the vegetable gardens, livestock watering, domestic water supply, and many household enterprises for which women use water from the system. Failure to recognize the value of those resources, or the customary rights of their users, can reduce both the productivity and equity of management systems.

In many cases, collective action by women has been instrumental in securing rights for women, either as a group or individually. Where women are blocked from holding land individually, they may be able to obtain a parcel for a women's group to use for a collective garden or nursery, for example. Women's irrigation groups in Bangladesh have invested in water pumps, enabling them to use or sell groundwater (van Koppen and Mahmud 1995). In addition, collective action may lead to a change in the rules, permitting individual women to obtain stronger rights over the resource (see Agarwal 1994 for good examples from South Asia). Although there is a large number of both nongovernmental organizations (NGOs) and local self-help groups, an important question in each area is the presence (or absence), shape, objectives, and membership of women's organizations as vehicles for learning more about women's roles and needs and as a potential base for women to assert their needs.

The other side of this issue concerns the integration of women as rights-holders and decisionmakers into traditionally male-dominated institutions for collective resource management. Policies that devolve authority for resource management from the state to local institutions make it all the more critical that the local institutions function effectively. But how does the gender composition of local institutions affect their strength and effectiveness? Male emigration or diversification out of agriculture makes this increasingly important in many parts of the world. A related question is whether women are better off integrating into existing male-dominated groups or setting up their own groups for resource management (nurseries, social forestry action)? The different roles and responsibilities of women can prejudice their ability to integrate successfully in mixed groups. A food producer cooperative set up in 1992 in Gbefi, Ghana, with grant funding conditioned on 50 percent of the members being female initially had a membership level of 59 percent women. Members were required to provide labor to the cooperative fields, receiving a profit share in proportion to their labor input. However, women's domestic responsibilities prevented them from supplying the labor needed when required, lowering their profit share and causing many women to withdraw. By 1995 the number of male farmers exceeded the number of women (Ahenkora and others 1995). However, groups made up of all men or all women do not necessarily imply homogeneity and the ability of all to participate. A credit program in Mali was targeted to women using group formation as a "collateral" mechanism through peer group pressure to repay. An evaluation indicated that women with preschool children were less able to use the credit to generate a positive return than women without preschool children (De Groote and others 1996). Homogeneity of group members in terms of the activity or its goal may be more important than gender per se (see, for example, Baland and Platteau 1996). This requires empirical determination and the lessons from successful and unsuccessful cases to be shared more widely.

Gains to Increasing Women's Incomes

The previous sections have shown how disparities in men's and women's human and physical capital and access to natural resources have costs in terms of forgone increases in agricultural output and the potential for sustainable natural resource management. Constraints to increasing women's incomes may also have important consequences for food security, individual welfare, and investment in the next generation's human capital. A growing body of evidence suggests that women spend a higher proportion of their incomes on expenditures related to household food and nutrition security. A number of studies conducted in the 1980s suggest that women and men spend income under their control in systematically different ways (Guyer 1980 and Dwyer and Bruce 1988). Women typically spend a high proportion of their income on food and health care for children as well as other goods for general household consumption. In contrast, men retain discretionary control over a higher proportion of their own incomes for personal expenditures. These studies have been confirmed by more recent studies that provide quantitative measures of the different effects of men's and women's incomes. Many of these studies have already attempted to control for unobservable household-level and community factors and the endogeneity of women's labor income.[4]

Evidence from Africa, Asia, and Latin America shows that women's income has a greater effect on household food security and preschooler nutrition than men's income (see table 15.2). In southwestern Kenya, for a given level of household income, female-controlled income share has a positive and significant effect on household consumption, while men's income has a negative effect (Kennedy 1991). Similarly, in Taiwan (China), women's income share has a significant and positive effect on household budget shares allocated to staples and education and a negative effect on budget shares allocated to alcohol and cigarettes (Thomas and Chen 1994). In Guatemala, the average yearly profits from nontraditional agricultural export crops would increase household food expenditures by twice as much if they were controlled by women rather than their husbands (Katz 1992). Finally, the effect of women's unearned income on child survival probabilities in Brazil is almost 20 times higher than that of men's unearned income (Thomas 1994).

More recent work summarized in table 15.3 has attempted to control for the possibility that government programs are placed as a result of community characteristics rather than random assignment (Pitt and Khandker 1996) as well as to derive alternative measures of nonlabor income based, for example, on asset ownership (Doss 1997). Pitt and Khandker's (1996) study of three NGO credit programs in Bangladesh finds that in the overwhelming majority of cases, the effects of women's credit compared with credit provided to men are as expected—that is, greater female labor supply, greater accumulation of women's nonland assets, greater use of contraceptives, and lower fertility. In addition, a strong result on household total consumption is found in two out of three programs. Indeed, for one of the programs run by the Grameen Bank, total household consumption increases by Tk18 for every Tk100 lent to a woman and by Tk11 for the same amount lent to a man (Pitt and Khandker 1995). Doss's (1997) study finds that, in rural Ghana, a larger share of assets controlled by women increases the budget shares allocated to food and improves child education outcomes; it decreases the budget shares of alcohol, recreation, and tobacco. Consistent with the earlier literature, Wang (1996) finds that in Zambia, mother's income increases weight-for-age more for infant girls than for infant boys, while father's income increases weight- and height-for-age more for boys than for girls.

Why do men and women tend to spend income differently? Societal and cultural norms may assign women the role of "gatekeepers," who ensure that household members, especially children, receive an adequate share of available food. Alternatively, women may prefer to spend more on children's daily needs because they spend more time with them. Women may also face different constraints than men. To minimize the competing demands on their time, for example, women may spend more on food because they purchase more expensive calories that take less time to prepare. Finally, women and men may have different income flows and thus different transaction costs. In other words, because women's income tends to come more frequently and in smaller amounts, it may be more readily spent on household daily subsistence needs than lumpier seasonal income, which tends to come to men and is likely to be spent on more expensive items (Hamilton, Popkin, and Spicer 1984). A recent study tested this hypothe-

Table 15.2. Effects of Men's and Women's Income or Assets on Household Welfare

Country and study	*Effect on*	*Effect of women's income or assets*	*Effect of men's income or assets*	*Ratio of effect of women's income or assets to that of men's income or assets*
Africa				
Kenya, Kennedy (1991)	Household calorie level	Positive	n.a.	n.a.
Asia				
Taiwan (China), Thomas and Chen (1994)	Household budget share for alcohol	Negative	Negative	1.3
Latin America				
Brazil, Thomas (1994)	Child weight-for-height	Positive	Positive	4.2
Brazil, Thomas (1994)	Child survival	Positive	Positive	18.2
Guatemala, Katz (1992)	Food expenditures	Positive	Positive	2

n.a. Not applicable.

Table 15.3. Summary of Recent Studies of the Effects of Female Income, Assets, and Credit on Selected Welfare Outcomes

Study	*Country (data year)*	*Estimation and test*	*Conclusion*	*Comments*
Rogers, Swindale, and Ohri-Vachaspati (1996)	Honduras (1993–94)	Ordinary least squares: effect of percent of income earned by mother on preschooler standardized height-for-age	"Controlling for income level, the percent of income earned by the child's mother has a positive effect on nutritional outcome … the positive effect … decreases as children get older." (p. 150)	No instrumenting of the share of women's income
Pitt and Khandker (1995, 1996)	Bangladesh (1991–92)	Weighted exogenous sampling maximum likelihood fixed effects: testing for differential impact of male and female borrowing from NGOs on eight outcomes—boy's and girl's schooling, women's and men's labor supply, total household expenditure, contraception use, fertility, and value of women's nonland assets	"The set of female credit variables is statistically significant in seven of eight cases at the 0.05 percent level. By contrast the set of male credit variables is significant in three out of eight cases … the hypothesis that male and female credit parameters are jointly equal for each of these three (NGO) programs is rejected in only four cases: women's labor supply, women's nonland assets, contraception, and fertility." (p. 41) Household consumption increases by Tk18 for every Tk100 lent to a woman and Tk11 for the same amount lent to a man (Pitt and Khandker 1995)	Estimation and sampling of households take care to identify the average impact of the credit program;the key is to sample households in program villages that are ineligible due to some exogenous constraint (such as landholding restrictions in the absence of a land market)
Ward-Batts (1996)	United Kingdom (1973–76 and 1980–82)	Ordinary least squares and Tobit: Are budget shares on 23 goods affected by dummy variable for policy shift in U.K. child benefit allocation (from men and women to women)?	Follows up on work of Lundberg, Pollak, and Wales (1997) Strong negative impact on men's clothing, tobacco, housing, strong positive impacts on children's clothing, fuel, and food at home	Total expenditure not instrumented; changes in other policies not fully accounted for (value added tax)
Doss (1997)	Ghana (1991–92)	Ordinary least squares and logit: budget share equations savings	For urban households, female asset share has a significant impact on seven out of eight budget share categories: food (+), alcohol (–),	Total expenditure not instrumented; assets = accounts and

			education (+), recreation (–), tobacco (–), household (+), and miscellaneous (–); for rural households, female farmland share has a significant impact on food (+), alcohol (–), recreation (–), tobacco (–), and completeness of vaccinations (+). For rural households female asset share significantly affects four out of eight budget share categories: food (+), alcohol (–), recreation (–), tobacco (–), and child education outcomes (+)	businesses controlled by household; assignation of ownership of assets assumes that knowledge and responsibility for business equal control
Wang (1996)	Zambia (1991–93)	Fixed-effects instrumental variables	Mother's income increases weight-for-age more for infant girls than for infant boys; father's income increases boys weight- and height-for-age more for boys than for girls; the difference between these differential effects is only significant for weight-for-age at the 10 percent level; for height-for-age, the differential effect of male and female income is significantly different	Because of fixed-effects methods used, can only look at the differential effect of mother's and father's income on child anthropometric outcomes; need to use panel IV methods that allow for instrumentation of time-invariant variables (for example, the Hausman-Taylor estimator)
Gammage (1997)	Nigeria (1996)	Univariate logit	Association of women's income contributions with decisions on reproductive health and fertility: subsets of these associations are significant and positive, while none is negative and significant	Needs multivariate analysis to control for other observable factors and for unobservable factors

sis by controlling for the flows of income earned by men and women in Niger (Hopkins, Levin, and Haddad 1994). The findings indicate that the timing of female income flows has a significant effect on both total household expenditures and food expenditures in a given season, while the timing of male income flows has no effect. This suggests that women have less access than men to resources that tend to even out consumption, such as credit and savings. Thus, both the timing of overall household income and the flow of income by gender influence seasonal food expenditures.

What are the policy and programmatic implications of the differential impacts of women's income? The implications for the design of transfer programs are straightforward: when the benefits to targeting women (in terms of the intervention's objective) outweigh the costs, then targeting should be undertaken. More generally and more importantly, these results stress the impact of women's economic status within the household on a number of important policy objectives: fertility reduction, food purchases, nutrition, and education to name a few. The previous sections suggested that excluding women from productive activity is a waste of talent that will result in lost economic productivity today. The results in this section suggest that the second-round impacts of improving economic opportunities for women could be profound in terms of improving economic productivity tomorrow—through lowered fertility, improved child survival, and increased human capital.

Toward Gender-Sensitive Project Design

The previous sections have documented the striking differences between men and women in the ownership of, access to, and control of assets and why that matters for a wide range of outcomes such as agricultural productivity, food security, environmental sustainability, and human welfare. If policymakers and project designers continue to implement programs that do not take account of gender roles, responsibilities, and resource constraints, neither sustainable economic growth nor successful development is likely.

Recognition of these differences, however, is not new. The Percy Amendment to the U.S. Foreign Assistance Act, passed by the U.S. Congress in 1973, mandated the integration of women into national economies, recognizing that doing so would improve development efforts as well as the status of women. In 1975, at the World Conference on Women in Mexico City, the United Nations declared that 1976–85 would be the U.N. Decade for Women: Equality, Development, and Peace. The 1995 Fourth World Conference on Women in Beijing set forth a platform of action for member states to undertake to improve women's status and welfare. Although much has been learned over the years, much remains to be done.

Not surprisingly, an early approach to reducing the asymmetries focused on targeting projects and programs to women. Unfortunately, in most cases it was no more successful at improving women's status than previous policies targeted to the household, and thereby often implicitly to men (Peña, Webb, and Haddad 1996). There are two key reasons for this: women are not isolated economic actors, and women per se are not a homogeneous group. There are numerous examples of project failure with respect to the former. For example, when irrigation was introduced in The Gambia in the early 1980s to raise yields, commercialize rice production, and increase women's share of household income, community initiatives ended up reducing women's income. Previously, women were the rice growers. Yield increases transformed rice from a private crop under the control of women to a communal crop under the control of men (von Braun and Webb 1989 and Dey 1981). Moreover, more recent attempts by donors to safeguard women's access to land were frustrated by managers of the project, who sided with male household heads in disputes over access to land (Dey 1992).

Neither are women, as a group, homogeneous. Heterogeneity is demonstrated by differences in class and caste, ownership of land versus landlessness, rental versus ownership, life cycle stage, marriage order (where there is polygyny), position as a female head of household or as part of a joint or male-headed household, and household composition, all of which are sources of variance that may be greater than their common interests as women (Meinzen-Dick and others 1997). Class and power relationships crosscut gender. In Mexico, privatization of land dominated by the well-to-do has marginalized poor men as well as poor women, with men migrating out (Goldring 1996). Ethnicity may be important, as in Ecuador, where indigenous and mestizo women have differential access to productive resources such as land (Ahlers 1995).

Many researchers and policymakers who are unfamiliar with gender issues use a "shortcut" approach to reaching women by identifying female-headed households to whom to provide benefits or promote equity.

Although these households may be perceived to have more autonomy, they are by no means more homogeneous than women in general. They are differentiated not only by the factors identified above but also by the process by which they became female headed: divorce, widowhood, or male outmigration. For those de jure female heads of household, inheritance patterns—patrilineal or matrilineal—are a significant determinant of the degree of resource access and ownership.

Successful project or policy formulation is unlikely without gender analysis.[5] This goes beyond a women-in-development approach. In the latter, the differences in resource access, ownership, and control between men and women are identified; in the former, the determinants of these asymmetries and their consequences for individuals, households, communities, and economic development are recognized (Murphy 1995). In essence, the women-in-development approach recognizes an outcome, whereas a gender-and-development approach recognizes both the outcome and its consequences as well as the process leading to those outcomes. The understanding gained from the gender-and-development approach facilitates understanding not only how a planned project or policy will affect men and women but also how it will affect the underlying processes that condition the allocation of rights, resources, and responsibilities within communities and households.

We recognize that gender analysis requires resources. However, the costs of failing to understand these asymmetries can be high in terms of wasted resources or negative effects on household welfare and resource use (von Braun, Puetz, and Webb 1989). Additionally, recognizing the process by which gender asymmetries are determined broadens the array of policy instruments available to tackle them. For example, changing legal frameworks or incentive structures can create an enabling environment to encourage project or policy success.

Although participatory approaches to project design have been recognized as a way of ensuring community involvement, commitment, and ownership of development projects, they are not synonymous with incorporating gender analysis. As shown here, community resource allocation is not necessarily equitable. Neither do participatory approaches to project design guarantee that women will participate in either their needs assessment or the proposed project design (Mosse 1995). A World Bank study finds that only 20 rural water supply projects in a portfolio of 121 were highly participatory in their design, and of those, only 10 successfully reached women (Narayan 1995). Respect needs to be given to local norms of behavior to ensure that women are given appropriate opportunities to identify their needs and participate in project discussions. This may involve holding separate meetings with men and women, structuring women's meetings to fit with their domestic work schedules, and identifying who the implicit female leaders are. Even where planned projects are thought to be relevant to either men or women, they are likely to be less successful if spouses are not involved in preproject community discussions. In an oilseeds processing project in Zimbabwe, men only permitted their spouses to become fully committed to project meetings and activities after the project had been discussed with them and they could appreciate the benefits to their wives and families (Mehra 1996).

Successful gender analysis at this early stage will ensure not only good technical design but also good socioeconomic design of projects: a design that takes account of local production systems, resource bases and access, distribution of labor, distribution of potential project benefits, and the bargaining power of men and women both within and across different classes and castes.

The results of gender analysis should then be operationalized. In the first instance, having completed the gender analysis stage, a planner should be in a position to determine the gender "design" or balance of the project. At one end of the spectrum, projects may be designed solely for women, with only female participants. At the other end of the spectrum, projects may be completely mainstreamed, with a gender-neutral project design giving equal opportunities for participation to men and women. Somewhere in-between these two options lie hybrid approaches—having a "women-only" component within a larger project or having separate budgeting and reporting of objectives by gender within a mainstream project. There is no "first-best" recommendation because each has its advantages and disadvantages and the appropriate choice depends on the community's characteristics and the project's goals.

Mainstreaming, for example, has the advantage of being able to draw on all project resources and is thus more likely to be sustainable. It also provides scope for capacity building. Unless the organization or implementing agency is sensitive to women's needs, however, the project itself can be gender insensitive, such that women may be marginalized. Moreover, it may be difficult to monitor performance with regard to gender. Mainstreamed projects are more likely to be successful if a high level of awareness of gender issues is reflected in project design. A strong commitment from local institutions, community leaders, and project directors is also essential if women are to reap benefits in proportion to their participation.

Another option is to have an identifiable budget and reporting structure within a mainstream project. In this model, attention to gender issues is more visible, gender-specific indicators are easier to monitor, and the project itself can raise awareness of the need to pay attention to gender issues. Although separate budgeting can help to maintain a high profile for women in the project, greater care needs to be taken to prevent tokenism: it may simply be used as an excuse to allocate a small budget for women's participation.

A separate women's project component allows the needs of women to be met more specifically than in a mainstream project. This enables the project to be visible and fine-tuned to women's needs, while being linked to the mainstream. It may also enable project implementers to tap other resources and access technical expertise. It also has a higher chance of guaranteeing women access to project benefits. However, such projects are often a low priority of policymakers and implementing organizations. They are typically underbudgeted and small in scale and, without sufficient commitment from project leaders, face the danger of being marginalized. They may also divert attention from gender issues in the broader project. Lastly, having a tendency to focus on welfare issues, they may neglect women's role as agricultural producers and managers. However, they are a good method of demonstrating to local communities the efficacy of addressing women's issues and can build a constituency for mainstreaming gender issues in the future.

Finally, organizations may choose to implement a free-standing women-in-development project. Such a project addresses women's specific needs, is visible, and is easily monitored. However, it is often difficult to get a constituency for the project from both community leaders and government. Because free-standing projects are typically small in scale, they can be invisible to local and national government ministries and thus may not raise awareness of gender issues for future projects. They often focus on welfare rather than broader issues of empowerment and resource access. Nevertheless, these projects can be highly successful in meeting women's needs in situations where there is no constituency for more gender-integrated projects. Striving to make the project visible but sensitive to community norms can work to build a constituency for the future. This can be the first step to achieving gender integration, first through a project subcomponent, after which it is fully mainstreamed.

Having decided on the appropriate gender balance of the project, it is important to ensure the ongoing participation of women. During gender analysis, the "critical path" should be identified—those aspects of the project that must be achieved to guarantee overall project success with respect to women's participation and accrual of benefits. These elements should form a central core of project monitoring activities. A system of indicators should be devised at the beginning of the project that will adequately track the performance of the project. This is particularly important where there is a lack of constituency for gender issues and where supervision is provided from outside the community.

Care should be exercised so that women can make the required and envisaged project commitments. For example, where "sweat equity" represents the community commitment to a project, it is necessary to ensure that the timing of such provision fits into women's domestic schedules; otherwise project timing may need to be modified. Where user groups are formed to manage natural resources, appropriate representation by women should be assured. Even where projects are targeted solely to women, such as reproductive health programs, they are less likely to be successful if program delivery does not assign value to women's time and recognize their significant economic roles and responsibilities (Mehra 1996).

Concluding Remarks

We have reviewed a literature suggesting that gender asymmetries and failure to acknowledge them have short-run and long-run costs. In the short run, gains in agricultural productivity and household food security are forgone. In the longer run, sustainability of agricultural production is compromised by less-than-optimal natural resource management, as are investments in schooling and fertility control.

If the gains to recognizing and mitigating gender asymmetries are so compelling, why do they persist? First, gender analysis is not yet a widely used tool for policy analysts and project designers and implementers, partly due to the confusion between the women-in-development goals and the gender-and-development goals. Second, gender analysis can be costly in terms of data collection (more disaggregated, though better targeted) and training. Third, there is confusion as to whether targeting women necessarily reduces asymmetries. Fourth, although the social and economic environment can be altered to reduce or eliminate these asymmetries, it is a difficult process that may involve changing the legal environment. Fifth, many institutions beyond the household (such as community groups, extension agents, credit groups, project managers) need to see the value to all their members of using the gender lens. This involves operationalizing gender analysis within ministries, NGOs, and projects.

The challenge for research is to improve estimates and documentation of the costs of gender asymmetries and to develop better operational tools for gender analysis. These tools are needed to enhance the ability of development actors to assess, a priori, the costs and benefits of gender analysis for themselves.

Notes

1. A classic example is Jones's (1986) study of Cameroon. Rice is considered a male crop, so any income generated from it was controlled by men, even if the crop was produced by women. Consequently, few women entered into improved rice cultivation. Instead, they continued to grown sorghum, which they controlled, despite its lower returns.

2. In Hindu law, women do not have the right to own, acquire, or dispose of property, although they may have the right to some personal property. In Islamic law, although women's landownership rights are explicit, the share given to female heirs is half that given to males.

3. This is related to the broader question of accounting for women's productivity, including both home and market production. See Beneria (1992) and Quisumbing (1996) for reviews of these issues.

4. Econometric approaches have involved using fixed-effects methods to control for unobserved household and community-level factors. The endogeneity of women's income has been addressed by instrumenting women's income or using nonlabor income instead of total income.

5. Good sources for gender analysis include Feldstein and Jiggins (1994), Thomas-Slayter, Esser, and Shields (1993), and Fong, Wakeman, and Bhushan (1996).

References

The word "processed" describes informally reproduced works that may not be commonly available through libraries.

Agarwal, Bina. 1994. *A Field of One's Own: Gender and Land Rights in South Asia.* Cambridge, U.K.: Cambridge University Press.

Ahenkora, Salome, Elizabeth Akpalu, Joanna Kerr, Kofi Marfo, and Francis Ulzen-Appiah. 1995. "Gender and the Implementation of Structural Adjustment in Africa: Examining the Micro-Meso-Macro Linkages. The Ghana Study." Report prepared for the Canadian International Development Agency for the Structural Adjustment and Gender in Africa (SAGA) Initiative. North-South Institute, Ottawa. Processed.

Ahlers, R. 1995. Untitled electronic mail communication submitted to the gender-prop e-mail conference, International Food Policy Research Institute, Washington, D.C., 13 October. Processed.

Alderman, Harold, John Hoddinott, Lawrence Haddad, and Christopher Udry. 1995. *Gender Differentials in Farm Productivity: Implications for Household Efficiency and Agricultural Policy.* Food Consumption and Nutrition Division Discussion Paper 7. Washington, D.C.: International Food Policy Research Institute.

Alderman, Harold, John Hoddinott, Lawrence Haddad, and Stephen Vosti. 1994. "Strengthening Agricultural and Natural Resources Policy through Intrahousehold Analysis: An Introduction." *American Journal of Agricultural Economics* 76(5):1208–12.

Appleton, Simon, David L. Bevan, Kees Burger, Paul Collier, Jan Willem Gunning, Lawrence Haddad, and John Hoddinott. 1991. "Public Services and Household Allocation in Africa: Does Gender Matter?" Center for African Studies, Oxford University, Oxford. Processed.

Baland, Jean-Marie, and Jean-Philippe Platteau. 1996. *Halting Degradation of Natural Resources. Is There a Role for Rural Communities?* Oxford: Clarendon Press and Food and Agriculture Organization of the United Nations.

Beneria, Lourdes. 1992. "Accounting for Women's Work: The Progress of Two Decades." *World Development* 20(11):1547–60.

Besley, Timothy. 1995. "Property Rights and Investment Incentives: Theory and Evidence from Ghana." *Journal of Political Economy* 103(5): 903–37.

Bindlish, Vishva, and Robert Evenson. 1993. *Evaluation of the Performance of T&V Extension in Kenya.* Technical Paper 208. Washington, D.C.: World Bank.

Boserup, Ester. 1970. *Women's Role in Economic Development.* New York: St. Martin's Press.

Brouwer, R. 1995. "Common Goods and Private Profits: Traditional and Modern Communal Land Management in Portugal." *Human Organization* 54(3):283–94.

Brown, Lynn R., and Lawrence Haddad. 1995. "Time Allocation Patterns and Time Burdens: A Gendered Analysis of Seven Countries." International Food Policy Research Institute, Washington, D.C. Processed.

Chambers, Robert, and Jon Morris, eds. 1973. *Mwea: An Irrigated Rice Settlement in Kenya.* Munich: Weltforum Verlag.

De Groote, Hugo, Eileen Kennedy, Ellen Payongayong, and Lawrence Haddad. 1996. "Credit with Education for Women in Mali: Self-Selection and Impacts on Women's Income, with Implications for Adult Female and Preschooler Nutrition." Report for the U.S. Agency for International Development. International Food Policy Research Institute, Washington, D.C. Processed.

Dey, Jennie. 1981. "Gambian Women: Unequal Partners in Rice Development Projects." *Journal of Development Studies* 17(3, April):109–22.

———. 1992. "Gender Asymmetries in Intrahousehold Allocation of Land and Labor: Some Policy Implications." Paper presented at the conference on intrahousehold resource allocation: policy issues and research methods, International Food Policy Research Institute and World Bank, Washington, D.C., February. Processed.

Dixon, Ruth B. 1982. "Women in Agriculture: Counting the Labor Force in Developing Countries." *Population and Development Review* 8(3, September):539–65.

Doss, Cheryl. 1997. "The Effects of Women's Bargaining Power in Household Health and Education Outcomes: Evidence from Ghana." Department of Economics, Williams College, Williamstown, Mass. Processed.

Dwyer, D., and J. Bruce. 1988. *A Home Divided: Women and Income in the Third World.* Palo Alto, Calif.: Stanford University Press.

Feldstein, Hilary Sims, and Janice Jiggins, eds. 1994. *Tools for the Field: Methodologies Handbook for Gender Analysis in Agriculture.* West Hartford, Conn.: Kumarian Press.

Fong, Monica S., Wendy Wakeman, and Anjana Bhushan. 1996. *Toolkit on Gender in Water and Sanitation.* Survey Toolkit Series 2. Washington, D.C.: World Bank, Poverty and Social Policy Department.

Freudenberger, M. S. 1994. "Tenure and Natural Resources in The Gambia: Summary of Research Findings and Policy Options." Land Tenure Center, University of Wisconsin, Madison. Processed.

Gammage, Sara. 1997. "Women's Role in Decision Making at the Household Level: A Case Study in Nigeria." International Center for Research on Women, Washington, D.C. Processed.

Goldring, L. 1996. "Irrigation, Privatization, and Migration." Paper submitted to gender-prop e-mail conference, International Food Policy Research Institute, Washington, D.C., 11 October. Processed.

Guyer, Jane. 1980. "Household Budgets and Women's Incomes." African Studies Center Working Paper 28. African Studies Center, Boston University, Boston. Processed.

Haddad, Lawrence, John Hoddinott, and Harold Alderman. 1997. *Intrahousehold Resource Allocation: Methods, Models, and Policy.* Washington, D.C.: International Food Policy Research Institute; Baltimore, Md.: Johns Hopkins University Press.

Hamilton, S., Barry Popkin, and D. Spicer. 1984. *Women and Nutrition in Third World Countries.* Praeger Special Studies. New York: Begin and Garvey.

Higgins, Paul A., and Harold Alderman. 1997. "Labor and Women's Nutrition: The Impact of Work Effort and Fertility on Nutritional Status in Ghana." *Journal of Human Resources* 22(3): 577–95.

Hopkins, Jane, Carol Levin, and Lawrence Haddad. 1994. "Women's Income and Household Expenditure Patterns: Gender or Flow? Evidence from Niger." *American Journal of Agricultural Economics* 76(5):1219–25.

Jamison, Dean, and Lawrence Lau. 1982. *Farmer Education and Farm Efficiency.* Baltimore, Md.: Johns Hopkins University Press for the World Bank.

Jha, Dayanatha, Behjat Hojjati, and Stephen Vosti. 1991. "The Use of Improved Agricultural Technology in Eastern Province." In Rafael Celis,

John T. Milimo, and Sudhir Wanmali, eds., *Adopting Improved Farm Technology: A Study of Smallholder Farmers in Eastern Province, Zambia.* Washington, D.C.: International Food Policy Research Institute.

Jodha, N. S. 1992. *Common Property Resources: A Missing Dimension of Development Strategies.* Discussion Paper 169. Washington, D.C.: World Bank.

Jones, Christine. 1986. "Intrahousehold Bargaining in Response to the Introduction of New Crops: A Case Study from Northern Cameroon." In Joyce L. Moock, ed., *Understanding Africa's Rural Households and Farming Systems.* Boulder, Colo.: Westview Press.

Katz, Elizabeth. 1992. "Intrahousehold Resource Allocation in the Guatemalan Central Highlands: The Impact of Nontraditional Agricultural Exports." Ph.D. diss. University of Wisconsin, Madison.

Kennedy, Eileen. 1991. "Income Sources of the Rural Poor in Southwestern Kenya." In J. von Braun and R. Pandya-Lorch, eds., "Income Sources of Malnourished People in Rural Areas: Microlevel Information and Policy Implications." Working Paper on Commercialization of Agriculture and Nutrition 5. International Food Policy Research Institute, Washington, D.C. Processed.

Lastarria-Cornhiel, Susana. 1997. "Impact of Privatization on Gender and Property Rights in Africa." *World Development* 25(8):1317–33.

Lundberg, Shelly J., Robert A. Pollak, and Terence J. Wales. 1997. "Do Husbands and Wives Pool Their Resources? Evidence from the United Kingdom Child Benefit." *Journal of Human Resources* 32(3):462–80.

McGuire, Judith S., and Barry M. Popkin. 1990. *Helping Women Improve Nutrition in the Developing World: Beating the Zero Sum Game.* Technical Paper 114. Washington, D.C.: World Bank.

Mehra, Rekha, ed. 1996. *Taking Women into Account: Lessons Learned from NGO Project Experiences.* Washington, D.C.: International Center for Research on Women.

Meinzen-Dick, Ruth S., Lynn R. Brown, Hilary S. Feldstein, and Agnes R. Quisumbing. 1997. "Gender, Property Rights, and Natural Resources." *World Development* 25(8):1299–302.

Moock, Peter. 1976. "The Efficiency of Women as Farm Managers: Kenya." *American Journal of Agricultural Economics* 58(5, December):831–35.

Moser, Caroline. 1989. "Gender Planning in the Third World: Meeting Practical and Strategic Gender Needs." *World Development* 17(11, November): 1799–825.

Mosse, David. 1995. "Authority, Gender, and Knowledge: Theoretical Reflections on Participatory Rural Appraisal." *Economic and Political Weekly*, 18 March.

Murphy, Josette L. 1995. *Gender Issues in World Bank Lending.* Washington, D.C.: World Bank, Operations Evaluation Department.

Narayan, Deepa. 1995. *The Contribution of People's Participation: Evidence from 121 Rural Water Supply Projects.* Environmentally Sustainable Development Occasional Paper 1. Washington, D.C.: World Bank.

Peña, Christine, Patrick Webb, and Lawrence Haddad. 1996. *Women's Advancement through Agricultural Change: A Review of Donor Experience.* Food Consumption and Nutrition Division Discussion Paper 10. Washington D.C.: International Food Policy Research Institute.

Pitt, Mark M., and Shahidur R. Khandker. 1995. "The Impact of Group-Based Credit Programs on Poor Households in Bangladesh: Does the Gender of Participants Matter?" Department of Economics, Brown University, Providence, R.I. Processed.

———. 1996. *Household and Intrahousehold Impact of the Grameen Bank and Similar Targeted Credit Programs in Bangladesh.* Discussion Paper 320. Washington, D.C.: World Bank.

Place, Frank, and Peter Hazell. 1993. "Productivity Effects of Indigenous Land Tenure Systems in Sub-Saharan Africa." *American Journal of Agricultural Economics* 75(1):10–19.

Quisumbing, Agnes R. 1995. *Gender Differences in Agricultural Productivity: A Survey of Empirical Evidence.* Food Consumption and Nutrition Division Discussion Paper 5. Washington, D.C.: International Food Policy Research Institute.

———. 1996. "Male-Female Differences in Agricultural Productivity: Methodological Issues and Empirical Evidence." *World Development* 24(10):1579–95.

Rocheleau, Diane, and Davids Edmunds. 1997. "Women, Men, and Trees: Gender, Power, and Property in Forest and Agrarian Landscapes." *World Development* 25(8):1351–71.

Rogers, Beatrice L., Ann Swindale, and P. Ohri-Vachaspati. 1996. "Determinants of Household

Food Security in Honduras." Department of Nutrition, Tufts University, Bedford, Mass. Processed.

Safilios-Rothschild, Constantina. 1985. "The Persistence of Women's Invisibility in Agriculture: Theoretical and Policy Lessons from Lesotho and Sierra Leone." *Economic Development and Cultural Change* 33(2, January):299–317.

Saito, Katrine, Hailu Mekonnen, and Daphne Spurling. 1994. *Raising the Productivity of Women Farmers in Sub-Saharan Africa.* Discussion Paper 230. Washington, D.C.: World Bank.

Saito, Katrine, and Daphne Spurling. 1992. *Developing Agricultural Extension for Women Farmers.* Discussion Paper 156. Washington, D.C.: World Bank.

Seager, Juri, and Ann Olson. 1986. *Women in the World: An International Atlas.* New York: Simon and Schuster.

Subbarao, Kalanidhi, and Laura Raney. 1996. "Social Gains from Female Education: A Cross-National Study." *Economic Development and Cultural Change* 44(1):105–28.

Thomas, Duncan. 1994. "Like Father, Like Son, or Like Mother, Like Daughter: Parental Education and Child Health." *Journal of Human Resources* 29(4):950–88.

Thomas, Duncan, and Chien-Liang Chen. 1994. "Income Shares and Shares of Income." Labor and Population Working Paper 94-08. Rand Corporation, Santa Monica, Calif. Processed.

Thomas-Slayter, Barbara P., A. Lee Esser, and M. Dale Shields. 1993. "Tools of Gender Analysis. A Guide to Field Methods for Bringing Gender into Sustainable Resource Management." ECOGEN Research Project, International Development Program, Clark University, Worcester, Mass. Processed.

Tisch, S., and Thelma Paris. 1993. "Labor Substitution in Philippine Rice Farming Systems: An Analysis of Gender." International Rice Research Institute, Manila. Processed.

Udry, Christopher. 1996. "Gender, Agricultural Production, and the Theory of the Household." *Journal of Political Economy* 104(5):1010–46.

van Koppen, Barbara, and Simeen Mahmud. 1995. "Women and Water-Pumps in Bangladesh: The Impact of Participation in Irrigation Groups on Women's Status." Department of Irrigation and Soil and Water Conservation, Wageningen Agricultural University, Wageningen, The Netherlands. Processed.

von Braun, Joachim, Detlev Puetz, and Patrick Webb. 1989. *Irrigation Technology and Commercialization of Rice in The Gambia: Effects on Income and Nutrition.* Research Report 75. Washington, D.C.: International Food Policy Research Institute.

von Braun, Joachim, and Patrick Webb. 1989. "The Impact of New Crop Technology on the Agricultural Division of Labor in a West African Setting." *Economic Development and Cultural Change* 37(20):513–34.

Wang, M. 1996. "Gender Differences in Intrahousehold Resource Allocations: An Empirical Analysis of Child Health in Zambia." Department of Economics, George Washington University, Washington, D.C. Processed.

Ward-Batts, Jennifer. 1996. "Out of the Wallet and into the Purse: Does Income Control Really Affect Intrahousehold Resource Allocation?" Department of Economics, University of Washington, Seattle. Processed.

Zwarteveen, M. 1997. "Water: From Basic Need to Commodity: A Discussion on Gender and Water Rights in the Context of Irrigation." *World Development* 25(8):1335–49.

Part III
Technical Issues and Perspectives

In parts I and II we stressed the need to put in place appropriate policy and institutional frameworks that are conducive to sustainable rural development. One of the ways in which these frameworks function is by creating the right incentives for farmers to innovate in their production technologies. Here in part III we focus mainly on technical issues because sustainable rural development must also be technically sound. We cover the following issues: soil conservation, mainstreaming of biodiversity, integrated pest management and other strategies to reduce pesticide use, economic and environmental aspects of irrigation and drainage, livestock-environment interactions, inappropriate deforestation, and agroforestry.

Chapter 16 by Lutz and others seeks to shed empirical light on the question of soil conservation. Soil degradation represents a reduction in the land's actual or potential uses. Many cultivation practices tend to degrade soil over time. For example, cultivation can expose soil to water and wind erosion, repeated tillage can weaken soil structure, crop production can remove nutrients, and use of machinery can compact the soil. Soil degradation, in turn, affects productivity. As soil is degraded, crop yields decline or input levels (and hence costs) rise in an effort to keep or restore productivity.

Despite long-standing concern about these problems, surprisingly little hard evidence exists on their magnitude. The degradation figures quoted in the literature are often extrapolated from very limited data and may exaggerate the problem because they often consider "moved soil" as "lost soil," even though much of it may be deposited on the same or other agricultural land. Soil degradation is predicted to have severe effects on agricultural productivity, yet evidence on the magnitude of these effects is hard to find. In fact, in many cases declines in productivity are claimed with little or no supporting evidence at all.

Degradation can be slowed or arrested by a large range of options, including cultural practices such as contour plowing and minimum tillage, vegetative practices such as grass strips, strip cropping, and vegetative barriers, and mechanical measures such as terraces and cutoff drains. Adoption of any of these techniques can be costly, either directly in investment requirements or indirectly in production forgone, and some measures are better suited to some conditions than to others. The critical question facing farmers, and society as a whole, is whether the benefits of a given conservation measure or set of measures are sufficient to make the costs worth bearing.

The problem of soil degradation and conservation can be examined from two perspectives. The first is that of society as a whole. From this perspective, all the costs and benefits of a given activity must be considered, including off-site effects. From the second perspective, that of the farmers themselves, only the costs and benefits that actually accrue to the agent making the decisions about resource use are considered.

Lutz and others, in chapter 16, examine the returns

to investment in conservation measures mainly from the farmers' point of view for two reasons. First, decisions about land use are ultimately made by the farmers themselves in light of their own objectives, production possibilities, and constraints. Understanding the incentives facing individual farmers is necessary, therefore, if patterns of resource use are to be understood and appropriate responses to problems formulated. Second, land use problems generally depend heavily on site-specific biophysical characteristics, which can vary significantly even within small areas. A farm-level approach also places the emphasis firmly on the effects that degradation has on farm productivity. In developing countries, where substantial numbers of people depend directly on agricultural production, the effect of degradation on yields is often critical. This is not to belittle the importance, in some situations, of off-farm effects, such as siltation of reservoirs and waterways.

In making their decisions about land use, farm households need to consider both the agroecological and the economic characteristics of the environment in which they operate. The household's problem can be summarized as one of maximizing the present value of the stream of expected net returns to agricultural production. With regard to the adoption of conservation measures, the issue is whether returns under the optimal path of the new, more conserving system are sufficiently greater than returns under the path of the current, more degrading system to justify the cost of switching.

The estimated productivity losses of soil erosion and degradation vary considerably, as do the estimated effects of conservation practices on yields. In some instances, yields are expected to recover once conservation measures are established—partly because soil regenerates once the processes of degradation are halted, partly because fertilizers are used more efficiently, and partly because improved cultivation practices are sometimes introduced together with conservation. Elsewhere, conservation measures might slow but not halt the decline in yield.

Introducing conservation measures often has an adverse effect on production because some of the area cultivated is turned over to use as diversion ditches, terraces, or hedges. Physical structures, in particular, usually reduce the area available for cultivation more than 10 percent. But conservation measures can also have beneficial impacts. Besides reducing soil loss and the rate of decline in yield, they can encourage the retention of moisture and stimulate improvements in the soil's physical structure.

Whether conservation measures are profitable for the farmers is an empirical and site-specific issue. Returns to conservation depend on the specific agroecological conditions faced, on the technologies used, and on the prices of inputs used and outputs produced. Generally, the farmers' decision to invest in conservation is based on normal considerations of benefit and cost: farmers tend to adopt conservation measures when it is in their interest to do so, unless some constraint is present. Cases in which returns to conservation are low or negative correlate well with low adoption rates. Profitability of conservation practices is a necessary, but not always a sufficient, condition for their adoption. Factors other than strict cost-benefit considerations also play a role. Institutional and motivational issues must be considered together with the results of the cost-benefit analysis.

Advocates of soil conservation often argue that subsidies are indispensable to induce farmers to adopt conservation measures. But such statements assume that conservation is inherently desirable whether or not there is concrete evidence that the benefits outweigh the costs. Lutz and others, in chapter 16, show that this may be far from the case; frequently, the benefits of specific conservation techniques (such as mechanical structures) do not justify their costs. Unless there are important off-site effects or the price signals received by farmers are significantly distorted, subsidies to induce adoption will not enhance economic efficiency. Whatever their justification, the use of subsidies encounters several difficulties. First, the divergence between social and private returns to conservation must be established, so that intervention can be directed where it will be most effective. Second, it is difficult to design appropriate incentive structures so that social objectives are met. Third, subsidization schemes often create perverse incentives for farmers.

One important way for governments to help is to make sure that constraints such as insecure tenure do not prevent farmers from adopting conservation measures. Also, governments already conduct research on soil conservation and provide, through extension services, some assistance to farmers who undertake conservation work. However, research in experiment stations tends to favor technical efficiency (including structural measures such as terraces) over cost-effectiveness. Further, government extension work is often ineffective. In many cases, nongovernmental organizations NGOs, such as Vecinos Mundiales in Central America, have proven to be more effective than government at presenting the range of options to farmers and delivering related technical assistance. Given the wide variety of conditions that farmers face, government extension services should also provide, explain, and demonstrate to farmers the corresponding variety of options available rather than, as has often happened in the past, pushing broadly for the adoption of specific techniques. And it may be innovative as well as effective for governments to decentralize decision-making and channel budgetary resources for soil conservation to the local level. This would allow communities to participate and contract assistance from those from whom the greatest contributions can be expected.

Chapter 14 discussed biodiversity conservation with a focus on protected areas. But much of the world's biodiversity exists in human-managed or -modified systems. That is the focus of Srivastava and others in chapter 17. Biodiversity has two important dimensions: the genetic variation within species and populations and the preservation of habitat. The significance of variation within a species, although poorly appreciated, is critical, particularly for agriculture. The continued productivity of existing crops and livestock hinges in large part on harnessing the genetic variation found within each species. Habitat conservation seeks to safeguard natural habitats for wild species and populations and to manage habitats that have been modified for human use, such as farmland. Agrobiodiversity is found in habitats that have been modified for crop and livestock production. Agrobiodiversity includes all plants and animals that contribute directly or indirectly to the raising of crops and livestock.

Although it is conceptually useful to differentiate agrobiodiversity from the larger array of species and habitats, the boundaries between biodiversity and agrobiodiversity are not clear-cut. All of biodiversity is potentially of use to agriculture, particularly with the emerging biotechnologies. Agriculture is highly dynamic, and the interface between wild and domesticated plants and animals is constantly shifting. This fact underscores the importance of conserving as much biodiversity as feasible for agricultural development in the future.

How agriculture can be intensified without damaging biodiversity is a critical question for rural development. Environmentally inappropriate intensification of agriculture has led to eutrophication of lakes and estuaries, loss of soil micro-organisms, accelerated soil erosion, contamination of groundwater, and draining of wetlands. All of these activities trigger a potentially dangerous loss of biodiversity. But wild species are essential for agricultural improvement because they are the source of new economic plants and animals and provide important services such as pollination and pest control.

Possible remedial measures can be adopted to address the loss of biodiversity associated with agricultural development: (a) minimizing habitat fragmentation by providing wildlife corridors along "bridges" of natural habitat, (b) shifting to integrated pest management (IPM) strategies, such as rotating crops and relying on biocontrol agents to check crop and livestock pests, and (c) eliminating fiscal and regulatory measures that promote homogeneity in crop and livestock production.

A balanced conservation strategy includes in situ conservation (for example, maintaining animal and plant genetic resources in places where they occur naturally) as well as ex situ conservation (for example, maintaining them in seed or field gene banks). In situ conservation of crops and livestock can be supported by (a) emphasizing the safeguarding of plants and animals for the future improvement of agriculture as a new dimension of existing wilderness parks and biological reserves, (b) creating world heritage sites for genes for agricultural development, (c) integrating agrobiodiversity in ecotourism where appropriate, (d) helping to find markets for lesser-known crops and local varieties under the motto "use it or lose it," and (e) helping to find ways for livestock owners to generate more revenues from threatened breeds.

Better in situ conservation can also be supported by rapid agrobiodiversity assessment teams. Many development organizations, including the World Bank, have procedures for screening projects for their environmental impacts before they are approved. For the most part, such screening focuses on off-site impacts of agricultural development and includes an assessment as to whether the project is likely to lead to the loss of forest or other "natural" habitats. But surveys of agrobiodiversity may also be needed before a development project is implemented.

An assessment of biodiversity before a project is approved and implemented would encompass biodiversity in relatively undisturbed habitats and agrobiodiversity. Rapid assessment methodologies have been tried and tested more fully with the first category. Much work remains to be done in designing methodologies for adequately assessing agrobiodiversity. Rapid assessments of agrobiodiversity should document the diversity of crops cultivated by local farmers, the varietal diversity of each crop, the number of livestock species raised, the number of breeds per livestock species, and the effect of the land use practice on soil biodiversity as an indicator of soil "health" (Srivastava and others, chapter 17).

The new vision for agricultural research adopts a holistic approach that is more sensitive to environmental concerns, while still addressing the need to boost the yields and incomes of rural producers and caretakers of the land. It includes, but is not restricted to, (a) integrated pest management, (b) a participatory approach with farmers, (c) better use of farmer knowledge, (d) greater support for research, development, and dissemination of lesser-known crops and animals, (e) support for research on new crops and livestock, (f) greater sensitivity to the value of a mosaic of land uses, (g) greater diversity of habitats within land use systems, (h) greater reliance on recycling of organic matter, (i) shift of the research focus from individual traits to lifetime and herd productivity characteristics, (j) determination of the critical number of breeds for conservation purposes, and (k) an effort to learn more about the genetic components of adaptation in livestock.

The notion of a new research paradigm has implications for institutional development and the exploration of new ways of doing business. Innovative institutional arrangements would include more effective partnerships among agricultural research centers, NGOs, growers associations, private companies involved in the manufacture and sale of agricultural technologies, universities and agricultural extension agencies, and development lending institutions. To some degree, all of these partnerships are being explored and tested.

Generating the gains in agricultural productivity necessary to secure food availability and livelihoods in the developing world over the coming decades requires an approach in which the intensification of agricultural systems is consistent with the conservation of the natural resource base. This approach requires less reliance on the intensive use of external inputs and greater dependence on management skills and location-specific knowledge of agroecosystems. Integrated pest management constitutes one such approach and is critical to sustainable rural development (Schillhorn and others, chapter 18; Pingali and Gerpacio, chapter 19).

Agricultural intensification—the movement from an extensive to an intensive production system or from a subsistence production system to a commercial one—has increased the use of agrochemicals. Intensive monoculture systems using high-yielding cereal varieties have resulted in an environment conducive to the buildup and infestation of pests, and the consequent use of pesticides has disrupted the natural balance between pest and predator. The rapid growth in pesticide use can also be attributed to the misinformation and risk aversion of both policymakers and farmers as well as to the profit motive, including sales techniques of the chemical industry. Further, indiscriminate and injudicious use of agrochemicals often has been encouraged by inappropriate or inadequate pesticide policies in developing countries. Finally, bilateral and multilateral aid for pesticide use, provided through grants or subsidized prices, has permitted governments to distribute pesticides to farmers at low or no cost.

Indiscriminate use of pesticides in the production of various crops has often impaired health due to direct or indirect exposure to hazardous chemicals, contaminated ground and surface waters through runoff and seepage, transmitted pesticide residues through the food chain, increased the resistance of pest populations to pesticides, thereby reducing their efficacy and causing pest outbreaks, and reduced the population of beneficial insects ("good bugs" like parasites and predators), thereby reducing the effectiveness of pest control strategies that attempt to minimize pesticide use.

A reappraisal of the role of pesticides in agriculture has been under way for some time. But only recently has the topic shifted from technological solutions (for example, better protection of workers and consumers, classification and phaseout of the most dangerous chemicals) to the agricultural policy agendas of government and donors. This has been associated with a wider shift in focus from agricultural production to environmentally sustainable management of production systems in which IPM is a critical element. Integrated pest management is a key component of integrated farming practices that are based on an understanding of ecology and the interaction between crops or animals and their pests, as well as an understanding of the environments in which pests operate. Yet, despite notable successes in some countries, widespread implementation of IPM remains an elusive goal in most parts of the world.

Host-plant resistance—the plant's ability to resist high levels of pest infestation—has been the cornerstone of the scientific strategy for developing sustainable pest control systems for cereal crops. Significant advances have been made in the last three decades in the development and dissemination of crop varieties with resistance to the major cereal pests. Much of the advance has come through the use of conventional breeding approaches, although substantial gains in the development of resistance could come through the use of modern biotechnology tools. The introduction of varieties with host-plant resistance has dramatically reduced the need for insecticides in rice and maize and the need for fungicides in wheat.

Given the extent of breeding efforts in producing pest-resistant crop cultivars, and evidence of insignificant productivity benefits of pesticide use, why are high levels of agrochemicals still used in crop production? First, the dissemination of crop varieties resistant to pest pressures has not been accompanied by extension messages on the reduced need for pesticides. Second, farmers' pest control decisions, scientists' research priorities, and policymakers' prescriptions, which are all based largely on perceived pest-related losses in yield (in turn often not related to actual losses) have led to the promotion of pesticide use. Consequently, high and injudicious applications of broad-spectrum pesticides have continued as before, causing the breakdown in varietal resistance. New varieties generated to replace cultivars with resistance breakdown have subsequently been overcome through further biotype changes in pest populations. This breeding treadmill could only be overcome through dramatic changes in crop management practices, especially in the use of pesticides.

Under low levels of pest infestation, natural control is the economically dominant strategy for pest management. (This is true for insects and diseases for rice, wheat, and maize.) Natural control relies on predator populations to control pest infestations under normal circumstances, when pest-resistant varieties are used. Pesticides may have to be used as a last resort in the rare instance of high infestations. Natural control does not imply doing nothing; rather, it is based on the premise of in-depth farmer knowledge of the pest-predator ecology and frequent monitoring of field conditions by the farmer. In this regard, natural control can be considered the ultimate goal of an IPM program, and farmers who are well versed in IPM techniques would converge toward it (Pingali and Gerpacio, chapter 19). This implies a paradigm shift in the traditional IPM strategies from when best to apply to when not to apply. Therefore, continued investments in IPM training would be essential for the successful adoption of reduced or minimum use of insecticides.

IPM is knowledge-based; it presents by far the most difficult challenge to traditional, small-scale farmers in developing countries who are making the transition to scientific farming. IPM requires farmers to grasp complex sets of data that are often anything but self-evident, unitary, and standardized or amenable to trial-and-error learning. The institutional and economic structure in the rural sector of developing economies also requires some policy intervention to reconcile long-term societal goals with short-term individual objectives in pest control. Promoting sustainable pest management within an IPM framework requires improved research and extension linkages, effective farmer training methods, community action, and an undistorted price structure.

The IPM concept is holistic; it requires farmers to take a systems view of the farm enterprise and to understand the interlinkages among various components of the system. Therefore, farmer training in sustainable pest management is an essential component of a strategy to achieve minimum use of insecticides. The eventual goal is to build farmers' capacity to identify and solve problems based on a thorough understanding of field ecology. The experience of the Food and Agriculture Organization's farmer field schools has shown that trained farmers use significantly lower levels of pesticides than untrained farmers. They are also more likely to experiment with other components of sustainable production systems, such as improved fertilizer management and more efficient water management.

As yet, several unresolved research and policy issues are related to intensive farmer training in IPM. The most important one is related to the costs and benefits of farmer training. The costs of training millions of farmers are enormous and need to be justified clearly in terms of farm-level benefits, both reduced expenditure on insecticides and reduced social costs of pesticide use. Attention also should be focused on the opportunities for reducing the overall costs of training. There are essentially two options, which are not mutually exclusive, for reducing the costs of training. The first is to train a core group of farmers within a geopolitical unit, such as a municipality, and then to rely on farmer-to-farmer training for disseminating the IPM message to a wider group of farmers. The second is to condense the complex message into simple rules that are easy for the farmer to implement. An example, from rice, of such a rule is "Do not spray insecticides against leaf-feeding insects for the first

40 days of crop growth." The rule is based on detailed pest ecology studies showing that the predominant insect pests during the first 40 days of crop growth are leaf-feeding insects and that even very high levels of infestations of these insects rarely lead to any loss in yield (Pingali and Gerpacio, chapter 19).

The "no spray for 40 days" rule is an example of a simple message, distilled from in-depth scientific investigation, that can be transmitted easily to farmers. A small set of such rules, which are mutually consistent, would go a long way toward improving insect pest management in tropical crop production. Simple rules ought not to be seen as a substitute for farmer training but rather as a strong complement to a training program. Simple messages can be transmitted rapidly even while investments are made in trying to reach all farmers with intensive training.

The success of pest management programs depends on, among other things, collective organization against infestations of migratory pests. The actions of individual farmers in the management of their pest problems could have detrimental effects on the community as a whole. In this regard, management of pests could be viewed as a common property problem and dealt with through effective collective action. In Asian agriculture, where farms are uniformly small and farmers are nearly homogeneous, collective action for pest control seems quite attractive. Synchronization of planting is the single most important community action strategy in pest management.

Even with a well-established IPM program, pesticides may have to be kept as a technology of last resort. Essentially, the idea of pesticide use in IPM is to spray only when imperative, using the smallest amount possible to do the job. However, agrochemicals continue to have a significant impact on productivity in the management of weeds; herbicides will continue to be the preferred alternative in the foreseeable future, even when the health costs of herbicides are explicitly taken into account.

A number of policy and regulatory instruments are available to governments to encourage environmentally sound and economically rational pest management practices. The more important of these instruments are (a) development of a system that increases the awareness of policymakers, consumers, and producers of the hazards of pesticide use, (b) development of a regulatory framework to ensure appropriate and safe production, distribution, and use of pesticides, (c) introduction of appropriate economic incentives, including taxes and special levies on pesticide use, to account for negative externalities and short-term subsidies to account for positive externalities in the use of IPM, (d) orientation of research and technology policies to generate a steady supply of relevant pest management information and technologies, including adequate budget allocations for research, extension, and training, and (e) signing of and adherence to international agreements and conventions (Schillhorn and others, chapter 18).

Over the past three decades, the dramatic spread of irrigated agriculture was mainly responsible for increases in food production that kept pace with growth in population. This was achieved in part through the large-scale development of new water resources and construction of new irrigation capacity (Chakravorty, chapter 20). The International Irrigation Management Institute has estimated that to meet future demand, irrigated agriculture may need to deliver output increases of more than 3.5 percent annually—a daunting task by any measure of performance. This growth must come primarily from increases in irrigation efficiency and not from additions to new irrigation capacity. Moreover, as is becoming increasingly common in the industrial countries and even in many developing countries, rising demands for residential use of water and rising environmental concerns are beginning to put serious limitations on new water development projects. Producing more food with less water—placing a smaller financial burden on the taxpayer and reducing the environmental costs—will be the major challenge for irrigation in the twenty-first century.

Despite major investments in irrigation during the past four decades—the World Bank alone has invested more than $31 billion and leveraged an additional $53 billion from partner countries and cofinanciers—irrigation projects around the world are in trouble for a multitude of reasons. Ex post project benefits have often been far lower than projected ex ante returns and full-cost pricing of water is rare. Most projects only recoup some fraction of their operation and maintenance (O&M) costs. Low tax collections and poor O&M of project structures have further led to high rates of water loss through seepage and percolation, and inadequate investments in drainage have rendered large tracts of prime agricultural land unusable because of salinity and waterlogging. The low price of irrigation water, which is often unrelated to water use, has led farmers to withdraw too much, giving rise to widespread salinity and waterlogging. Investments in the irrigation sector in the developing countries have relied too much on bureaucratic federal and state agencies that have paid little attention to the economic pricing of irrigation services, the creation of reliable delivery systems, and the participation of users in the O&M of projects. Adverse health and environmental effects of water use have received little attention, partly because of a lack of coordination and interaction between water and public health agencies.

These problems call for a new approach that is more comprehensive and integrates the elements of irrigation policy, which include the economic and environmental effects and the associated intersectoral linkages (Chakravorty, chapter 20). This approach must be built on pricing of resources that reflects opportunity costs whenever possible and creation of delivery infrastructure that promotes efficient allocation and, if feasible, trading of the resource among users.

Optimal resource allocation is best done by means of a smoothly functioning market system. In the case of water, these conditions are not met on several counts. For example, the mobility of water makes it difficult to measure and hold, thereby making the establishment of property rights difficult. Economies of scale in water generation and conveyance often create conditions for a natural monopoly. The joint product nature of water, which enables multiple users to access a given stock of water sequentially, and the externalities created as a result of the degradation of quality through use make the creation and functioning of water markets extremely difficult. For these reasons, the water sector in both industrial and developing countries has been characterized by public investment in the construction of irrigation projects, state ownership and operation of facilities, and regulated distribution and allocation of water.

The introduction of markets that allow trading among users would lead to a more efficient allocation of water. Water markets function in different parts of the world, often without help from outside parties. However, they only work under very selective conditions: (a) the presence of effective water users associations or water authorities, (b) the removal of legal and institutional regulations that prohibit or limit water trading, and (c) the creation of infrastructure (canals) and control systems for the transfer and measurement of water.

In the long run, properly functioning water markets may allocate scarce supplies to their most valued use. They may also improve the efficiency of water in the agricultural sector and thus release water for alternative uses such as to meet urban needs. However, in many developing countries, the transition from the short to the long run is of profound importance and is beset by serious problems relating to the redistribution of rents accruing to beneficiaries. Most irrigation systems in developing countries are heavily subsidized by the taxpayer, with substantial rents from water accruing to large landowners and the financially better-off farmers who are usually located in the upper reaches of the distribution system.

Many governments, policymakers, and environmental groups are beginning to realize that a more efficient irrigation system means more water for everyone. This has induced changes in the way water rights have been allocated, and restrictions to trading in water have often been removed. In developing countries, governments are now motivated and willing to undertake major institutional reforms as a precondition to initiating new multilaterally funded irrigation projects.

New "holistic" thinking is also reflected in the current emphasis on "basinwide" management promoted by agencies such as the International Irrigation Management Institute (Chakravorty, chapter 20). These agencies argue that water efficiency must be measured not within a project but over an entire basin so as to include the reuse of drainage water from seepage and percolation. This suggests that policies such as the adoption of sprinkler irrigation within an irrigation project may improve the efficiency of water use within the project but reduce the availability of water elsewhere in the basin. Similarly, salt and chemical pollution of water through the application of inputs may adversely affect agriculture downstream of the project. These issues can only be handled if the basin is adopted as the relevant unit of analysis and the relevant externalities are internalized.

Livestock can damage global natural resources in a number of ways, but they can also contribute to environmental balance (Steinfeld and others, chapter 21). About 34 million square kilometers, or 26 percent of the world's land area, are used for grazing livestock. In addition, 3 million square kilometers, or about 21 percent of the world's arable area, are used for cereal production for livestock feed. Livestock produce 13 billion tons of waste per year. A large part of this is recycled, but where animal concentrations are high, waste poses an enormous environmental hazard. Livestock grazing can affect the water balance in certain areas. Water is needed to produce fodder and feed concentrate, to provide drinking water for animals, and to drain surplus waste and chemicals. Livestock interact directly and indirectly with biodiversity.

Policies need to be designed to correct the environmental effects of livestock production. These policies should address the underlying causes of environmental degradation and must be flexible, site-specific, and well targeted. Instruments to enhance positive and mitigate negative environmental impacts include pricing, regulations, and institutional development. The collective purpose of these instruments is to establish feedback mechanisms to ensure that the use of livestock is consistent with overall social objectives.

Within an enabling policy framework, a wide range of technologies are available, while others still need to be developed. Technologies can be grouped into four different sets (Steinfeld and others, chapter 21). Although there is some overlap, these categories help to provide a good picture of livestock-resource interactions:

- *Technologies that reduce the environmental damage by alleviating the direct pressure on natural resources or by reducing the pollution load through modifying the chemical or physical characteristics of products.* In the arid and semiarid grazing areas, careful water development can help to prevent environmental damage. Investments in market technology may also reduce environmental pressure by encouraging greater off-take. In addition, new and more benign methods are available to control diseases. Environmental damage of intensive systems can be reduced significantly by focusing on emissions from manure by, for example, improving collection and storage. The main focus should be on reducing nitrogen losses, most of which are in the form of ammonia from manure. Nutrient losses during and after application of manure on soils can be significantly reduced by injecting or applying manure into the subsoil. Better timing of application in response to crop requirements avoids further losses and enhances the nutritive value of manure. In tanneries, dairies, and slaughterhouses, anaerobic systems can purify wastewater and reduce by half the biological oxygen demand (BOD), while more sophisticated anaerobic systems reach 90 percent BOD purification.
- *Technologies that enhance natural resources by making them more productive or richer.* For grazing systems in arid zones, "deferred grazing," which has been a traditional practice in many Middle Eastern countries, and, in the semiarid zones, overseeding or planting of adapted fodder may regenerate the vegetation. The introduction of a multispecies grazing pattern will often encourage better use of vegetation and may have positive effects on plant and animal biodiversity. Livestock, mainly through their input function within a mixed crop-livestock system, enhance the main natural resource—land. Animal manure and traction make the land more productive than would be the case in their absence. Thus, all technologies that reduce nutrient losses from manure, and improve the efficiency of their application, enhance land productivity.
- *Technologies that save natural resources by allowing farmers to get more revenue from the same resource or to get the same from less.* The livestock sector possesses and continues to develop technologies that increase the efficiency of natural resource use. In particular, these technologies target feed conversion because feed typically accounts for 60–70 percent of production costs. Better feed conversion saves land used for its production, while reducing the animals' waste load. Technologies also provide solutions to saving and sparing other natural resources such as water and biodiversity.
- *Technologies that turn waste into products by closing cycles.* Historically, livestock were kept because they used resources for which there was no alternative use. This explains why efficiency per animal was not, and in many low-input systems still is not, a major concern. The conversion of organic waste into livestock products, although associated with livestock waste, reduces the environmental hazards associated with crop and agroindustrial waste. Also, livestock consume food wastes and increasingly so, as urban agglomeration and changing eating habits offer a window of opportunity for the collection of food waste from catering units to be recycled as feed. Large amounts of straw, otherwise burned on the fields or slowly decomposing with little nutritional benefit to the crops, may be turned into quality feed, for example through urea treatment. The cycles of matter can also be closed by using livestock waste as feed, energy, or fertilizer.

Novel concepts are being developed to integrate crop and livestock production in a farming area rather than on a mixed farm (Steinfeld and others, chapter 21). This method of area-wide crop-livestock integration allows individual enterprises to operate separately while linking the flows of energy and organic and mineral matter through markets and regulations. This allows for highest efficiencies at the enterprise level, while maximizing social benefits. Thus, the motto for the more densely populated parts of the developing world is to intensify, but not concentrate, animal production. A policy framework is needed to organize crop and livestock production in such a way.

Grazing systems will remain a source of extensively produced animal products. To some extent, these systems can intensify production by incorporating new technologies, especially in the higher-potential areas (subhumid and highland areas). Often this can be facilitated by stronger organizations, local empowerment, and regulation of access to resources. In the semiarid and subhumid tropics, where there is potential for mixed farming, policies need to facilitate the transition of grazing systems into mixed-farming systems by integrating crops and livestock (manure management, animal draft, residue feeding, and fodder crops).

Mixed-farming systems will see continued intensification and important growth. Smallholder and family mixed farming will remain predominant for some time to come, with livestock based on crop by-products and surplus (Steinfeld and others, chapter 21). Important productivity gains can be achieved by enhancing the flow of nutrients and energy between the two components. Livestock's role, in addition to production, is to enhance and substitute natural resources. The environmental and economic stability of this system makes it the prime focus for the continuing transfer and development of technology.

Industrial systems in areas of high animal densities will face the challenge of coping with higher production costs as a result of more stringent regulations and pollution levies. This will remove, in some cases, the competitive edge that industrial production has over land-based production. Potentially, this will also raise the prices for livestock products, reduce demand, and provide incentives for intensification.

As a result of the interaction between livestock production systems and natural resources, coupled with factors such as market access, there are development opportunities as well as threats to sustainability. A comprehensive perspective is needed to ensure an enabling policy framework in which to introduce effective technologies (Steinfeld and others, chapter 21). Technology remains a key component because future development, including that of the livestock sector, will depend on technology to substitute for natural resources. This trend toward knowledge-intensive systems is already widely observed.

Expansion of livestock production has often been named as one of the main causes of tropical deforestation. Contemporary concern about deforestation focuses on tropical countries because that is where the majority of forest cover is being removed (Kaimowitz and others, chapter 22). In the period 1980–90, 137.3 million hectares of tropical forests were cleared, about 7.2 percent of the total that existed in 1980. Any claims regarding the magnitude of global deforestation must be taken with caution, however, because there are serious problems with the data and definitions used.

Deforestation often implies:

- The loss of livelihoods for forest-dependent people, many of whom may not wish or be able to find other sources of employment
- Decreasing stocks of fuelwood and nontimber forest products as well as of industrial timber
- Greater soil erosion and river siltation
- Substantial loss of species and genes, in view of the high level of endemic biodiversity in tropical forests
- Substantial emission of carbon dioxide, which contributes to global warming
- Other types of local and regional climate change.

The concern is not only with deforestation but also with forest degradation, which can be defined as a decrease of density or increase of disturbance in forest classes. In the long run, this may be just as important as deforestation itself.

The process of deforestation must be analyzed at two levels: agents and causes. The agents of deforestation are the people who physically (or through decisions over their labor forces) convert forests to nonforest uses: small farmers, plantation and estate owners, forest concessionaires, infrastructure construction agencies, and so forth. In Latin America, most deforestation is associated with medium- and large-scale operations (resettlement schemes, large-scale cattle ranching, hydroelectric dams) and is characterized by transitions from closed forest to nonforest land uses. In Africa, deforestation is related largely to the expansion of small-scale farming and rural population pressure (the growing number of smallholders), associated with conversion from closed forest cover to short fallow farming. In Asia, deforestation is associated with both relatively large operations (as in Latin America) and rural population pressure (as in

Africa). There are no simple, universal single-cause explanations. Different agents coexist and closely interact in many countries, and their relative importance varies over time and between regions. Given such complex relationships, there are no clear guilty or innocent parties, and no one should expect neat, simple solutions.

Individuals and businesses deforest because it is their most profitable alternative. To get them not to deforest in situations where forest clearing is inappropriate, deforestation must be made less profitable or other alternatives (either based on retaining forests or completely outside forest areas) must be made more profitable. This is, in fact, the main thesis of Kaimowitz and others (chapter 22).

The "causes" of deforestation refer to the multiple factors that shape agents' actions and, in particular, their decision to deforest. These include market forces (international price fluctuations of agroexport commodities), economic policies (currency devaluation), legal or regulatory measures (a change in land tenure laws), institutional factors (the decision to deploy more forest rangers to a particular area), and political decisions (a change in the way forest concessions are allocated), among others. A key question is how certain causes can be manipulated to influence the behavior of agents, so as to lessen the rate of inappropriate deforestation (Kaimowitz and others, chapter 22).

Some deforestation is inappropriate for two reasons. First, deforestation generally causes negative externalities that generate costs to society that are not reflected in existing prices and has long-term consequences that individual producers rarely consider. Second, the relative importance of the negative externalities tends to grow over time if an increasing proportion of deforestation occurs in areas that have only marginal value for agriculture but sequester large amounts of carbon, have fragile soils, or are high in biodiversity.

Any decision regarding which deforestation is appropriate or not is ultimately political in nature and cannot be justified on purely technical grounds. Nevertheless, forest clearing is more likely to be inappropriate when it involves the following types of areas:

- Areas that have little value for agriculture by virtue of the quality of soil or gradient of land
- Areas that have large amounts of biodiversity, particularly endemic biodiversity, that is poorly represented in existing protected areas
- Areas that have large numbers of forest-dependent people who show no inclination to abandon their existing livelihood strategies
- Areas that, by virtue of their rich potential and comparative advantage for timber production, make this the most profitable use of land, even after the area has been logged for the first time
- Fragile areas where the ecological cost of conversion resulting from "downstream" effects outweighs any economic gain from nonforest land uses
- Periurban areas where forests play a key role in conserving aquifers, providing fuelwood, and supporting recreational and tourism activities.

It is important to distinguish inappropriate from appropriate deforestation for three reasons. First, it is often assumed (at least implicitly) that all tropical deforestation is inappropriate, and this is not necessarily the case. Second, to the extent that inappropriate deforestation can be defined, the geographical areas (and sometimes socioeconomic groups) can be specified that should be the targets of policy designed to reduce inappropriate deforestation. Third, by clearly defining the areas appropriate for conversion, pressure can conceivably be eased on forests where conversion is inappropriate.

Kaimowitz and others (chapter 22) offer a rudimentary conceptual framework for distinguishing inappropriate from appropriate deforestation. It takes into account three cross-cutting categories of valuation: biophysical, economic, and political. Decisionmakers must take all of them into account simultaneously.

Deforestation can be made less profitable by

- Reducing the demand or prices for products produced from newly cleared land
- Increasing the unit costs and riskiness of activities associated with deforestation
- Eliminating speculative gains in land markets.

Alternatives to deforestation can be made more profitable by

- Increasing the income stream to be obtained by maintaining forests
- Reducing the costs of maintaining forests
- Increasing the opportunity costs of labor and capital that might otherwise be used in activities associated with deforestation (Kaimowitz and others, chapter 22).

All the policies analyzed are evaluated on the basis of the following six criteria: effectiveness, ability to be targeted, equity, political viability, direct cost, and indirect cost. There is no perfect or generalizable policy for reducing inappropriate deforestation. There are no "first-best" options. Each national situation is different, much uncertainty remains about key cause-and-effect relations, and there are usually tradeoffs among effectiveness, ability to be targeted, political viability, and direct and indirect costs of policies. Most of the policy instruments discussed are very blunt instruments in regards to stopping deforestation—governments will be forced to choose from a mix of measures specifically crafted to local conditions.

Logging will usually lead to increased forest conversion by an influx of migrants, if the following conditions apply simultaneously:

- Roads are constructed that open up new areas of forests.
- The use of nonforest land is much more profitable than the retention of forests (in part due to policy distortions).
- Forest boundaries are poorly enforced by government agencies (such as the forest service or the national parks service), and given the institutional or legal context, people expect that the land they occupy, claim, or "stake out" will eventually be recognized, even legalized, by the government, for example, by creating an open-access "frontier."
- A large pool of unemployed or landless people, or with very low incomes and prospects, constitute potential migrants (whether moving spontaneously or sponsored by the state or private agoindustries). The pace of colonization might be related to the difference between current incomes of potential migrants and the amount they expect to earn by colonizing forest areas.

This assessment, if correct, suggests that the answer to the forest conversion issue is not necessarily to stop logging per se, or to stop logging in all new areas, or to ban all new road construction in forest areas, but rather to reform those policies and institutions that at present make forest colonization seem more attractive than the potential migrants' current activities. This might include the pull factors (to reduce the profitability of illegally clearing forests or of speculating in land that was supposed to be kept as forest) or the push factors (to increase the limited livelihood options outside of forests). The evidence from the rapid economic growth of Asian economies before 1997 is that as employment and income prospects outside the agricultural sector improved, fewer people wanted to undertake the dangerous, illegal, difficult, and often unprofitable activities of temporary agriculture in forestlands. However, if the new land use is profitable (growing cocoa, coffee, cinnamon, rubber, or fruit trees, or even timber trees like eucalyptus, teak, or gmelina) and if the potential capital gains from "capturing" some real estate from the government forests are high, it might be very difficult to slow the rate of forest conversion by such people.

Thus the principal reforms that could reduce inappropriate tropical deforestation are likely to be a combination of the following government policies:

- Eliminate subsidies to agricultural and pastoral industries that encourage deforestation
- Eliminate legal and institutional incentives or requirements to clear forests as a basis for gaining recognized land tenure
- Reform forest industry concessions and licenses to provide incentives for long-term sustainable management
- Develop innovative institutional arrangements for devolving more decisionmaking authority and responsibility to those whose livelihoods and quality of life are directly linked to the extent and quality of tropical forests
- Encourage voluntary market differentiation by consumers that discriminates positively toward products that have been sustainably produced from forests
- Facilitate the recognition of and compensation for environmental services provided by forests and ensure that transfer payments are received by the persons making the decisions at the forest frontiers.

Most of the major environmental problems in developing countries are due not to the pursuit of economic development, but rather to incorrect economic policies: poorly defined property rights, underpricing of resources, state allocations and subsidies, and neglect of nonmarketed social benefits. Instead of devising new policies to stop further resource and environmental deterioration while promoting development, one should first try to eliminate those (legal, social, political, and institutional) factors that cause or exacerbate the problems (a point made repeatedly in part I of this volume).

One response to deforestation has been to encourage agroforestry: a land use system in which trees, shrubs, palms, and bamboos are cultivated on the same land as agricultural crops or livestock for economic and environmental reasons. The system includes managing natural regrowth, seeding, planting, and maintaining trees as border plantings, and interplanting them in agricultural crops, in woodlots, in home gardens, or in other systems. Agroforestry's special characteristics are that it includes a large number of species, configurations, and management intensities, has longer gestation than most agricultural crops, and produces outputs with multiple uses.

Numerous projects have been established in the past two decades throughout Central America and the Caribbean to promote communal and individual tree plantings and agroforestry systems (Current and others, chapter 23). The main emphasis used to be tree planting for fuelwood. But many of those projects were unsuccessful. Subsequently, the emphasis shifted from fuelwood production to the concept of multipurpose tree species in agroforestry systems. The establishment of multipurpose tree species on farmland can provide a wide range of benefits. Conflicts do arise between the cultivation of trees and agricultural crops, but some agroforestry systems allow farmers to integrate trees into their farming systems, with only small drawbacks for crop production or even an increase in farm productivity overall.

The shift in emphasis toward multipurpose tree species was accompanied by a corresponding realization that a project is more likely to succeed if project planners consult local communities about their perceived needs and design projects to meet those needs instead of imposing schemes that they may not consider a priority.

In addition to the estimated financial return, farmers attach considerable importance to how an agroforestry system fits into the overall farm production system and the existing land, labor, and capital constraints (Current and others, chapter 23). Many agroforestry systems are profitable to farmers under a considerable range of economic conditions, and various types of (low-intensity) traditional agroforestry are practiced in many areas. Even when agroforestry is profitable, smallholders will and should adopt agroforestry incrementally and gradually because of management and resource constraints; poorer farmers, in particular, are often hampered by limited land, labor, and capital resources and their need to ensure food security and reduce risks.

Local scarcity of wood products is, as might be expected, a key motivator in adopting nontraditional agroforestry. Projects must begin by assessing the scarcity of wood as well as the existence of local markets for products. Taungya, perennial intercrops, trees on contours, and tree lines have proved to be the easiest systems to introduce. Results are mixed for alley cropping, home gardens, windbreaks, green manuring, dispersed trees in cropland, and tree-pasture systems. Farmers are willing to invest in rehabilitating their land where systems also produce products or income, and they prefer less-intensive systems.

The demonstration effect of fast-growing tree species on farms, and of benefits on demonstration plots, has helped to expand agroforestry activities, reducing the costs of extension and increasing its effectiveness. Rather than offering standard designs, programs serve farmers best if they offer a broad selection of species and systems from which to choose those most suitable to their household's needs and resources. Involving local people as paratechnicians is often a successful, low-cost approach to promoting technology.

Financial incentives and subsidies should be kept to a minimum. Agroforestry technologies promoted by extension should be financially profitable, and thus adoptable, for the farmer without subsidies. The possible exceptions are time-limited financial incentives for early adopters of unfamiliar technologies. In-kind, material inputs encourage farmers to experiment with and adopt agroforestry, but the experiences with food-for-work incentives are mixed.

16

The Costs and Benefits of Soil Conservation in Central America and the Caribbean

Ernst Lutz, Stefano Pagiola, and Carlos Reiche

Soil degradation can be defined as a reduction in the land's actual or potential uses (Blaikie and Brookfield 1987). Many cultivation practices tend to degrade soil over time. For example, cultivation can expose soil to water and wind erosion, repeated tillage can weaken soil structure, crop production can remove nutrients, and use of machinery can compact the soil. Central America's often mountainous environment and heavy rainfall make much of the region particularly vulnerable to degradation—a problem exacerbated by population pressures that have opened up new areas only marginally suited to agriculture. Soil degradation, in turn, affects productivity. As soil is degraded, crop yields decline or input levels (and hence costs) rise in an effort to restore productivity.

Despite long-standing concern about these problems, surprisingly little hard evidence exists on their magnitude. The degradation figures quoted in the literature are often extrapolated from very limited data and may exaggerate the problem because they often consider "moved soil" as "lost soil," even though much of it may be deposited on other agricultural land. For instance, in a recent assessment of the extent of human-induced soil degradation, the International Soil References and Information Centre estimated that 56 percent of the land in Central America has experienced moderate degradation (implying that productivity has been substantially reduced) and that 41 percent has experienced strong degradation (implying that agricultural use has become impossible; Oldeman, Hakkeling, and Sombroek 1990). Aggregate measures such as these, however, often have a weak empirical basis. Studies that directly measure erosion rates and the factors that influence them are few and have generally been scattered and unsystematic. Even less effort has been devoted to studying other forms of land degradation, such as nutrient depletion, damage to physical and chemical properties of soil, or reductions in its capacity to retain moisture. Table 16.1 presents an overview of available estimates of erosion rates in Central American countries. The figures presented in the table were obtained in a variety of ways and are therefore not always strictly comparable, but they do give some idea of the great diversity of erosion rates present within the region.

Predictions abound of the catastrophic effects that soil degradation will have on agricultural productivity. Evidence on the magnitude of these effects, by contrast, is hard to find. In fact, in many cases claims of declines in productivity are made with no supporting evidence at all (see Biot, Lambert, and Perkin 1992 for some African examples). Leonard (1987), for example, simply asserts that a pattern of extensive land use leading to soil loss or decline in fertility is apparent in the Caribbean areas of Central America. Speaking of the highland areas, he points to increasing reports of localized desertification in areas of western Honduras and Costa Rica. He also mentions that cotton yields are reportedly declining where severe erosion has been experienced. But nowhere does he indicate the size or rate of fertility loss. More

Table 16.1. Empirical Evidence on Soil Erosion in Central America and the Caribbean

Country and region or area	Rainfall (millimeters)	Slope (percent)	Farming system	Average rate of erosion: Metric tons per hectare per year	Average rate of erosion: Millimeters
Dominican Republic					
Taveras	—	—	—	275	—
Northcentral	—	36	Various	24–69	—
Southwest	—	30	Various	2–1,254	—
El Salvador					
Metapán	1,600	—	Corn, beans	49	—
Haiti					
Camp-Perrin	2,000	30	Hedges	4–45	—
Papaye	1,200	25	Grass hedge	8	—
Honduras					
Tatumbla, Morazán	2,000	45	Corn, beans	42	3
	900–1,500	15–40	—	18–30	—
Nicaragua					
Cristo Rey	1,700	30–40	Cotton	40	—
Panama					
Cuenca del Canal	1,200	35	Rice	153	—
	1,200	35	Corn	137	—
	1,200	35	Rice	118	—
Coclé	1,937	—	Rice, corn	340	17
Chiriquí	1,500–2,800	—	Cassava, beans	35	5
	1,500–2,800	—	Pasture	77	11
	1,500–2,800	—	Coffee	183	27

— Not available.
Note: Figures are rounded.
Source: Case studies in Lutz, Pagiola, and Reiche 1994.

generally, the assumption that fertility must be declining rapidly is usually left implicit from statements about high rates of erosion.

But erosion rates, even where they are significant, may have very little effect on productivity under certain conditions. Erosion rates in the Tierra Blanca area of Costa Rica's Cartago Province, for example, are extremely high, but the effect on productivity is minor because soils in that region are deep (up to 1 meter in some places) and have high organic matter throughout the soil profile. Moreover, the subsoil that underlies these soils is itself productive, although less so than the topsoil. The Chiriquí region in Panama provides another example of this. Conversely, areas with shallow soils or unfavorable subsoils, such as the Turrubares area in Costa Rica, can be very sensitive to even limited rates of erosion. The same is true of other forms of soil degradation. The impact of nutrient loss on productivity, for example, depends on the initial stock of nutrients and on their rate of regeneration.

Given the different effects of soil degradation on productivity, a specific soil conservation technique—

particularly an expensive one—may not necessarily be worthwhile from the perspective of a farm household or society. Degradation can be slowed or arrested by a large range of options, including cultural practices such as contour plowing and minimum tillage, vegetative practices such as grass strips, strip cropping, and vegetative barriers, and mechanical measures such as terraces and cutoff drains. Adopting any of these techniques can be costly, either directly in investment requirements or indirectly in production forgone, and some measures are better suited to some conditions than to others. The critical question facing farmers and society as a whole is whether the benefits of a given conservation measure or set of measures are sufficient to make the costs worth bearing.

Conceptual Issues

The problem of soil degradation and conservation can be examined from two perspectives. The first is that of society as a whole. From this perspective, all the costs and benefits of a given activity must be considered. If agricultural production leads to siltation of reservoirs, for example, this represents a real cost to society that should be considered together with the value of the output obtained and any effects on fertility. In addition, valuation of the resources used and obtained from agricultural production should be adjusted for any distortions resulting from policy interventions or market failures, in order to measure their true opportunity cost. From the second perspective, that of the farmers themselves, only the costs and benefits that actually accrue to the agent making the decisions about resource use are considered. These costs and benefits are valued at the prices these agents actually face, with no attempt to adjust for distortions.

This chapter examines the returns to investment in conservation measures mainly from the farmers' point of view for two reasons. First, decisions about land use are ultimately made by the farmers themselves and not by social planners or government agencies. Farmers decide how to use their land in light of their own objectives, production possibilities, and constraints and not on the basis of any theory of the social good. Understanding the incentives facing individual farmers is necessary, therefore, if patterns of resource use are to be understood and appropriate responses to problems formulated. Second, land use problems generally depend heavily on site-specific biophysical characteristics, which can vary significantly even within small areas (Pagiola 1993). Analysis at the farm level is the most apt to incorporate site-specific effects.

A farm-level approach also places emphasis firmly on the effects of degradation on farm productivity. In developing countries, where substantial numbers of people still depend directly on agricultural production, the effect of degradation on yields is often critical. This is not to belittle the importance, in some situations, of off-farm effects of soil degradation, such as siltation of reservoirs and waterways. But even where such off-farm effects are the primary concern, considering them first at the farm level is appropriate because conservation measures have to be implemented on farms.[1]

In making their land use decisions, farm households need to consider both the agroecological and the economic characteristics of the environment in which they operate. In addition, they often face numerous constraints, such as tenure problems, liquidity constraints, and the need to meet consumption requirements and to compensate for missing or incomplete markets. Moreover, many farm decisions are made in the context of considerable risk and uncertainty. A complete analysis of land use decisions, therefore, requires that one look at the issue in the context of overall decisionmaking of the household (Singh, Squire, and Strauss 1986 and Reardon and Vosti 1992).

The farm household's problem can be formulated as one of maximizing the utility of consumption over time, subject to a budget constraint imposed by the returns from agriculture over time and from nonfarm activities and subject to any other constraints it might face. Singh, Squire, and Strauss (1986) show that if markets exist for all goods and services, the problem of maximization is separable, in the sense that production decisions are made independently of consumption decisions. Even when production decisions are not separable, however, they can be analyzed independently as long as the "prices" of goods for which markets are missing are interpreted as shadow prices that reflect the farm household's perception of the severity of the constraints it faces (de Janvry, Fafchamps, and Sadoulet 1991).

The household's problem, then, can be summarized as one of maximizing the present value of the stream of expected net returns to agricultural production (Pagiola 1993). In practice, data are generally not available to estimate complex maximization models. But for empirical analysis, the model can be reformulated to fit a cost-benefit analysis framework. The

Table 16.2. Sites of Country Studies

Sites	*Biophysical environment*	*Degradation problem*	*Conservation measures proposed*
Costa Rica			
Barva area, Province of Heredia	Important coffee-producing region; relatively deep soil, but vulnerable to erosion because of topography	Soil loss affecting nutrients available to coffee	Diversion ditches
Tierra Blanca–San Juan Chicoá, Province of Cartago	Important vegetable-producing area; deep volcanic soils	Deep soils so decline in yield is not significant; erosion washing away seed and fertilizer and exposing rocks	Diversion ditches recommended, but interfere with prevalent cultivation practices
Turrubares, Central Pacific region	Previously used for pasture, now converted to production of cocoa yam for export	Very high rates of erosion; thin soils vulnerable to erosion	Diversion ditches or terraces
Dominican Republic			
El Naranjal subwatershed, Peravia Province	Subsistence agriculture; steep slopes, soils of moderate natural fertility	High rates of erosion	Diversion ditches at 10-meter intervals, live barriers, and cropping on the contour
Guatemala			
Patzité, Department of Quiché	Small-farmer area; strongly undulating topography; soils of medium depth and fertility	Heavily affected by soil erosion	Terraces with a protected embankment
Haiti			
Maissade Watershed, Central Plateau region	Hilly area; generally less degraded and more productive than most other hilly regions of Haiti	Erosion	*Ramp pay* (indigenous technique: crop residue placed along the contour, held in place by stakes), hedgerows along the contour, and contour rock walls
Honduras			
Tatumbla, Department of Francisco Morazán	Subsistence agriculture predominant; thin topsoil, with low levels of organic material and of many nutrients	Susceptible to water erosion, especially in the high areas	Diversion ditches protected by live barriers
Yorito, Department of Yoro	Small-scale subsistence agriculture, still largely forested; shallow, easily erodible soils, of medium to low natural fertility	Cleared plots vulnerable to erosion	Diversion ditches with live barriers
Nicaragua			
Santa Lucía valley, watershed of Malacatoya River	Subtropical foothills, moderately deep soils; one of the most productive areas in the country	High risk of erosion due to steep slopes, scarce vegetation cover, and intense precipitation; deforestation on upper slopes	Manually constructed diversion ditches with stone barriers
Panama			
Coclé	Subsistence agriculture using slash-and-burn techniques, with plots cultivated one year every five; shallow soils, generally low in organic matter and nutrients, on steep slopes	Rapid yield decline on cleared plots; deforestation	Combination of erosion prevention measures (planting on the contour, live and dead barriers, diversion ditches) and improved cultivation practices

household's choice can be thought of as selecting between two or more alternative cropping systems. For concreteness, one might think of a household deciding whether to replace its traditional cultivation system, in which conservation measures are limited to practices such as contour plowing, with a more conserving system, which might include, for example, terraces or reduced-tillage techniques. Each system is characterized by a distinct production function and soil growth function and, therefore, generates a different optimal path. From the household's perspective, the problem is whether returns under the optimal path of the new, more conserving system are sufficiently greater than returns under the optimal path of the current, more degrading system to justify the cost of switching.

Basically, it is in the farm household's financial interest to adopt the new, more conserving system if the net present value (NPV) of the incremental returns from switching is positive (NPV > 0). This formulation is equivalent to a standard cost-benefit analysis formulation and lends itself particularly well to empirical analysis because data are often available in a suitable form. Observing practices in use allows time paths of yield and use of inputs to be constructed; these are then used to project costs and revenues over time. The method can also be used if the only data available are on total costs and revenues in each period, and it lends itself well to the incorporation of lumpy investments and other discontinuities in cropping practices (Walker 1982 and Taylor and others 1986).[2]

The discussion so far has assumed that the only constraints on behavior are those imposed by the properties of the biophysical system. The NPV > 0 criterion is thus a necessary, but not a sufficient, criterion for adoption of a new production system, because other factors might prevent adoption of a new system even if the NPV estimate is positive. In principle, these other constraints can be built into the optimization framework. The effect of insecure tenure might be included, for example, by limiting the length of the time horizon. In practice, however, it generally proves easier to compute the profitability of a system assuming that no constraints hold and then verifying whether other constraints are binding. The cost-benefit calculations themselves often provide insight into whether particular constraints are likely to prove binding. The length of time it takes for an investment to be repaid, for example, indicates whether tenure problems are likely to pose problems. If the investment is repaid very rapidly, insecurity of tenure is unlikely to affect adoption. Of course, if adopting a new production system is unprofitable for the farm household, the question of whether other constraints might prevent its adoption does not arise.

Methodology

Cost-benefit analysis techniques lend themselves well to the evaluation of soil conservation measures because they provide a coherent framework for integrating information on the biophysical and economic environments facing farmers. Variants of these techniques have been used to examine a number of soil conservation cases: for example, in the Dominican Republic (Veloz and others 1985), in India (Magrath 1989), and in Kenya (Pagiola 1992). Other simple techniques, such as calculating the value of lost nutrients (Repetto and Cruz 1991), can only provide rough indicators of the severity of the problem. They cannot provide guidance in selecting the best response.

The basic principles of the analysis are straightforward. First, the effects of continued erosion (or other types of soil degradation) on productivity are estimated for the time horizon of interest. These are then used to estimate returns at each point in time. Second, the calculations are repeated under the conditions that would be experienced if a specific conservation measure were adopted. The returns to the investment in this measure are then obtained by taking the difference between the streams of discounted costs and benefits in the cases with and without conservation. It must be stressed that this method estimates the returns to the specific conservation measures being examined, not to conservation per se. A finding that certain conservation practices are not profitable does not imply that all conservation measures are not profitable—often, numerous measures designed to reduce degradation rates are already being practiced.

As was argued in the previous section, when the analysis is carried out at the farm level using prices actually facing farmers, a positive NPV estimate for a given conservation measure can be interpreted as showing that adoption of the measure is profitable for the farmer. Farmers should, in principle, be willing to adopt such a measure voluntarily. But, as with all cost-benefit analysis, there is no guarantee that other, unexamined options would not be preferable. When several options are known to exist, the analysis can be repeated for each in turn, and the most profitable among them found.

For this article and the larger work from which it is drawn, the availability of data dictated the choice both of the sites studied (see table 16.2) and of the aspects of the problem analyzed—erosion and mechanical methods of conservation.[3] Research efforts have focused almost exclusively on problems arising from erosion, to the neglect of other forms of soil degradation, and most conservation projects in the past have tended to emphasize mechanical conservation structures. Consequently, the case studies do not present a comprehensive overview of soil conservation problems and practices in the region; they do, however, illustrate the wide diversity of conditions encountered, help explain farmers' behavior, and indicate appropriate policy responses.

The country studies, with the exception of Haiti, were conducted by local practitioners. In most cases, teams were composed of economists, agronomists, and soil scientists from relevant government agencies. This collaborative and participatory approach to the research proved successful in drawing on local knowledge and expertise. It also developed local analytical capacity.

The data needed on the nature and rate of degradation caused by current practices, on the effects of degradation on future productivity, and on the effects of conservation practices were very scarce. Different methods for estimating the required relationships were chosen, depending on the nature of the available data. Econometric techniques were sometimes employed to estimate the effect on yield of certain observed conditions (such as the presence or absence of certain conservation measures). Disentangling the impact of soil degradation is very difficult (Capalbo and Antle 1988). For our purposes, however, estimating a time trend of yields with and without a given conservation measure was usually sufficient, although even this limited objective encounters problems such as bias in sample selection when nonconserved and conserved fields are compared. In addition, many of the case studies had to rely on farmer recall data and were able to control for other sources of yield variation, such as weather, only to a limited extent. In other cases, simple models of the physical environment—such as the Universal Soil Loss Equation and, in Haiti, the Soil Changes under Agroforestry model—were employed, using a mixture of experimental and observational data. This modeling approach is more flexible because it allows parameter values to be drawn from a variety of data sources. But it requires detailed knowledge (both qualitative and quantitative) of the biophysical environment; building and validating a complete and realistic model are complex endeavors. Even calibrating existing models is far from easy.

Obtaining the required economic data was generally less problematic. Crop production budgets, used to estimate returns, were generally widely available, although rarely at the degree of disaggregation needed. Fortunately, preliminary budgets built from available secondary data were easy to confirm, supplement, and correct during fieldwork. The most important task was to ensure that the crop production budgets accurately reflected practices and prices in the area. Inputs provided by the households themselves, such as family labor, were priced at their cost in the nearest market. Output and input prices used in the analysis were meant to represent long-run real price trends for outputs and inputs. Assessing the discount rate is crucial, given the intertemporal nature of the problem, but beset by controversy. Here, because the analysis examines the profitability of conservation from the farm household's viewpoint, the appropriate discount rate to use should be the farmers' cost of borrowing or their rate of time preference. Little empirical evidence exists on either, however (Holden, Shiferaw, and Wik 1998). Therefore, and to facilitate comparability of results across study sites, a common real discount rate of 20 percent was used in each case study. In addition, the internal rate of return (IRR) was computed in each case. If the appropriate discount rate, assuming it is known, is smaller than the IRR, the proposed conservation measures would be profitable.

Effects of Degradation on Productivity

The estimated productivity losses vary considerably across the case studies; table 16.3 presents findings for some of the crops analyzed. In several cases, the data point to rapid rates of decline in yield. In the Maissade watershed of Haiti, for example, yields of corn and sorghum would decline as much as 60 percent over a decade in the absence of conservation measures. In the Tatumbla region in Honduras, corn yields would decline almost 50 percent in 10 years if no conservation measures were used. Elsewhere, estimated declines would be minor. Coffee yields in the Barva region of Costa Rica, for example, would decline just over 10 percent in 10 years, and there is reason to believe that this rate of decline is overestimated. In Costa Rica's Tierra Blanca region, declines in potato yield caused by erosion would be compen-

Table 16.3. Estimated Impact of Soil Degradation on Productivity of Select Crops in the Areas Studied
(percentage of initial yield)

		Years					
Country and area	*Crop*	*10*	*20*	*30*	*40*	*50*	*Projected shutdown year*
Costa Rica							
Barva	Coffee	89	78	67	56	46	20
Turrubares	Coco yam	0	0	0	0	0	4
Dominican Republic							
El Naranjal	Pigeon peas	58	16	0	0	0	
	Peanuts	100	100	100	100	100	16[a]
	Beans	77	53	30	0	0	
Guatemala							
Patzité	Corn	0	0	0	0	0	10[b]
Haiti							
Maissade	Corn, sorghum	41	22	10	1	0	25
Honduras							
Tatumbla	Corn	53	39	39	39	39	8
Yorito	Corn	82	65	47	41	41	11

Note: Projected year for production shutdown is in the absence of conservation measures.

a. The 16-year shutdown period applies to the pigeon peas-beans-peanuts intercrop system. Because the three crops are cultivated together, peanut cultivation is assumed to cease when the yields of the other crops make production uneconomic.

b. Corn can be produced in years one through nine, but decline in yield is so rapid that it reaches zero in year 10.

Source: Case studies in Lutz, Pagiola, and Reiche 1994.

sated easily by small increments in fertilizer use; indeed, potato production has been increasing steadily despite high rates of erosion. The effects of degradation can also vary significantly across crops, even in the same area, as shown by the data from El Naranjal in the Dominican Republic.

If no conservation measures were adopted, returns to agricultural production would gradually decline in each of the cases studied. Eventually, production would become uneconomic and cease, although exactly when this will happen will vary, depending on the rate of decline in yield, the cost of production, and the price of the output. (Because farmers are likely to adjust their production practices as yields decline, the time before production becomes unprofitable is likely to be overestimated.) The very high rates of decline experienced in Turrubares mean that the production of coco yam would shut down in four years if no conservation measures were adopted; by contrast, in Tierra Blanca the production of potatoes would remain profitable more or less indefinitely even without conservation.

Not all the damage caused by soil degradation takes the form of yield losses. In Tierra Blanca, for example, the effects of degradation on agricultural production are reflected primarily in higher costs arising from the need to apply higher rates of fertilizer, from the lower efficiency of fertilizer (because some is washed away), and from the need to remove stones that accumulate on fields as soil is eroded. In Panama's Coclé Province, agricultural production could only be sustained for a very short time on a given plot if no conservation measures were used. The costs of degradation, therefore, are reflected primarily in the need to clear new plots of land at frequent intervals.

These examples, together with the diverse effects on yield, reinforce the need for site-specific information for understanding degradation problems and devising effective ways of helping farmers respond to them. Note, however, that these case studies are by no means a random sample of degradation conditions in the region; they are drawn from sites for which data were available, and therefore primarily from areas where degradation problems were serious enough to warrant data collection. Consequently, they probably represent high-case scenarios of the degree and rate of degradation in the region.

The estimated effects on yields of conservation practices are likewise varied. In some instances, yields would recover once conservation measures were established—partly because soil regenerates once the processes of degradation are halted, partly because fertilizers are used more efficiently, and partly because improved cultivation practices are sometimes introduced together with conservation. In the Tatumbla area of Honduras, for example, corn yields would increase about 145 kilograms annually if diversion ditches were built and improved planting practices were adopted, up to a maximum set by the local agroecological conditions and the technology employed by the farmers. Elsewhere, conservation measures might slow but not halt the decline in yield. In the Turrubares area of Costa Rica, for example, diversion ditches would halve the rate of decline in yield; the much more expensive terraces, in contrast, would reduce the rate to one-tenth. Again, the diversity of conditions is evident.

As well as reducing soil loss and hence the rate of decline in yield, conservation measures can affect yields by encouraging the retention of moisture and stimulating improvements in the soil's physical structure (English, Tiffen, and Mortimore 1994). In Haiti's Maissade area, land treated with conservation structures was found to produce an average of 51 percent more corn and 28 percent more sorghum than plots without conservation structures in 1988 (a year of poorly timed rainfall) and an average of 22 percent more corn and 32 percent more sorghum in 1989 (a more normal year). In dry areas, therefore, soil conservation can often reduce the risk of crop failure by improving moisture retention.

Although introducing conservation measures can bring long-term benefits, it also can often result in adverse effects on production because some of the area cultivated was turned over to use as diversion ditches, terraces, or hedges. Physical structures, in particular, usually reduced the available area for cultivation by more than 10 percent. Construction of cutoff drains in Tierra Blanca, for example, reduced the effective area by about 14 percent, while terrace construction in the Patzité region of Guatemala led to a 15 percent reduction. Further, terracing often entails the movement of earth, which can bring unproductive soil to the surface. In Tierra Blanca, diversion ditches had the additional disadvantage of interfering with the prevailing production practices, which rely heavily on mechanical equipment. Such drawbacks heavily influence the ultimate profitability of these conservation measures.

Because some of the productivity estimates are based on weak or incomplete data, extensive sensitivity analyses were incorporated into each of the case studies. The results are robust to changes in the estimated effects on yield in several cases, but in others results are affected significantly by changes in assumed rates of decline in yield. In such instances, the premium to additional research would be high. In the Santa Lucía case study in Nicaragua, data were insufficient to estimate the effects of degradation on productivity. Simulation analysis was used, therefore, to examine returns to the proposed conservation measures (manually constructed diversion ditches with stone barriers) under a range of assumptions about the effect of degradation and conservation on yield. The results of the simulations show that the proposed conservation measures are likely to be profitable only if the yield benefits of conservation are substantial.

Farm-Level Returns to Soil Conservation Measures

Effects on yield are not the only factors to consider in analyzing the costs and benefits of investing in a given conservation measure. The cost of constructing and maintaining conservation measures must also be considered. The cost of constructing and maintaining conservation measures must also be considered. Table 16.4 summarizes the results of a full economic analysis of each of the case studies where data were sufficient to allow adequate assessment.

The most profitable conservation measure studied, in terms of rate of return, was found in Maissade, Haiti. This indigenous technique, known as *ramp pay*, consists of crop stubble laid out along the contour, supported by stakes, and covered with soil. It is cheap to construct and very effective in halting erosion. Moreover, without conservation measures, yield would decline particularly rapidly in that area. High rates of return were also estimated for conservation measures in Turrubares, Costa Rica, where highly profitable export crop production was threatened by

Table 16.4. Estimated Returns to Investments in Conservation in the Case Study Areas

Country and area	*Conservation measure*	*Crop*	*Net present value (U.S. dollars)*	*Internal rate of return (percent)*	*Number of years to break even*
Costa Rica					
Barva	Diversion ditches	Coffee	–920	< 0	> 100
Tierra Blanca	Diversion ditches	Potatoes	–3,440	< 0	> 100
Turrubares	Diversion ditches	Coco yam	1,110	84.2	2
	Terraces	Coco yam	4,140	60.2	3
Dominican Republic					
El Naranjal	Diversion ditches	Pigeon peas, peanuts, beans	–132	16.9	> 100
Guatemala					
Patzité	Terraces	Corn	–156	16.5	> 100
Haiti					
Maissade	Ramp pay	Corn, sorghum	1,180	Undefined	0
	Rock walls	Corn, sorghum	956	Undefined	1
Honduras					
Tatumbla	Diversion ditches	Corn	909	56.5	4
Yorito	Diversion ditches	Corn	83	21.9	18
Panama					
Coclé	Terraces	Rice, corn, yucca, beans	34	27.2	8

Note: Net present value is computed over 50 years, using a 20 percent real discount rate.

rapid rates of yield decline, and in the Tatumbla area of Honduras, where yield decline would also have been very rapid if no conservation measures had been taken.

The least profitable conservation measures studied were found in Barva and Tierra Blanca, Costa Rica. The Tierra Blanca case is particularly interesting, because rates of erosion are very high. But the region's deep volcanic soils mean that degradation has very little effect on productivity. In fact, production would be higher without the proposed conservation measures—diversion ditches—because their construction would reduce the effective cultivated area and, by interfering with current production practices, would increase the costs of production. It is not surprising that farmers in the area had little interest in adopting these conservation measures.

In Maissade in Haiti, Turrubares in Costa Rica, and Patzité in Guatemala, data were available to examine the returns to different forms of conservation. In Maissade, *ramp pay* is clearly superior to rock walls, which are more expensive and lack the agronomic advantages of *ramp pay*. In Turrubares, the choice is less clear: terraces slow erosion much more effectively than diversion ditches, but they are also more expensive to construct and entail a greater reduction in effective cultivated area. The tradeoff that must be made between effectiveness and cost is fairly easy in this case because the greater effectiveness of terraces more than compensates for their additional cost. But this is not always true. In Patzité, for example, a combination of diversion ditches and live barriers appears to be substantially more profitable than terraces, even if they are much less effective.

This case appears to be more representative of conditions encountered in Central America: in analyses of 20 conservation techniques in Mexico, for example, McIntire (1994) finds that cultivation and cropping practices, including vegetative barriers, are superior to structural measures. Only when crop production is very profitable but extremely vulnerable to degradation (as in the case of Turrubares) are expensive conservation measures likely to be justified.

Unfortunately, data were insufficient to examine differences in returns *within* the study areas. Evidence from Kenya (Pagiola 1994) suggests that returns to conservation can vary considerably even within narrowly defined agroecological zones. Farmers on different slopes, for example, experience different rates of erosion. They also face different costs of conservation; the optimal spacing of terraces and diversion ditches, for example, is a function of slope. Whether these differences are significant in any given instance is an empirical matter.

In each case, adoption rates correlate well with the estimated profitability of conservation. The profitability of *ramp pay* is confirmed by its widespread adoption in Maissade. Conservation measures were also adopted at high rates in the Tatumbla region of Honduras and the Turrubares region of Costa Rica; not surprisingly, adoption rates were very low in Tierra Blanca. Adoption rates were also low in Yorito, Honduras; there, the studies estimated the conservation measures to be marginally profitable, but the estimates were based on particularly weak data and were fairly sensitive to changes in assumptions. Thus it may be perfectly rational for farmers not to adopt the proposed conservation measures. In some cases—in Tierra Blanca, for instance—degradation simply is not a significant problem for productivity. In others, the costs of the proposed conservation measures are too high relative to their benefits. The case of Patzité in Guatemala illustrates this best: although degradation is relatively rapid and, if left untreated, will make production uneconomic within a decade, the proposed terraces are very expensive and take a lot of the land out of cultivation. Again, this is not to say that *all* conservation measures are unprofitable. Visits to Tierra Blanca show, for example, that although farmers have not adopted diversion ditches, they do plant along contours and, on steeper slopes, construct temporary bunds on their fields. (The effects of these measures are implicit in the estimates of degradation and of impact on productivity for the "without conservation" case.)

Obstacles to Adopting Conservation Measures

Profitability of conservation practices is a necessary but not always a sufficient condition for their adoption. Factors other than strict cost-benefit considerations also play a role (Murray 1994). Some of these factors are reflected in the cost-benefit analysis to the extent that they affect the prices facing farmers. The effect of imperfect factor markets, for example, would be reflected in higher prices for inputs, which would affect the profitability of production activities. Most often, however, institutional issues must be considered together with the results of the cost-benefit analysis. The analysis carried out for the case studies does not always provide conclusive evidence on these, but it does provide some insights.

It has often been argued that insecure property rights dissuade farmers from undertaking long-term investments, such as investments in soil conservation, because they may not themselves be able to reap the benefits (Ervin 1986 and Wachter 1992). To respond to this problem, numerous efforts have been made to make tenure more secure by providing farmers with legal title to their land. The U.S. Agency for International Development (USAID), for example, has funded titling projects in several countries, including El Salvador and Honduras. But equating land titles with secure tenure and thus with increased investment is too simplistic. Unless numerous improvements are made to the legal system and governmental institutions, land titles often prove to be too costly to obtain or enforce for most farmers, and unless access to credit is improved for farmers holding titles, the desired effect on investment may not materialize (López 1996).

The time required for investments in conservation measures to break even provides an important indicator of the severity of tenure insecurity. Farmers with insecure tenure may doubt that they will be able to profit from conservation measures if the benefits will be reaped in the distant future. Table 16.4 shows that in most of the case studies, profitable conservation measures had relatively short payback periods. Where long payback periods were forecast, the measures were either unprofitable or only marginally profitable and thus unlikely to be adopted even in the absence of tenure problems. Other evidence from the case studies also suggests that tenure insecurity is not as significant a problem as is sometimes thought. About 80 percent of the farmers in the Tatumbla area in Honduras own land by occupation, that is, they do not have legal title, yet most have adopted the recommended conservation measures. In the Patzité region in Guatemala, the proportion of farmers without title is similar; only 10 percent of farmers have title to their land. Although erosion is a significant problem, adoption of conservation measures in this area has been relatively slow. At first sight, this might appear to be evidence for the importance of titling. But negative profitability of the recommended conservation measures is more likely to account for low adoption rates than insecure tenure or lack of land titles.

Another oft-cited obstacle to adoption is the lack of capital markets. If credit markets fail, adoption of conservation will be limited by the farmers' ability to self-finance the required investments (Pender 1992). The research carried out for this project did not bring to light any direct evidence on the functioning of capital markets in the areas studied. The estimated rates of return for investments in conservation measures shown in table 16.4 give some indication of the maximum rates that could be supported before the investments would become unprofitable. Several are encouragingly high.[4]

Conclusions and Policy Implications

Whether conservation measures are profitable for the farmers is an empirical and site-specific issue. Returns to conservation depend on the specific agroecological conditions faced, on the technologies used, and on the prices of inputs used and outputs produced. Hard data on the extent of soil degradation and its effects on productivity remain extremely scarce despite several decades of soil conservation efforts (Lal 1988). More systematic research is needed on soil degradation and its consequences, and there is considerable scope for collaboration on such research, because all countries within Central America include a large number of agroecological regions, and many agroecological regions are found in more than one country. Regional organizations such as CATIE (Centro Agronómico Tropical de Investigación y Enseñanza, Tropical Agricultural Research and Higher Education Center) have an obvious coordinating role to play. The payoff is likely to be high, because the approach to soil conservation is more targeted, with efforts concentrated where they are needed most.

The results of the case studies carried out in the region show that conservation is profitable in some cases, but not in others. In view of the small number of cases studied and weakness of the available data, broad lessons must be drawn with care. It seems safe to say, however, that except when high-value crops are planted on very fragile soils (like the coco yam in Turrubares), expensive mechanical structures are unlikely to be profitable for the farmers. Conservation measures are particularly likely to be profitable either when they are cheap and simple or when they allow the adoption of improved practices.

Generally, the farmers' decision to invest in conservation is based on normal considerations of benefit and cost: farmers tend to adopt conservation measures when it is in their interest to do so, unless some constraint is present. Cases in which returns to conservation are estimated to be low or negative correlate well with low adoption rates.

A full examination of the role of government policy in conservation would require a broader analysis than that undertaken here; in particular, off-site effects of degradation would have to be explicitly included and allowance made for distortions in observed price signals resulting from government policies or market failures. Nevertheless, several important points emerge from the present analysis.

Advocates of soil conservation often argue that subsidies are indispensable to induce farmers to adopt conservation measures. But such statements assume that conservation is inherently desirable whether or not there is concrete evidence that the benefits outweigh the costs. The results presented in this article show that this may be far from the case; frequently, the benefits of specific conservation techniques (such as mechanical structures) do not justify their costs. Unless there are important off-site effects or the price signals received by farmers are significantly distorted, subsidies to induce adoption would therefore not bring increased economic efficiency.

When off-site effects are present, the rationale for intervention is potent, because the farmers' estimation of returns to conservation will pay inadequate attention to its social benefits. In the Santa Lucía Milpas Altas watershed in Guatemala, for example, a USAID project uses subsidies (so-called *pago social*) to induce farmers to build terraces and thus reduce flooding in the historic town of Antigua. In the same watershed, farmers who do not receive subsidies generally use less costly conservation methods such as vegetative barriers and live fences. Although these measures are profitable to the farmers, they may not control floods.

The effect of price distortions is more difficult to

establish. The many factors that affect the profitability of a given conservation measure and their complicated interactions make it hard to predict whether a distortion encourages or discourages conservation (Pagiola 1996). Recent evidence suggests that typical policy distortions in developing countries tend to encourage degradation (Panayotou 1993), but more work is needed to substantiate this point. The best way of dealing with policy distortions or market failures is to attempt to eradicate the distortions themselves; subsidization should be resorted to only in the rare instances that such direct action is virtually impossible.

Whatever their justification, the use of subsidies encounters several difficulties. First, the divergence between social and private returns to conservation must be established, so that intervention can be directed where it will be most effective. However, subsidies are often used where no off-site effects are present, wasting scarce budgetary resources in areas where they are not justified by any social benefits. In Costa Rica, for example, the soil conservation service (SENACSA) subsidizes half the cost of establishing conservation measures on the fields of small farmers, irrespective of location. Subsidies are also provided in cases such as Turrubares, where individual farmers already have sufficient incentive to conserve purely on productivity grounds. Conversely, subsidies are not always provided in cases where off-site effects are present. More commonly, subsidies are provided to construct, but not maintain, the conservation measures, so farmers sometimes allow them to decay. In Nicaragua, for example, terraces were built on fields in the Lake Xolotlán watershed above Managua in an effort to reduce flooding in the city and sedimentation in its reservoirs. Built at no cost to the farmers, these terraces interfered with cultivation practices and did not result in private net benefits to the farmers; most were soon destroyed. Similar experiences have occurred in the Tierra Blanca area of Costa Rica.

The second problem in using subsidies, then, is the difficulty of designing appropriate incentive structures so that social objectives are met. The case of the Lake Xolotlán watershed illustrates a situation in which subsidies are insufficient to overcome the divergence between private and social returns to conservation. The El Naranjal watershed in the Dominican Republic provides another example. Here, the USAID-funded Management of Natural Resources Project (MARENA) provided subsidized credit to participating farmers. Consequently, adoption rates were initially very high, even though the evidence suggests that the measures were unprofitable from the farmers' perspective. In 1985, more than 90 percent of the area's farms practiced soil conservation. Five years later, however, only half of these farms continued to do so. Subsidies can only convince farmers to modify their behavior as long as they continue to be paid. In contrast, MARENA's successor seems to have stimulated considerable use of conservation techniques without offering subsidies; in fact, the cost of participation was quite high because conservation was tied to access to irrigation. Although sufficient data were not available to analyze the new practices fully, they appear to be highly profitable.

Another risk in designing subsidization schemes is that of creating perverse incentives for farmers. In Costa Rica, for example, a reforestation credit system unintentionally encouraged farmers to deforest their land so that they might qualify for the credit. The expectation that subsidies are forthcoming to fund conservation efforts may also encourage farmers to delay conservation, even when such measures are privately profitable, in the hope that the government will bear part of their cost. Even when they are justified, then, subsidies must be used with great care.

Governments should also ensure that constraints such as insecure tenure do not prevent farmers from adopting conservation measures. But such efforts require substantiating research if they are to be effective. Too often the existence of tenure problems and the effectiveness of titling as a solution are simply taken as given.

Governments already do some research on soil conservation and provide, through extension services, some assistance to farmers who undertake conservation work. However, research in experiment stations has tended to favor technical efficiency (including structural measures such as terraces) over cost-effectiveness. Further, government extension work is often ineffective. In many cases, nongovernmental organizations, such as Vecinos Mundiales in Central America (López and Pío Camey 1994) have proven to be more effective at presenting the range of options to farmers and delivering related technical assistance. Given the wide variety of conditions that farmers face, government extension services should also provide, explain, and demonstrate to farmers the variety of options available rather than, as has often happened in the past, pushing specific techniques. And it may be innovative as well as effective for governments to decentralize decisionmaking and channel budgetary resources for soil conservation to the local level. This would allow communities to participate and contract assistance from those from whom the greatest contributions can be expected.

Research is not likely to produce a "breakthrough technology" that will solve all conservation problems. Improvements are likely to be more marginal. But, alone or in combination, improved techniques can have a significant impact on productivity. Modifications in the *ramp pay* technique used in Haiti are an example. Here, the traditional practice of gathering crop stubble along the contour was improved by more exact placement and by covering the structure with upslope soil, thus discouraging rat infestations and encouraging surface flow infiltration. These changes made the practice much more effective in halting degradation and more acceptable to farmers. Similar improvements in techniques arising from research have been successful in West Africa (Reij 1992).

The conflict between conservation and production noted in many of the case studies often affects the returns to conservation very significantly. Attempts to develop practices that reduce or eliminate this conflict—overlap technologies, in the terminology of Reardon and Vosti (1992)—should be especially encouraged. And to make the research truly useful, it should be carried out primarily on the farm and in close consultation with farmers.

Notes

The data used in the case studies were collected and analyzed by Mauricio Cuesta and Héctor Manuel Melo Abreu (Costa Rica); José Abel Hernández (Dominican Republic); José Bueno Alferez, José Roberto Hernández Navas, and Rafael Lazo Meléndez (El Salvador); Luis Eduardo Barrientos, Saúl Adolfo Lima, and Pedro Antonio Rosado (Guatemala); Jon L. Jickling and Thomas A. White (Haiti); José Wilfredo Andino, Carlos Awad Ramírez, Gabino López Vargas, Gilberto Palma, Augustino Pío Camey, and Antonio Valdéz (Honduras); Danilo Antonio Montalbán and Miguel Obando Espinoza (Nicaragua); and César Isaza, Julio Santamaría, and Tomás Vásquez (Panama). The authors are indebted to Hans Binswanger, John English, John McIntire, Augusta Molnar, Alfredo Sfeir-Younis, Stephen Vosti, and several members of the Editorial Committee for valuable comments and suggestions.

1. Off-farm effects and another important land degradation problem—the inappropriate use of common property lands—are outside the scope of this research. For a discussion of off-farm effects, see Magrath and Arens (1989); for common property issues, see Bromley (1992).

2. Combined investments from households in a village or watershed area are sometimes required to manage land degradation problems effectively. For analysis of such problems in the same area as the Haiti case study described below, see White and Runge (1992).

3. In addition to the sites listed in table 16.2, research was carried out at other sites in several of the countries listed and at several sites in El Salvador. Data on these sites—in particular, on the effects of degradation on yields—are insufficient to allow a full analysis of the returns to conservation measures.

4. Even when rates of return to investment in conservation are high, however, conservation might not be undertaken if even higher rates of return can be obtained by investing in off-farm income opportunities. Southgate (1994), for example, argues that high returns to urban employment in Ecuador encourage farmers to depreciate their land assets and then move to urban areas. Similarly, Schneider and others (1993) argue that perceptions of limitless land resources in the Amazon prompt farmers to "mine" their soils and then move on.

References

The word "processed" describes informally reproduced works that may not be commonly available through libraries.

Biot, Yvan, Robert Lambert, and Scott Perkin. 1992. *What's the Problem? An Essay on Land Degradation, Science, and Development in Sub-Saharan Africa.* Discussion Paper 222. East Anglia: University of East Anglia, School of Development Studies.

Blaikie, Piers, and Harold Brookfield. 1987. "Defining and Debating the Problem." In Piers Blaikie and Harold Brookfield, eds., *Land Degradation and Society.* London: Methuen.

Bromley, Daniel W., ed. 1992. *Making the Commons Work: Theory, Practice, and Policy.* San Francisco: ICS Press.

Capalbo, John M., and Susan M. Antle. 1988. *Agricultural Productivity: Measurement and Explanation.* Washington, D.C.: Resources for the Future.

De Janvry, Alain, Marcel Fafchamps, and Elisabeth Sadoulet. 1991. "Peasant Household Behavior with Missing Markets: Some Paradoxes Explained." *The Economic Journal* 101:1400–17.

English, John, Mary Tiffen, and Michael Mortimore. 1994. *Land Resource Management in the Machakos District, Kenya: 1930–1990.* Environment Paper 5. Washington, D.C.: World Bank.

Ervin, David E. 1986. "Constraints to Practicing Soil Conservation: Land Tenure Relationships." In Stephen B. Lovejoy and Ted L. Napier, eds., *Conserving Soil: Insights from Socioeconomic Research.* Ankeny, Iowa: Soil Conservation Society of America.

Holden, S. T., B. Shiferaw, and M. Wik. 1998. "Poverty, Market Imperfections, and Time Preferences: Of Relevance for Environmental Policy?" *Environment and Development* 3(1): 83–104.

Lal, Rattan C. 1988. "Preface." In Rattan C. Lal, ed., *Soil Erosion Research Methods.* Ankeny, Iowa: Soil and Water Conservation Society.

Leonard, H. Jeffrey. 1987. *Natural Resources and Development in Central America: A Regional Environmental Profile.* New Brunswick, N.J.: Transaction Books for the International Institute for Environment and Development.

López, Gabino Vargas, and Augustín Pío Camey. 1994. "Practical Experiences and Lessons Learned by Vecinos Mundiales from Soil Conservation Work in Rural Communities of Honduras." In Ernst Lutz, Stefano Pagiola, and Carlos Reiche, eds., *Economic and Institutional Analyses of Soil Conservation Projects in Central America and the Caribbean.* Environment Paper 8. Washington, D.C.: World Bank.

López, R. 1996. "Land Titles and Farm Productivity in Honduras." Department of Agriculture and Resource Economics, University of Maryland, College Park, Md. Processed.

Lutz, Ernst, Stefano Pagiola, and Carlos Reiche, eds. 1994. *Economic and Institutional Analyses of Soil Conservation Projects in Central America and the Caribbean.* Environment Paper 8. Washington, D.C.: World Bank.

Magrath, William B. 1989. "Economic Analysis of Soil Conservation Technologies." Divisional Working Paper 1989-4. Environment Department, World Bank, Washington, D.C. Processed.

Magrath, William B., and Peter Arens. 1989. "The Costs of Soil Erosion on Java: A Natural Resource Accounting Approach." Environment Department Working Paper 18. World Bank, Washington, D.C. Processed.

McIntire, John. 1994. "A Review of the Soil Conservation Sector in Mexico." In Ernst Lutz, Stefano Pagiola, and Carlos Reiche, eds., *Economic and Institutional Analyses of Soil Conservation Projects in Central America and the Caribbean.* Environment Paper 8. Washington, D.C.: World Bank.

Murray, Gerald. 1994. "Technoeconomic, Organizational, and Ideational Factors as Determinants of Soil Conservation in the Dominican Republic." In Ernst Lutz, Stefano Pagiola, and Carlos Reiche, eds., *Economic and Institutional Analyses of Soil Conservation Projects in Central America and the Caribbean.* Environment Paper 8. Washington, D.C.: World Bank.

Oldeman, L. R., R. T. A. Hakkeling, and W. G. Sombroek. 1990. *World Map of the Status of Human-Induced Soil Degradation: An Explanatory Note*, 2d ed., rev. Wageningen: International Soil Reference and Information Centre.

Pagiola, Stefano. 1993. "Soil Conservation and the Sustainability of Agricultural Production." Ph.D. diss. Food Research Institute, Stanford University, Palo Alto, Calif.

———. 1994. "Soil Conservation in a Semi-Arid Region of Kenya: Rates of Return and Adoption by Farmers." In T. L. Napier, S. M. Camboni, and S. A. El-Swaify, eds., *Adopting Conservation on the Farm: An International Perspective of the Socioeconomics of Soil and Water Conservation.* Ankeny, Iowa: Soil and Water Conservation Society.

———. 1996. "Price Policy and Returns to Soil Conservation in Semi-Arid Kenya." *Evironmental and Resource Economics* 8:225–71.

Panayotou, Theodore. 1993. *Green Markets: The Economics of Sustainable Development.* ICEG Sector Studies 7. San Francisco: ICS Press.

PCEO (Proyecto de Control de Erosión de Occidente). 1981. "Internal Report." Irena, Nicaragua. Processed.

Pender, John L. 1992. "Credit Rationing and Farmers' Irrigation Investments in Rural South India: Theory and Evidence." Ph.D. diss. Food Research Institute, Stanford University, Palo Alto, Calif.

Reardon, Thomas, and Stephen A. Vosti. 1992. "Issues in the Analysis of the Effects of Policy on Conservation and Productivity at the Household

Level in Developing Countries." *Quarterly Journal of International Agriculture* 31(4, October-December):380–96.

Reij, Chris. 1992. "Building on Traditions: The Improvement of Indigenous SWC Techniques in the West African Sahel." Paper presented at the international symposium on soil and water conservation, Soil and Water Conservation Society, Honolulu, Hawaii, 19–21 October. Processed.

Repetto, Robert, and Wilfrido Cruz. 1991. *Accounts Overdue: Natural Resource Depreciation in Costa Rica.* Washington, D.C.: World Resources Institute.

Schneider, Robert, Gunars Platais, David Rosenblatt, and Maryle Webb. 1993. "Sustainability, Yield Loss, and *Inmediatismo*: Choice of Technique at the Frontier." LATEN Dissemination Note 1. Latin America Technical Department and Environment Division, World Bank, Washington, D.C. Processed.

Singh, Inderjit, Lyn Squire, and John Strauss, eds. 1986. *Agricultural Household Models: Extensions, Applications, and Policy.* Baltimore, Md.: Johns Hopkins University Press for the World Bank.

Southgate, Douglas. 1994. "The Rationality of Land Degradation in Latin America: Some Lessons from the Ecuadorian Andes." In T. L. Napier, S. M. Camboni, and S. A. El-Swaify, eds., *Adopting Conservation on the Farm: An International Perspective of the Socioeconomics of Soil and Water Conservation.* Ankeny, Iowa: Soil and Water Conservation Society.

Taylor, Daniel B., Douglas L. Young, David J. Walker, and Edgar L. Michalson. 1986. "Farm-Level Economics of Soil Conservation in the Palouse Area of the Northwest: Comment." *American Journal of Agricultural Economics* 68(2, May):364–65.

Veloz, Alberto, Douglas Southgate, Fred Hitzhusen, and Robert Macgregor. 1985. "The Economics of Erosion Control in a Subtropical Watershed: A Dominican Case." *Land Economics* 61(1, May): 145–55.

Wachter, Daniel. 1992. "Land Titling for Land Conservation in Developing Countries." Divisional Working Paper 1992-28. Environment Department, World Bank, Washington, D.C. Processed.

Walker, David J. 1982. "A Damage Function to Evaluate Erosion Control Economics." *American Journal of Agricultural Economics* 64(4, November):145–55.

White, Thomas A., and C. Ford Runge. 1992. "Common Property and Collective Action: Cooperative Watershed Management in Haiti." Working Paper P92-3. Center for International Food and Agricultural Policy, University of Minnesota, St. Paul. Processed.

17

Toward a Strategy for Mainstreaming Biodiversity in Agricultural Development

Jitendra Srivastava, Nigel Smith, and Douglas A. Forno

The conservation of biodiversity has emerged as a major priority. It is no longer just the concern of bird watchers and a handful of field botanists. Citizens and politicians alike have rallied to the cause of saving biodiversity, which is harnessed by cultures in various ways to produce food and other products. Much of the planet's remaining biodiversity will be lost unless future needs can be met from areas already cultivated or grazed. Demand for food and other agricultural products is likely to increase significantly over the next few decades. As societies on every continent become more urbanized and income levels rise, more livestock products are consumed, and greater demands are placed on landscapes to produce feed for cattle, pigs, and other domestic animals. This process is likely to exert further pressure on biodiversity unless a concerted effort is made to adopt more environmentally sound agricultural practices.

Agriculture: Friend or Foe of Biodiversity?

Some dramatic changes will be needed in the ways that people raise crops and livestock if much biodiversity is to survive the next 50 years. How agriculture is transformed and intensified in a sustainable manner will be the key to how many species and how much genetic variation are still around in the next century. A focus on conserving biodiversity in "protected areas" alone will not work (box 17.1). The main purpose of this chapter is to make the case that agriculture and biodiversity are intimately connected and that one cannot survive without the other. Continued progress in raising and sustaining agricultural yields hinges on better protecting and harnessing the planet's biological riches.

Agriculture is often seen as the "enemy" of biodiversity, rather than as part of it. This perception arises because the raising of livestock and crops inevitably alters vast expanses of the earth's surface. Population growth and other factors encourage deforestation, pastoralists are squeezed into ever-diminishing spaces and sometimes overgraze the land, and high-input, modern farming practices frequently pollute the water and soil with chemicals. All of these activities trigger a potentially dangerous loss of biodiversity. However, some land use systems and agricultural practices enhance biodiversity within managed landscapes. For example, the judicious use of livestock waste as organic manure enhances the species diversity of macrofauna (Bohac and Pokarzhevsky 1987). Also, inappropriate agricultural practices can be modified to mitigate adverse impacts on the environment.

Agricultural Intensification: A Bane or Blessing for Biodiversity?

To many, agricultural intensification means more purchased inputs, such as fertilizers, pesticides, herbicides, and machinery. Eutrophication of lakes and estuaries, loss of soil micro-organisms, accelerated soil erosion, contamination of groundwater, and draining of wetlands—to mention a few of the adverse impacts of some agricultural practices—attest to the dangers inherent in intensifying agriculture without regard to the long-term consequences for the natural resource base.

This chapter does not purport to provide all the answers. It is only one step in that direction. But sustainable agricultural intensification would include approaches such as:

- More rational use of nutrients, space, and energy in all land use systems
- Greater recycling of nutrients
- Better use of biological resources to raise and maintain yields of crops and livestock, for example, by improving the use of genetic resources and biocontrol agents of crop pests
- Greater appreciation for, and use of, indigenous knowledge, especially concerning numerous neglected crops that could help improve livelihoods and the environment
- More effective measures for soil and water conservation
- The deployment of "environmental corridors" in landscapes that have been transformed by agriculture and livestock raising.

Box 17.1. A Holistic Approach to Biodiversity Conservation

Protection of a sample of natural habitats is neither sufficient nor desirable to conserve biodiversity for two simple reasons: (a) most of the world's biodiversity exists in human-managed or -modified systems, and (b) land use patterns and sociopolitical factors in areas adjacent to parks and reserves have major implications for the integrity of biological diversity in "protected" areas (Pimentel and others 1992). This relationship has clearly been demonstrated by the fate of 62 bird species in an 86-hectare woodland in West Java. After several square kilometers of surrounding woodland were destroyed, 20 bird species disappeared, four declined almost to extinction, and five more declined noticeably (Diamond, Bishop, and Van Balen 1987). The remaining species appeared to be unaffected. This example highlights the need for regional conservation (Ricklefs 1987) and the need for biodiversity conservation in both protected areas and agricultural ecosystems.

The Core Issue

How agriculture can be intensified while enhancing biodiversity is the critical question tackled in this chapter. Our goal is to identify some of the critical dimensions to this issue, to illuminate their multiple facets, and to suggest policies that mitigate the adverse impacts of agriculture on the environment. Our concern here is not only to highlight ways in which agricultural practices can be tailored so that they are more environmentally friendly, however. We are especially concerned with incorporating greater biodiversity within agricultural production systems. New approaches to agricultural research and development are being tried in various places around the world, and virtually all of them emphasize harnessing and managing biological resources better than in the past. Instead of excessive reliance on an arsenal of potent chemicals to improve soil fertility and thwart the attacks of insects and disease-causing organisms, agricultural research is geared increasingly to the manipulation of genes and release of predators of crop pests, among other biological assets. When crops and livestock are bred so that they can thrive under the incessant onslaught of challenges to productivity, agricultural production systems become more resilient.

Agricultural intensification does not automatically trigger greater harm to the environment. On the contrary, it can save and enhance biodiversity. Benign policies and practices that enhance agricultural productivity as well as biodiversity conservation are possible. This chapter pinpoints ways in which this has already been accomplished in certain areas and suggests measures that might accelerate the wider adoption of sound practices. Such information will be useful to persons engaged not only in designing and implementing agricultural development projects but also in establishing priorities for agricultural research. Individuals involved primarily in the conservation of biodiversity as it is most widely understood—the safeguarding of wildlife and "nat-ural" habitats—might also find use in the discussion on the complementarities between agriculture and environmental conservation.

Biodiversity and Agrobiodiversity Defined

Before exploring the complex issues surrounding agricultural development and biodiversity, it is useful to define what we mean by biodiversity and agrobio-

Figure 17.1. Agrobiodiversity as a Subset of Biodiversity

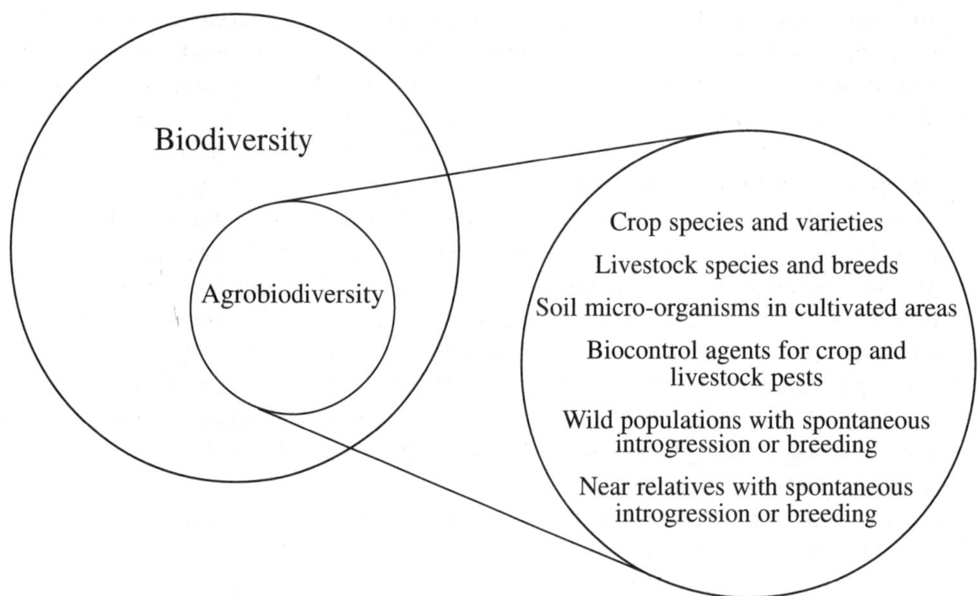

diversity, the subset of biological resources that support agricultural production (figure 17.1).

Biodiversity has three main dimensions: the genetic variation within species and populations, the number of species, and the preservation of habitat (Srivastava, Smith, and Forno 1996). The significance of variation within a species is not widely appreciated but is critical, particularly for agriculture. The continued productivity of existing crops and livestock hinges in large part on harnessing the genetic variation found within each species.

The second dimension of biodiversity is fairly straightforward: it is an index of species richness, or the number of distinct plants and animals in a given environment. Thus tropical rain forests are especially rich in species, and their fate has major implications for many crops important for subsistence and cash income in the tropics and subtropics (box 17.2). In order to protect species and genetically distinct populations of each species, it is necessary to safeguard their environments.

The issue of habitat conservation has two parts: safeguarding natural habitats for wild species and populations and managing habitats that have been modified for human use, such as farmland. The second item is poorly understood among the general public and many policymakers. Economic growth and poverty alleviation hinge in large measure on managing biodiversity in habitats transformed by humans, such as farmland, rangeland, and forests exploited for lumber and nontimber forest products. How farmers and livestock owners manipulate natural resources is

Box 17.2. Biodiversity and the Tropics

The conservation and management of biodiversity are particularly critical in lower latitudes because they contain many more species than are typically found in temperate areas. Agrobiodiversity is also exceptionally high in the humid tropics. Tropical rain forests cover only 7 percent of the earth's surface but more than half of the world's biota (Wilson 1988). A couple of examples suffice to illustrate the extraordinary levels of biodiversity found in tropical rain forests and to underscore their importance for conservation and the need for careful management of their biological riches. A 13.7 square kilometer portion of the La Selva reserve in Costa Rica contains almost 1,500 plant species, more than are found in all of the United Kingdom (Myers 1988). The Malayan peninsula is only about a third as large as the United Kingdom, but it contains five-and-a-half times as many plant species (Whitmore 1985). Tropical forests contain wild populations of hundreds of crops, such as mango, coffee, and rubber, as well as candidates for new crops and livestock.

therefore critical to the question of conserving biodiversity.

It is in habitats that have been modified for crop and livestock production that one finds agrobiodiversity, which includes all plants and animals that contribute directly or indirectly to the raising of crops and livestock.

Although it is conceptually useful to differentiate agrobiodiversity from the larger array of species and habitats, it is worth emphasizing that the boundaries between biodiversity and agrobiodiversity are not clear-cut. All of biodiversity is potentially of use to agriculture, particularly with the emerging biotechnologies. Agriculture is highly dynamic, and the interface between domesticated plants and animals and wild species is constantly shifting. A wild plant or animal of little or no current market value could eventually provide significant employment and income in the future. The fact that it is not easy to draw a firm line between biodiversity and agrobiodiversity only underscores the importance of conserving as much biodiversity as feasible to maintain its future value for agricultural development, among other reasons.

Agriculture has a direct stake in safeguarding the wider biodiversity. Wild species are essential for agricultural improvement because they are a source of new economic plants and animals and provide important services such as pollination and pest control. Moreover, advances in biotechnology are pushing back the limits to exchanging genes between unrelated organisms.

Given the ultimate importance of biodiversity in its broadest sense to agriculture, a strategy for mainstreaming biodiversity in agricultural development needs to address the off-site impacts of land use systems. Steps in this direction are outlined in the call for a new agricultural research and development paradigm. Work is already under way to address a range of issues related to off-site impacts, including the reduction or elimination of agricultural pollutants in groundwater and in runoff and the placement of greater emphasis on integrated pest management (IPM) strategies.

To many, biodiversity conservation is essentially synonymous with protecting showcase mammals, birds, and wild and spectacular landscapes. Although efforts to safeguard the habitats of wildlife warrant support, this conventional approach to conservation of biodiversity is insufficient. Most of the earth's surface has now been modified in various ways by human activities. On a global scale, less than 5 percent of the land is in nominally protected parks or reserves, and some 70 percent of the total land surface is in agriculture or managed forests (Pimentel and others 1992). Considerable biodiversity is often found in areas managed for agricultural production or extractive products.

How and why rural people conserve, enhance, and use biodiversity have rarely been taken into account when designing management interventions and devising policy for agricultural development and natural resource management. But the active participation of farmers, ranchers, and pastoralists—and especially resource-poor operators—is essential in the design and implementation of biodiversity and agricultural development projects (Thrupp, Cabarle, and Zazueta 1994 and Wilcox and Duin 1995).

The Role of the Business Community

One of the leitmotifs of this chapter is that market forces can be harnessed to ferret out and promote crops or old varieties that are in danger of slipping into extinction. The same idea applies to ancient, but dwindling, livestock breeds and some unconventional livestock species, such as iguanas. If markets can be found for "forgotten" crops and livestock breeds, they are more likely to survive. Gene banks and embryo stores cannot save all the varieties and breeds that are no longer commercially viable. Markets are constantly changing, and new opportunities are emerging for tapping some of these dwindling pockets of agrobiodiversity to generate income for locals and enrich the diets of people around the world.

Many of the innovative approaches to marketing the untapped wealth of biodiversity in Latin America are likely to come from relatively small-scale enterprises. One example is Kapok International, based in Chagrin Falls, Ohio. Kapok International has a Brazilian subsidiary in Manaus, Brazil, and currently markets some Amazonian fish. It soon expects to begin marketing unusual Amazonian fruits for the juice and candy industry in North America (box 17.3).

With assistance from the International Finance Corporation, an innovative trust fund has been established to spur greater private sector investments in improving the use and management of the wealth of biological resources in Latin America. The Biodiversity Enterprise Fund for Latin America is designed to help companies start up or expand business activities linked to sustainable use of the continent's biodiversity, which is both underused and, in

Box 17.3. Business Opportunities in Agrobiodiversity: The Case of Kapok International

Turning the tide on the rapid destruction of species-rich rain forests can only occur if the economic needs of the local people are addressed. To that end, Kapok International is seeking to develop new markets for some intriguing Amazonian fruits that are little known outside the region. Based in Chagrin Falls, Ohio, Kapok International is exploring the potential of some nontimber forest products and some promising crops that can be grown or collected on a sustainable basis. In this manner, economic value will be added to forests, and locals can improve their standards of living.

Market opportunities in the United States abound for "cause-oriented" foods. Consumers in the United States, as well as many other countries, are increasingly interested in novel foods, especially if they come from exotic locations such as the Amazon. And some consumers are willing to pay a premium for extractive and agricultural products that have been obtained without damaging the environment or the cultures of local people, especially indigenous groups.

One product that Kapok International expects to market in the near future is *cupuaçu* cocoa. Pronounced *coo-poo-a-su*, this relative of cacao renders a fine chocolate-like drink with a hint of citrus. *Cupuaçu* makes a creamy-tasting cocoa, but without the milk, a major plus for the millions of people who suffer from lactose intolerance. Fine *cupuaçu* chocolate bars may soon grace supermarket shelves. Several drinks containing *cupuaçu* pulp have appeared recently in several U.S. markets.

A lack of infrastructure in most parts of the Amazon is a major impediment to exploring the full potential of the region's numerous tropical fruits. Poorly maintained roads impede the timely delivery of fruits to processing plants. Unsanitary conditions and inadequate refrigeration at the region's few agro-industrial plants also hinder the export of frozen pulps and purees to demanding markets in North America and Europe. With supportive agricultural policies and an improved investment climate, such barriers can be overcome to the benefit of consumers, local people, and biodiversity.

Box 17.4. Biodiversity and Sustainability of Agricultural Systems

As with any ecosystem, the functioning and sustainability of agricultural systems depend greatly on biological diversity. Through an experimental study, Tilman, Wedin, and Knops (1996) demonstrate that more diverse plant communities use and retain nutrients more efficiently, thereby attaining greater productivity and reducing the loss of nutrients leaching from the ecosystem. Various soil microbes also effectively recycle nutrients (Lal 1991). Earthworms, insects, and fungi all play a vital role in the functioning and sustainability of agricultural systems. Biological diversity also enhances natural pest control mechanisms in agroecosystems. In fact, every single species that exists in agroecosystems has intrinsic value. Insects provide a virtually untapped source of food (Defoliart 1989), dyes (Metcalf, Flint, and Metcalf 1962), and pharmaceutical products (Eisner 1990). The elimination or addition of even one species can have profound effects. Pimentel and others (1992) cite a remarkable example to illustrate this. Until recently, oil palm trees in Malaysia were pollinated manually—an inefficient and expensive way of performing the task. Ten years ago, the government introduced a tiny weevil from West Africa's forests that is associated with palm pollination. The pollination of palm trees in Malaysia is now accomplished entirely by the weevil, with annual savings of $140 million (Greathead 1983).

some cases, overexploited. Compared with traditional investments in agriculture, fisheries, and forestry, the $20 million to $30 million fund for biodiversity-based enterprises is a small but promising start. As experience accumulates as to what works and what does not with such specialized investment funds, the expansion of such efforts should be considered in Latin America and in other regions (see box 17.4).

The private sector has critical roles to play at various steps of the process of making better use of agrobiodiversity. In the case of planting material, multinational seed companies dominate the production and sale of major cereals, but opportunities exist for small-scale private seed companies to exploit the growing demand for unconventional crops, particularly in areas where the crops have recently been introduced. Although the private sector is still a relatively minor source of new varieties in most developing countries, this picture is gradually changing (Dalrymple and Srivastava 1994). Policies at the national and international levels are needed to facilitate this encouraging trend.

Major Recommendations

This section brings together major policy recommendations within a framework that both highlights the multiple dimensions of agriculture-biodiversity interactions and suggests some ways to proceed. We explore fresh approaches on in situ conservation of plant and animal genetic resources, underscore the value of work on the systematics of plants and animals to help us better understand variation within species and relationships between species, and make a case for assembling interdisciplinary teams to assess agrobiodiversity before agricultural development projects are implemented. Major policy recommendations span issues such as promoting quality pricing, extending credit to traditional varieties as well as modern cultivars, and reducing agricultural subsidies, which often exacerbate environmental problems. Dimensions to the emerging agricultural research paradigm that stresses sustainability are also examined.

A strategy is needed to help policymakers reconcile the task of preserving biodiversity while at the same time increasing agricultural productivity. In the first section, we explored some of the important relationships between agriculture and biodiversity and argued that biodiversity conservation and agricultural development can be complementary activities. Here we explore some tentative steps toward an overall strategy for mainstreaming biodiversity in agricultural development.

Agenda for Action

Various policies and practices can be devised to promote biodiversity conservation in managed landscapes and to enhance the greater use of bioiversity in agriculture (table 17.1). The major points for policy consideration are summarized here.

Toward a balanced conservation strategy. Too often conservation is posed as an either/or proposition: some argue that resources should be concentrated in maintaining animal and plant genetic resources in places where they occur naturally (in situ), while others suggest that genetic variation is more useful, and often safer, if it is maintained in seed or field gene banks (ex situ). Both strategies are needed to sustain the productivity of agricultural systems.

Promising in situ approaches. The following approaches to in situ conservation of crops and livestock warrant particular support:

- Emphasize a new dimension to existing wilderness parks and biological reserves by safeguarding plants and animals for the future improvement of agriculture. Some existing parks and reserves contain wild populations or near-relatives of crops and livestock; highlighting such attributes would strengthen the case for improving their protection and management. A major problem here is that many areas under consideration for protection have not been adequately inventoried from the botanical standpoint.
- Create world heritage sites for genes for agricultural development. Most parks and reserves have been set up to protect spectacular animals or rare and endangered plants and habitats. Although such efforts deserve continued support, the agrobiodiversity aspect of environmental conservation has been largely overlooked. Many areas are "hot spots" for agrobiodiversity and warrant better conservation and management in their own right.
- Integrate agrobiodiversity in ecotourism where appropriate. A virtually untapped market exists for showing tourists who visit nature reserves and major archaeological sites "traditional" villages and farmlands in the surrounding area.
- Help find markets for lesser-known crops and local varieties under the motto "Use it or lose it." If a highly localized crop that is on the decline suddenly ignites interest in a distant market, growers will be motivated to look around for highly productive varieties. This search for planting material might save varieties that would otherwise be abandoned.
- Help find ways for livestock owners to generate more revenues from threatened breeds. If such breeds can "pay" for themselves, they are less likely to be replaced by the more widespread breeds or to lose their identity in continuous crossbreeding.

Greater involvement of the private sector. Too often, biodiversity conservation is seen as a public

Table 17.1. Toward a Strategy for Mainstreaming Biodiversity in Agricultural Development

Objective	*Proposed action*	*Responsible organizations*
Commitment to enhancing biodiversity in agriculture		
Increase government commitment to mainstreaming biodiversity in agriculture and rural development	Conduct substantive dialogue with client countries on the need to conserve and manage biodiversity resources	Multilateral development banks, bilateral aid agencies, NGOS
Increase farmer commitment to incorporate greater diversity in farming systems	Support efforts to establish property rights so that farmers are more willing to invest in long-term strategies to incorporate more biodiversity, such as intercropping with perennials; identify other incentives that encourage farmers to promote agrobiodiversity in their cropping systems	National governments
Generate greater understanding of how to implement policies and sound practices to enhance biodiversity in agricultural development	Prepare training manuals and packets for human resource development	Multilateral development banks, bilateral aid agencies, international agricultural research centers, NGOS, agricultural extension agencies
Diversify farming operations with viable alternative crops	Support more research in lesser-known or more localized crops	National agricultural research programs
Collect sufficient data on the potential impact of agricultural development projects on biodiversity and agrobiodiversity	Develop methodologies and performance indicators for assessing agrobiodiversity in areas targeted for agricultural development	Multilateral development banks, bilateral aid agencies, national agricultural research programs, NGOS, farmers
New agricultural research and extension paradigm		
Increase participation of farmers in the design and implementation of agricultural development projects to enhance agrobiodiversity	Conduct substantive dialogue with developing-country institutions on the need to incorporate, not just consult, local farmers in agricultural research and development projects	Farmers, multilateral development banks, international agricultural research centers, NGOS
Increase the deployment of genetically diverse (heterozygous) populations of crop varieties and animal breeds on managed landscapes	Offer greater support for developing polygenic resistance, multilines, high-yielding, open-pollinated varieties instead of hybrids	International agricultural research centers, national agricultural research programs

Improve productivity of farming and livestock raising in marginal environments	Conduct more research on undervalued crops and livestock that are better adapted to difficult environments than the "mega" crops and livestock	Farmers, international agricultural research centers, national agricultural research programs
Create greater awareness of appropriate practices for enhancing biodiversity in agricultural development	Develop and disseminate "sound practices" tailored to task managers in developing-country institutions and international development agencies	Multilateral development banks, bilateral aid agencies, NGOS, national agricultural research programs, agricultural extension
Create rich landscape mosaics with a variety of habitats for wildlife as well as crop and livestock production	Develop an understanding of factors that motivate farmers to retain or destroy buffer strips and woodlots or to plant a wider variety of crops; suggest policies to promote landscape mosaics	*Research:* universities, national agricultural research programs; *implementation:* agricultural extension
Develop understanding of how to implement policies and sound practices to enhance biodiversity in agricultural development	Prepare training manuals and packets for human resource development	Multilateral development banks, bilateral aid agencies, international agricultural research centers, NGOS, agricultural extension agencies
Approaches to conservation of agrobiodiversity		
Achieve appropriate balance between in situ and ex situ approaches by crop	Develop methodology for estimating the value of in situ conservation of genetic resources as a basis for allocating limited funds	International agricultural research centers (especially International Plant Genetic Resources Institute), universities, Food and Agriculture Organization
Create incentives for in situ conservation of crop and livestock genetic resources	Identify and implement a new system of intellectual property rights applicable to traditional varieties and breeds to benefit local people	Food and Agriculture Organization
Create incentives to maintain materials in ex situ collections through recognition of their tangible economic value	Explore the possibility of recognizing intellectual property rights for some genetic materials in collections with royalties to be paid to gene banks and groups that provide the germ plasm	Food and Agriculture Organization

sector responsibility. Experience has shown that reliance on parks and reserves alone to conserve biodiversity does not work because local populations often do not receive any benefits from them. Indeed, they may even suffer damages from wildlife-damaging crops or livestock or be denied access to resources they once depended on for part of their subsistence or income.

To counteract the reliance on formal protected areas, community empowerment and management of natural resources have become a new rallying cry among donors and organizations that promote conservation and rural development. The decentralization of control over natural resources can bring many benefits, but "community" management alone may not work in all cases.

Both protected areas and community or local initiatives to manage and conserve biodiversity warrant continued support, but they are not enough. A third, parallel approach is called for, particularly in the case of agrobiodiversity: greater involvement of the private sector. Where possible, market forces can and should be harnessed to conserve and better use biodiversity.

Tapping farmer knowledge. As a matter of policy, all agricultural research programs and development projects should include components that seek to incorporate farmers in the design and implementation of research and development efforts. The benefits of such collaboration far outweigh the "costs" of taking the time and trouble to improve the channels of communication between the research establishment and the intended beneficiaries: farmers and livestock managers.

Indigenous knowledge has made specific contributions to sustainable agriculture:

- Crop varieties adapted to harsh growing conditions
- Cropping patterns that minimize the buildup of diseases
- Breeds of livestock that tolerate poor feed or resist diseases and pests that afflict imported breeds
- Intimate knowledge of botanical resources in forest and other habitats that could provide leads for plant domestication or other uses.

Greater support for systematics. The task of conducting in situ conservation and agrobiodiversity surveys prior to agricultural development will be made much easier if competent specialists are available to identify and sort materials. Currently, not nearly enough taxonomists are available to analyze the plant and animal materials collected, particularly in the tropics. This shortage stems in part from the pull of glamorous fields, such as microbiology and biotechnology. Development organizations and donors could help fill this need by providing:

- Support to expand the room for botanical and zoological collections in museums, universities, and other research-oriented institutions in developing countries
- Scholarships for students from the developing world to undertake taxonomic training in both developing and industrial countries.

Rapid Agrobiodiversity Assessment Teams

Many development organizations, including the World Bank, have procedures for screening projects for their environmental impacts before approval. For the most part, such screening focuses on the off-site impacts of agricultural development and assesses whether the project is likely to lead to the loss of forest or other "natural" habitats. But surveys of agrobiodiversity are also needed before a development project is implemented.

An assessment of biodiversity before a project is approved and implemented would encompass two main topics: biodiversity in relatively undisturbed habitats and agrobiodiversity. For both situations, biodiversity indicators would need to be established and their significance for decisionmaking assessed. Rapid assessment methodologies have been tried and tested more fully with the first category. Much work remains to be done in designing methodologies for adequately assessing agrobiodiversity.

Biodiversity in relatively undisturbed habitats. The implications of the agricultural development project for the expected shift in land use patterns would need to be analyzed, paying special attention to the likelihood that the project might exert greater pressure on remaining relatively intact habitats. Using geographic information system techniques, one could begin delineating habitats with minimal disturbance where much of native biodiversity is concentrated.

For the different "natural" habitats, an inventory of the plant and animal species would be called for, noting any species with restricted distributions (this is already part of the environmental assessment process) and identifying any wild populations of domesticated plants and animals that could be of special interest to breeders. Efforts should be made to establish whether spontaneous introgression is occurring between weedy populations and crops, because such exchange of genes often improves the hardiness of crops.

Agrobiodiversity. A survey is warranted to document:

- The diversity of crops cultivated by local farmers
- Varietal diversity of each crop
- The number of livestock species raised
- The number of breeds per livestock species
- Assessment of land use practices on soil biodiversity as an indicator of soil "health."

Performance indicators. Performance indicators can serve as "warning bells" when agricultural practices threaten to impair the integrity of both "natural" and cultural habitats. Dangers signs include:

- Natural habitat loss
- Habitat fragmentation
- Species loss even when natural habitat remains intact
- Decline in biodiversity of crop species on-farm
- Decline in biodiversity within species.

The following is a sample of remedial measures that can be adopted to address the loss of biodiversity associated with agricultural development:

- Minimize habitat fragmentation by providing wildlife corridors along "bridges" of natural habitat
- Shift to integrated pest management strategies, such as crop rotation and reliance on biocontrol agents, to check crop and livestock pests
- Eliminate fiscal and regulatory measures that promote homogeneity in crop and livestock production
- Support research on traditional varieties that can achieve high yield
- Support research on modern varieties that are genetically resistant to pests and diseases.

More work is needed, however, to operationalize an indicator matrix, particularly with regard to methodologies for measuring indicators.

Adjustment of the policy environment. A wide range of policy areas are involved, such as credit, trade, intellectual property rights, and land tenure. The following policy levers would help conserve and improve the use of agrobiodiversity:

- *Promote quality pricing.* If farmers receive premium prices for unusual, but attractive, varieties of crops, they will be encouraged to grow them. A price grading system that rewards farmers not only for producing "clean" produce but also for offering a diverse array of fruit and leaf types will help to generate heterogeneity on agricultural landscapes. Quality pricing is especially important for fruits, vegetables, coffee, and cacao because a discriminating pricing structure would signal farmers that some of the lesser-known varieties also have market value and that it is worth their trouble to process and handle produce with care.
- *Open credit to nonrecommended varieties and livestock.* For the most part, farmers can only obtain credit for government-approved varieties or "super" breeds. Such discrimination accelerates the process of abandoning traditional varieties and breeds and contributes to genetic erosion.
- *Continue to encourage the reduction of farming subsidies.* This process is well under way in many countries, in part because of international trade agreements. It should be encouraged because agricultural subsidies often trigger biodiversity loss. Such losses occur because farmers often use more purchased inputs, such as environmentally damaging pesticides and fertilizers or irrigation water, than they would if they had to pay market prices.

A new agricultural research and development model. The old research model emphasized maximizing output, almost at any cost. Also, research tended to focus on commodities rather than on production systems as a whole. The interrelatedness of the parts of farming systems was neither appreciated nor understood, with sometimes adverse affects on the landscape and peoples' livelihoods. The new vision for agricultural research adopts a more holistic approach that is sensitive to environmental concerns, while still addressing the need to boost yields and incomes of rural producers and caretakers of the land.

The evolving agricultural research paradigm includes, but is not restricted to:

- *Adopt integrated pest management.* IPM strategies include the release of biocontrol agents, deployment of genetically resistant cultivars and breeds, more judicious use of pesticides and herbicides, alteration of cropping patterns to thwart the buildup of pests and diseases, and placement of greater emphasis on crop rotation, where economically feasible, to retard soil degradation and reduce pest pressure.
- *Use a participatory approach with farmers.* Two types of on-farm research are typically found: demonstration plots on farmers' land and experi-

mental work that involves farmers and other stakeholders in the design of models from the "ground up." Much more of the second type of on-farm research is needed that involves farmers, pastoralists, and other "clients" of agricultural research and development from the inception of the study design. In this manner, research would be more demand driven.

- *Improve the use of indigenous knowledge.* How and why local people use natural resources can provide important information for more appropriate agricultural research and development efforts.
- *Support research, development, and dissemination of lesser-known crops and animals.* Neglected traditional varieties and breeds, many of which are particularly well suited to difficult environments, would be included in this broad research effort. But sustained support for research on the major food and industrial crops as well as livestock remains essential.
- *Support research on new crops and livestock.* Scope exists for new crops and livestock to fill specialty market and environmental niches. In some cases, natural vegetation communities could be managed for the production of new domesticated animals. A deeper commitment to research on crop and livestock candidates would thus underscore the value of conserving biodiversity and natural habitats.
- *Be sensitive to the value of a mosaic of land uses.* Even land uses that are "desirable" from the biodiversity viewpoint can be promoted too far. Biodiversity in managed landscapes is often best served by promoting a mixture of land uses that provides varied habitats for wildlife adapted to altered areas.
- *Achieve a greater diversity of habitats within land use systems.* Biodiversity within a land use system, such as intensive cereal cropping, can be achieved by allowing for a variety of habitats, such as riparian buffer strips, shelter belts, windbreaks, strip cropping, and wetlands. Diversity of habitats on the landscape creates more niches for wildlife, some of which are beneficial in controlling crop pests. More diverse habitats, including managed ones, also promote the more efficient use of nutrients and create microclimates that buffer crops from inclement weather.
- *Recycle organic matter.* Measures such as incorporating livestock or green manure—no-till or minimum-till farming—help to sustain the diversity of soil micro-organisms, which are so important in nutrient recycling.
- *Focus research on lifetime and herd productivity characteristics.* Deterministic simulation models and live animal experimentation can be used in some cases to achieve this goal.
- *Determine the critical number of breeds for conservation purposes.* Analysis of genetic variation can be used to highlight the genetic "spacing" between breeds and to identify those breeds that are significantly different or unique from others.
- *Learn more about genetic components of adaptation in livestock.* A better understanding of traits such as resistance to ticks and use of body reserves would aid breeding efforts and would likely underscore the importance of safeguarding so-called "minor" breeds.

Institutional development. The notion of a new research paradigm has implications for developing new institutional arrangements and exploring new ways of doing business. Innovative institutional arrangements would include more effective partnerships between:

- Agricultural research centers and NGOS
- Agricultural research centers and growers' associations
- Agricultural research centers and private companies involved in the manufacture and sale of agricultural technologies
- Universities and agricultural extension agencies
- Development lending institutions and all of these entities.

To some degree, all these connections are being explored and tested. Results are mixed. Initial suspicion between some national agricultural research centers and NGOS has gradually subsided, while in other cases misgivings about interinstitutional collaboration have been borne out by experience. Links between agricultural extension agencies and generators of technology are, with few exceptions, weak. To facilitate future collaborative ventures in the management of agrobiodiversity, organizations' roles and responsibilities will have to be established more clearly at the outset.

References

The word "processed" describes informally reproduced works that may not be commonly available through libraries.

Bohac, J., and Pokarzhevsky. 1987. "Effect of Manure and NPK on Soil Macrofauna in Chernozem Soil." In J. Szegi, ed., *Soil Biology and Conservation of Biosphere*, 2 vols. Proceedings of the ninth international symposium. Budapest: Akademiai Kiado.

Dalrymple, D. G., and J. P. Srivastava. 1994. "Transfer of Plant Cultivars: Seeds, Sectors, and Society." In J. R. Anderson, ed., *Agricultural Technology: Policy Issues for the International Community*, pp. 180–207. Wallingford, U.K.: C.A.B. International.

Defoliart, G. R. 1989. "The Human Use of Insects as Food and as Animal Feed." *Bulletin of the Entomological Society of America* 35(1):22–35.

Diamond, J. M., K. D. Bishop, and S. Van Balen. 1987. "Bird Survival in an Isolated Javan Woodland: Island or Mirror?" *Conservation Biology* 1(2):132–42.

Eisner, T. 1990. "Prospective for Nature's Chemical Riches." *Issues in Science and Technology* (National Academy of Sciences) 6(2):31–34.

Greathead, D. J. 1983. "The Multi-million Dollar Weevil That Pollinates Oil Palm." *Antenna* (Royal Entomological Society of London) 7(3):105–07.

Lal, Rattan. 1991. "Soil Conservation and Biodiversity." In D. L. Hawksworth, ed., *The Biodiversity of Microorganisms and Invertebrates: Its Role in Sustainable Agriculture*, pp. 89–104. Wallingford, U.K.: C.A.B. International.

Metcalf, C. L., W. P. Flint, and R. L. Metcalf. 1962. *Destructive and Useful Insects*. New York: McGraw-Hill.

Myers, N. 1988. "Threatened Biotas: 'Hotspots' in Tropical Forests." *The Environmentalist* 8(autumn): 1–20.

Pimentel, D., U. Stachow, D. A. Takacs, H. W. Brubaker, A. R. Dumas, J. J. Meaney, J. A. S. O'Neil, D. E. Onsi, and D. B. Corzilius. 1992. "Conserving Biological Diversity in Agricultural/ Forestry Systems." *Bioscience* 42(5):354–62.

Ricklefs, R. E. 1987. "Community Diversity: Relative Role of Local and Regional Processes." *Science* 235:167–71.

Srivastava, Jitendra, Nigel J. H. Smith, and Douglas Forno. 1996. *Biodiversity and Agriculture: Implications for Conservation and Development*. Technical Paper 321. Washington, D.C.: World Bank.

Thrupp, L. A., B. Cabarle, and A. Zazueta. 1994. "Participatory Methods in Planning and Political Processes: Linking the Grassroots and Policies for Sustainable Development." *Agriculture and Human Values* 11(2-3):77–84.

Tilman, D., D. Wedin, and J. Knops. 1996. "Productivity and Sustainability Influenced by Biodiversity in Grassland Ecosystems." *Nature* 379(6567):718–20.

Whitmore, T. C. 1985. *Tropical Rain Forests of the Far East*. Oxford: Clarendon Press.

Wilcox, B. A., and K. N. Duin. 1995. "Indigenous Cultural and Biological Diversity: Overlapping Values of Latin American Ecoregions." *Cultural Survival Quarterly* 18(winter):49–53.

Wilson, E. O. 1988. "The Current State of Biological Diversity." In E. O. Wilson and F. M. Peter, eds., *Biodiversity*, pp. 3–18. Washington, D.C.: National Academy Press.

18

Integrated Pest Management: Strategies and Policies for Effective Implementation

Tjaart W. Schillhorn van Veen, Douglas A. Forno, Steen Joffe, Dina L. Umali-Deininger, and Sanjiva Cooke

Generating the gains in agricultural productivity necessary to secure food availability and livelihoods in the developing world over the coming decades requires an approach in which the intensification of agricultural systems is consistent with conservation of the natural resource base. This approach requires less reliance on the intensive use of external inputs and greater dependence on management skills and location-specific knowledge of agroecosystems. Integrated pest management (IPM) constitutes one such approach.

The IPM concept is far from new. Farmers used integrated pest control long before scientists coined the term. In traditional farming systems, pest management is inseparable from sound farm management. It involves farm practices that promote good plant and animal health and keep pest losses in check without the use of manufactured chemicals. Changes in farming systems during the past half century lost sight of this approach, and chemical control methods became the pillar in the control of pests and diseases in modern agriculture. During the past decade, however, growing concerns about the risk and negative effects of chemical methods have spurred agriculturists, environmentalists, and economists to explore pest management strategies that have fewer side effects on public health and the environment. The most well-known among these strategies is IPM.

IPM can best be described as a farmer's knowledge and use of an optimal mix of pest control tools and tactics, taking into account a variety of factors, including yield, profits, risk, sustainability, safety, and pest population dynamics. IPM is a key component of integrated farming practices that are based on an understanding of ecology and the interaction between crops or animals and their pests as well as an understanding of the environments in which pests operate.

An essential aspect of IPM is its integration of technical and social knowledge. This integration requires a sound understanding of key pest constraints and biological and farm management systems. These are often highly location-specific, and farmer participation and networking are essential in the design of modern science-based IPM schemes. The major limiting factor, however, is insufficient backup of farmers with adequate and applicable IPM-related technologies. Another problem is a lack of appropriate farm-focused research.

Development agencies, notably bilateral programs sponsored by Germany, the Netherlands, and the United States, as well as those sponsored by the Food and Agriculture Organization, have been advocating and supporting efforts to implement IPM for more than 20 years. The World Bank is now a significant IPM supporter and has invested some $80 million in implementation projects since 1988. However, despite these efforts and notable successes in some countries, widespread implementation of IPM remains an elusive goal in most parts of the world.

If IPM is to become widespread, farmers must have the appropriate incentives, relevant knowledge, and practical techniques to make use of nonchemical-based approaches. To apply IPM, farmers need to accept a practice that is usually more management- and labor-intensive than the use of chemical agents. Hence farmers will need to see a demonstrable economic payoff. Ultimately, the choice of pest management technology will be influenced by the costs, benefits, and availability of competing alternatives, as well as by any rules or other social norms governing their use.

Governments influence the prospects for widespread implementation of IPM through the incentive structures and regulations affecting the choice of pesticides or alternative approaches. Government influence is also strong through support for research, extension, education, and training initiatives. This dimension of IPM implementation, including the influence that donor organizations can have on government polices, is one of the most important but least-documented aspects of the challenge facing the international community in implementing the recommendations of the United Nations Conference on Environment and Development (UNCED) on IPM.

Consequently, this chapter reviews how governments can support and encourage the implementation of IPM and how the World Bank and other development agencies can assist this process. The first section considers the factors that affect farmers' choice of pest management technology. The second section reviews issues related to development strategies and government policies and shows how many countries have created a policy environment that favors the use of chemical pesticides. The third section assesses the policy instruments that governments can use to create a more level playing field for alternative pest management practices and to promote IPM where appropriate. The fourth section discusses national IPM policy and the need for more farmer-centered research, extension, and training.

Farmers and Pest Management

Farmers have available a range of pest management techniques, from natural controls based primarily on cultural, physical, and mechanical techniques to the use of biological control agents or chemical pesticides. In simple terms, the choice of pest management technique is a function of *costs* (purchased inputs, other variable costs such as labor, and fixed costs such as the cost of sprayers or information) and *returns* (labor-saving prevention of crop loss in monetary or subsistence terms). Poor farmers are limited in their choices and may especially benefit from cost-effective IPM approaches.

Because agricultural pests are rarely limited to single farms but occur regionally or beyond, there are limits to farmer-centered approaches. Most pest control programs also need a regional or national base as a guide, as exemplified by national and regional quarantine regulations, by the highly successful eradication of screwworm disease in the United States and more recently in northern Africa, and by the beneficial effects of nationally or regionally harmonized pesticide use.

Pest Management, Farmers, and the Chemical Industry

The profit motive of the chemical industry has been partly responsible for the detrimental environmental and public health effects of pesticides. Initial increases in pesticide use after World War II were fueled by extra supplies and industrial capacity. In later years, the industry has been responding to the demand, albeit somewhat manipulated, of its clients. Farmers embraced pesticides because of labor savings, increased crop security, higher quality and homogeneity of product, and persuasive messages from research, extension, and industry advertising activities. Consumers increasingly demanded visibly blemish-free agricultural and horticultural products; the industry responded by developing an array of pest control chemicals.[1] Consumers, however, are becoming increasingly interested in food safety and are demanding wholesome products and production practices that are not detrimental to the environment.

Just as public attitudes have changed, so too has the pesticide industry. Insecticide use in agriculture, initially the major culprit in public health concerns, has been stagnant or declining in the past two decades in industrial countries (National Research Council 1989), although herbicide use has increased. "Cure-all" drugs and agrochemicals are very slowly being replaced by target-specific chemicals, albeit at higher costs. Current research in the agrochemical industry is increasingly directed at highly specific interventions and at disease- or weed-resistant genetics (as well as herbicide-resistant crops that could increase future herbicide use). Following stricter regulation, as demanded by society, considerable improvements have been made in the safe handling of pesticides, although this has been applied more widely in the industrial world than in the developing world.

Economic Threshold in Pest Management Decisions

The concepts of *action thresholds* and *economic thresholds* are often used, albeit in a simplified way, to explain decisionmaking in pest management. Thus Stern (1973: 261) defines the action threshold level as "the population density at which control action will result in little likelihood of the pest population exceeding the economic threshold level." The economic threshold level is the "break-even point at which the value of loss in yield quantity or quality is equal to the cost of a control method that successfully eliminates pest damage and yield loss." The action and economic thresholds overlap, but they are not necessarily the same; farmers' actions may derive from a lack of specific knowledge (for example, "seeing many bugs" rather than assessing potential damage) or from advertising campaigns that promote use at levels where costs may be higher than benefits. Too often farmers, because they have only limited understanding of pest ecology, assume that all bugs are harmful.

In some cases it may indeed pay to take action very early against emerging pest populations, thereby preventing later buildup. In other cases it may not pay to take any action at all; many farmers may underestimate the recuperative ability of their plants and animals. Cotton plants, for example, can compensate well for early defoliation by leafworms; yet Sudanese cotton growers still spray early, inadvertently eliminating beneficial pests and thus creating subsequent problems with aphids and whiteflies.

Uncertainty about the degree of crop losses caused by pests creates output risks for farmers. The manner in which farmers respond to these risks (or their degree of risk aversion) in turn influences pest management activity. If they perceive the level of pest infestation and damage to be uncertain, farmers may apply pesticides at a lower threshold level to reduce the risk of crop losses. This decision contributes to more frequent applications and greater total volume applied. Uncertainty about the effectiveness of a particular technique could also lead to increased pest management activity to reduce risk at the margin (Feder 1979).

Policy Issues in Pest Management

In traditional farming systems a dynamic equilibrium (or endemic stability) exists between many pests and beneficial organisms. A degree of loss is accepted but a varied and ecologically well-balanced mix of crops (and livestock and wildlife) helps to limit the losses. The past half century has seen a worldwide trend toward more intensive and often mechanized agricultural systems that increase productivity and food supplies.

Agricultural Development Strategies: Changing Incentives for Pesticide Use

Production intensification, however, has been associated with a trend toward monoculture and chemical-based approaches to pest management. One side effect has been to create conditions conducive to the rapid growth of pest populations and to increase the risk that such populations will reach epidemic proportions. Formal agricultural research and extension systems have responded with the development of high-yielding plant (and livestock) varieties, combined with technology packages of pesticide- or drug-based approaches to counteract losses. Although an effective and economic course of action in the short term, the use and continuing development of cheap and quick-acting pesticides and veterinary pharmaceuticals may exacerbate the long-term pest problem by eliminating the population of pest predators ("good bugs," such as ladybirds, earthworms, and birds) that help to keep populations of undesirable bugs in check. Pest predators thus improve the natural defense systems of plants and animals.

Less emphasis has been placed on maintaining genetic diversity, including the maintenance of beneficial organisms, and on breeding for disease and pest resistance in plants and animals. A widely pursued development policy of industry-led growth has been accompanied by the taxation of agriculture (keeping food prices low), which has been partly compensated through input subsidies (including pesticides). Rising labor costs associated with the commercialization of agriculture have discouraged further the use of more labor-intensive traditional pest management practices.

Governments, Donors, and Excessive Use of Chemical Pesticides

The increased dependence on pesticides fostered by agricultural development strategies has led over time to the entrenchment of a chemical mode of pest control as the dominant paradigm in pest management. Today, many developing-country governments have incentive frameworks that favor pesticide use over the adoption of more environmentally benign IPM approaches.

Pesticide use tends to be promoted through both direct and indirect government and industry measures (table 18.1) and through the mechanism of prices (Waibel 1994 and Farah 1994). Direct government price subsidies reduce the farmer's costs of using pesticides and encourage increased use. Weak or inappropriate legislation and regulations and their inadequate enforcement contribute further to improper use. Bilateral and multilateral aid for pesticide use, provided through grants or subsidized prices, permits governments to distribute pesticides to farmers at low or no cost. Moreover, lack of information on alternative, more environmentally benign approaches to pest management often biases extension service toward pesticide use.

Governments indirectly subsidize the cost of pesticides through overvalued exchange rates that make pesticide imports more available, through public sector investments in pesticide research and development that are not recouped from pesticide users, and

Table 18.1. Direct and Indirect Measures Fostering the Use of Pesticides

Direct measures		*Indirect measures*	
Price factors	*Nonprice factors*	*Price factors*	*Nonprice factors*
Government price subsidies	Public sector policies, such as direct involvement of the public sector in marketing	Indirect public sector subsidies	Public sector policies, such as taxation
Public sector production and distribution	Weak or inappropriate pesticide legislation	Overvaluation of exchange rates that facilitate pesticide imports	Technology policy restricting the entry of more environmentally benign technologies (import ban, lack of intellectual property protection)
Cost subsidies to private pesticide manufacturers	Weak (or lack of) enforcement of pesticide-use regulations; conflicts of interest	Pesticide bias in government research investments	Weak (or lack of) environmental policy
Pesticide sales tax exemption	Public sector extension service promotion of pesticides	Foreign and domestic direct-investment incentives, including for the manufacture of pesticides	Farm input trade guarantee
Subsidized credit and insurance tied to pesticide use	Government pest management policies, such as support for promotion or pesticide research (pest eradication; specific legislation regarding prophylactic treatments)	Minimum wage policy that discriminates against more labor-intensive and environmentally benign production techniques	Inadequate curricula of agricultural extension
Preferential import duties and exchange rates for pesticide imports	Prochemical bias of industry information and extension advice	High import duties on alternative technologies	
Plant protection outbreak budget	Strong pesticide lobby of interest groups		
Donor price subsidies			
Donor pesticide grant or aid at subsidized costs			

Source: Adapted from Waibel 1994.

through special foreign and domestic investment incentives (for example, duty-free imports of capital goods and tax incentives) that extend to pesticide manufacturing. Policies that create barriers to the entry of new technologies (for example, lengthy, cumbersome, and costly registration procedures for seeds of improved disease-resistant plant varieties or lack of intellectual property protection) also limit farmers' pest management technology options. Banks may also encourage pesticide use by linking credit to certain agronomic practices, including the use of pesticides (Thrupp 1990).

National governments sometimes also use pesticide regulation to protect their own pesticide industry. In some cases such protection is provided to outdated manufacturing processes or to unsafe manufacturing plants or products, which prevents new and safer products from entering the market. Pesticide industries often use governments to expand the market for their products by encouraging them to promote technology packages that combine seed, fertilizer, and pesticide; by supporting applied research using pesticides; or by supporting extension services that endorse pesticide use in general.

Integrated Pest Management and Public Policy Tool Box

A reappraisal of the role of pesticides in agriculture has been under way since the publication of Rachel Carson's *Silent Spring* 40 years ago (Carson 1962 and Farvar and Milton 1972). But only in the last few years has the topic shifted from technological solutions (for example, better protection of workers and consumers, classification and phaseout of the most dangerous chemicals) to the mainstream of government and donor agricultural policy agendas. This has been associated with a wider shift in focus from agricultural production per se to environmentally sustainable management of production systems, as reflected in the UNCED Agenda 21 agreements.

Government intervention to influence farmers' choice of technology between pesticide- and nonpesticide-based approaches can be justified by the environmental and public health implications of pesticide use. These include harmful effects on public health, ranging from acute poisoning and a high risk of cancer among pesticide-handling farmworkers to contamination of food with pesticide residues (see chapter 19). There are also wider environmental effects on water, soil, and a variety of nontargeted organisms. These hazards give rise to negative externalities that impose costs on society and the environment. But these costs are rarely reflected in the price that users pay, and other forms of compensation are not applied. In this situation there is divergence between the private costs to the user and costs to society, the true "social costs." Thus the level of pesticide use is not optimal from society's perspective, and intervention is warranted.

A number of policy and regulatory instruments are available to governments to encourage environmentally sound and economically rational pest management practices. The more important of these instruments are outlined here (their implementation is discussed later):

- Development of a system that increases the awareness of policymakers, consumers, and producers of the hazards of pesticide use
- Development of a regulatory framework to ensure appropriate and safe production, distribution, and use of pesticides
- Reorientation of agricultural and environmental policies to introduce appropriate economic incentives, including taxes and special levies on pesticide use, to account for negative externalities, and short-term subsidies, to account for positive externalities in the use of IPM
- Orientation of research and technology policies to generate a steady supply of relevant pest management information and technologies, including adequate budget allocations for research, extension, and training
- International agreements and conventions.

Information Programs and Education

Many of the policies that encourage excessive pesticide use were developed at a time when little information was available on the health hazards of pesticide use and on sustainability. Greater awareness of these issues among policymakers, producers, and consumers, which can be achieved by improving the flow of information, encourages reducing pesticide use by helping to make decisions that better reflect true social costs. Informed consumer preferences can become a powerful factor influencing pest management practice. Informed consumers can make their preferences clear through pressure on public authorities to regulate the development, handling, and use of pesticides; through the market, by paying a premium for pesticide-free produce; and directly through dialogue with producer groups. Public investment in education and communications media to provide unbiased information on the benefits and risks of pesticide use is necessary as part of a general effort to level the playing field between pesticide dependency and IPM.

Regulation

Establishment of a pesticide regulatory system is the primary means used to ascertain pesticide quality, to minimize exposure, and to determine a proper balance between the risks and benefits of using agrochemicals. The major objectives are to (a) protect pesticide users and handlers, (b) protect public health, (c) prevent negative ecological effects, (d) prevent the buildup of pesticide resistance in target and nontarget populations, and (e) protect beneficial pests. Appropriate regulatory frameworks are necessary for pesticide regulation; this is an area where deregulation is misplaced (see box 18.1).

Regulatory systems operate through a variety of compliance provisions and mechanisms (table 18.2). Regulatory authorities may restrict, tax, or ban the production, import, sale, or use of certain pesticides. Use criteria may be determined by geographical zones (for example, sensitive habitats), according to different classes of user (for example, institutions or individual farmers), or by commodity-production system and pesticide resistance management. Introduction of such a regulatory framework requires an understanding of the feasibility of various regulatory options under prevailing conditions in a particular country.

A pesticide registration scheme requires manufacturers to demonstrate, by providing relevant data, the safety and efficacy of specific pesticides for particular uses when used according to the registered instructions. Often allied to the registration process is a requirement for accurate labeling, including information about safe use of the product and potential hazards and associated risks of individual pesticides. This requirement usually carries over to advertising and extension messages. Monitoring and evaluation of specific products may also be required. Through a licensing and certification scheme, pesticide handlers may be required to demonstrate adherence to set standards and completion of relevant training. The registration of pesticides in the United States and Europe is often used as a baseline for suitability of use in other countries.

Box 18.1. Pesticide Deregulation and Development of Resistance in China and India

Dramatic declines in cotton production in China in the past few years have occurred as major pests, such as the cotton bollworm (*Heliothis armigera*), have become resistant to many insecticides. The economic losses have been enormous. In some counties of the Yellow River valley, yields have declined 50–60 percent through a combination of resistance and drought. In Hebei, Henan, and Shandong, overall yields have been reduced 30–50 percent. In addition, the number of sprays used has risen from between 8 and 10 to between 15 and 25, and the seasonal cost of sprays has doubled or tripled over a short period. In the worst-affected areas, cotton production has become completely uneconomic. A report by the International Organization of Pesticide Resistance Management (IOPRM 1993) finds that the single most important factor behind the development of resistance was the unregulated pesticide market.

A similar picture has emerged in India. Here deregulation has led to an increase in the number of smaller companies selling (subsidized) pesticides with no effective oversight of the product quality or of the advice given to farmers. Throughout India, there are more than 86,000 dealers. Under these circumstances of uncontrolled marketing of pesticides, it is likely to be very difficult, if not impossible, to implement an IPM policy (Gillham and others 1995).

Economic and Legal Disincentives

One approach to internalizing the negative externalities associated with pesticide use is called the polluter-pays principle. According to this principle, polluters should be made to pay the costs of the damage they inflict on society. In the case of pesticides, strict implementation of polluter payment is complicated by the difficulty and the cost of clearly identifying who the polluters are, to what degree each user contributed to the damage, and the need to estimate the optimal level of pesticide use and the associated optimal tax. Using the polluter-pays principle to control pesticide use is being seriously considered in the European Union and elsewhere (Bonny and Charles 1993).

Implementation of the polluter-pays principle has included the imposition of licensing fees, user fees, and a sales tax. These measures can be tailored to the

Table 18.2. Common Pesticide Regulatory Activities

Type of regulation	*Enforcement process*	*Target group*
Pesticide registration	Quality control of new agrochemicals; risk assessment of new pesticides	Manufacturers, formulators, and distributors
Pesticide reregistration (of new formulations or other use)	Continuous risk assessment of existing agrochemicals	Manufacturers, formulators, and distributors
Bans on unsafe pesticides	Banning of sales and manufacturing	Manufacturers, formulators, distributors, and end users (no indemnification for the farmer)
Food safety	Development of quality standards; monitoring of food quality; prosecution of violators	Producers, processors, market chains, and researchers
Pesticides in groundwater	Monitoring of application and groundwater residue; prosecution of violators	Producers and manufacturers
Storage and disposal of pesticides	Certification and inspection; prosecution of misuse	Producers, distributors, and pesticide applicators
User or farmworker safety	Certification and training	Producers, applicators, and farmworkers
Resistance prevention	Voluntary or mandatory participation	Producers, applicators, and distributors

characteristics of individual pesticides. Such levies are usually further supported by penalties for noncompliance, negligence, and failure to meet relevant standards. Overall, taxation of pesticide use offers a more flexible tool than the all-or-nothing pesticide registration requirement (that is, the pesticide is either approved for use or not), but taxation requires more sound (and probably costlier) enforcement.

Economic Incentives

Positive incentives could also be used to promote environmentally benign pest management practices. These incentives include support for alternative practices, either through temporary grants and tax breaks or through research and extension. For example, to encourage the adoption of environmentally friendly technologies, governments could set up short-term, time-bound programs that provide matching grants to farmers who adopt IPM practices. This subsidy may be economically justified because it rewards early adopters who are paving the way for more widespread adoption through their demonstration effect; for example, an initial matching grant to farmer groups may initiate a collective pest scouting scheme or other means of enhancing pest monitoring or farmer networks.

Research and Extension

A bias in favor of pesticides in the funding of research, extension, and training in many countries is a response to distorted economic incentives that artificially boost the returns to pesticide-related research. This bias also reflects the economic strength of pesticide industries, both domestic and foreign, that have been able to promote and support the use of proprietary packages. One means by which governments can encourage a socially nearly optimal mix of research and related downstream activities is the judicious use of the economic and regulatory provisions outlined above. This would have the effect of bringing the costs of pesticide use closer to the social costs, thus stimulating demand for

Table 18.3. Examples of Successful Integrated Pest Management Applications

Crop	*Country*	*Impact*
Soybean	Brazil	Pesticide use on soybeans reduced 80 percent in seven years
Rice	India (Orissa)	Insecticide use on rice reduced
Rice	Indonesia	Pesticide use on rice reduced more than 60 percent
Cotton	Nicaragua	Pesticide use reduced one-third during the 1970s
Various	Cuba	Pesticide use reduced 80 percent during the 1980s

Source: Postel 1987.

alternative techniques including IPM (table 18.3 shows some examples of a successful switch from pesticides to IPM). It would also encourage industry to invest in the supply of environmentally benign technologies.

International Agreements and Conventions

Since the early 1980s the United Nations system has taken the lead in the establishment of international agreements covering pesticides. The Food and Agriculture Organization, for example, has issued a series of detailed guidelines on the procurement, trade, and handling of pesticides under the umbrella of the International Code of Conduct on the Distribution and Use of Pesticides (FAO 1990). These guidelines cover all aspects, including regulatory provisions and industry practice. Acceptance of the code has been slow, often because implementation is in the hands of agricultural plant protection departments, many of which are also involved in the distribution of pesticides.

A relatively recent addition to internationally accepted practice is the prior informed consent provision that is being jointly implemented by FAO and United Nations Environment Programme. Under the informed consent procedures, listed products (those that have been banned or severely restricted in a number of countries) can be exported to another country only if that country has full information and gives explicit consent. The original list of 12 chemicals is being expanded to 39. Currently, 120 countries are signatories to the agreement.

Another important international standard, established by the World Health Organization, groups pesticides according to their oral toxicity in mammals. These data provide a reference for relevant standards including the maximum residue levels of allowable pesticides in the Codex Alimentarius agreement, which governs the quality of agricultural produce.

Implementation Strategies for IPM

Experiences from Africa, Asia, Latin America and the Caribbean, and the United States (California) are yielding valuable lessons about implementation of IPM-based programs in developing countries (Teng and Heong 1988, Wiebers 1993, Kiss and Meerman 1991, and Murray and Nathan 1994). These experiences suggest that there are two key strategic elements in implementing IPM on a wide scale: (a) creation of a level playing field for different modes of pest management, which means eliminating policies that promote environmentally unsustainable pest management techniques and strengthening regulatory institutions and (b) implementation of positive measures to promote IPM through support for public awareness, research, extension, and training, with an emphasis on decentralized, farmer-centered initiatives.

The first of these measures is necessary to create an enabling environment for IPM but is rarely sufficient to promote implementation. The second measure is also necessary but fails to have a significant impact unless the enabling environment is in place. The political commitment and management needs in terms of policy and institutional change are complex in that they cross sectoral boundaries and need to cover both macro- and micro-level initiatives. Careful sequencing and targeting of specific measures are also needed to minimize the economic costs associated with policy shifts, in particular where this will create or strengthen a constituency opposed to change.

National Integrated Pest Management Policy

Establishment of a national IPM policy framework provides a useful first step in implementing an IPM strategy at the national level. Several countries have

Box 18.2. Integrated Pest Management in Indonesia

In the early 1980s extensive areas of rice production in Indonesia were infested with the brown plant-hopper *Nilaparvata lugens*. It was no longer economically possible to control the pest, despite ongoing resistance breeding by the International Rice Research Institute and intensive pesticide applications by farmers.

The Indonesian national IPM policy was announced in 1986 by presidential instruction and supported by the Food and Agriculture Organization of the United Nations (FAO; see Kenmore 1991 and World Bank data). This followed research evidence that insecticides were not increasing rice yield significantly and, by destroying natural enemy populations, were destabilizing the system of rice production. Many insecticides were banned; subsidies were reduced or removed; crop protection and associated research and extension institutions were strengthened. A major IPM training effort began. By mid-1992 about 1,000 pest observers, 3,000 extension staff, and at least 150,000 farmers had been trained to observe and understand the local ecology of the planthopper and its natural enemies and to follow simple threshold rules.

During 1987–90, the volume of pesticides used on rice fell more than 50 percent, while yields increased about 15 percent. Farmers' incremental net profits are estimated at $18 per IPM-trained farmer per season. This compares with an estimated average training cost of about $4 per farmer. The federal government has benefited from a $120 million a year (85 percent) reduction in pesticide subsidies and has begun to realign these funds into further IPM initiatives. In 1993 the government of Indonesia, the World Bank, and the U.S. Agency for International Development committed $53 million to a project that will build on and extend the use of IPM throughout the country.

This Indonesia project is an often-quoted example, but India's irrigated rice ecosystem is also well suited to natural pest control. Complete IPM approaches in other production systems and ecosystems may require more information and different approaches. However, the field school and farmer training and empowerment concept has positive effects beyond pest control.

already moved to do this, such as Cuba (Rosset and Benjamin 1994), Indonesia (Kenmore 1991), the Netherlands (Vijftigschild 1996), and the United States (U.S. Department of Agriculture, Agricultural Research Service 1993). The advantages stem from overseeing the disparate package of measures needed to implement IPM, both upstream policy elements and on-farm implementation, within a single, coherent decision framework. The same decision framework can provide the basis for consultations with all relevant stakeholders to secure broad institutional support for what may involve a significant shift (see the example of Indonesia; box 18.2). Finally, government's explicit statement of intent relative to IPM provides a strong signal to all relevant actors.

A broad consultation process using workshops and seminars can be used to define targets and priorities and to define appropriate institutional roles and responsibilities for different elements of implementation, thus helping to bring the various public, private, and government organization stakeholders in IPM together within a concerted and open program. This process also helps to raise awareness and understanding by both consumers and producers about the health and environmental risks associated with pesticides and to create a broad constituency in favor of implementing IPM and mobilizing public funds to support implementation.

Information, Analysis, and Priorities

Policies that support IPM will increase aggregate social welfare, but they may also negatively affect the productivity and livelihoods of some groups in the short term. Thus Zilberman and others (1991) find that bans on pesticide imports, when no provisions have been made for effective substitutes, result in lower production levels, higher prices, a substantial loss of discretionary income for consumers, and a redistribution of income toward agricultural producers who do not use pesticides. To achieve a smooth transition to environmentally sound pest management practices and to maintain progress toward wider agricultural sector growth targets, some targeting and phasing of policy changes are necessary.

A first step is to determine the status and availability of suitable IPM technologies and the prospects for farmer adoption. This determination will be specific to the system of production and requires field data based on a multidisciplinary assessment. For example, in some high-input systems, direct measures to limit pesticide use, through either the pricing mechanisms or regulation, may be the single most important way to influence technology choice toward IPM. In theory, agricultural service providers, both public and private, will adjust the supply of information and technologies to meet farmers' requirements for alternative techniques. In practice, a variety of market imperfections and technical limitations and constraints on the flow of information could slow the pace of any adjustment.

A concerted effort to support farmers' transition to IPM through on-farm adaptive research, demonstration, and training can be a powerful and effective adjunct to economic and regulatory measures. This has been effective in irrigated rice production systems in Indonesia and other Asian countries (see box 18.2). Although such an intensive training effort can be costly, a growing body of evidence suggests that it yields high economic returns (Pincus 1994).

Although measures to ensure undistorted pricing for pest management technologies continue to be important, a critical constraint to IPM implementation is likely to be the availability of effective and competitive alternative (nonchemical) techniques. For example, there remains a shortage of low-cost IPM technologies that are relevant to the mixed-farming systems prevalent in much of Sub-Saharan Africa (Kiss and Meerman 1991). In the absence of viable alternatives, any measures that would raise the costs of pesticides may be perceived as placing unwarranted limitations on growth of productivity and livelihoods, with the long-run social benefits being heavily discounted by both producers and policymakers. A relatively substantial investment in the generation of new knowledge and practical IPM techniques is therefore essential, along with improved infrastructure and services. The aim is to ensure that as farming systems evolve, farmers continue to have access to relevant pest management technologies that make minimal use of chemical pesticides.

Economic and Regulatory Measures

Many developing countries face significant constraints to implementing the economic and regulatory provisions required to support IPM. Surveys indicate that 25 percent of developing countries lack the legislation and 80 percent lack the resources to implement or enforce such legislation (FAO 1993). This is especially the case in Africa, with 76 percent of countries lacking pesticide-control statutes (FAO 1993 and Tobin 1994). Several factors explain this situation. A comprehensive and rigorous network of registration, legislation, standards, training, monitoring, and information systems appears to be more expensive than alternative public sector programs with a shorter-term and more visible direct impact. These costs are raised by limitations of infrastructure and communications. At the same time, the financial impact of IPM policies, such as lower subsidies, taxes, and licensing and registration fees, falls most heavily on producers and marketers of pesticides and initially on farmers engaged in the production of cash crops, who use the most pesticides. These are both powerful interest groups.

Given such constraints, there is a need to:

- Develop a streamlined regulatory system with low administrative costs (registration procedures, for example, can often use data from other countries with comparable or higher standards)
- Ensure that the institutions charged with implementing economic and regulatory measures have adequate authority to push through and enforce relevant measures
- Seek to recover implementation costs with revenue (for example, through registration fees)
- Combine implementation of policies with public information and education on the benefits
- Encourage a transparent and open process that includes consultation with industry and other interest groups for possible resolution of conflicting interests.

Farmer-Centered Research, Extension, and Training

The involvement of farmers in helping to generate locally specific techniques suitable for particular farming systems appears to be an important factor determining the success of efforts to implement IPM, a finding that is likely to apply equally to other sustainable production technologies and resource management approaches. The field implementation of IPM requires that farmers, extension workers, and local crop protection technicians have a practical understanding of the ecology and life cycle of major pests and their natural enemies and that this knowledge be translated into appropriate decisionmaking tools (for example, action thresholds) and practical control tactics.

Training of farmers, pest observers, agricultural development officers, and extension agents should therefore represent a major component of any implementation initiative. This training should take place close to the field; much of the information will be locally specific and generated in collaboration with farmers through adaptive research on farmers' fields. Such a decentralized initiative, involving the establishment of farmer field schools, has been very successful in a number of Asian countries (Asian Development Bank 1993).

A related feature of successful initiatives has been the role of pilot projects as the platform for demonstrating and validating IPM techniques before more widespread extension and training. This training is backed by a continuous flow of information from pest scouts, farmers, and others to decisionmakers and is consequently knowledge-, information-, and research-intensive. Such information can be transmitted daily (weather reports, for example) or weekly (by radio, television, or newspapers). Support for IPM implementation is inconsistent with a top-down technology-transfer approach.

An alternative mode, which has been used in Indonesia, is the so-called facilitation model (box 18.3). Rather than transferring ready-made packages of technology and blanket recommendations, this approach seeks to enhance the capacity of farmers to learn, to make use of their experience, and to adapt technologies as required (Röling and van de Fliert 1994). The use of such a different approach has implications for investment design, the design and management of institutions, training of staff, and links between research and extension, including those with centers of the Consultative Group on International Agricultural Research and other international research groups.

Box 18.3. Lessons Learned in Indonesia

According to van de Fliert, Pontius, and Röling (1995), experience in Indonesia has taught the following lessons:

- Introduction of IPM cannot be compared to the transfer of technology from expert to nonexpert; rather, it requires the facilitation of a learning process.
- The critical issue in IPM training for rice is understanding the ecology of the rice field.
- Training IPM instructors also requires a learning process and is related directly to daily farming experiences.
- Extension workers will inevitably be involved in IPM training.
- Farmers are considered the best primary trainers in IPM for other farmers.

Note

1. The global market for agrochemicals is currently estimated at $21 billion; nearly 27 percent of this is in North America, 30 percent in Western Europe, and 14 percent in Japan. The market growth in OECD (Organisation for Economic Co-operation and Development) countries has flattened out, but use in the developing world is expected to grow considerably. In 1990 developing countries accounted for 20 percent of world agrochemical use, but this percentage is expected to double by 2000. However, use in these countries is limited largely to a few crops: cacao, cotton, coffee, flowers, fruits (mainly banana), maize, rice, tobacco, and vegetables. See OECD (1995).

References

The word "processed" describes informally reproduced works that may not be commonly available through libraries.

Asian Development Bank. 1993. *IPM Guidelines*. Manila.

Bonny, S., and R. Charles. 1993. "Perspectives d' evolution de l'emploi des engrais et des phytosanitaires dans l'agriculture française." *Cahiers d'economie et sociologie rurale* 26(1):29–62.

Carson, Rachel. 1962. *Silent Spring*. Cambridge, Mass.: Riverside Press.

FAO (Food and Agriculture Organization of the United Nations). 1990. *International Code of Conduct on the Distribution and Use of Pesticides*. Amended version. Rome.

———. 1993. *Safe and Efficient Use of Pesticides in Africa*. Rome.

Farah, Jumanah. 1994. *Pesticide Policies in Developing Countries: Do They Encourage Excessive Use?* Discussion Paper 238. Washington, D.C.: World Bank.

Farvar, M. Taghi, and John P. Milton, eds. 1972. *The Careless Technology: Ecology and International Development*. Garden City, N.J.: Natural History Press.

Feder, Gershon. 1979. "Pesticides, Information, and Pest Management under Uncertainty." *American Journal of Agricultural Economics* 61(1):97–103.

Gillham, Fred E. M., Thomas M. Bell, Tijen Arin, Graham A. Matthews, Claude Le Rumeur, and A. Brian Hearn. 1995. *Cotton Production Prospects for the Next Decade*. Technical Paper 287. Washington, D.C.: World Bank.

IOPRM (International Organization for Pesticide Resistance Management). 1993. "The Management of Resistant Cotton Pests in China." Paris. Processed.

Kenmore, Peter E. 1991. "Indonesia's Integrated Pest Management: Policy, Production, and Environment." Paper presented at U.S. Agency for International Development environment and agriculture officers conference, Colombo, Sri Lanka, 11 September. Processed.

Kiss, Agnes, and Frans Meerman. 1991. *Integrated Pest Management and African Agriculture*. Technical Paper 142. Washington, D.C.: World Bank.

Murray, D. L., and E. L. Nathan. 1994. "From Pesticides to Integrated Pest Management: Prospects for an Alternative Development Paradigm in Central American Agriculture." Paper presented at the eighteenth international congress of the Latin American Studies Association, Atlanta, Georgia, 10–12 March. Processed.

National Research Council. 1989. *Alternative Agriculture*. Washington, D.C.: National Academy Press.

OECD (Organisation of Economic Co-operation and Development). 1995. "No. 6 Guidelines for Aid Agencies on Pest and Pesticide Management." In *Guidelines on Environment and Aid*. Paris.

Pincus, J. 1994. "Farmer Field School Survey: Impact of IPM Training on Farmers." Pest Control Behavior Internal Report. National IPM Program, Jakarta. Processed.

Postel, Sandra 1987. *Defusing the Toxics Threat: Controlling Pesticides and Industrial Waste*. Worldwatch Paper 79. Washington, D.C.: Worldwatch Institute.

Röling, Niels, and Elizabeth van de Fliert. 1994. "Transforming Extension for Sustainable Agriculture: The Case of Integrated Pest Management in Rice in Indonesia." *Agriculture and Human Values* 11 (2-3):97–108.

Rosset, Peter, and Medea Benjamin. 1994. *Two Steps Back, One Step Forward: Cuba National Policy for Alternative Agriculture*. Gatekeeper Series 46. London: International Institute of Environment and Development.

Stern, V. M. 1973. "Economic Thresholds." *Annual Review of Entomology* 18:259–80.

Teng, P. S., and K. L. Heong. 1988. "Pesticide Management and Integrated Pest Management in Southeast Asia." Consortium for International Crop Protection, Beltsville, Md. Processed.

Thrupp, Lori Ann. 1990. "Inappropriate Incentives for Pesticide Use: Agricultural Credit Requirements in Developing Countries." *Agriculture and Human Values* 7(3-4):62–69.

Tobin, R. J. 1994. *Bilateral Donor Agencies and the Environment: Pest and Pesticide Management*. Arlington, Va.: Winrock International Environmental Alliance.

U.S. Department of Agriculture, Agricultural Research Service. 1993. "USDA Programs Related to Integrated Pest Management." USDA Program Aid 1506. Beltsville, Md. Processed.

van de Fliert, Elizabeth, J. Pontius, and Niels Röling. 1995. "Searching for Strategies to Replicate a Successful Extension Approach: Training of IPM Trainers in Indonesia." *European Journal of Agricultural Education and Extension* 1(1):41–63.

Vijftigschild, R. A. N. 1996. "The Relationship between Pesticide Use and Environmental Burden for Arable Farms in the Netherlands." In H. Waibel and J. C. Zadocks, eds., *Institutional Constraints to IPM*. Pesticide Policy Project. Hanover, Germany: Institut für Garten-bauokonomie.

Waibel, H. 1994. "Towards an Economic Framework of Pesticide Policy Studies." In S. Agne, G. Fleischer, and H. Waibel, eds., *Proceedings of the Workshop on Pesticide Policies*. Gottingen, Germany: Institut für Agrarokonomie der Universitat Gottingen.

Wiebers, Uwe-Carsten. 1993. *Integrated Pest Management and Pesticide Regulation in Developing Asia*. Technical Paper 211. Washington, D.C.: World Bank.

Zilberman, David, A. Schmitz, G. Casterline, E. Lichtenberg, and J. B. Siebert. 1991. "The Economics of Pesticide Use and Regulation." *Science* 253(5019):518–22.

19

Toward Reduced Pesticide Use for Cereal Crops in Asia

Prabhu L. Pingali and Roberta V. Gerpacio

Intensive crop monoculture systems, popularized by the green revolution, created an environment that was conducive to pest growth. Although pest-related losses in yield were small in percentage terms, they were highly visible and led policymakers to fear major outbreaks (Rola and Pingali 1993). This desire to avoid large-scale pest outbreaks resulted in policies that made pesticides easily accessible and affordable at the farm level (Pingali and Rola 1995). Pesticides were seen both as a guarantee against crop failure and as a means of realizing the full potential yield of the crops (Waibel 1986). Although frequent and indiscriminate use of pesticides is the norm in developing countries today, farm-level promotion of pesticide use is accompanied by little, if any, training in judicious use, farmer safety, and adverse ecological and health consequences.

Indiscriminate pesticide use in the production of various crops has been documented to result in the following: (a) health impairment due to direct or indirect exposure to hazardous chemicals, (b) contamination of ground and surface waters through runoff and seepage, (c) the transmittal of pesticide residues through the food chain, (d) increased resistance of pest populations to pesticides, thereby reducing their efficacy and causing pest outbreaks, and (e) the reduction of the population of beneficial insects like parasites and predators, thereby reducing the effectiveness of pest control strategies that attempt to minimize the use of pesticides (see Pingali and Roger 1995 for rice, Cornwell 1995 for tobacco, Crissman and Cole 1994 for potato, and Forget, Goodman, and de Villiers 1993).

Technological change is dramatically reducing the need for pesticide use in cereal crop production systems. Today, with improved host-plant resistance, improved crop management, and advances in integrated pest management, along with a reformed policy environment that is beginning to discourage the use of pesticides, the link between pesticides and cereal crop productivity has become less clear. Certainly in the case of insecticides for rice and fungicides for wheat, recent evidence indicates that the productivity benefits of applying chemicals are marginal at best. In the case of maize, chemical use under subsistence production systems has always been low, but commercialization has led to increased pesticide use. However in recent decades, maize varieties with improved resistance to insects and diseases have been developed. Herbicides are the one uncertainty for the future of chemical use. Adequate nonchemical controls for weeds are not yet available, and herbicide use is increasing dramatically as the opportunity cost of labor rises across the developing world.

This chapter describes the opportunities currently available for achieving a dramatic reduction in the use of agrochemicals in the production of cereal crops in Asia. The first section summarizes past trends and future prospects for pesticide use in Asia for the three major cereals: rice, wheat, and maize. The second section identifies the factors that contribute to the rise in farm-level demand for pesticides and the factors that could reduce that demand. The third section highlights the advances that have been made in the generation and dissemination of cereal crops that are less dependent on chemicals because they are resistant to insects and diseases. The final section discusses the integrated management approaches that are necessary for maximizing and sustaining the productivity gains offered by resistant varieties.

The Pesticide Market and Use in Asian Cereal Production

From $900 million in 1960, the global agricultural pesticide market was valued at $27.5 billion in 1993 and at $30.3 billion in 1995, indicating an annual growth rate of almost 11 percent. Global pesticide sales in 1993 were equivalent to almost 2 million metric tons of active ingredients, of which 497,000 metric tons—or 25 percent—were consumed in Asian agriculture (FAO 1997). Cereal crop production across the world is reported to be the second largest consumer of agricultural pesticides, next only to fruit and vegetable production. In 1995 rice and maize production consumed about 23 percent of all agrochemicals, while the production of fruits and vegetables consumed about 26 percent (IFPRI 1996). Global rice production, for instance, consumed about $3.2 billion worth of agricultural pesticides (12 percent of the 1993 pesticide sales), of which $2.6 billion (or 81 percent) were spent in Asia.

Before the 1960s, pesticides were mostly consumed in industrial nations. With agricultural modernization in developing countries in the mid-1960s, developing-country exports became an important part of the global pesticide industry. In the 1960s and 1970s, France, Germany, Italy, Japan, Switzerland, United Kingdom, and the United States accounted for 75 percent of world chemical production and provided approximately 85 percent of all agrochemicals used in Asia (Boardman 1986 and United Nations 1990, as cited in Pingali and Rola 1995).

A large portion of the pesticides exported to the developing countries are products no longer used in the industrial countries on experimental and health grounds. Weir and Shapiro (1981), for instance, point out that at least 25 percent of U.S. pesticide exports are products that are banned, are heavily restricted, or have never been registered for use in the United States. Warburton, Palis, and Pingali (1995) also report that the vast majority of insecticides used in developing Asian countries are highly hazardous category I and category II chemicals, as classified by the World Health Organization, many of which have been banned for agricultural use in most industrial countries. These insecticides include organochlorine endosulfan; organophosphates such as methyl parathion, monocrotophos, and chlorpyrifos; carbamates such as BPMC, carbaryl, and carbofuran; and pyrethroids such as cypermethrin and deltamethrin. They continue to be widely used in tropical Asia due either to ignorance or to substantially lower costs than the safer alternatives.

There is considerable variation from country to country and from crop to crop in the type and quantity of pesticides used. Herbicides dominate North American and European domestic markets, but insecticides are more common elsewhere in the world, especially in the Asia-Pacific region. The share of East Asia in global pesticide consumption is 14 percent but accounts for 54 percent of Asia's pesticide use. Wood Mackenzie Consultants, Ltd. (1993 as cited in IRRI 1995) reports that, in 1993, Japan alone accounted for 62 percent of all pesticides consumed in Asia's rice production: 46 percent of the insecticides, 71 percent of the herbicides, and 72 percent of the fungicides used on rice in the region (table 19.1). The next large users of rice pesticides in Asia were the Republic of Korea, China, and India, at 12, 11, and 7 percent, respectively.

Aggregate figures, though, tend to mask the intensity of pesticide use. Malaysia used just 9 percent of Asia's total pesticide consumption but was the region's most intensive user of pesticides, applying 23.42 kilograms of active ingredients per hectare of arable land. In contrast, India consumed 14 percent of all agricultural pesticides in Asia but applied only about 0.43 kilogram of active ingredients per hectare of arable land.

In Asian cereal crop systems, insecticides are used predominantly in rice and commercial maize production; almost none is used in wheat production. Fungicides are generally not important for cereal crops, except for rice production in the East Asian countries. The use of fungicides for rice may increase in the rest of Asia if the current push toward higher rice yields is pursued (Pingali, Hossain, and Gerpacio 1997). In the case of wheat, the plant's ability to resist high levels of disease pressure has made fungicide use almost obsolete in Asian wheat production systems. However, the use of herbicides is increasing for all three cereals, as rising agricultural wages, due to increasing off-farm employment opportunities,

Table 19.1. Share in Consumption of Pesticides Used in Rice Production in Select Asian Countries, 1993 (computed based on rice pesticide sales)

Country	*Insecticides*	*Herbicides*	*Fungicides*	*All pesticides*
Bangladesh	1.03	0.54	1.38	0.95
China	16.55	8.67	7.59	11.22
India	12.41	2.49	4.14	6.58
Indonesia	4.14	0.65	0.69	1.94
Japan	45.60	71.40	72.41	61.81
Korea, Rep. of	13.24	11.92	11.31	12.17
Myanmar	1.03	0.54	0.69	0.76
Philippines	1.55	1.19	0.41	1.10
Thailand	2.48	1.95	0.69	1.79
Vietnam	1.96	0.65	0.69	1.14

Source: For basic data, IRRI 1995.

reduce the cost-effectiveness of hand weeding (Naylor 1997).

Although Asia's rice production remains the major consumer of agricultural pesticides, the composition of its pesticide market has been changing in recent years. The proportion of insecticides used in Asian rice production, for example, decreased from about 45 percent (of all pesticides) in 1984 to 37 percent in 1993 (table 19.2). Total insecticide use, for rice production, actually decreased 10 percent from 1990 to 1993 (table 19.3). In the same time period, the proportion of herbicides in total chemical use increased from 27 to 35 percent. Herbicide use grew at a rate of about 4 percent a year during 1987–92.

In 1984, rice production in China used 64 percent of all insecticides used, 8 percent of herbicides, and 28 percent of fungicides; by 1993, insecticide use had dropped to 54 percent, while herbicide use had risen to 27 percent of all pesticides used. India and the Philippines also demonstrated the same trend in rice production. In contrast, rice economies with cheap agricultural labor, such as Myanmar and Vietnam, still show increasing use of insecticides, but decreasing use of herbicides, in rice production (table 19.2).

Indiscriminate and injudicious use of agrochemicals is the direct result of inappropriate or inadequate pesticide policies in developing countries. Enlightened policy environments along with the introduction of crop varieties resistant to insects and diseases, as well as the promotion of integrated pest management, generally result in less use of pesticides, more use of safer chemicals, and improved farmer safety and environmental sustainability.

Agricultural Modernization and the Demand for Agrochemicals

In the developing world, agricultural modernization, which includes agricultural intensification and increased market orientation, led to rapid growth in the demand for agrochemicals. The rapid spread of the green revolution for cereal crops, especially in Asia during the late 1960s and the 1970s, led to an upward shift in the demand for modern agricultural inputs, including pesticides. Although the demand for pesticides as a whole grew rapidly, the trends for chemical subgroups, such as insecticides, herbicides, and fungicides, varied substantially by country or region and by crop. Pingali and Rola (1995) provide a stylized representation of the factors leading to regional differences in pesticide use. The following summarizes their discussion.

Agricultural intensification—the movement from an extensive to an intensive production system or from a subsistence production system to a commercial one—has been extensively documented to promote the use of agrochemicals. Table 19.4 shows the differential effects of increasing land scarcity and increasing market orientation on the demand for chemicals. Consider first a sparsely populated subsistence society with limited access to markets. Agriculture in such societies is characterized by extensive land use and almost complete reliance on nontraded inputs, such as farmyard manure. Pest pressure is low in such cultivation systems and is kept that way through a variety of management practices, such as crop rotations and the use of traditional cultivars with known resistance

Table 19.2. Proportion of Insecticides, Herbicides, and Fungicides Used in Rice Production in Select Asian Countries, 1984–93
(percent)

Country and year	*Insecticides*	*Herbicides*	*Fungicides*
Bangladesh			
1984	61.3	19.4	19.4
1988	42.6	27.7	29.8
1993	40.0	20.0	40.0
China			
1984	64.3	7.8	27.9
1988	77.9	4.2	17.9
1993	54.2	27.1	18.6
India			
1984	79.1	9.4	11.5
1988	77.1	12.1	10.8
1993	69.4	13.3	17.3
Japan			
1984	29.8	37.4	32.8
1988	32.1	40.7	27.1
1993	27.1	40.6	32.3
Philippines			
1984	59.5	24.3	16.2
1988	37.5	37.5	25.0
1993	51.7	37.9	10.3
Myanmar			
1984	36.4	45.5	18.2
1988	33.3	33.3	33.3
1993	50.0	25.0	25.0
Vietnam			
1984	50.0	33.3	16.7
1988	30.8	38.5	30.8
1993	63.3	20.0	16.7
All Asia			
1984	44.5	26.8	27.6
1988	42.1	31.4	25.7
1993	36.8	35.1	27.6

Source: For basic data, IRRI 1995.

to chronic pests.

In subsistence societies, increasing land scarcity due to population growth leads to agricultural intensification—that is, increased intensity of land use (Boserup 1965 and Pingali and Binswanger 1987). Where the opportunity costs of family labor are low, food production continues to rely predominantly on nontraded inputs. Pest pressure increases with intensification due to increased spatial and temporal carryover of pests. Although increased weed pressure is handled by family labor, increased insect pressure is no longer amenable to traditional management practices, and small amounts of insecticides begin to be used for cereal crops, even in subsistence societies. Low to moderate amounts of fungicides also tend to be used for crops such as cotton, tobacco, and horticultural products, especially fruits and vegetables (see table 19.5).

The contrasting scenario is one of a sparsely populated area that has excellent market access. In this

Table 19.3. Percentage Change and Annual Growth Rate of Sales in Insecticides, Herbicides, and Fungicides Used in Rice Production in Select Asian Countries, 1980–93
(percent)

Country and	*Percentage change*			*Annual growth rate*	
chemical	*1980–84*	*1985–89*	*1990–93*	*1981–86*	*1987–92*
Bangladesh					
Insecticides	—	–37.5	–66.7	—	–6.3
Herbicides	—	0.0	–66.7	—	–6.1
Fungicides	—	33.3	–47.4	—	6.8
China					
Insecticides	109.1	17.9	–8.6	12.0	4.4
Herbicides	40.0	–37.5	—	9.0	—
Fungicides	31.6	0.0	22.2	1.2	2.1
India					
Insecticides	175.0	13.8	–28.6	8.5	4.4
Herbicides	62.5	53.3	–17.9	9.3	9.4
Fungicides	14.3	33.3	25.0	0.4	11.9
Japan					
Insecticides	62.3	51.9	1.4	14.6	–1.1
Herbicides	6.8	53.4	14.6	11.6	1.0
Fungicides	3.1	16.6	34.6	11.0	1.9
Philippines					
Insecticides	37.5	–40.0	–21.0	7.4	0.0
Herbicides	125.0	40.0	–21.4	17.6	2.5
Fungicides	0.0	83.3	–78.6	1.1	12.0
Myanmar					
Insecticides	4.0	167.9	36.2	7.3	15.5
Herbicides	0.0	83.3	155.8	16.0	31.6
Fungicides	0.0	191.7	–6.8	8.7	9.9
Vietnam					
Insecticides	20.0	–16.7	58.3	3.3	10.4
Herbicides	—	20.0	–50.0	—	0.0
Fungicides	—	150.0	–58.3	—	13.6
All Asia					
Insecticides	77.9	36.4	–10.4	11.5	3.3
Herbicides	23.0	46.0	24.4	13.0	3.7
Fungicides	12.8	35.5	10.2	9.4	4.2

— No data available.
Source: For basic data, IRRI 1995.

Table 19.4. Intensification, Market Orientation, and Demand for Agrochemicals

Land/labor ratio	*Low market orientation*	*High market orientation*
High	Low demand for all chemicals	Herbicides predominate
Low	Insecticides predominate	High demand for all chemicals

Source: Pingali and Rola 1995.

Table 19.5. Agrochemical Use, by Stage of Agriculture and Cropping System

Farm characteristics	*Cereals (rice)*	*Other field crops (cotton, tobacco)*	*Horticulture (fruits, vegetables)*
Land-abundant subsistence	All chemicals, none	All chemicals, none to low	All chemicals, none to low
Land-scarce subsistence	Insecticides, low to moderate	Insecticides, fungicides, low to moderate	Fungicides, low to moderate
Land-abundant market-oriented	Herbicides, moderate to high	Herbicides, fungicides, moderate to high	Fungicides, herbicides, moderate to high
Land-scarce, market-oriented	Insecticides, herbicides, high	All chemicals, high	All chemicals, high

Source: Rola and Pingali 1995.

case, agricultural intensification is high due to high land values, but unlike the subsistence case, the opportunity cost of labor is also high, hence the high levels of traded input use. Although increasing insect pressures can be controlled through appropriate crop rotations and seasonal fallows, weeds are the dominant constraint. High levels of herbicide use are the norm. High fungicide use can also be observed for horticultural crops, especially fruits and vegetables. Where market access and land scarcity are both high, high opportunity costs of land and labor result in agricultural intensification with high use of traded inputs. The demand for all chemicals is high in such societies.

Modern cereal crop production systems further aggravate pesticide use. Intensive monoculture systems using high-yielding cereal varieties result in an environment conducive to the buildup and infestation of pests, and the consequent use of pesticides disrupts the natural balance between pests and predators. The risk of insect- and disease-related losses grew because early modern varieties were often highly susceptible to local pests and because crops were no longer heterogeneous. Regular prophylactic application of pesticides promoted by extension services and supported by government subsidies became a standard part of the early green revolution package.

In addition, because early formulations of pesticides were often nonselective, pesticides proved equally lethal to beneficial pests that preyed on crop pests. In many cases, pesticide use actually lowered yields, as crop pests, freed of their natural predators, multiplied without constraints (Pagiola 1995 and Rola and Pingali 1993). In the case of rice in tropical Asia, a large number of pest outbreaks have been associated with injudicious pesticide applications rather than with the use of modern high-yielding cultivars, high cropping intensity, or high use of chemical fertilizers (Heinrichs and Mochida 1984, Kenmore and others 1984, Joshi and others 1992, and Schoenly and others 1996). Outbreaks of secondary pests of rice, notably the brown planthopper, that were previously of minor significance began to occur in regions adopting modern varieties and using agrochemicals (Pingali and Gerpacio 1997). High and injudicious applications of pesticides disrupted the rich diversity of pest and predator populations, where the species richness and abundance of predator populations are often greater than those of pest populations.

The rapid growth in pesticide use can also be attributed to the misinformation and risk aversion of both policymakers and farmers. In an extensive review of rice case studies from across Asia, Rola and Pingali (1993) provide documentation that policymakers' perception of yield losses is higher than farmers' perception of yield losses, which in turn is substantially higher than actual yield losses. Both farmers' and policymakers' perceptions of pest-related yield losses are anchored in exceptionally high losses during major infestations, even when the probability of such infestations is low. In cases of minor infestations, physical damage to the plant is easily visible but does not always result in a loss of yield. Heong, Escalada, and Mai (1994) have shown that leaf damage, even as much as 50 percent, has insignificant effects on rice yields, yet 80 percent of insecticide applications for rice in Asia are targeted toward leaf-feeding insects.

Although in aggregate terms cereal crops account for the bulk of pesticide use in Asia, on a per hectare basis their use is quite small when compared with the levels applied on fruits, vegetables, and other high-value commodities. Pesticide application in high-value crops is related to consumer demand for aesthetically appealing agricultural products. Because these high-value agricultural products enjoy a substantial price premium for unblemished physical appearance, risk-averse farmers tend to apply pesticides, beyond the technical optimum, in order to capture this price differential. Pagiola (1995), for example, reports that in Bangladesh where 70 percent of pesticides are used on rice, the amounts used per unit of area and the total area affected are both relatively small. On vegetables, the use of insecticides follows a pattern almost diametrically opposed to that found in rice. Whereas rice is sprayed only two to three times a season, vegetables such as eggplant and country beans are often sprayed several times a week. A survey of eggplant producers in Jessore indicates a range of application from 17 to 150 times for one crop cycle (Kabir and others 1994, as cited in Pagiola 1995). Other vegetables have rates of use that are lower, but still generally higher than rice. Cauliflower and cabbage, for instance, are commonly sprayed three to four times. In the case of wheat, maize, and rice, which are the most important cereal crops in Asia, pesticides do not enhance physical quality in any way, and there is no price differential to capture (Rola and Pingali 1993). Given the positive income elasticity of demand for fruits and vegetables, the long-term prognosis for developing economies is one of increasing areas under these crops and, in the absence of alternative pest control strategies, increasing share of the agrochemical market (Pingali and Rola 1995).

With recent advances in developing the cereal crop's ability to tolerate insects and diseases, and in integrated pest management, the outlook for future reductions in pesticide use is very promising. Modern science has reduced the dependence on insecticides for rice and maize and the dependence on fungicides for wheat. The dependence on herbicides for cost-effective management of weeds continues to be problematic, with little scientific advance either in genetic or in crop management.

Breeding for Host-Plant Resistance as a Technological Alternative to Pesticides

Host-plant resistance—the plant's ability to resist high levels of pest infestation—has been the cornerstone of the scientific strategy for developing sustainable pest control systems for cereal crops. Significant advances have been made in the last three decades in the development and dissemination of crop varieties with resistance to the major cereal pests (table 19.6). Much of the advance has occurred through the use of conventional breeding approaches, although substantial future gains in resistance development could come through the use of modern biotechnology tools. Host-plant resistance ought to be seen as an essential building block in an integrated pest management strategy. Attempts to promote resistant varieties without a commensurate change in pesticide management practices have generally failed due to the early break-

Table 19.6. The Evolution of Host-Plant Resistance for the Major Cereals, 1960s–90s

Period	*Rice*	*Wheat*	*Maize*
1960s	Striped stemborer	Stem rust	European corn borer
1970s	Brown planthopper, green leafhopper, rice whorl maggot	Septoria tritici blotch	Earworms, tropical borers, southwestern corn borer
1980s	Yellow stemborer, white-backed brown planthopper, thrips	Leaf rust	Fall armyworm
1990s	*Bacillus thuringiensis*	Spot blotch, fusarium scab, stripe rust	*Bacillus thuringiensis*

down in resistance.

Conventional Breeding for Host-Plant Resistance

Although, technically tedious and time-consuming, conventional breeding techniques have led to significant progress in generating varieties resistant to major insect pests and diseases. Early work at the International Rice Research Institute (IRRI) in the Philippines reported that most of the important sources of resistance to major diseases and pests had been incorporated into modern rice varieties (IRRI 1972). The emphasis on releasing rice varieties with improved host-plant resistance to major insect and disease pressures has continued through subsequent decades. Lines considered promising enough have been named as varieties, while others have proven to be good parents in new crosses. Named varieties and promising breeding lines provide resistance to important rice insect pests such as green leafhoppers, brown leafhoppers, stemborers, and gall midge and to diseases such as blast, bacterial leaf blight, bacterial leaf streak, tungro, and grassy stunt (IRRI 1985).

For wheat and maize, the Centro Internacional de Mejoramiento de Maíz y Trigo (CIMMYT, International Maize and Wheat Improvement Center) based in Mexico has an active program that develops maize and wheat germ plasm with desirable levels of resistance to insect pests and diseases. CIMMYT combines "shuttle breeding" and "hot spot" multilocation testing within Mexico and abroad to obtain multiple disease resistance for the different agroecological zones in mega-environments where wheat and maize are grown (Dubin and Rajaram 1996). Insect problems are not as important in wheat as in the other two cereals. "Shuttle breeding" refers to a method in which breeding generations from the same crosses are alternately selected in environmentally contrasting locations as a way of combining desirable characteristics. "Hot spots" are locations where significant variability for a pathogen exists and where screening plant generations exposes the genotypes to as broad a range of virulence genes and combinations as possible. This screening, together with multilocation testing, increases the probability of developing durable resistance.

In wheat, considerable genetic progress has been achieved in developing host-plant resistance, with regard to leaf rust, septoria leaf blotch, fusarium scab, and bacterial leaf streak in the high-rainfall areas of tropical highlands. Some progress has also been achieved on resistance to the second-priority diseases, such as barley yellow dwarf, tan spot, and septoria nodorum blotch (Dubin and Rajaram 1996).

In the case of maize, most of the released hybrids and composites have high resistance to foliar diseases, and breeding for resistance to prevalent insect pests and diseases has made considerable progress (Commonwealth Agricultural Bureaux 1980). More recently, Kumar and Mihm (1995) have found that CIMMYT has developed potentially useful single cross hybrids and varieties (for example, CML 139 x CML 135, CML 139 x Ki3, CML 67 x CML 135) that, because of improved resistance, suffer less damage and percentage loss in grain yield to fall armyworm, southwestern corn borer, and sugarcane borer than earlier varieties. In 1993, it was reported that screening of CIMMYT bank accessions and elite germ plasm has continued to find new or better sources of maize resistance to stemborers, armyworms, maize weevils, spider mites, rootworms, and corn earworms (Mihm and others 1994, as cited in Edmeades and Deutsch 1994).

For decades, plant breeders have ensured that each new elite cultivar and breeding line possesses a wide range of genes for resistance to pests and diseases. However, breeding programs are challenged by the fact that genetic variation is not confined to the host plant but is also a feature of the pest or parasite. With certain types of resistance, selection for a resistant population of the host is followed closely by natural selection of the pest or parasite for those variants that can overcome the resistance. The risk is greater when large areas are planted year-long to a single, homogeneous cultivar in association with indiscriminate use of pesticides. Pest and pathogen populations change over time, and genetic resistance breaks down, while some pest and disease problems remain intractable. To keep ahead of evolving and changing pest populations, breeding activities constantly have to aim for increased genetic diversity against pests and cultivars with more durable resistance. Recent advances in biotechnology can help to improve the durability of host-plant resistance.

Contributions of Biotechnology to Breeding for Host-Plant Resistance

Biotechnology adds a new dimension to the research on improved host-plant resistance by enhancing conventional breeding efforts through the use of wide hybridization, molecular marker technologies, and genetic engineering. Biotechnology enhances the efficiency of conventional breeding efforts by reducing the length of time involved and the inherent trial-and-error nature of conventional breeding. Biotechnology may also make resistance more durable by incorporating alien and novel genes into the cultivated plant species. In the short to medium term, the contributions of biotechnology are likely to be most significant in improving the durability of pest resistance rather than in improving yield and grain quality. The following paragraphs discuss the prospects of biotechnology for the major cereal crops.

Wide hybridization allows breeders to use genetic diversity drawn from the wider gene pool of a crop's entire genus. A crucial tissue culture technique, embryo rescue, enables breeders to cross cultivated species with wild species or distant relatives and to stabilize select breeding lines faster in the laboratory than in the field. Pieces of the wild genome are integrated into the chromosomes of the elite parents, useful genes are transferred from different wild species into cultivated varieties, and the progeny are readily stabilized.

Embryo rescue has produced hybrids between cultivated rice and 12 wild species. Genes for resistance to brown planthopper, white-backed brown planthopper, bacterial blight, and blast have already been transferred from wild species into elite rice breeding lines (IRRI 1993a).

Diversity for resistance to or tolerance of several stresses is now available within the wide hybridization wheats under prevailing conditions in Mexico. Resistance to Karnal bunt, septoria tritici blotch, and spot blotch from wide hybridization wheats is being used to improve bread wheat based on disease screening data obtained over several years. Resistant wide hybridization genetic stocks of wheat have now been crossed onto elite cultivars, and the resulting lines are expressing diversity for resistance to these three diseases. This germ plasm is being incorporated into advanced bread wheat lines to diversify the genetic base of resistance to or tolerance of other stresses as well (Pingali and Rajaram 1997).

Molecular marker technology increases the efficiency of combining specific desirable genes into improved breeding lines by facilitating the selection of genes. A genetic map is used to show the relative position of genes on a chromosome and the distance between genes. The position of useful and important genes on the chromosomes is roughly identified or tagged by the use of molecular markers, allowing plant breeders indirectly to select genes of interest. Instead of selecting progeny plants having the gene of interest by checking for the action of the gene, scientists use marker-aided selection to identify the presence of the linked marker in the plant's DNA. Such markers are especially useful in breeding for resistance to nonendemic pests and diseases or for genes with overlapping effects that can contribute to complex and more durable resistance.

In the case of rice, more than 20 single genes for disease and insect resistance have been located relative to the restriction fragment length polymorphism (RFLP) markers. Among them are several genes for resistance to rice blast fungus and major genes for resistance to bacterial blight and three insects (brown planthopper, white-backed brown planthopper, and gall midge). A current project in wheat based on the use of marker-assisted selection aims to develop resistance to barley yellow dwarf, a viral disease that affects wheat as well as rice and barley.

Genetic engineering refers to a process that includes the identification and incorporation of economically valued genes into a specific crop (with genes originating from any living organism, from a virus, or even from chemical synthesis), the delivery systems that introduce the desired gene into the recipient cells, and the expression of the new genetic information in the recipient cells. Plant genetic engineering techniques provide breeders with new opportunities to improve the efficiency of production and increase the utility of agricultural crops. Modern varieties could be further improved by inserting foreign genes into plant cells and regenerating viable, stable, and fertile plants that possess useful traits such as disease and insect resistance, better-quality grain, and resistance to certain herbicides. Perhaps the biggest advantage of these techniques is that they have the potential to shorten the time required to develop improved traits in wheat plants.

The first group of useful foreign genes to be put into rice included soybean trypsin inhibitor for insect resistance and two barley genes, chitinase and ribosome inhibitor protein (RIP), for resistance to the fungal diseases sheath blight and blast (IRRI 1993b). When introduced into rice, novel genes such as the *Bacillus thuringiensis* (Bt) gene for insect resistance and coat protein genes for tungro resistance should impart high levels of resistance. Coat protein–induced virus resistance is directed against the rice tungro virus (Beachy and others 1989), the most important disease problem for rice (Herdt and Riely 1987). A coat protein–mediated resistance to rice stripe virus (an insect-transmitted virus) has been introduced in Japonica rice (Hayakawa and others 1992). The Bt genes are targeted against stemborers (Joos and Morrill 1989), which are the most important insect pest problem for rice (Herdt and Riely 1987). Insect-resistant rice generated by introducing modified endotoxin gene of Bt has been shown to produce Bt protein and to enhance insect resistance in Japonica rice plants (Fujimoto and others 1993).

Genes for resistance to viral, bacterial, fungal, and other diseases have been identified and transferred successfully into many crops (Ahl Goy and Duesing 1995 and Persley 1990). The first genetically engineered Indica rice based on protoplast system (Datta and others 1990) now provides resistance to sheath blight in Indica rice (Lin and others 1995). Several chitinase genes in combination with Bt genes have been introduced in rice to enhance protection against pathogenic fungi, bacteria, and insects. The identification of many other genes that may enhance the genetic yield potential of rice or confer resistance to biotic and abiotic stresses is still at an early stage. After their identification, these genes have to be isolated and constructed in a manner suitable for their introduction and expression into the rice genome.

In the case of maize, genetically engineered plants for resistance to the corn borer have been produced using the Bt gene. Although the transformation of wheat plants is still in its infancy, there are significant hopes of developing additional transgenic plants with fungal resistance and abiotic stress tolerance.

Although progress in biotechnology is rapid, only in a few cases have agronomically viable plants been developed. Although some progress has been made, gene mapping techniques remain slow and in need of refinement to facilitate marker-aided gene selection (N. Huang, IRRI, personal communication). The procedure is expensive and requires highly trained personnel. Traditional plant breeding problems are resolved, but challenges facing plant improvement programs remain essentially the same. The germ plasm still needs to be screened for desirable genes, which then have to be transferred to the crop, and, finally, the performance of the improved plant has to be tested in the field. In addition, assessing the usefulness of genetic engineering techniques requires understanding the possible negative consequences. One example is the incorporation of a gene for herbicide resistance into cereal crops. A herbicide-resistant plant could sustain crop productivity in the face of increasing competition from weeds, especially in direct-seeded and minimum-tillage systems. Widespread use of such plants could, however, increase the risk of cross-pollination of the gene into weed and wild species, thus aggravating, rather than reducing, the problem of weed competition. To date, however, there is no evidence that such cross-pollination has occurred. A herbicide-resistant plant could also encourage high levels of herbicide use and thereby contribute to increased environmental pollution. However, a herbicide-resistant plant could encourage the switch to smaller doses of safer herbicides. Moreover, a host of intellectual property issues have to be dealt with in the transfer of genes and advanced biotechnology tools from advanced-country laboratories to developing-country research programs where they would be used to develop improved cereal crops.

Pesticide Productivity, Yield Variability, and Host-Plant Resistance: Evidence from Experimental Trials and Farmers' Fields

On-farm experiments and examination of farmers' yields do not suggest a yield or profitability response to pesticide applications on resistant crop varieties. The introduction of varieties with host-plant resistance has dramatically reduced the need for insecticides in rice and maize and the need for fungicides in wheat. Evidence for the three major cereals indicates that the extent of crop loss due to insects and diseases has dropped over the last two decades and that the extent of yield loss due to the failure to apply chemicals has declined significantly.

Rola and Pingali (1993) summarize the data on rice crop losses due to insects in the Philippines and indicate that estimated crop losses have decreased since the late 1960s and early 1970s. Pathak and Dyck (1973) recorded a crop loss of 22.5 percent during 1968–72, but more recently, both Litsinger (1991) and Waibel (1986) recorded losses of only 8.6 and 8.9 percent, respectively. The drop in insect-related crop losses can be attributed to the widespread use of resistant cultivars in farmers' fields after the mid-1970s. Both Litsinger and Waibel observe no significant differences in yield between insecticide-treated and -untreated plots in more than half the trials. In other words, half the time farmers do not need to spray against rice pests, especially on varieties with notable host-plant resistance. Pingali and Gerpacio (1997) report that the rice yield response, on farmers' fields, to insecticide application has been negligible with the adoption of resistant varieties. Prophylactic application of insecticides on resistant varieties tends to increase, rather than decrease, the variability in crop yields by disrupting the balance between pest and predator and by inducing the breakdown in resistance.

Host-plant resistance has similarly been successful in controlling the major corn insect pests, such as the corn borer, corn earworm, and fall armyworm. The use of insecticides on insect-resistant corn varieties has generally not been economically worthwhile. Diseases, such as rusts of the stem and leaf in wheat, that used to cause significant loss in yields have been controlled in the last two decades through improved resistance. Sayre and others (1997) report that there is no significant difference in wheat yields between plots sprayed with fungicides and those not sprayed when resistant varieties are used.

The widespread availability of insect- and disease-resistant varieties for the major cereals has reduced the productivity benefits and the profitability of applying insecticides and fungicides. Discussion on productivity benefits of pesticides revolves around the presumption that pesticides are a risk-reducing input—that is, they reduce the variability of yield. One needs to question this presumption, both for insecticides and for fungicides (Pingali and Rola 1995). In the case of insecticides, Rola and Pingali (1992, 1993) provide empirical evidence for rice production showing that insecticide applications increase variability of yield. They estimate an insecticide response function and evaluate the returns to four alternative pest management strategies for the Philippines, using both farm panel data and data from on-farm experiments conducted over several growing seasons. Response function estimates indicate very modest effects of insecticides on the mean and variance of yield distribution. Comparing the returns to prophylactic control, economic threshold levels, farmers' practice, and natural control, Rola and Pingali find, for lowland tropical rice systems, natural control to be the economically dominant pest management strategy. Natural control, in association with varietal resistance, is consistently more profitable in an average year than prophylactic treatment and economic thresholds. The dominance of natural control is found to hold even when farmers' risk aversion is taken into account. It becomes even greater when the health costs of exposure to insecticides are taken explicitly into account, because the positive production benefits of applying insecticides are overwhelmed by the higher health costs. The value of crop loss to pests is invariably lower than the cost of pesticide-related illness (Rola and Pingali 1993) and the associated loss in farmer productivity (Antle and Pingali 1994). Although such studies have not been conducted for wheat and maize in developing countries, similar results could be anticipated.

Agrochemicals continue to have a significant impact on productivity in the management of weeds, where herbicides will continue to be the preferred alternative in the foreseeable future. In intensive cereal crop production systems, the use of herbicides is more cost-effective than the use of human labor, because the seasonality of weeding often creates labor scarcity so that wage rates for weeding are higher than those for other crop management operations. The use of herbicides has also gotten a boost from the general rise in farm wages due to overall economic growth and the growth in nonfarm employment opportunities, particularly in Asia. As cheap herbicides become available and as farm wages rise, cereal crop farmers increasingly substitute herbicides for human labor, and the savings in labor cost more than compensate for the additional cost of herbicides. Pingali and Márquez (1997) document the productivity benefits of herbicide use, even when the health costs of herbicides are taken explicitly into account.

There are few genetic and management alternatives to herbicides, and those that exist are generally not very cost-effective. The use of more competitive cereal varieties can avert the effect of weed competition and the consequent use of more herbicides, but there appears to be a tradeoff between yield and the plant's ability to compete with weeds (Moody 1991). Research on varietal improvement for weed management in cereals is still at a very early stage (Khush 1996). Among the management options for weed control that minimize labor use are the manipulation of water in the paddy, in the case of rice (Pingali, Hossain, and Gerpacio 1997), and, in the case of wheat and maize, the use of ridge tillage systems (Sayre 1996) and the use of cover crops and intercropping. However, none of these options has proven to be economically as attractive as the use of herbicides, and the challenge for the research and policy community is to find cost-effective mechanisms for reducing herbicide use in cereal crop production.

The development of genetic resistant to *Striga*, the most important weed in Africa, is an exception to the lack of success in using genetic means for controlling weeds. Approximately 50 million hectares devoted to cereal production in Africa are in areas infested with *Striga*. Although three species of *Striga* are dominant, maximum loss is caused by *Striga hermonthica*. Both tolerance (plants that have lower yield loss than susceptible plants) and resistance (plants that prevent attachment of the parasite) are known to exist in maize and other crops. In a recent study (Kim, Adetimirin, and Akintunde 1997), *Striga* reduced maize yields by 25 percent in tolerant hybrids and 50 percent in susceptible hybrids over a range of nitrogen levels. Genetics seem to be a viable option to be pursued in the pesticide-free management of *Striga*.

Policies for Sustaining a Minimum-Pesticide Strategy

Given the extent of breeding efforts in producing pest-resistant crop cultivars and evidence of insignificant productivity benefits of pesticide use, why are high levels of agrochemical use so prevalent in crop production? First, the dissemination of crop varieties resistant to pest pressures is not accompanied by extension messages on the reduced need for pesticides. Second, farmers' pest control decisions, scientists' research priorities, and policymakers' prescriptions, which are all based largely on perceived pest-related yield losses (in turn often unrelated to actual yield losses), promote the use of pesticides (Rola and Pingali 1993). Consequently, high and injudicious applications of broad-spectrum pesticides continue as before, causing the breakdown in varietal resistance. New varieties generated to replace cultivars with resistance breakdown are subsequently overcome through further biotype changes in pest populations. This "breeding treadmill" can only be overcome through dramatic changes in crop management practices, especially in the use of pesticides. However, whenever government policies, such as subsidies and credit programs, keep pesticide prices artificially low, the incentives are also low for farmers to invest in knowledge that improves their perceptions and to adopt alternative pest control strategies with a broad ecological approach.

Under low levels of pest infestation, natural control is the economically dominant strategy for managing pests (Rola and Pingali 1993). This is true for insects and diseases for rice, wheat, and maize. Natural control relies on predator populations to control pest infestations under normal circumstances, when pest-resistant varieties are used. Pesticides may have to be used as a last resort in the rare instance of high pest infestations. Natural control does not imply "do nothing"; rather, it is based on the premise of in-depth farmer knowledge of the pest-predator ecology and frequent monitoring of field conditions by the farmer. In this regard, natural control can be considered the ultimate goal of an integrated pest management (IPM) program, and farmers who are well versed in IPM techniques would converge toward it. This implies a paradigm shift in the traditional IPM strategies from when best to apply to when not to apply. Therefore, continued investments in IPM training are essential for the successful adoption of reduced or minimum use of insecticides.

Integrated Pest Management

Although it is generally recognized that the sustainable use of host-plant resistance requires a more integrated approach to pest management, farm-level practices of pesticide management have not changed significantly over the past 20 to 30 years. Farmers have been slow to adopt IPM practices, although they readily accept seeds and other tangible crop production technologies. However, IPM is knowledge-based and presents by far the most difficult challenge to traditional, small-scale farmers in developing countries as they make the transition to scientific farming (IRRI 1994 and Goodell 1984). IPM requires farmers to grasp complex sets of data that are often anything but self-evident, unitary, and standardized or amenable to trial-and-error learning. The institutional and economic structure in the rural sector of developing economies also requires some policy intervention to reconcile long-term societal goals with short-term individual objectives in pest control. Promoting sustainable pest management within an IPM framework requires improved research-extension linkages, effective farmer training methods, community action, and an undistorted price structure.

The IPM concept is holistic; it requires farmers to take a systems view of the farm enterprise and to understand the interlinkages among the various components of the system. Disseminating such a holistic message requires an extension system that has traditionally been geared toward promoting component technologies, such as improved seeds and fertilizer, to adopt new skills. Pest and pesticide problems are intrinsically local in nature. National policy ought to nourish the rural community's capacity to handle pest problems effectively, profitably, and equitably. To be successful, the IPM concept needs to be adapted to particular local situations. Such adaptation has to be done with close collaboration of researchers, extension personnel, and farmers. Most IPM success stories have been preceded by research done in farmer fields with the farmer actively participating in all stages of the research process (Escalada and Heong 1993).

The wider dissemination of locally validated research results requires a decentralized extension system in which the formulation of the extension message is determined at the subprovincial or municipal level. In such a system, the extension worker acts as a local-level researcher adapting research and technology to suit local agroecological conditions rather than as an agent transmitting messages that are barely understood and often inappropriate to the situation at hand. Such a decentralized extension setup will only come about with a paradigm shift from the current emphasis on the top-down transfer of information.

Farmer training in sustainable pest management is an essential component of a strategy toward minimum use of insecticides. The eventual goal is to build farmer capacity in identifying and solving problems based on a thorough understanding of field ecology. The experience of the Food and Agriculture Organization's farmer field schools has shown that trained farmers use significantly lower levels of pesticides than untrained farmers (Kenmore 1991). Trained farmers are also more likely to experiment with other components of sustainable production systems, such as improved fertilizer management and more efficient water management.

There are as yet several unresolved research and policy issues related to intensive farmer training in IPM. The most important one concerns the costs and benefits of farmer training. The costs of training the 120 million farmers in Asia are enormous and need to be justified in terms of farm-level benefits, both reduced expenditures on insecticides and reduced social costs of pesticide use. Attention also should be focused on opportunities for reducing the overall costs of training. There are essentially two options, which are not mutually exclusive, for reducing training costs. The first is to train a core group of farmers within a geopolitical unit, such as a municipality, and then rely on farmer-to-farmer training for disseminating the IPM message to a wider group of farmers. There are definite scale economies to the farmer-to-farmer training approach if the quality of the message transmitted does not deteriorate as it gets passed down the line.

The second option is to condense the complex message into several simple rules that are easy for the farmer to implement. An example, from rice, of such a rule is: Do not spray insecticides against leaf-feeding insects for the first 40 days of crop growth. The rule is based on detailed pest ecology studies, which have shown that the predominant insect pests during the first 40 days of crop growth are leaf-feeding insects and that even very high levels of infestations of these insects rarely lead to any loss in yield (Heong 1990). Leaf-feeding insects are very visible, and farmers tend to attach great importance to controlling them (Heong, Escalada, and Lazaro 1995). Controlling leaf-feeding insects, however, comes at a cost that substantially exceeds the value of yield saved, if any. Early-season insecticide applications tend to wipe out the leaf-feeding insects as well as the beneficial predator populations that are building up in the paddy. Rice paddies receiving one or two insecticide applications within the first 40 days tend to be susceptible to secondary pest infestations, especially the brown planthopper, which build up unchallenged due to the lack of natural controls. Controlling the growth of secondary pests requires further applications, thus spiraling the use of insecticides. By not applying insecticides early in the crop season, farmers can reduce the need for them later on due to the abundance of predator populations in the paddy fields.

The "no spray for 40 days" rule is an example of a simple message, distilled from in-depth scientific investigation, that can be transmitted easily to farmers. A set of such rules, which are mutually consistent, could go a long way toward improving the management of insect pests in tropical cereal production. Simple rules ought to be seen not as a substitute for farmer training but rather as a strong complement to a training program. Simple messages can be transmitted rapidly even while investments are being made to reach all tropical rice farmers with intensive training.

Community Action

The success of pest management programs depends on, among other things, collective organization against infestations of migratory pests. Actions of individual farmers in the management of their pest problems could have detrimental effects on the community as a whole. In this regard, management of pests can be viewed as a common property problem that is best dealt with through effective collective action. In Asian agriculture, where farms are uniformly small and farmers are nearly homogeneous, collective action for pest control seems quite attractive. Synchronization of planting is the single most important community action strategy in pest management.

Asynchrony is significantly related to the buildup and field-to-field carryover of pest populations. Because synchronous planting can prevent the build up of pests and thereby reduce the damage to crops, it could greatly improve yields. The extent of synchrony would be based on the minimum period in which pests can complete one life cycle. Rice fields in a contiguous area have to be planted within three to four weeks in order to capture the benefits of synchrony (Heinrichs and Mochida 1984).

Several factors may constrain synchronization of planting schedules in a particular location and for particular crops. Loevinsohn (1985, 1987) investigates the causes, extent, and effects of synchronous rice cultivation in Nueva Ecija, Philippines. These include irregularities in water distribution and drainage or variation in the arrival of irrigation water at farms, unavailability of a tractor or custom hire work during land preparation, unavailability of labor during transplanting, and lack of access to credit.

On the whole, treating pests as a common property resource (or a public liability) would entail collective action by constituents and support from the public sector. Recognizing both the positive and the negative externalities of pest control agents, group action could be much more effective than individual action. The challenge to governments is to create an environment that promotes these strategies in ways that achieve growth, equity, and environmental sustainability.

Pesticide Pricing Policies and IPM

Even with a well-established IPM program, pesticides may have to be kept as a technology-of-last-resort. Essentially, the idea of using pesticides in IPM is to spray only when imperative, using the smallest amount possible to do the job. To make IPM more attractive, pesticides ought not to be subsidized. In fact, farmers will bother to learn and to apply IPM techniques only when the cost savings associated with insecticide application make it worth their while. The highly acclaimed IPM training in Indonesia was preceded by a comprehensive reform of pesticide regulation and the removal of all subsidies for rice insecticides (Rola and Pingali 1993). In addition to the removal of subsidies, a sustainable pest management program may warrant the taxation of agrochemicals to account for the social costs associated with indiscriminate pesticide use. Antle and Pingali (1994) have shown that a 100 percent tax on insecticides for rice could actually improve productivity by improving farmer health through a reduction in exposure and an increase in labor productivity. For industrial-country agriculture, Carlson and Wetzstein (1993) and Zilberman and others (1991) argue that pesticide taxation has the potential to eliminate many of the problems associated with pesticide use.

References

The word "processed" describes informally reproduced works that may not be commonly available through libraries.

Ahl Goy, P., and J. Duesing. 1995. "From Pots to Plots: Genetically Modified Plants on Trial." *Bio/Technology* 13(5):454–58.

Antle, J. M., and Prabhu L. Pingali. 1994. "Pesticides, Productivity, and Farmer Health: A Philippine Case Study." *American Journal of Agricultural Economics* 76(3):418–30.

Beachy, R. N., B. L. Subba Rao, K. Gough, P. Shen, and M. Kaniewska, with H. H. Hibino. 1989. "Characterization of the Viruses Associated with the Tungro Disease." Paper presented at the third annual meeting of the Rockefeller Foundation International Program on Rice Biotechnology, Colombia, Mo., 8–10 March. Processed.

Boardman, R. 1986. *Pesticides in World Agriculture: The Politics of International Regulation.* New York: St. Martin's Press.

Boserup, Ester. 1965. *Conditions of Agricultural Growth.* Chicago: Aldine Publishing Company.

Carlson, Gerald, and M. E. Wetzstein. 1993. "Pesticides and Pest Management." In G. Carlson, D. Zilberman, and J. A. Miranowski, eds., *Agricultural and Environmental Resource Economics.* New York: Oxford University Press.

Commonwealth Agricultural Bureaux. 1980. *Perspectives in World Agriculture.* London: Unwin Brothers; Old Woking, U.K.: The Gresham Press.

Cornwell, J. 1995. "Risk Assessment and Health Effects of Pesticides Used in Tobacco Farming in Malaysia." *Health Policy and Planning* (United Kingdom) 10(December):431–37.

Crissman, C., and D. Cole. 1994. "Economics of Pesticide Use and Farm Worker Health in Ecuadorian Potato Production." *American Journal of Agricultural Economics* 76(August):593–97.

Datta, S. K., A. Peterhaus, K. Datta, and I. Potrykus.

1990. "Genetically Engineered Fertile Indica Rice Recovered from Protoplasts." *Bio/Technology* 8(8):736–40.

Dubin, H. J., and S. Rajaram. 1996. "Breeding Disease-Resistant Wheats for Tropical Highlands and Lowlands." *Annual Review of Phytopathology* 34:503–26.

Edmeades, G. E., and J. A. Deutsch, eds. 1994. *Stress Tolerance Breeding: Maize That Resists Insects, Drought, Low Nitrogen, and Acid Soils.* Maize Program Special Report. Mexico D.F. : Centro Internacional de Mejoramiento de Maíz y Trigo.

Escalada, M. M., and K. L. Heong. 1993. "Communication and Implementation of Change in Crop Protection." In D. J. Chadwick and J. Marsh, eds., *Crop Protection and Sustainable Agriculture*. Ciba Foundation symposium on world food production by means of sustainable agriculture: the role of crop protection, held in collaboration with the Centre for Research on Sustainable Agricultural and Rural Development, Madras, India, 30 November–2 December 1992. Chichester, U.K.: John Wiley and Sons.

FAO (Food and Agriculture Organization of the United Nations). 1997. "Statistics on Pesticide Use." Internet copy. Statistical Analysis Service, Statistics Division, Rome. Processed.

Forget, G., T. Goodman, and A. de Villiers. 1993. *Impact of Pesticide Use on Health in Developing Countries*. Proceedings of a symposium held in Ottawa, Canada, 17–20 September 1990. Ottawa: International Development and Research Centre.

Fujimoto, H., H. Itoh, M. Yamamoto, J. Kyozuka, and K. Shimamoto. 1993. "Insect Resistant Rice Generated by Introduction of a Modified Delta-Endotoxin Gene of *Bacillus thuringiensis*." *Bio/Technology* 11(10, October):1151–61.

Goodell, G. 1984. "Challenges to International Pest Management Research and Extension in the Third World: Do We Really Want IPM to Work?" *Bulletin of the Entomological Society of America* 30(3):18–25.

Hayakawa, T., Y. Zhu, K. Itoh, Y. Kimura, T. Izawa, K. Shimamoto, and S. Toriyama. 1992. "Genetically Engineered Rice Resistant to Rice Stripe Virus, an Insect-Transmitted Virus." *Proceedings of the National Academy of Sciences* 89(20):9865–69.

Heinrichs, E. A., and O. Mochida. 1984. "From Secondary to Major Pest Status: The Case of Insecticide-Induced Rice Brown Planthopper, *Nilaparvata lugens*, Resurgence." *Protection Ecology* 7(2-3):201–18.

Heong, K. L. 1990. "Feeding Rates of the Rice Leaffolder, *Chaphalocrocis medinalis* (Lepidoptera: Pyralidae), on Different Plant Stages." *Journal of Agricultural Entomology* 7(2):81–90.

Heong, K. L., M. M. Escalada, and A. A. Lazaro. 1995. "Misuse of Pesticides among Rice Farmers in Leyte, Philippines." In Prabhu L. Pingali and P. A. Roger, eds., *Impact of Pesticides on Farmer Health and the Rice Environment*. Norwell, Mass.: Kluwer Academic Publishers; Los Baños, Philippines: International Rice Research Institute.

Heong, K. L., M. M. Escalada, and Mai Vo. 1994. "An Analysis of Insecticide Use in Rice: Case Studies in the Philippines and Vietnam." In *Rice IPM Network Workshop Report on Message Design for a Campaign to Encourage Farmers' Participation in Experimenting with Stopping Early Insecticide Spraying in Vietnam*. Network workshop, Ho Chi Minh City, Vietnam, 25–28 May. Los Baños, Philippines: International Rice Research Institute.

Herdt, R. W., and F. Z. Riely. 1987. "International Rice Research Priorities: Implications for Biotechnology Initiatives." In *Proceedings of the Rockefeller Workshop on Allocating Resources for Developing-Country Agricultural Research, 6–10 July, Bellagio, Italy*. New York: Rockefeller Foundation.

IFPRI (International Food Policy Research Institute). 1996. *Pest Management: The Expected Trends and Impact on Agriculture and Natural Resources to 2020*. 2020 Vision Food, Agriculture, and the Environment Discussion Paper. Washington, D.C.

IRRI (International Rice Research Institute). 1972. *Rice, Science, and Man*. Papers presented at the tenth anniversary celebration of the International Rice Research Institute, 20–21 April. Los Baños, Philippines.

———. 1985. *Twenty-fifth Anniversary Celebration*. Papers presented at the twenty-fifth anniversary celebration of the International Rice Research Institute, 6–8 June. Los Baños, Philippines.

———. 1993a. *IRRI Program Report for 1992*. Los Baños, Philippines.

———. 1993b. *Rice in Crucial Environments: IRRI 1992–1993 Corporate Report*. Los Baños, Philippines.

———. 1994. *IRRI Program Report for 1993*. Los Baños, Philippines.

———. 1995. *World Rice Statistics 1993–94*. Los Baños, Philippines.

Joos, H., and W. Morrill. 1989. "Control of Insect

Pests on Rice Using *Bacillus thuringiensis* Genes." Paper presented at the third annual meeting of the Rockefeller Foundation International Program on Rice Biotechnology, Colombia, Mo., 8–10 March. Processed.

Joshi, R. C., B. M. Shepard, P. E. Kenmore, and R. Lydia. 1992. "Insecticide-Induced Resurgence of Brown Planthopper (BPH) on IR62." *International Rice Research Newsletter* 17(3, June):9–10.

Kabir, K. H., F. M. A. Rouf, M. E. Baksh, N. A. Mondal, M. A. Hossain, M. A. Satter, and M. S. M. Asaduzzaman. 1994. "Insecticide Usage Pattern on Vegetables at Farmers' Level of Jessore Region: A Survey Report." Research report. Jessore Regional Agricultural Research Stations, Jessore, Bangladesh. Processed.

Kenmore, P. E. 1991. *Indonesia's Integrated Pest Management: A Model for Asia.* Intercountry IPC Rice Programme. Manila, Philippines: Food and Agriculture Organization of the United Nations.

Kenmore, P. E., F. O. Cariño, C. A. Pérez, V. A. Dyck, and A. P. Gutiérrez. 1984. "Population Regulation of the Rice Brown Planthopper (*Nilaparvata Lugens*) within Rice Fields in the Philippines." *Journal of Plant Protection in the Tropics* 1(1): 19–37.

Khush, G. 1996. "Rice Varietal Improvement for Weed Management." in R. Naylor, ed., *Herbicide in Asian Rice: Transition in Weed Management.* Palo Alto, Calif.: Stanford University, Institute of International Studies; Los Baños, Philippines: International Rice Research Institute.

Kim, Soon-Kwon, V. O. Adetimirin, and A. Y. Akintunde. 1997. "Nitrogen Effects on *Striga Hermonthica* Infestation, Grain Yield, and Agronomic Traits of Tolerant and Susceptible Maize Hybrids." *Crop Science* 37(3):711–16.

Kumar, H., and J. A. Mihm. 1995. "Antibiosis and Tolerance to Fall Armyworm, *Spodoptera Frugiperda (J. E. Smith)*, Southwestern Corn Borer, *Diatraea Grandiosella Dyar*, and Sugarcane Borer, *Diatraea Saccharalis Fabricius*, in Selected Maize Hybrids and Varieties." *Maydica* 40(3):245–51.

Lin, W., C. S. Anuratha, K. Datta, I. Potykus, S. Muthukrishnan, and S. K. Datta. 1995. "Genetic Engineering of Rice for Resistance to Sheath Blight." *Bio/Technology* 13(7):686–91.

Litsinger, J. A. 1991. "Crop Loss Assessment in Rice." In E. A. Heinrichs and others, eds., *Rice Insects: Management Strategies.* New York: Springer-Verlag.

Loevinsohn, M. 1985. "Agricultural Intensification and Rice Pest Ecology: Lessons and Implications." Paper presented at the International Rice Research conference, International Rice Research Institute, Los Baños, Philippines, 1–5 June. Processed.

———. 1987. "Insecticide Use and Increased Mortality in Rural Central Luzon, Philippines." *Lancet* 1(8546):1359–62.

Mihm, J. A., J. Deutsch, D. Jewell, D. Hoisington, and D. G. de Leon. 1994. "Improving Maize with Resistance to Major Insect Pests." In G. E. Edmeades and J. A. Deutsch, eds., *Stress Tolerance Breeding: Maize That Resists Insects, Drought, Low Nitrogen, and Acid Soils.* Maize Program Special Report. Mexico D.F.: Centro Internacional de Mejoramiento de Maíz y Trigo.

Moody, K. 1991. "Weed Management in Rice." In D. Pimentel, ed., *Handbook of Pest Management in Agriculture*, 2d ed., pp. 301–28. Boca Raton, Fla.: CRC Press.

Naylor, R. 1997. *Herbicides in Asian Rice: Transitions in Weed Management.* Palo Alto, Calif.: Stanford University, Institute for International Studies; Los Baños, Philippines: International Rice Research Institute.

Pagiola, Stefano. 1995. *Environmental and Natural Resource Degradation in Intensive Agriculture in Bangladesh.* Environmental Economic Paper 15. Washington, D.C.: World Bank.

Pathak, M. D., and V. A. Dyck. 1973. "Developing an Integrated Method of Rice Insect Pest Control." *PANS* 19(4):534–44.

Persley, G. J. 1990. *Beyond Mendel's Garden: Biotechnology in the Service of World Agriculture.* Wallingford, U.K.: C.A.B. International.

Pingali, Prabhu L., and Hans P. Binswanger. 1987. "Population Density and Agricultural Intensification: A Study of the Evolution of Technologies in Tropical Agriculture." In G. Johnson and R. Lee, eds., *Population Growth and Economic Development.* Washington, D.C.: National Research Council.

Pingali, Prabhu L., and R. V. Gerpacio. 1997. "Living with Reduced Insecticide Use in Tropical Rice." *Food Policy* 22(2):107–18.

Pingali, Prabhu L., M. Hossain, and R. V. Gerpacio. 1997. *Asian Rice Bowls: The Returning Crisis?* Wallingford, U.K.: C.A.B. International.

Pingali, Prabhu L., and C. B. Márquez. 1997. "Herbicides and Rice Farmer Health: A Philippine

Study." In R. Naylor, ed., *Herbicides in Asian Rice: Transitions in Weed Management*. Palo Alto, Calif.: Stanford University, Institute for International Studies; Los Baños, Philippines: International Rice Research Institute.

Pingali, Prabhu L., and S. Rajaram. 1997. "Technological Opportunities for Sustaining Wheat Productivity Growth." Paper presented at the Illinois World Food and Sustainable Agriculture Program conference on meeting the demand for food in the twenty-first century: challenges and opportunities for Illinois agriculture, Urbana-Champaign, Ill., 27 May. Processed.

Pingali, Prabhu L., and P. A. Roger, eds. 1995. *Impact of Pesticides on Farmer Health and the Rice Environment*. Norwell, Mass.: Kluwer Academic Publishers; Los Baños, Philippines: International Rice Research Institute.

Pingali, Prabhu L., and A. C. Rola. 1995. "Public Regulatory Roles in Developing Markets: The Case of Pesticides." In Prabhu L. Pingali and P. A. Roger, eds., *Impact of Pesticides on Farmer Health and the Rice Environment*. Norwell, Mass.: Kluwer Academic Publishers; Los Baños, Philippines: International Rice Research Institute.

Rola, A. C., and Prabhu L. Pingali. 1992. "Choice of Crop Protection Technologies under Risk: An Expected Utility Maximization Framework." *Philippine Journal of Crop Science* 17(1):45–54.

———. 1993. *Pesticides, Rice Productivity, and Farmers' Health: An Economic Assessment*. Washington, D.C.: World Resources Institute; Los Baños, Philippines: International Rice Research Institute.

Sayre, K. D. 1996. "The Role of Crop Management Research at CIMMYT in Addressing Bread Wheat Yield Potential Issues." In M. P. Reynolds, S. Rajaram, and A. Macnab, eds., *Increasing Yield Potential in Wheat: Breaking the Barriers*. Mexico D.F.: Centro Internacional de Mejoramiento de Maíz y Trigo.

Sayre, K. D., R. P. Singh, J. Huerta-Espino, and S. Rajaram. 1997. "Genetic Progress in Reducing Losses to Leaf Rust in CIMMYT-Derived Mexican Wheat Cultivars." CIMMYT Wheat Program, Centro Internacional de Mejoramiento de Maíz y Trigo, Mexico D.F. Processed.

Schoenly, K., J. E. Cohen, K. L. Heong, G. Arida, A. T. Barrion, and J. A. Litsinger. 1996. "Quantifying the Impact of Insecticides on Food Web Structure of Rice-Arthropod Populations in Philippine Farmers' Irrigated Fields: A Case Study." In G. Polis and K. Winemiller, eds., *Food Webs: Integration of Patterns and Dynamics*, chap. 32. New York: Chapman and Hall.

United Nations. 1990. *Commodity Trade Statistics*. Statistical Paper Series D, vol. 38. no. 1-19. New York.

Waibel, H. 1986. *The Economics of Integrated Pest Control in Irrigated Rice: A Case Study from the Philippines*. Crop Protection Monograph. Berlin: Springer-Verlag.

Warburton, H., F. G. Palis, and Prabhu L. Pingali. 1995. "Farmer Perceptions, Knowledge, and Pesticide Use Practices." In Prabhu L. Pingali and P. A. Roger, eds., *Impact of Pesticides on Farmer Health and the Rice Environment*. Norwell, Mass.: Kluwer Academic Publishers; Los Baños, Philippines: International Rice Research Institute.

Weir, D., and M. Shapiro. 1981. *The Circle of Poison: Pesticide and People in a Hungry World*. San Francisco: Third World Publications for the Institute for Food and Development Policy.

Wood Mackenzie Consultants Ltd. 1993. "Global Rice Pesticide Market." A report submitted to the Plant Protection Department of Ciba-Geigy Ltd., Basil, Switzerland. Processed.

Zilberman, D., A. Schmitz, G. Casterline, E. Lichtenberg, and J. B. Siebert. 1991. "The Economics of Pesticide Use and Regulation." *Science* 253(August 2):518–22.

20

The Economic and Environmental Impacts of Irrigation and Drainage in Developing Countries

Ujjayant Chakravorty

The world's population is expected to grow from approximately 6 billion today to nearly 10 billion by the middle of the next century (United Nations Population Fund 1993). Most of this growth will take place in the developing world, increasing its share of the earth's population to 90 percent. Over the past three decades, the dramatic spread of irrigated agriculture, which enabled widespread adoption of high-yielding production technology, was mainly responsible for enabling increases in food production to keep pace with the growth in population. This was achieved through the large-scale development of new water resources and the construction of new irrigation capacity.

Both rising population and rising incomes due to economic development are expected to increase substantially the demand for food in the coming years. The International Irrigation Management Institute (IIMI 1992) has estimated that to meet this demand, irrigated agriculture will need to increase output more than 3.5 percent annually—a daunting task by any measure of performance. However, the major problem is that this growth must come primarily from more efficient irrigation rather than from new capacity. Moreover, as is becoming more common in the industrial countries and even in many developing countries, higher residential demand for water and rising environmental concerns are beginning to restrict new water development projects. Environmental demands for water to preserve instream uses, wetlands, freshwater fisheries, and other recreational purposes are in direct competition with the diversion of water for irrigation in the United States and in many other industrial countries. The same pattern is likely to repeat itself in the developing countries with the rise in environmental consciousness and the concomitant degradation of environmental resources.

Thus the irrigation sector in developing countries will be expected to deliver higher agricultural output with reduced supplies of water, because there is little hope that new investment in irrigation will match the global trend of diverting water from agriculture to urban and environmental uses. The capital cost of new irrigation development rose 70–116 percent per hectare in the 1980s alone, according to figures available from the World Bank (Dinar 1997), and new investment in irrigation has declined substantially in countries such as China and India. Producing more food with less water—placing a smaller financial burden on the taxpayer and reducing the environmental costs—will be the major challenge for irrigation in the twenty-first century (Lenton 1994).

However, in spite of its significant contribution to agricultural production and food security, the current state of irrigation is not encouraging. Despite major investments in irrigation during the past four decades—the World Bank alone has invested more than $31 billion and leveraged an additional $53 billion from partner countries and cofinanciers—irrigation projects around the world are in serious trouble for a multitude of reasons: ex post project benefits have often been far lower than projected ex ante returns, and full-cost pricing of water is rare, with most projects recouping only a fraction of their operation and maintenance (O&M) costs. The low price of irrigation water, which is often unrelated to water use, has led farmers to withdraw too much, giving rise to widespread salinity and waterlogging. For example, Biswas (1991) suggests that more than a quarter of global irrigated acreage has been damaged by salinization.

On a broader scale, investments in the irrigation sector in developing countries have relied too much on bureaucratic federal and state agencies that have paid little attention to the economic pricing of irrigation services, the creation of reliable delivery systems, and the participation of users in the operation and maintenance of projects. Adverse health and environmental effects of water use have received little attention, partly because of the lack of coordination and interaction between water and public health agencies.

These problems call for a new approach that is more comprehensive and that integrates the different elements of irrigation policy, which include the economic and environmental effects and the associated intersectoral linkages. This approach must be built on pricing of resources that reflects opportunity costs whenever possible and the creation of delivery infrastructure that promotes the efficient allocation and, if feasible, trading of the resource among users.

This chapter reviews current thinking and practice on the economics of irrigation and drainage in developing countries. In particular, it examines recent experiences with the reform of irrigation water policy and attempts to draw broad conclusions and policy recommendations for which approaches work and which do not and under what conditions. The chapter is addressed to policymakers and practitioners and aims to complement their specific expertise on resource management issues with a more global perspective on issues concerning the economics and management of agricultural water resources.

The second section discusses critical issues related to the economics of irrigation in the developing countries, followed by a section highlighting problems related to rainfed agriculture, which is still the primary means of food production in the developing world. Next, a section deals with the environmental problems that have arisen as a result of inadequate drainage and have resulted in large irrigated areas losing their productive potential. This is followed by a section highlighting a few important factors that contribute to the success or failure of irrigation projects. A final section concludes.

The Economics and Pricing of Irrigation Water: Critical Issues

Economic principles dictate that under certain conditions regarding the characteristics of the resource and of the market in which it is allocated, resource allocation is best done by means of a smoothly functioning market system (Young and Haveman 1985). In the case of water, these conditions are not met on several counts. For example, water is mobile, which makes it difficult to measure and hold, thereby making difficult the establishment of property rights. Economies of scale in water generation and conveyance often create conditions for a natural monopoly. The joint product nature of water, which enables multiple users to access a given stock of water sequentially, and the externalities created as a result of the degradation of quality through use make the creation and functioning of water markets extremely difficult. For these reasons, the water sector both in industrial and in developing countries has been characterized by public investment in the construction of irrigation projects, state ownership and operation of facilities, and regulated distribution and allocation of water.

Irrigation water projects in developing countries face several important issues that arise from the existence of insufficient or outmoded institutional mechanisms and economic policies. Many of the policies that allocate water among users were introduced at a time when there was no perceived scarcity of water and the primary goal was to provide cheap and abundant water supplies to encourage people to occupy and farm new settlement areas. These rules are mostly some variation of the first-come, first-served principle that allowed settlements to occur at low enforcement and transaction costs. These distributional mechanisms are still in use in most irrigation projects, where water allocation is based on the seniority of water rights, which prevents the trading and transfer of water among users. Low tax collections and poor O&M of project structures have further led to high rates of water loss through seepage and percolation, and inadequate investments in drainage have rendered large tracts of prime agricultural land unusable because of salinity and waterlogging.

Conveyance Losses

One major problem in irrigation management is the continued use of traditional design principles that implicitly assume that the opportunity cost of water is zero. These techniques were appropriate in an era when water was plentiful. For example, the design capacity at the source and the allocation of water to each farmer are often based on exogenously determined crop water requirements, not on the shadow price of water. This implies the overuse of available water resources to irrigate a project area that is smaller than economically optimal (Chakravorty and Roumasset 1994). Frequently, politicians, in an attempt to maximize the number of beneficiaries, exert pressure to extend the project area beyond what is specified in the design (Repetto 1986). Over time, large transmission losses and excessive withdrawals of water by large and influential farmers (who typically are located in the upstream reaches of the project canals) shrink the irrigated area covered by the project.

Wade (1984) provides a fascinating description of this phenomenon for an irrigation project in South India. He points out that canal performance can be substantially improved by better operation and maintenance and that output can be expanded significantly with as little as 50–100 percent more outlays for maintenance and without expensive upgrading of facilities. Wade describes efficiency in an irrigation project as akin to stretching a membrane. Higher levels of canal performance require the water supply to be stretched through better management. Poor operation and maintenance, in contrast, lead to a contraction of the system, limiting water allocation to upstream locations.

Wade further compares two water delivery schedules in northern and southern India that have very different impacts on irrigation efficiency and water distribution. In northwestern India, irrigation is characterized by elaborate rotational delivery schedules that require all farmers to share the available water. This is enforced by allowing each field to withdraw all the water available in the channel for a fixed number of minutes each week. This system ensures that both head- and tail-enders receive their allocations. In contrast, in southern India, paddy irrigation is characterized by continuous flow. Water flows out of all outlets at the same time, and the Irrigation Department is supposed to ensure that the head-enders do not take more than their allocated amount and leave water in the canal for the tail-enders. This system also supplies the full water requirement to certain blocks, while depriving lands not in the service area of any water at all. Wade provides a detailed account of the opportunity for corruption that this system promotes, with officials taking bribes for supplying more water to certain villages and with the bribery permeating every level of the bureaucratic and political apparatus. The system has a well-developed scheme of rewards and punishment: bureaucrats who participate in corruption are transferred to "lucrative" posts, while those who do not are shunted to undesirable or remote locations.

In a simulation model of this phenomenon, Chakravorty, Hochman, and Zilberman (1995) find that under plausible conditions, better management of canal water may expand the service area of the project by as much as 200–300 percent, with similar increases in agricultural output. This can be done, for example, by water users associations that collect water charges as well as operate and maintain the distribution system. Reducing water losses through better maintenance allows the irrigated area in their model to expand. However, if the demand elasticity for agricultural production is low, then the gains in service area and output may be smaller. Rigorous empirical evidence of the impact of better canal management is not available, however, and is an important problem for further research.

Salinity and Waterlogging

The complex relationship between transmission losses and the buildup of salinity and waterlogging is revealed in an ongoing study conducted by the Government of Pakistan and the IIMI (Lenton 1994). The study finds that because the delivery of water in the lower reaches of the distribution system was unreliable, farmers leaned heavily on tubewell systems for irrigation. Because the tubewell water was of poor quality, intensive use led to high salinity in the project area. This discovery was a complete surprise to the project researchers, who had initially assumed that salinity was the direct result of a rising water table. In this project, researchers found no evidence of waterlogging, but they did find a high incidence of salinity. They also observed that poor maintenance of distribution canals, unauthorized withdrawals of water, and capacity constraints led to spatial inequity between farmers at the head and tail ends. Farmers at the tail end received much less water than those at the head end and faced significantly more variable water supplies. According to the author, these findings represent a radical departure from the traditional approach to the problem of salinity, which focuses on waterlogging and overwatering. The new approach to controlling salinity in the project focuses on the adoption of crops with a higher tolerance for salinity and better management of water distribution, which helps to reduce the high reliance on tubewell irrigation.

Maintenance of Public Irrigation Facilities

One of the most serious issues facing policymakers and international agencies like the World Bank and the regional banks is the need to maintain project structures. Service in public irrigation facilities tends to deteriorate rapidly a few years after commissioning, mostly because of poor maintenance. Systems are frequently operated so that the availability of water has little to do with the seasonal pattern of water demand by farmers (Repetto 1986). The *warabandi* rotation schedule, long popular in South Asia as a means of delivering canal water to farmers on a rotation schedule, is sometimes unreliable, with farmers never knowing when they will receive their allocation. This is mainly because of poor maintenance in upstream regions and the natural instability of the canal bed, which in turn affects the water level and the amount of water withdrawn from each individual outlet. This uncertainty leads farmers to underinvest in important production inputs, such as fertilizer and seeds.

In the public tubewell irrigation system in the state of Uttar Pradesh in India, steady decreases are observed in both the hours of operation and the area irrigated per tubewell. So although the number of tubewells has increased over the years, the total acreage served has declined. These systems were put in place to provide irrigation for low annual cropping intensity and limited water application. However, the surface distribution system has failed to distribute the limited water over the target area. Incomplete distribution canals have meant that farmers farther away from the source do not receive enough water. Inadequate operation and maintenance have led to frequent breakdowns, delayed repairs, and interrupted service (Cunningham 1992). Despite the poor performance of public irrigation systems, the government has continued to develop public groundwater schemes, mainly because such schemes have relatively short construction periods and can reach beneficiaries who either cannot afford private tubewells or inhabit inaccessible regions.

Groundwater irrigation is characterized by high operating costs relative to capital costs, unlike surface water irrigation. The government also provides subsidized capital costs by providing farmers with cheap electricity and diesel. In Uttar Pradesh, the government charges a flat rate for electricity use, which is one reason why there has been a rapid growth in private tubewells run on electricity. The flat rate induces farmers to overirrigate and to select larger-capacity pumps so that they can irrigate quickly, which is useful in an environment of random and fluctuating power supply.

A major study of a sample of World Bank–funded projects in irrigation and drainage between 1961 and 1985 reveals some interesting results. Jones (1994) finds that estimates of economic return from projects average 15 percent, lower than the 21 percent predicted at the project appraisal stage. The study finds that two of the biggest factors affecting returns from irrigation are those over which the project has no control—macroeconomic distortions and commodity prices. However, among the factors that *can be controlled* is the scale of the project—the benefit-cost ratio seems to be greater in bigger projects. The study does not find a direct relationship between benefits from an irrigation project and O&M, mainly because water charges are usually paid directly into the central treasury, although some countries have enacted legislation requiring collections to be used directly for O&M (Radosevich 1986). However, other studies, such as the Asian Development Bank's assessment of irrigation user charges, suggests a direct link between cost recovery and the financial autonomy of the organization undertaking O&M. In projects where O&M activities are financed out of water fees, cost recovery is achieved up to the costs of operation and maintenance.

Water Trading and Markets

Economists generally believe that the introduction of markets and trading among users lead to efficient allocation. Water markets function in different parts of the world, often without help from outside parties. However, they only work under very selective conditions and do not seem to work when any of those conditions are not satisfied. For example, in Alicante, Spain, irrigation systems in the *huerta* (irrigation service area) consist mostly of small farmers cultivating up to 1 hectare of land or less. Water supply is scarce, irregular, and sufficient only for about a quarter of the land (Reidinger 1994). The *huerta* has been in use for hundreds of years and receives water from a reservoir, two interbasin canals, and a local groundwater source. The water users association, which runs the irrigation system, owns only part of the water and buys the rest on a regular basis. It allocates water by holding auctions every week in which tickets to a certain amount and duration of water flow are traded. This system allows for scarce water in the *huerta* to be allocated to its most valued use.

In the northern Indian states of Punjab and Haryana, a limited water market, locally called *warabandi*, exists in which farmers trade water turns. This happens mainly because the operating system of the irrigation canals operates on a different schedule than the water scheduling needs of the farmers, resulting in long delays between irrigation (Reidinger 1974).

In California, where surface water rights are based on a first-come, first-served queuing system, users are allowed to divert water for "beneficial" use but are not allowed to sell the water or to trade it among themselves. These systems generally allow users to draw as much water as they need, often paying a diversion fee that is much lower than the true cost of water delivery. The California drought of 1986–92 reduced the state's water supplies by nearly half. The water districts responded to this situation through several measures, including creating a water bank that bought and sold water. The limited water trading induced by the creation of the water bank was highly successful and enabled the transfer of water from surplus to deficit plots. The reallocation of water to higher-valued crops and fallowing of lands that were being used to cultivate water-intensive crops (such as alfalfa) resulted in only a 20 percent reduction in gross revenue, using only half the amount of water as before. Although drought conditions no longer exist in California, there is a move to continue encouraging water trading in the state, partly because lawmakers are under pressure to allocate more water to environmental uses. Sophisticated trading mechanisms have been introduced, such as an electronic bulletin board for the exchange of information among water suppliers and demanders in the huge Westlands Water District in the Central Valley, an experiment that is now being emulated by 23 other districts (Chakravorty and Zilberman 1997).

These experiences in limited water trading produce several important clues to the conditions that need to be created for trading to occur:

- The presence of effective water users associations or water authorities. Such an organization can be effective in resolving disputes over water trades, including their third party impacts. These associations can also help organize trades and provide information on them.
- Removal of legal and institutional regulations that prohibit or limit water trading. For example, water rights could be made distinct from land rights and should be specified in volumetric terms.
- Creation of infrastructure (such as canals) and control systems for the transfer and measurement of water. However, the capital, operation, and maintenance costs of these systems must be lower than the incremental benefits from water trading.

Effect on Income Distribution

In the long run, properly functioning water markets may allocate scarce supplies to their most valued use. They may also improve the efficiency of water in the agricultural sector and thus release water for alternative uses, such as to meet urban needs and protect aquatic habitat and coastal areas. However, in many developing countries, the transition from the short run to the long run is of profound importance and is beset with serious problems related to the redistribution of rents accruing to beneficiaries. For example, as pointed out by Repetto (1996), Wade (1984), and others, most irrigation systems in developing countries are heavily subsidized by the taxpayer, with substantial rents from water accruing to large landowners and financially better-off farmers, who are usually located in the upper reaches of the distribution system. Any policy reforms that introduce water markets and improve conveyance structures that facilitate water trading will move water prices closer to opportunity cost. As Chakravorty, Hochman, and Zilberman (1995) show, these reforms would allow the improved project to serve more beneficiaries but would also significantly reduce the rents accruing to beneficiaries located closer to the upstream regions. Reform proposals must take this into account, possibly by subsidizing upstream farmers or making them shareholders of the water utility, so that they receive a share of the aggregate rents from the project.

Another problem is the possible capitalization of infrastructure improvements into land values and the displacement of small and marginal farmers by more resourceful farmers. For example, if upstream farmers are allowed to sell their water, downstream farmers will be subject to even smaller supplies. In that case, policies need to be designed to ensure that downstream farmers can obtain rights over use of the return flows or are compensated with some portion of the proceeds of the trade. The large inequities prevalent in most developing-country irrigation systems make it imperative that income distribution issues of irrigation reform be studied thoroughly before they are implemented.

Marginal-Cost Pricing

Pricing of irrigation services has probably been the most troublesome issue to deal with for both international agencies involved in irrigation as well as national and local governments. In most developing countries, the revenue collected is not even enough to cover O&M costs, let alone capital expenditures. Water pricing techniques vary considerably between countries and often between projects and regions within the same country. The pricing principles applied in most projects are based on financial criteria such as recovery of O&M costs rather than on any economic concept such as marginal-cost pricing. Not only do O&M costs vary across projects, but also there is considerable uncertainty regarding their magnitude, and irrigation authorities in developing countries have been known to inflate their estimates to secure funds for increased construction activity. In many countries, the inability of the irrigation agency to charge for water services reduces its claim on government resources. For example, in Egypt, employee salary scales for revenue-generating sectors are higher than for nonrevenue-generating sectors such as irrigation (Perry 1996).

Pricing schemes for irrigation water vary considerably from place to place. The most commonly implemented schemes are not based on the quantity of water withdrawn but on a flat rate that is not related to actual use, often leading farmers to use water until its marginal value is zero. In many countries, water rates are levied on a per hectare basis and depend on the type and extent of crop irrigated. Often land taxes are levied to tax the increased benefits created by improvements in irrigation. Betterment levies are also used to tax the higher land value capitalization from irrigation. In some countries, such as Peru, volumetric fees are charged based on minutes of flow, while in Israel, where water is distributed through closed pipes, block prices are in effect, which consist of a basic low price for an initial volume and higher prices for additional blocks (Sampath 1992).

Various surveys have shown that marginal-cost pricing principles are rarely applied in water projects in developing countries (see Small 1987). A reason often cited for departures from marginal-cost pricing is the existence of externalities from irrigation to consumers outside the project, for example, in the form of lower food prices and increased food security or through spillover effects in the regional economy. If these externalities were incorporated in the marginal-cost calculations, it might well be optimal to charge farmers below marginal cost. There are other problems with marginal-cost pricing. Farming activities vary from season to season and across locations within the same project and climate. To charge marginal-cost prices each time would require not only perfect information on the parameters of demand and cost but also the use of complex pricing formulas. This may mean high administrative and political costs that could well offset the gains from marginal-cost pricing (Sampath 1992). However, as the scarcity value of water rises with time, and the cost of computing time- and space-dependent prices decreases with the advent of low-cost microcomputers, it will become economically feasible to use complex pricing policies for irrigation water as is the practice for other spatially distributed commodities such as electricity and natural gas.

Implementation of marginal-cost pricing implies that farmers farther from the water source must be charged higher prices to account for conveyance costs than those located closer to the source. This also implies that farmers located farther away will withdraw less water than farmers in the head reaches of the distribution system. A pricing policy that discriminates between farmers at different locations may be difficult to implement. It has been suggested that if conveyance losses are reduced by lining canals and maintaining control structures, the cost differences in water delivery between locations will also be reduced, so that a spatially uniform pricing policy may be a reasonable approximation to marginal-cost pricing (see Chakravorty, Hochman, and Zilberman 1995).

As Sampath (1992) points out, many water economists preach the virtues of marginal-cost pricing without realizing that social and political objectives may support a deviation from pricing water at marginal cost. Pricing models developed by economists need to incorporate these multiple objectives. Solving these second-best models may yield water prices that are closer to those observed in practice. The difference between these second-best prices and the realized prices may provide a more accurate picture of the suboptimality of water prices. However, although recognizing diverse social and political objectives (such as equitable distribution of benefits among beneficiaries) is important and provides important clues to current pricing practices, it may be difficult to justify those objectives because the pricing policies they promote cause a range of problems that were not originally anticipated, such as adverse environmental impacts and rent-seeking. What may be needed is analysis showing that these noneconomic objectives within the irrigation sector may actually cause economic and environmental impacts that result in serious efficiency and equity effects in the economy as a whole.

The IIMI has examined irrigation cost recovery and water pricing policies in USAID (U.S. Agency for International Development) projects in Egypt and the impact of alternative pricing mechanisms on the agriculture sector (Perry 1996). The study evaluates three policies: (a) a flat rate, independent of type of crop or cropping intensity, (b) a crop-based charge linked to water consumption, and (c) a volumetric charge. The results show that although a flat rate charge has no impact on crop selection, full recovery of allocated costs to agriculture would reduce farm incomes by about 4.5 percent. More important, the study finds that a crude crop-based charge in which water charges are proportional to farm demand is a good approximation for volumetric pricing because it induces farmers to grow water-efficient crops. This finding suggests that volumetric pricing, which requires water measurement, may not be cost-effective under conditions of relatively plentiful supply.

Further modeling to simulate the impact of a 15 percent reduction in water supplies found that service charges equal to 25 percent of farm incomes would be required to reach the targeted reduction in demand. However, under uniform rationing of water, production falls only 4 percent. If only tail-end farmers are given reduced supplies of water, the losses increase to 7 percent. These figures suggest the importance of pricing policies that allow for optimal water conservation through crop and technology choice at the farm level.

No discussion of water pricing mechanisms is complete without referring to the case of Israel, which not only is located in one of the most arid and water-scarce regions of the world but also has one of the highest water use efficiencies in irrigated agriculture. Between 1951 and 1985, total output from irrigated land grew tenfold, with only a fivefold increase in irrigated acreage. However, aggregate water use increased only threefold, and water use per hectare actually dropped about 40 percent (Bhatia, Cestti, and Winpenny 1995). Crop production per cubic meter of water almost tripled during this period. The Water Commission of the Ministry of Agriculture has implemented a complete package of measures to ensure the efficient use of water. For example, each farmer's allocation of water is based on the area cultivated, crop mix, and water requirements for each crop. Wasteful practices come with penalties in the form of reduced acreage or fines for excessive consumption of water. Water charges are based on a progressive block rate structure with a basic rate for 80 percent of the allocation and higher rates for additional amounts. Water supplied during the peak irrigation season carries a 40 percent premium. All the water supplied is metered, and the government subsidizes the installation of more efficient irrigation technology and other control devices. As a result, more than 90 percent of the irrigated area is under sprinkler irrigation, and the rest is under drip technology (Tahal Consulting Engineers 1991).

Water Management in Rainfed Areas

In spite of phenomenal increases in acreage under irrigation, rainfed agriculture continues to be a major source of food production in many developing countries. An estimated 70 percent of the cropped land in Asia is rainfed. These production systems are complex and diverse and are constantly affected by the uncertainties of monsoon, drought, and flood cycles. Rainfed farming systems are also characterized by wide variations in soil quality, ranging from the dry and sterile soils of arid zones to the deep, humus-rich, water-soaked soils of tropical rain forests (FAO 1982).

Both the deficient and the excess supply of rain contribute to the instability of production. The high degree of uncertainty regarding availability leads to a less-than-optimal investment in agricultural inputs, especially in the new green revolution package based on high-yielding and fertilizer-responsive seed varieties that have benefited farmers who have access to irrigated lands. Thus farmers who crop rainfed, often marginal, lands generally tend to be from the economically poorer sections of society, and their inability to invest in modern agricultural technologies serves to widen the gap between rich and poor in the rural areas of developing countries.

In certain regions, such as South and Southeast Asia, rainfall is highly variable. About 80 percent of rain falls in a period of four to six months, and the rest of the year is mostly dry. Thus cropping periods have to be adjusted to make optimal use of the rainy season, but if the rain is delayed, or there is a prolonged dry spell, the cropping schedule is often severely disturbed. Empirical data suggest a close relation between rainfall and crop productivity (Saleh and Bhuiyan 1995).

Rainfed agriculture is affected by three major types of problems. These can be classified as soil-related (undulating and sloping lands, poor fertility), socioeconomic (fragmented holdings, poor economic condition of farmers), and climatic (water deficit). A number of relevant technologies can be targeted toward rainfed environments. According to Katyal and Venkateswarlu (1993), these include:

- *Land treatments that promote maximum preservation of rainwater where it falls, that is, in situ moisture preservation.* For example, off-season tillage increases the use of rainwater.
- *Collection of water during high-intensity rains and recycling it during periods of water scarcity, which could lead to stable production from drylands.* The runoff can be collected in ponds and used for life-saving irrigation of the rainy-season crop or irrigation before sowing or after the season.
- *Efficient use of harvested water.* Experiments have been conducted to determine the optimal timing of supplemental irrigation for the rainfed crop. For example, moisture stress during the flowering of rainy-season crops causes considerable damage. Supplemental irrigation at this stage may be highly beneficial.

These technological fixes must be undertaken in relation to other economic and institutional reforms and must be done for the catchment as a whole. Thus, land and water conservation measures need to be adopted along with crop diversification, use of relevant agroecological information, and access to credit and inputs by rainfed farmers. Here too, farmer participation and agency commitment are critical. For example, there may be asymmetry in the distribution of benefits because farmers owning land in the catchment area may need to be compensated by farmers in the command area who are the direct beneficiaries of water conservation. This, in turn, requires substantial cooperation and negotiation between upstream and downstream farmers (Pandey 1991).

Environmental Impacts of Irrigation and Drainage

There are severe environmental impacts of irrigation projects, including waterlogging and salinity, and their resulting adverse impacts on health include the incidence of vector-borne diseases such as schistosomiasis and malaria. The principal causes of waterlogging and salinity are irrigation without drainage, excessive irrigation, and large leakages from the canal system (such as a delivery efficiency of 35–40 percent from canal head to root zone). Many of these practices are intricately linked to water pricing and management policies. For example, pricing irrigation water too low leads to excessive irrigation as well as lower incentives to adopt modern irrigation technologies (Caswell and Zilberman 1986), resulting in low water efficiencies and relatively large water losses.

The problem is particularly severe in Pakistan where semiarid climatic conditions encourage the accumulation of salts in the root zone. In addition, irrigation water supplies add 1.2 metric tons of salt per hectare to the root zone every year. Unregulated pumping of groundwater further aggravates the situation by mobilizing salts dissolved in the groundwater aquifer. All this has contributed to more than a third of the gross command area being waterlogged and 14 percent of it being saline. The twin problems of waterlogging and salinity are especially severe in Sindh Province, where the water table is less than 8 feet deep in more than a third of the cropped area and more than half the area is moderately saline. High groundwater tables inhibit root growth and therefore reduce crop yields. Salinity retards plant growth, and its impact in reducing Pakistan's agricultural output is estimated to be on the order of nearly 25 percent.

Donor agencies and the Pakistan government agree that the lack of an effective drainage system is the principal threat to the sustainability of agriculture in the Indus basin. The steps to be taken require that the drainable surplus must be removed from the source but also that internal drainage must be sufficient and the Indus Basin Irrigation System must be provided a drainage outlet to the Arabian Sea. In the past, half-hearted attempts to program drainage failed to address some of the key institutional and economic factors affecting water use and drainage practices. Important components were neglected, such as O&M, public participation, drainage research, and the application of research results to the project. Recently, the Pakistan government agreed to initiate a comprehensive approach to river basin management that will include the reform of water markets and individual property rights, the restructuring of government agencies, and the formation of water users associations.

This integrated approach to drainage management is expected to contribute not only to better drainage but also to a wide variety of direct and indirect benefits. In particular, it is expected to benefit tenants, smallholders, and tail-enders by improving the efficiency of service delivery, improving the availability of irrigation water and drainage, and lowering costs. The improvements in drainage structures and O&M will lead to less salinity and waterlogging and higher output of crops. Preliminary estimates suggest that farmers can expect to increase their incomes as much as 30 percent as a result of the anticipated drainage improvements, although the exact magnitude will depend on a wide variety of factors, including macroeconomic factors, the initial level of salinity and waterlogging, and agroecological conditions. Thus farmers will benefit not only because the steady decline in yields will stop but also because output will increase gradually and the associated water management measures will increase the availability of water, especially for tail-enders.

In many regions, the diversion of water for irrigation has resulted in the reduction of available water supplies to sensitive ecosystems such as wetlands that support vital plant and animal populations. In the Aral Sea region, the diversion of water from lakes to support irrigated rice and cotton cropping in remote desert areas of Kazakhistan and Uzbekistan has resulted in a drop in water level in the lake from 53 to 39 meters above sea level and a tripling of salinity levels as a result of the intrusion of salt water from the Aral Sea. This has decimated the fish population in the lake and caused severe climatic and other changes in the region's microenvironment (Dinar 1997).

Success Stories and Lessons for Policy Reform

Several important lessons can be gleaned from the cumulative experience with irrigation over the past four decades. Some critical ones are as follows. It is universally acknowledged that the present system of water pricing (or the lack of it) cannot continue forever and that some form of water pricing reform must be implemented. It is also clear that the straightforward introduction of water markets will not work. The introduction of market mechanisms must be preceded by the creation of conditions that ensure the proper functioning of markets. For example, investments must be made in infrastructure such as conveyance facilities capable of transporting water between locations and in control structures for volumetric measurement of water flows. Legal impediments to water trading must be removed, although they do not seem to hamper trading in water when other conditions are favorable. Transferable water rights need to be created to induce water users to use water efficiently.

The World Bank recently undertook a review of 64 active irrigation and drainage projects in 28 countries (World Bank 1996). The review highlights the main issues and problems with the irrigation portfolio and suggests how to improve its quality. The Bank report identifies several key problems in its irrigation portfolio:

- Excessive optimism regarding the commitment of the host government during project preparation and appraisal
- Miscalculation at the appraisal stage of what may be achieved in terms of hydraulic efficiencies and cropping intensities, which reduces the projects' cash returns and lowers their economic outcome as well as their economic rate of return
- Performance of economic analysis of projects based on a single probability of water availability through rainfall or other sources.

The report concludes that improving irrigation lending depends on several factors, some of which are external to the projects being implemented. For example, the slump in commodity prices in the 1980s led to a drop in rice prices and to lower returns for the entire portfolio of rice-based irrigation projects. Several measures are now being considered for improving the rate of project success in meeting targeted development goals:

- Elimination of the strict targets on professional input, overall cost, and speed in project preparation, which are causing a decline in the quality of projects at the entry level. The two years typically given for a project to receive funding from the Bank are too short to complete a series of tasks including land acquisition, field investigations, development of stakeholder participation, and detailed engineering design of irrigation projects. This has led to poor project quality and a general preference for projects that require less time and budget to prepare.
- Better understanding of the technical, economic, social, and organizational aspects of irrigation projects. Among other things, for example, the Bank has called for improved understanding of the interaction between project design and water management, so that rehabilitation and maintenance improvements lead directly to an improvement in service. It also has called for designs that not only provide improved service but also are simpler to operate.

A few irrigation projects have shown exemplary

performance mainly because of strong government commitment, user participation in the management of structures, and, in many cases, institutional reform and commitment of the irrigation authority to project implementation. For example, modernization of the Office du Niger, one of the oldest and largest schemes in Sub-Saharan Africa, resulted in the tripling of average paddy yields to about 5 tons per hectare, which compares well with yields obtained under the green revolution in Asia. Similarly, the Mexican water authority the, Comisión Nacional de Aguas, completed two successful projects that involved decentralization of its O&M activities in irrigation districts through the promotion of water users organizations. This resulted in an increase in the cost recovery component of the O&M budget from a low 18 percent to about 80 percent nationwide. In Colombia, legal reforms gave water users associations a key role in the management of irrigation districts and required full recovery of O&M costs and partial recovery of investment costs. As a result of changes created by the new law, O&M recovery targets were met in three of the four districts financed by the irrigation project.

Concluding Remarks

Several key lessons emerge from this discussion. From all accounts, the task of improving irrigation management has become more complex. This is partly because as populations have increased, so have demands for urban water use and for better environmental quality. The low marginal product of water in agriculture and the relatively high value of water in urban and environmental uses make it likely that more and more water will be siphoned away from agriculture in the future. This again implies that less water will be available to meet the agricultural needs of an expanding population. This situation is beginning to induce a structural change in the nature of irrigated agriculture. In spite of the presence of rent-seeking interests, restrictive property rights regimes, and a general belief that water, like air, should not be bought and sold, fundamental changes are already beginning to happen.

Many governments, policymakers, and environmental groups, such as in California, are beginning to realize that a more efficient irrigation system means more water for everyone. This has induced changes in the way water rights have been allocated, and, in many cases, restrictions on trading in water have been removed. In developing countries, governments are beginning to be motivated and willing to undertake major institutional reforms as a precondition to initiating new multilaterally funded irrigation projects.

However, major impediments remain. In many countries, the physical and institutional structures must be improved before sound economic and management policies can be implemented. User participation in operation and maintenance is now an essential component of any project, but the factors essential to forming effective water users associations are still being understood. The cultural and historical setting also plays a key role in determining beneficiary participation.

Another important insight obtained from irrigation projects is the importance of adopting an integrated approach to project rehabilitation. For example, if a project aims to improve drainage in a particular irrigation project, it needs to consider existing water institutions, prevailing prices of water, the issue of surface and groundwater allocation, and the adoption of technology by farmers. Simply creating better drainage structures may not work in the long run. The same is true in rainfed agriculture. A watershed-based approach that considers the impacts of management policies on both upstream and downstream beneficiaries has a better chance of working than policies that do not consider linkages across sectors or beneficiary groups.

This new "holistic" thinking is reflected in the current emphasis on "basinwide" management promoted by agencies such as the International Irrigation Management Institute (Seckler 1996). They argue that water efficiency must be measured not within a project but over an entire basin so as to include the reuse of drainage water from seepage and percolation. This suggests that policies such as the adoption of sprinkler irrigation within an irrigation project may improve the efficiency of water use within the project but also reduce the availability of water elsewhere in the basin. Similarly, salt and chemical pollution of water through application of inputs may adversely affect agriculture downstream of the project. These issues can only be handled if the basin is adopted as the relevant unit of analysis and the relevant externalities are internalized. These issues become more complicated if the basin spills across political boundaries, as in the case of the Imperial Irrigation District (IID) in California. A study by the California Department of Water Resources calculated that the IID could save nearly half a million acre-feet of water through canal lining and other measures, which could be used to augment municipal water supplies. However, these "savings" are also claimed by Mexico, which argues that its aquifers receive recharge through seepage from IID canals (Cummings and Nercissiantz 1992).

References

The word "processed" describes informally reproduced works that may not be commonly available through libraries.

Bhatia, Ramesh, Rita Cestti, and James Winpenny. 1995. *Water Conservation and Reallocation: Best-Practice Cases in Proving Economic Efficiency and Environmental Quality*. A World Bank–Overseas Development Institute Joint Study. Washington, D.C.: World Bank.

Biswas, Asit K. 1991. "Inaugural Address." Presented before the seventh world congress on water resources, International Water Resources Association, Rabat, Morocco, May. Processed.

Caswell, Margriet, and David Zilberman. 1986. "The Effects of Well Depth and Land Quality on the Choice of Irrigation Technology." *American Journal of Agricultural Economics* 68:798–811.

Chakravorty, Ujjayant, Eithan Hochman, and David Zilberman. 1995. "A Spatial Model of Optimal Water Conveyance." *Journal of Environmental Economics and Management* 29:25–41.

Chakravorty, Ujjayant, and James Roumasset. 1994. "Incorporating Economic Analysis in Irrigation and Management." *Journal of Water Resources Planning and Management* 120(6):819–35.

Chakravorty, Ujjayant, and David Zilberman. 1997. "California's Water Experience: Lessons of the Drought." In *Down to Earth*. New Delhi: Center for Science and Environment.

Cummings, Ronald, and Vahram Nercissiantz. 1992. "The Use of Water Pricing as a Means for Enhancing Water Use Efficiency in Irrigation: Case Studies in Mexico and the United States." *Natural Resources Journal* 32:731–55.

Cunningham, John. 1992. "Public and Private Groundwater Development in India: The Case of Uttar Pradesh." In Guy Le Moigne, Shawki Barghouti, and Lisa Garbus, eds., *Developing and Improving Irrigation and Drainage Systems: Selected Papers from World Bank Seminars*. Technical Paper 178. Washington, D.C.: World Bank.

Dinar, Ariel. 1997. "Irrigated Agriculture and the Environment: Problems and Issues in Water Policy." Paper presented at the workshop on sustainable management of water in agriculture: issues and policies, Organisation for Economic Co-operation and Development, Athens, November. Rural Development Department, World Bank, Washington, D.C. Processed.

FAO (Food and Agriculture Organization of the United Nations). 1982. *Rainfed Farming Systems in Asia and the Pacific: Cropping Patterns in Agro-Ecological Zones of Southeast Asia*. Bangkok.

IIMI (International Irrigation Management Institute). 1992. "Improving the Performance of Irrigated Agriculture: IIMI's Strategy for the 1990s." Colombo. Processed.

Jones, W. I. 1994. "A Review of World Bank Irrigation Experience." In Guy Le Moigne, K. William Easter, Walter J. Ochs, and Sandra Giltner, eds., *Water Policy and Water Markets: Selected Papers and Proceedings from the World Bank's Ninth Annual Irrigation and Drainage Seminar, Annapolis, Maryland, December 8–10, 1992*. Technical Paper 249. Washington, D.C.: World Bank.

Katyal, J. C., and B. Venkateswarlu. 1993. "Management of Water Resources in Rainfed Areas of India: The Critical Issues." *ICID Bulletin* 42(2):1–9.

Lenton, Roberto. 1994. "Research and Development for Sustainable Irrigation Management." *Water Resources Development* 10(4):417–24.

Pandey, Sushil. 1991. "The Economics of Water Harvesting and Supplementary Irrigation in the Semi-Arid Tropics of India." *Agricultural Systems* 36:207–20.

Perry, Chris. 1996. *Alternative Approaches to Cost Sharing for Water Service to Agriculture in Egypt*. Research Report 2. Colombo: International Irrigation Management Institute.

Radosevich, G. 1986. "Legal and Institutional Aspects of Irrigation Water Management." In K. C. Nobe and R. K. Sampath, eds., *Irrigation Management in Developing Countries: Current Issues and Approaches*. Boulder, Colo.: Westview Press.

Reidinger, Richard. 1974. "Instituting Rationing of Canal Water in Northern India: Conflict between Traditional Patterns and Modern Needs." *Economic Development and Cultural Change* 23(1):79–104.

———. 1994. "Observations on Water Markets for Irrigation Systems." In Guy Le Moigne, K.

William Easter, Walter J. Ochs, and Sandra Giltner, eds., *Water Policy and Water Markets: Selected Papers and Proceedings from the World Bank's Ninth Annual Irrigation and Drainage Seminar, Annapolis, Maryland, December 8–10, 1992*. Technical Paper 249. Washington, D.C.: World Bank.

Repetto, Robert. 1986. *Skimming the Water: Rent-Seeking and the Performance of Public Irrigation Systems*. Research Report 4. Washington, D.C.: World Resources Institute.

Saleh, A. F. M., and S. I. Bhuiyan. 1995. "Crop and Rain Water Management Strategies for Increasing Productivity of Rainfed Lowland Rice Systems." *Agricultural Systems* 49:259–76.

Sampath, Rajan. 1992. "Issues in Irrigation Pricing in Developing Countries." *World Development* 20(7):967–77.

Seckler, David. 1996. *The New Era of Water Resources Management: From "Dry" to "Wet" Water Savings*. Research Report 1. Colombo: International Irrigation Management Institute.

Small, Leslie. 1987. *Irrigation Service Fees in Asia.* ODI/IIMI Irrigation Management Network Paper 87/1C. London: Overseas Development Institute.

Tahal Consulting Engineers, Ltd. 1991. "Israel Water Sector Review." Paper presented at the international workshop on comprehensive water resources management policy, World Bank, Washington, D.C., June. Processed.

United Nations Population Fund. 1993. *State of the World's Population*. New York.

Wade, Robert. 1984. "Irrigation Reform in Conditions of Populist Anarchy: An Indian Case." *Journal of Development Economics* 14:285–303.

World Bank. 1996. "Portfolio Improvement Program: Irrigation Subsector." Agriculture and Natural Resources Department, Washington, D.C. Processed.

Young, Robert, and Robert Haveman. 1985. "Economics of Water Resources: A Survey." In Allen Kneese and James Sweeny, eds., *Handbook of Natural Resource and Energy Economics*. Vol. 2, pp. 465–529. Amsterdam: North-Holland.

21

Livestock and the Environment: Issues and Options

Henning Steinfeld, Cornelis de Haan, and Harvey Blackburn

Livestock interact with the environment in many ways. In the first section, this chapter provides the overall context and discusses the main issues. In the second section, it suggests policy and project interventions that can address those issues, including the tradeoffs between productivity and the environment.

The Context and Issues

Livestock can damage global natural resources in a number of ways, but there are also many examples of environmental balance and positive contributions. Livestock interact on a global scale with land, water, air, and plant and animal biodiversity. About 34 million square kilometers, or 26 percent of the world's land area, are used for grazing livestock. In addition, 3 million kilometers, or about 21 percent of the world's arable area, are used for cereal production for livestock feed. Livestock produce 13 billion tons of waste a year. A large part of this is recycled, but where animal concentrations are high, waste poses an enormous environmental hazard. Livestock and livestock waste cause gaseous emissions that have important local and global impact on the environment. Livestock grazing can affect the water balance in certain areas. Water is needed to produce fodder and feed concentrate, to provide drinking water for animals, and to drain surplus waste and chemicals. Livestock interact directly and indirectly with biodiversity; although biodiversity often is compromised, there are also examples of mutual benefits.

The Dynamics of Livestock Production Systems

For a large part, livestock interact with the environment within the confines of a production system. These production systems are evolutionary responses to population pressure, resource endowment, and marketing opportunities. Sére and Steinfeld (1996) distinguish three main production systems: grazing, mixed-farming, and industrial systems.

Grazing systems are mainly based on native grassland, with no or only limited integration with crops. These systems often do not involve imported inputs and generally have a low caloric output per hectare (Jahnke 1982). In arid grazing systems, the potential for increasing productivity is extremely limited, and livestock resources are best managed by allowing for variability—or "at disequilibrium" (Behnke, Scoones, and Kerven 1993)—and by adapting animal pressure continuously to the highly variable rainfall and availability of feed. The number of stock is allowed to fluctuate in response to unfavorable "bust" or drought conditions and favorable "boom" conditions. Where infrastructure and water development or crop encroachment interfere with resources and patterns of movement, ecologically sound systems are often disrupted, overgrazing and land degradation occurs, and the livelihood of pastoral communities is invariably threatened. In semiarid zones, there is some scope for intensification and transition into mixed-farming systems, because these areas have some potential for cropping. Parts of southern Asia have developed along this path, and the sudano belt in Africa is following the same course. Some of the humid and subhumid zones have been precluded from livestock development because of problems of disease (African sleeping sickness, which is transmitted by tsetse flies in Africa) or accessibility. With increasing influx of population and better methods of controlling disease, the threat of African sleeping sickness will become manageable, and these as well as other grazing systems (subhumid savannas of Brazil, for example) have excellent potential and may develop into mixed-farming systems.

Historically, the integration of livestock and crop activities into *mixed-farming systems* represents the main avenue for intensification. As by-products (crop residues, manure) of one enterprise serve as inputs into the other, this system is, in principle, environmentally friendly. In many places of the world, closed-cycle smallholder farms have developed and continue to be a predominant feature.

When land scarcity emerges as a pressure point for farming systems, farmers respond by specializing in crop production, by following a path of intensifying the crop-livestock interaction (Christiaensen, Tollens, and Ezedinma 1995), or by abandoning land-based production and moving into *industrial livestock production* where market demand is strong.

Industrial production systems are detached from immediate land in terms of feed supply and waste disposal. Where the demand for animal products increases rapidly, land-based systems fail to respond, leading to animal concentrations that are out of balance with the capacity of the land to absorb waste and supply feed. High concentrations of animals close to human agglomerations bear enormous pollution problems and associated human health risks. Because of water pollution and health hazards, most of these systems are moved out of city boundaries as soon as infrastructure permits. The decision of Singapore to abandon livestock production altogether is a case in point.

The opportunities that arise from strong market demand for meat and milk conflict with the limited potential to expand the conventional resource base on which livestock traditionally relied: that is, grazing land and feed resources of low or no alternative use. Increasingly the world livestock sector resorts to external inputs, notably high-quality feed but also more productive breeds, better animal health, and general husbandry inputs. Grazing systems offer only limited potential for intensification, and livestock production is becoming increasingly crop-based. Thus the importance of roughage as feed is decreasing at the expense of cereals and agroindustrial by-products. There is an important species shift toward monogastric animals, mainly poultry and pigs. Ruminant meat accounted for 54 percent of total meat production in the developing countries in 1970, decreased to 38 percent in 1990, and is projected to decrease further to 29 percent in 2010 (Alexandratos 1995). This species shift reflects the use of, and better conversion rates for, concentrate feeds by monogastric animals.

Livestock production is becoming separated from its land base, is urbanized, and is beginning to assume features of industrial production. In recent years, industrial livestock production globally grew at twice the rate (4.3 percent) of production in mixed-farming systems (2.2 percent) and more than six times the rate in grazing systems (0.7 percent; Sére and Steinfeld 1996). This trend continues to accelerate. In agroecological terms, livestock production is growing more rapidly in humid and subhumid zones than in arid tropical zones or highlands, essentially following patterns of human population growth.

Grazing and Overgrazing

About 34 million kilometers, or 26 percent of the world's land area, are used for grazing livestock. Grazing animals can improve soil cover by dispersing seeds with their hoofs and through their manure, while controlling shrub growth, breaking up soil crusts, and removing biomass that otherwise might provide fuel for uncontrolled bush fires. These impacts stimulate grass tillering, improve seed germination, and thus improve land and vegetation. However, heavy grazing causes soil compaction and erosion and decreases soil fertility, organic matter content, and water infiltration and storage. Overgrazing in hilly environments can accelerate erosion.

The arid rangelands have generally been related to widespread degradation. Dregne, Kassas, and Rozanov (1992) and World Resources Institute (1992) argue that the majority of the world's rangelands are moderately or severely desertified. The United Nations Environment Programme (UNEP; see Oldeman, Hakkeling, and Sombroek 1990) estimates that, since 1945, about 680 million hectares, or 20 percent of the world's grazing lands, show significant soil degradation.

However, if irreversibility and declining productivity are taken as the main characteristics of degradation (Nelson 1990), then the actual situation of the world's arid grasslands and rangelands is better than generally reported. For example, in the Sahel, satellite imageries (Tucker, Dregne, and Newcomb 1991) show vegetation at the same northern limits as before the big droughts of the 1970s and 1980s. They point to a fluctuating pattern, or a "contracting and expanding Sahara," rather than a continuously expanding desert. This resilience is confirmed by our own analysis showing that animal production per head and per hectare has increased, rather than declined, over the past three decades (see box 21.1). This has occurred in spite of a large increase in the number of livestock and a decrease in the area of rangelands.

In the light of this evidence, arid rangelands with less than a 90-day growing season are now seen as containing dynamic and highly resilient ecosystems, especially under traditional management of continuous adjustments to highly variable rainfall, both in time and in space. Arid vegetation is extremely resilient, and most of the changes observed are the result of severe dry spells and are therefore likely to be temporary. The continuous "disequilibrium" protects land, especially in arid areas with annual vegetation, because the low grazing pressure after a drought facilitates recuperation, and, although with a time lag, overall grazing pressure is adjusted to the amount of feed available (the theoretical bases for range management under those conditions have recently been well described by Behnke, Scoones, and Kerven 1993 and Scoones 1994). Flexibility and mobility are therefore key requirements to achieve sustainable use of rangeland in these areas.

Where this mobility is impaired and customary practice impeded, degradation often occurs. The nationalization of arid rangelands that was introduced by many governments in the postcolonial period in Africa and Asia undermined the intricate fabric of customary practice by replacing an ecologically well-balanced system of communal land use with a free-for-all open-access system (see box 21.2). Policies to "settle" the pastoralists, to promote ranches, and to regulate stocking rates, which were major principles of arid rangeland development of the 1960s and 1970s, reduced the critical mobility and flexibility in that system. In the Middle East and Central Asia, state farms are being privatized and cut up, thereby impeding pastoralists' mobility.

Box 21.1. Productivity Trends in the Sahel

An analysis of livestock production in five Sahelian countries over a 30-year period, carried out as part of this study, shows a 93 percent increase in the meat produced per hectare and a 47 percent increase in the meat produced per head. At the same time, there was a 22 percent increase in the animal population (from 14.5 to 17.6 million total livestock units).

This increase in productivity occurs in both cattle and small ruminants. Part of the increased productivity might result from the progressive move of the livestock population to more humid areas and the increased use of crop residues. The long-term trend points to sustained increases in productivity and a stable resource base.

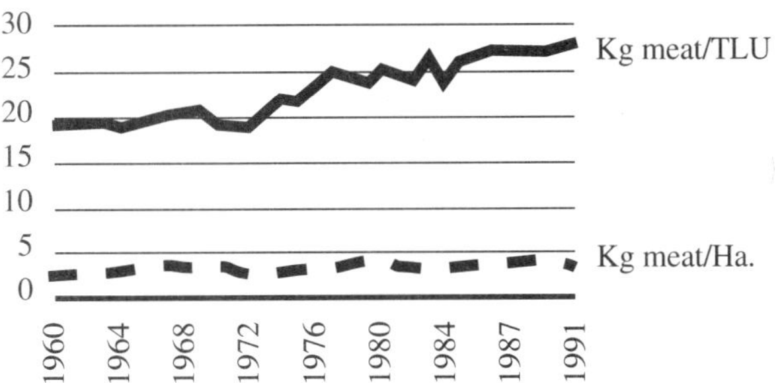

Installation of water points. The development of water points for human and livestock use often opens up arid lands and contributes to land degradation (see box 21.3). Around the water points, animal trampling causes bare surfaces, and degraded land typically accounts for 5 to 10 percent of total area. Normally, soil fertility is quite high, but the impact of hoofs, including soil compaction, and heavy grazing impede normal patterns of vegetation. Long-term studies in Senegal and Sudan (Thomas and Middelton 1994) do not find a significant expansion of the sacrifice areas of individual water points. More important, water development in arid areas might upset entire ecosystems by changing the relationship between traditional wet- and dry-season grazing areas and converting traditional dry-season grazing into year-around grazing.

In the semiarid zones, those with more than a 90-day growing season, land degradation caused by grazing livestock is much more serious than at the fringes of the desert. Data from a transect in Mali show that land degradation in areas with 600–800 millimeters of rainfall was significantly greater than in areas with 350–450 millimeters. In areas with higher rainfall, the percentage of barren soil increased from 0 to 10 per-

Box 21.2. The Deterioration of Common Property Resources in India

In a study of 75 villages in seven rather arid states of India, Jodha (1992) finds that the common property resource area declined by as much as 30 to 50 percent between the 1950s and 1982 and that cattle declined 20 percent in favor of small ruminants. The traditional communal management of the common property resource had broken down:

- In 1950, 70 villages had formal or informal rules for managing the common property resource; in 1982 only eight villages maintained those rules.
- In 1950, 55 villages levied formal or informal taxes to maintain the common property resource; none did so in 1982.
- In 1950, 65 villages had formal or informal obligations to maintain their common property resource; in 1982 only 13 had them.

As traditional common property resource management shifted to open access, common property resources degraded significantly, as shown by the 75 percent decline in the number of trees in the common property resource and the decline in the number of grazing days (Jodha 1992).

Box 21.3. Positive Effects of Intensification: The Machakos Case

Human pressure and intensification can also work positively. English, Tiffen, and Mortimore (1992) show that in the semiarid Machakos district in Kenya the natural resource base improved despite a 500 percent growth in population during the past 60 years, an increase in cropland from 35 percent of the higher-potential area in 1948 to 81 percent in 1992, and an increase in the livestock population per farm from 5.4 animal units in 1940 to 16.2 in 1980. Dynamic market development made farming profitable, and off-farm employment generated enough capital for investments in soil and water conservation. Horticulture and smallholder dairy production are the main activities generating the cash for resource conservation. The famine predicted in the 1930s for the Machakos district never occurred.

cent over the period 1950–90, whereas in arid areas, there was no significant change (Mainguet and others 1992). A similar situation is reported in parts of North Africa and the Middle East (Sidahmed 1996), although quantitative data are not available.

Land. This degradation is linked to growing population. Crop encroachment, fuelwood collection, and overgrazing are the interlocking factors causing land degradation in semiarid zones. Crop encroachment not only exposes the soil directly to the erosive effects of winds and downpours but also progressively hampers the movement of animals because it obliterates the passage between wet- and dry-season grazing areas. Fuel and fertilizer subsidies, especially in the Middle East and North Africa, often exacerbate the conversion into marginal cropland of sites with higher potential within arid rangelands. Drought emergency programs, which hand out subsidized concentrate feed, probably also contribute to range degradation, allowing too many animals to be maintained on the range and preventing an ecologically normal regeneration of range vegetation after the drought. These feed subsidies have now become a structural phenomenon, especially in North Africa and the Middle East, leading to continuous growth of the livestock population, which is no longer constrained by the poor productivity of pasture in low-rainfall regions.

Water. Data on the effect of grazing on water infiltration are site-specific. However, light-to-moderate grazing can improve or maintain infiltration, whereas heavy grazing almost always reduces it. Long-term observations on the Edwards plateau in Texas have shown that both heavy grazing and no grazing cause lower infiltration and higher erosion rates than moderate grazing (Grazingland Management Systems 1996). Water pollution by agrochemicals can be caused by the control of livestock disease vectors, such as ticks (the carrier of many diseases, such as anaplasmosis and east coast fever) and tsetse flies (the carrier of African sleeping sickness in man and animals), and of weeds. Ticks have been controlled traditionally by cattle dips or sprays using organochlorines. Several donor-supported projects promoted this technology in the 1970s and early 1980s. Fauna has been damaged both by inappropriate dosing, leading to increased resistance of ticks, as well as by inappropriate drainage, causing the liquid in dips to be discharged frequently in open water after use.

Plant and animal biodiversity. Arid vegetation is extremely resilient. Most of the changes observed are a result of dry spells and therefore are most likely temporary. The decrease in woody and perennial species observed in recent decades in the Sahel (Boudet and others 1987) is most likely the result of droughts. Resting an area brings the original flora back again (Hiernaux 1996), which indicates that the loss of these species is not irreversible. In effect, an extensive review of grazing and production data of 236 sites worldwide, including many sites in the arid zone, shows that there is no difference in biomass production, species composition, and root development in response to long-term grazing in the field (Milchunas and Lauenrot 1993). Grazing by animals can even improve plant biodiversity.

Complementarity can be observed between wildlife and livestock. There is increasing evidence that, in low-rainfall areas, the combination of wildlife and livestock can result in greater biodiversity and higher income for pastoralists and ranchers. Usually, livestock-wildlife combinations do not require significant reductions in livestock stocking rates. For example, a reduction of only 20 percent of the cattle stocking rate would create the "niche" for most wildlife species to prosper (D. Western, 1996, personal communication). Byrne, Staubo, and Grootenhuis (1996) estimate a larger dietary overlap, and thus the need for a greater reduction, but agree that there is substantial complementarity. There is also fairly general agreement that wildlife ranching on the basis of meat alone is not financially viable. Game meat is now attractive because it occupies a niche market. Combining the two activities, with adequate sharing of the benefits of common goods, is therefore considered the best strategy.

In humid areas, clearing forest and savanna to establish pastures causes soil nutrients to leach out rapidly. Weeds displace grasses, and artificial pastures can only be sustained for a period of up to 10 years. More than 50 percent of the pasture areas in the Amazon area have now been abandoned in a degraded state. Natural regeneration of forests is quite difficult, especially when areas are large. Under good management and using modern technology, however, the establishment of pasture and the introduction of cattle can be the second-best option for maintaining soil fertility. For example, the Centro de Investigación en Agricultura Tropical (CIAT 1994) developed a stable land management system, using grass and nitrogen fixing, that also proved attractive to smallholders.

Livestock and Deforestation

Tropical rain forests cover about 720 million hectares and contain approximately 50 percent of the world's biodiversity. Since 1950, more than 200 million hectares of tropical rain forests have been lost, as a result of ranching, crop cultivation, and forest exploitation. Ranching-induced deforestation is one of the main causes of loss of some unique plant and animal species in the tropical rain forests of South and Central America, the world's richest source of biodiversity. Since 1950, the area in Central America under pasture has increased from 3.5 million to 9.5 million hectares, and cattle populations have more than doubled from 4.2 million to 9.6 million head (Kaimowitz 1995). In Asia and Africa, deforestation is the result mainly of crop expansion.

Land speculation, titling procedures, and government-provided financial incentives have been the main reasons for ranch-induced deforestation (Kaimowitz 1995). Land speculation is heavily influenced by road construction because land prices are closely correlated with the distance to an all-weather road. In many countries, land titling procedures still require that land be under agricultural use before a title can be given. Such procedures, of course, encourage deforestation. Incentive policies, in the late 1960s and 1970s, which provided subsidized interest rates with lenient reimbursement conditions, and beef export subsidies played an important role in the expansion of ranching. They have been phased out, and this has reduced investment in large ranches by absentee owners. This is reflected in a decline in deforested area. In Central America in the 1980s, rain forest disappeared at an annual rate of 430,000 hectares a year. This decline slowed to 320,000 hectares annually over the period 1990–94, although this should be seen in context: by 1990, only about 19 million hectares of rain forest remained.

Crop-Livestock Interactions: Intensification and Involution

The integration of crop and livestock still represents the main avenue for intensification of food production. Mixed farming provides farmers with an opportunity to diversify risk from single-crop or livestock production, to use labor more efficiently, to have a source of cash, and to add value to low-value or surplus feed. To varying extents, mixed-farming systems allow the waste products of one enterprise (crop byproducts, manure) to be used as inputs to the other enterprise (as feed or fertilizer). Mixed farming is, in principle, beneficial for land quality because it helps to maintain soil fertility. In addition, the use of rotations between various crops and forage legumes replenishes soil nutrients and reduces soil erosion (Thomas and Barton 1995).

Adding manure increases the soil's nutrient retention capacity (or cation exchange capacity), improves the physical condition by increasing the water-holding capacity, and improves the structure. This is a crucial contribution because, in many systems, it is the only avenue available to farmers for improving the soil's organic matter. It is also substantial in economic terms. Approximately 20 million tons, or 22 percent, of total nitrogen fertilization out of 94 million tons in world agriculture (FAO 1997) and 11 million tons, or 38 percent, of phosphate are of animal origin, representing about $1.5 billion worth of commercial fertilizer. Not only does animal manure replenish soil fertility, but it also helps to maintain or create a better climate for soil micro-flora and fauna.

However, mixing crops and livestock neither generates new nutrients (with the exception of nitrogen fixation by leguminous plants) nor reduces nutrient surpluses. But livestock, even with low levels of technology, allow for (a) the spatial and temporal allocation of nutrients from areas with lower returns from cropping to those with higher returns, (b) the acceleration of the turnover of nutrients in the production cycle, and (c) the reduction of nutrient losses within the cycle compared to agricultural production without livestock.

The key issue is the nutrient balance. Most mixed-farming systems of the developing world have a negative nutrient balance. Deficits are partially covered by a flow of nutrients from grazing areas to cropland. As population pressure changes the ratio of cropland to grazing land, and when other sources are not available, fertility gaps tend to widen. Reported deficits range from about 15 kilograms of nitrogen per hectare a year in Mali to more than 100 kilograms of nitrogen per hectare a year in the highlands of Ethiopia (de Wit, Westra, and Nell 1996). The result is that crop yields continue to decline. Resource degradation, poverty, and population pressure carry a high risk of conflict, as the recent events in Rwanda have proven. However, positive trends—intensifying and diversifying production—also occur. A key factor facilitating such positive trends is access to markets.

Animal draft is still a main component of many smallholder farms in developing countries, substituting for human labor and drudgery. An estimated 250 million working animals provide the draft power to approximately 28 percent of the world's arable land, equivalent to 52 percent of total cropping land in developing countries. In southeastern and eastern Asia, animal draft is decreasing in importance as rapid mechanization takes place. In contrast, many areas in Africa are developing into mixed farming, with draft animals playing a key role in providing tillage for expanded crop areas and higher production. As a renewable form of energy, animal draft substitutes for fossil energy and the use of other natural resources.

Past policies have sometimes limited the synergistic effect of crops and livestock in nutrient-deficient situations. Imposing high import duties to protect domestic cereal production pushed cropping into marginal areas and upset the equilibrium between crops and livestock. Subsidization of inorganic fertilizer reduced the incentives for nutrient cycling using livestock.

In many places of the world, subsistence farms with crop, livestock, and household closely interlinked have developed and continue to be a predominant feature. With human population pressure increasing further, the need for intensification brings livestock more and more into cropping areas and integrates them in nutrient and energy cycles (see box 21.4). Almost throughout the world, the family-based and diversified mixed-farming system has come under pressure either from novel technologies and market forces or from resource degradation and poverty. Two major features emerge. One scenario leads to specialization, where market forces and technological requirements force mixed-farming systems to grow in unit size and to specialize. With specialization, there are fewer opportunities for on-farm integration of crops and livestock. Another significant trend is what has been described as "involution" or a collapse of the mixed-farming system. In virtually all tropical highland areas the relatively high densities of human population are traditionally sustained by rather complex, mixed-farming systems. Population pressures may decrease farm sizes to a point where associated land pressures are no longer compensated by commensurate increases in land productivity, resulting in disintegration of the system. Livestock, often large ruminants, can no longer be maintained on the farm. This results in a greater deficit of nutrients and energy and leads to natural resource degradation and loss of investment. There is mounting evidence that human population pressures, poverty, and resource degradation, aggravated by lack of access to markets and employment opportunities, are cause-and-effect fac-

Box 21.4. The Feed-Food Controversy

One-third of the world's grain harvest ends up in the digestive tracts of livestock, where the conversion of vegetable protein into animal protein is incurring losses of 60 to 90 percent of edible protein. This fact concerns large parts of the public.

Among the 996 million tons of concentrates used in 1994 (FAO 1996), all cereals, roots and tubers, pulses, and some feedstuffs of animal origin (milk powder) can be classified as edible. Edible feedstuffs provide livestock with 74 million tons of protein.

In the same year and by contrast, livestock produced 199 million tons of meat, 532 tons of milk, and 45 million tons of eggs, altogether yielding 53 million tons of protein. Leaving aside the differences in nutritional value, the world's livestock sector consumes more edible protein than it produces. The input-output ratio is 1.41 and continues to grow because livestock production increasingly relies on grains, and this reliance is only partially offset by gains in the efficiency of feed conversion. However, such input-output considerations overlook two important aspects: first, ruminants have the capacity to produce protein without being fed protein because of rumen flora activity. This allows for large ruminant populations to produce without being fed high-quality feed. Second, the quality of different proteins needs to be considered. Proteins of animal origin have a much higher digestibility and nutritive value than most vegetable proteins.

However, the question of feed-food competition is somewhat ill-posed. Rather than looking at what potential food goes to livestock, one should look at the resource requirements. For example, a highly intensive dairy production system may rely on alfalfa fodder produced under irrigation and with heavy use of external inputs including fossil fuel, fertilizer, and pesticides. Alfalfa is not edible, but the land and other resources could be used for food production. Apparently the food versus feed debate has to do with the value that food has beyond its commodity price—a less tangible religious and cultural dimension.

A recent FAO study (1996) shows that the increasing use of feed grains has not had an adverse effect on the provision of cereals for human consumption. In times of food shortages, such as in 1974–75, adjustments are made in the use of feed, and food consumption of cereals remains largely unaffected.

Box 21.5. Reducing Methane Emissions from Digestive Fermentation through Strategic Supplementation in South Asian Countries

Urea and other supplemental nutrients are mixed with molasses to make them palatable to livestock. In addition, molasses provide the energy needed to realize the improved microbial growth that can result from enhanced ammonia levels. These multinutrient blocks have been used in many countries including Bangladesh, India, Indonesia, and Pakistan (Habib and others 1991, Hendratno, Nolan, and Leng 1991, Leng 1991, and Saadullah 1991). Typical results have been increases of 20–30 percent in milk yield, increases of 80–200 percent in growth rate, and increases in reproductive efficiency. In addition, methane emissions per unit of product have declined as much as 40 percent. Bowman, Croucher, and Picard (1992) estimate that strategic supplementation of dairy animals would reduce methane emissions 25 percent, while increasing milk production 35 percent.

Efficient digestion in the rumen requires a diet that contains essential nutrients for the fermentative micro-organisms. Lack of these nutrients lowers the animal's productivity and raises methane emissions per unit of product. For animals on low-quality feed, the primary limitation to efficient digestion is the low concentration of ammonia in the rumen. Supplying ammonia can therefore greatly enhance digestive efficiency and the use of available feed energy. Ammonia can be supplied by urea, chicken manure, or soluble protein that degrades in the rumen. Urea is broken down in the rumen to form ammonia, and adding urea to the diet is the most effective method of boosting rumen ammonia levels.

tors of the involution (Himalayan hills, African highlands, Andean countries, Java).

This involution of previously well-integrated mixed farming is another livestock environment "hot spot." Here, it is not the interaction between livestock and natural resources that creates a degradation problem but rather the socioeconomic context that leads to a diminishing interaction that eventually ceases altogether.

Concentrate Feed Production

Cereals are the major component of livestock concentrate feed: 32 percent of the world's cereal production is consumed by livestock, and this has become an issue of debate (see box 21.5). This cereal is produced on roughly 21 percent of the total land devoted to cereal. All totaled, cereals, oilseeds, and roots and tubers used to feed livestock added up to an average of 750 million tons over the 1990–92 period. An additional 252 million tons (about 24 percent) of all concentrates were processing by-products (brans and oilcakes) for which there is little alternative use (see figure 21.1).

Averages based on 1990–92 data indicate that global cereal production was 1.8 billion tons. Of this, 600 million tons were used for livestock feed. Maize is the most important feed grain, accounting for 55 percent of total feed grain used, followed by barley and wheat. Soybean is the most important oil meal because it supplies more than half the requirements for these high-protein feeds.

By the mid-1980s, concentrate feed accounted for about a quarter of all feeds for livestock; and this proportion was growing at about 0.2 percentage point annually. Concentrates comprise about 40 percent of all feeds in the industrial countries and 12 percent in the developing world.

The current expansion of cropland is globally very small—0.1 percent annually, which compares with growth in crop production of 1.9 percent a year (FAO 1996). This suggests that the bulk of the increased demand for crops is being met by the intensification of cropland rather than expansion.

Crops differ in the degree of depletion of soil moisture and water resources, in their relative demands on soil nutrients, and in their typical need for applications of pesticides. In general, cereal crops, and in particular maize, have the potential to cause greater environmental damage than other crops. This is due to heavy use of fertilizers and pesticides, high demand for water, and poor ground cover in the early stages of plant development. On the contrary, potential impacts are generally lowest for legume crops, such as soybeans and pulses. Environmental risks due to nitrate and phosphate losses are greatest in maize and wheat, while the risks of soil nutrient depletion are greatest in cassava and sweet potato.

Waste from Processing

Like waste from animal production, the processing of animal products results in environmental damage where it is concentrated and unregulated. This is the case in urban and periurban environments, particularly in developing countries. Slaughtering requires large amounts of hot water and steam for sterilization and cleaning. Therefore, the main polluting component is wastewater (Verheijen and others 1996).

The concentration of organic compounds in

Figure 21.1. Global Production of Livestock Concentrate Feed, 1965–92
(millions of tons)

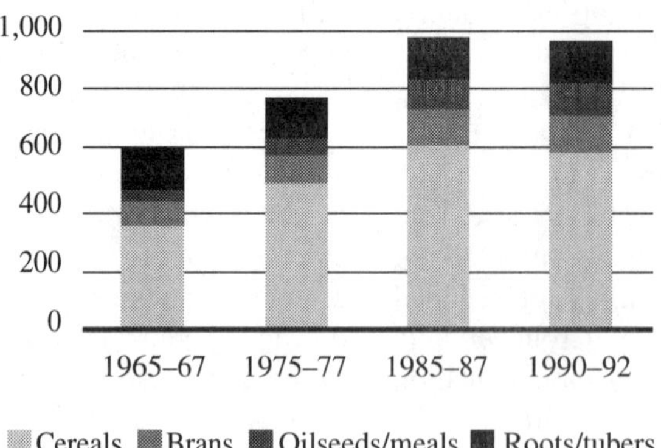

wastewater leads to oxygen demand, usually expressed as biological oxygen demand (BOD). Wastewater contains fat, oil, protein, carbohydrates, and other biodegradable compounds. Degradation of these organic substances requires oxygen. In addition, wastewater usually contains insoluble organic and inorganic particles called suspended solids. In most developing countries, tannery effluent is sewered, discharged into inland surface waters, or irrigated on the land (Verheijen and others 1996). High concentrations of salt and hydrogen sulfide present in tannery wastewater greatly affect the quality of water. Suspended matter such as lime, hair, and fleshings make the surface water turbid and settle to the bottom, thereby affecting fish. Chromium tanning is toxic to fish and other aquatic life. When mineral tannery wastewater is applied on the land, soil productivity is adversely affected and some part of the land may become completely infertile. Due to infiltration, groundwaters are also adversely affected. Discharge of untreated tannery effluent into sewers causes deposition of calcium carbonate and blocks sewers.

Gaseous Emissions of Livestock and Waste

Livestock and livestock waste produce gases. Some are local, such as ammonia. Others, such as carbon dioxide (CO_2), methane (CH_4), ozone (O_3), and nitrous oxide (N_2O), affect the world's atmosphere, by causing global warming or global climate change.

Carbon dioxide. There are three main sources of livestock-related carbon dioxide emissions. First, all domesticated animals emit carbon dioxide as part of their basic metabolic function or respiration, estimated at a total of 2.8 billion metric tons annually. Second, carbon dioxide emissions result from biomass burning, part of which can be attributed to the clearing of land and setting of bush fires to create pasture and enhance pasture growth. Third, carbon dioxide is released in relation to livestock-related consumption of fossil fuel for heating, manufacturing machinery, and producing feed.

Carbon dioxide is the least aggressive of the greenhouse gases but is emitted in large quantities. Annually, savanna and deforestation fires pump approximately 1.6 billion tons into the atmosphere, between 20 and 25 percent of the total anthropogenic CO_2 emission of 7.1 billion tons (Houghton, Callander, and Verney 1995). However, unlike the emission from fossil fuels, the same grasslands and forests recapture part of these emissions. In effect, technologically improved savannas accumulate 50 tons (or the annual emission of about 15 cars) more emissions per hectare than the traditional unimproved savannas (Fisher and others 1995).

Leaving aside the respiration of domesticated animals, livestock can be associated with about 15 to 20 percent of total carbon dioxide emissions. When the carbon sequestration capacity of natural and improved grassland is considered, the contribution to the net increase in carbon dioxide concentrations in the atmosphere is probably much smaller. There are, however, notable differences between the systems. Grazing systems, with little fossil fuel inputs and often considerable sequestration, are likely to have a positive carbon balance, whereas this balance is likely to turn negative as intensity and commercial inputs increase, such as in industrial systems.

Methane. Methane gas is much more aggressive (24 times) than carbon dioxide in causing global climate change. It is the product of animal production and manure management, rice cultivation, production and distribution of oil and gas (pipelines), coal mining, and landfills. Livestock production and manure management contribute about 16 percent of total annual production of 550 million tons. Methane gas is produced mainly as a by-product of the feed digestion of ruminants (box 21.5). On average, about 6 percent of the feed energy is lost in methane (U.S. Environmental Protection Agency 1995). Methane emissions are the direct result of the capacity of ruminants to digest large amounts of fibrous grasses and other feeds that cannot be used for human consumption. Pigs and poultry cannot digest these fibrous feeds, which is why emissions from these animals are relatively low. Methane emissions per unit of product are highest when the quality of feed and level of production are low, conditions that prevail in the arid and humid tropics and subtropics. This offers the opportunity for a *win-win* situation in which improved feeding leads to more efficient and profitable production and lower methane emissions.

Of methane emanating from animal production, 20 percent comes from manure stored under anaerobic conditions (U.S. Environmental Protection Agency 1995). Here the picture is reversed: high levels of methane emissions from manure management, as well as from large production units, are usually associated with high levels of productivity and intensity.

Despite growing livestock populations, global methane emissions from livestock remain static for two reasons. First, increases in productivity lower emission levels per animal and per unit of product. Advances in feed resources and nutrition and breed development are significant contributing factors. Second, monogastric production is growing at a much faster pace than ruminant production. About 80 percent of the total growth of the livestock sector is attributed to pigs and poultry, which emit comparatively small amounts of methane.

Any reduction in methane production, however, is likely to result in an increased emission of other gases, notably carbon dioxide and nitrous oxide, because fossil fuels and fertilizer will be required in the intensification process.

Nitrous oxide. This is the most aggressive greenhouse gas (320 times CO_2). It is produced in animal manure, which contributes about 400,000 tons of nitrogen a year or 7 percent of total global emissions. Indirectly, livestock are associated with N_2O emissions from grasslands and, through their concentrate feed requirements, with emissions from arable land and nitrogen-fertilizer use.

What Is to Be Done?

Policies need to be designed to correct the environmental effects of livestock production, which are not reflected in product and input prices. These policies should address the causes of environmental degradation and must be flexible, site-specific, and well targeted. Instruments to enhance positive and mitigate negative environmental impacts include pricing, regulations, and institutional development. The collective purpose of these instruments is to establish feedback mechanisms that ensure that livestock's use is consistent with overall social objectives.

Within an enabling policy framework, a wide range of available technologies can be employed, while others still need to be developed. Their purpose is not simply to reduce the environmental damage by reducing the polluting load. They should also be designed to enhance and spare natural resources through the use of livestock and to turn waste into useful products.

Policy Design

The collective role of all policy instruments aimed at reducing environmental damage associated with livestock is to establish feedback mechanisms that ensure that livestock's use is consistent with overall social objectives.

There is ample evidence that economywide policies, including fiscal and monetary policies or structural adjustment programs, have a strong impact on the natural resource base. These complicated interactions are poorly understood, and generalizations are difficult to make. However, where market failures occur, or policy distortions and poverty cause environmental damage, broad policy reforms that promote efficiency or reduce poverty can generally be regarded as beneficial to the environment.

The objective is not necessarily to modify the original broader policies that have conventional economic or social goals; this would be an unrealistic proposition, particularly in developing countries. Rather, complementary measures with specific or localized foci may need to be designed to mitigate the negative effects or enhance the positive effects of the original policies on the environment (Munasinghe 1994). The challenge is to design policies that correspond to the intended social and economic objectives but still comply with environmental sustainability.

Institutions

Institutions are required to develop environmental policies and laws and to assist line agencies in incorporating environmental concerns into economic development planning and budgeting. There must also be institutional capacity to monitor compliance and enforce regulations. Institutions are essential to managing access to common property resources, such as grazing land or forest, or ensuring security of tenure.

For most traditional grazing systems under pressure by changing property rights, institutions need to be improved by decentralizing decisionmaking. Local empowerment is required in all situations, but especially where governmental institutions are weak. The need to transfer authority and responsibility for resource management to the lowest level at which it can be exercised effectively is increasingly recognized. Often, consultation and direct participation of the community in decisionmaking enable local knowledge to be harnessed and responsibility for identifying problems and finding solutions to be taken at the local level. Institutions need to establish stewardship for environmental resources and to regulate access. This, in particular, refers to extensive systems where a large part of the environmental damage can be traced to a regulatory vacuum left by eroding traditional institutions.

Community-based wildlife management is becoming generally accepted as an essential component of sustainable wildlife management, particularly livestock-wildlife interactions. There has been considerable progress in decentralizing wildlife management, especially in East Africa, and in sharing the benefits from wildlife with the involved population (Ottichilo 1996). Key issues involve maintaining the strength and cohesion of the traditional social fabric and achieving an equitable distribution of revenues from wildlife among the local population.

Property rights instruments have the potential to grant resource users a tangible interest in the environmental consequences of their actions (Young 1996). Property rights need to be enforced by governments and, in a subsidiary fashion, by traditional institutions. Lack of security of tenure has often been identified as a prime cause of land degradation. In the Brazilian Amazon, for example, security of tenure has been conditional on land clearance, with the result that significant areas of rain forest have been cleared solely as a means to obtain access to wealth. Land tenure arrangements are a key factor in facilitating long-term investment in sustainable resource use, for both grazing and mixed systems.

Government institutions become critical in establishing and enforcing a regulatory framework for the protection of areas or the control of wastes, such as for industrial production systems or animal product processing units. To strengthen the pollution control and enforcement mechanisms in the developing world will be an important task in the future.

The establishment of financial institutions may alleviate pressure on the environment where significant numbers of animals are kept primarily for investment purposes, such as by absentee herd owners in Africa or Latin America (see box 21.6). Positive interest rates on safe deposits are providing better alternatives for investment if the primary purpose for keeping livestock is not production.

Incentive Policies

Incentive policies rely on market forces. The more a production system is exposed to the market, the more susceptible it will be to price changes. In particular, intensive production depends on inputs that contain a high component of natural resources, which are rarely reflected in their market price. Examples include concentrate feed, fossil fuel, inorganic fertilizer, livestock products, land, mechanization, and genetic material. These should be priced higher by abolishing subsidies or, in some situations, imposing taxes. This, in addition to the quantitative effect of reduced consumption, will induce a more efficient use of natural resources, with both economic and environmental gains. It will also favor a more even spatial distribution of livestock production and promote land-based systems. Correspondingly, subsidies or tax relief can be provided where natural resources are saved or used wisely, such as by using renewable energy (methane) or protecting biodiversity. Essentially, the recommendation is for full-cost pricing of the provision of all goods and services used by livestock producers and, also, those sectors that compete with the livestock sector for resources and markets.

There are many examples of incentive policies that reduce grazing pressures. Levying grazing fees for communal areas is common. Fees can also be progressive, with larger herds paying more per head (Narjisse 1996). Full cost recovery, especially for water and animal health services, is another key element of good pastoral policies. Water has often been a free good supplied publicly. Full cost recovery would reduce the number of large boreholes and therefore reduce local degradation around these water points. Appropriate benefit-sharing systems for the protection of biodiversity, on a local and international scale, will need to be devised. For intensive systems, levies on the discharge of waste and tradable manure quota schemes have been established in many industrial countries (Brandjes and others 1996).

Box 21.6. How Rational Are Pastoralists?

The perception that pastoralists maintain unproductive animals in their herds for prestige rather than economic reasons is still widespread. This, in the eyes of many, is one of the main reasons for overstocking and land degradation. However, almost all studies on pastoral and agropastoral systems show that there are very few unproductive animals in traditional herds (ILCA 1994). Animals are sold when they have their optimum market weight. Unproductive animals are sometimes found in "investment herds" owned by traders or civil servants who, in the absence of reliable and remunerative banking systems in Sub-Saharan Africa, invest in livestock.

Regulations

Regulations can be used to promote more sustainable use of resources by prescribing quantitative limits (of emissions or number of animals or use of inputs), technical methods, and access to resources. A further regulatory instrument is zonation, to promote a more even spatial distribution of animal production. For grazing systems in the humid and subhumid zones, the establishment of protected areas for the preservation of biodiversity is often the only appropriate measure. For intensive mixed-farming systems and industrial systems, the control approach focuses on regulating manure management. Enforcement of regulations within reasonable social costs remains a major challenge.

Training and Extension

Knowledge transfer, in many instances, is the key to keeping a balance between livestock and the environment. A considerable array of technologies are available to address environmental issues in mixed-farming and industrial systems. Livestock extension services in the developing world have traditionally focused on animal health services at the expense of production issues, and there has been an almost complete neglect of the environmental aspects of livestock production. The introduction of environmentally more benign practices often requires complex social and institutional changes, which front-line extensionists are poorly equipped to handle. Alternative technology is badly needed in grazing systems, even though no ecologically and socially viable alternatives to the traditional pastoral systems of arid zones have yet been identified. Although the current efficiency of those traditional systems makes this a formidable task, population pressure in these areas makes it imperative as well.

Technologies

Technological change occurs as a response to changes in the availability of inputs and in demand. Over the past decades the livestock sector has responded mainly by increasing efficiency and by undertaking the major structural shifts outlined earlier. Four sets of technologies, which overlap somewhat, can work to the benefit of the environment:

- Technologies that reduce the environmental damage by alleviating the direct pressure on natural resources or by reducing the pollution load through modifying the chemical or physical characteristics of work products
- Technologies that enhance natural resources by making them more productive or richer
- Technologies that save natural resources by making it possible to generate more revenue from the same resource or the same revenue from less
- Technologies that turn waste into products by closing cycles.

Reduction of Environmental Damage

In the arid and semiarid grazing areas, careful water development can help to prevent environmental damage. Investments in market technology may also reduce environmental pressure by encouraging greater off-take. In addition, new and more benign methods are now available to control diseases. Vaccines are available for the tick-borne diseases spread by a number of subspecies, although not yet for east coast fever. Long-duration tick sprays now exist that can be applied on individual animals, with negligible environmental effect. A number of environmentally benign control methods, using extremely low concentrations of easily degradable insecticides, also exist for tsetse control.

Environmental damage of intensive systems can be significantly reduced by focusing on emissions from manure by, for example, improving collection and storage techniques. The main focus should be on reducing nitrogen losses, most of which are in the form of ammonia from manure. Most losses can be avoided by manure collection and covered storage facilities. Minimal amounts of ammonia are emitted when manure is collected under solid floors, and 80 to 90 percent reductions can be achieved by covering storage tanks (Voorburg 1994). A further reduction of odor and ammonia loads can also be achieved through "bio-filters" or "bio-washers" that absorb odors and ammonia from polluted air. This is done by oxidizing ammonia into NO_2 and NO_3. Up to half of the ammonia can be eliminated through air washing systems, which are, however, costly in investment and operation (Chiumenti, da Borso, and Donanton 1994).

Nutrient losses during and after application of manure on soils can be significantly reduced by injecting or applying manure into the subsoil (Brandjes and others 1996). Better timing of application in response to crop requirements avoids further losses and enhances the nutritive value of manure.

Nitrification inhibitors can be added to slurry to reduce leaching from the soil under wet conditions.

In tanneries, dairies, and slaughterhouses, anaerobic systems can purify wastewater and reduce by half the biological oxygen demand, while more sophisticated anaerobic systems reach 90 percent BOD purification (Verheijen and others 1996). Wastewater treatment first separates solids from the liquid, followed by biological treatment under anaerobic conditions (lagoons). Nutrients, such as phosphorus, are then removed by chemical or physical processes such as adsorption, stripping, or coagulation. The same process removes the remaining BOD as well as pathogens.

Resource-Enhancing Technologies

For grazing systems in arid zones, "deferred grazing," a traditional practice in many Middle Eastern countries, may regenerate vegetation, while in the semiarid zones, this may be accomplished by overseeding or planting adapted fodder. The introduction of a multispecies grazing pattern will often encourage better use of the vegetation and may have positive effects on plant and animal biodiversity. Such technologies may be part of overall "management" approaches, which should explicitly acknowledge the efficiency of the current mobile systems and their nonequilibrium status. For the humid tropics, perennial grasses and legumes have been developed that maintain soil fertility better than any other crop. Biodiversity may also be enhanced through careful management of wildlife-livestock interactions in pastoral systems as well as through prevention of bush encroachment.

Livestock, mainly through their input function within a mixed crop-livestock system, enhance the main natural resource: land. Animal manure and traction make the land more productive than would be the case in their absence. Thus, all technologies that reduce nutrient losses from manure and improve the efficiency of their application enhance land productivity. For example, stall feeding doubles the effective availability of nitrogen and phosphorus. Fodder shrubs and trees reduce soil erosion and improve soil fertility. In Indonesia, agroforestry systems were successfully introduced with three strata, including grass, fodder shrubs, and tree crops such as oil and coconut palms or cashew nuts. Raising the productivity of land through crop-livestock interaction protects other land from being brought into cultivation.

Box 21.7. Alternatives to Cereal Feeding

Countries in the humid and subhumid tropics are cereal-deficit. Livestock production, in particular monogastric production, is thus faced with high prices for feed concentrates. This has spurred the development of sugarcane-based feeding systems in a number of tropical countries (Colombia, Cuba, Vietnam, Philippines; see Preston and Leng 1994). Sugarcane is one of the highest yielders of biomass per unit of time and area. The juice can be used to feed monogastrics, while the tops can be used to feed ruminants. As a perennial crop, sugarcane production has very low rates of erosion and can be produced with low external input. In the past, the association of sugarcane and livestock production has been problematic because sugarcane was traditionally produced on large plantations, geographically separated from livestock production. Recent developments in the diversified use of sugarcane may lead to more village-based, intensive systems of monogastric production in the humid tropics.

Resource-Sparing Technologies

The livestock sector continues to develop technologies that increase the efficiency of natural resource use. In particular, these technologies target feed conversion because feed typically accounts for 60 to 70 percent of production costs. Better feed conversion saves land, while reducing the animals' waste load (see box 21.7).

A wide array of technologies improve feed conversion. They are mainly used in industrial systems in industrial countries, but their use is quickly picking up in developing countries as well. The most important ones are:

- Introducing multiphase feeding, whereby feed composition is better suited to the needs of animal classes, which means that less waste is produced and less nitrogen and phosphates appear in waste and in the environment
- Improving the accuracy of determining nitrogen and phosphate requirements, followed by improving the balance of feeds with these essential nutrients. In this respect, the balance of essential amino acids, the building blocks of feed proteins, has improved in the rations of pigs and poultry. For example, a combination of more balanced feed,

improved digestibility, and inclusion of synthetic amino acids substantially reduces the protein content in feed and hence reduces the excretion of nitrogen and phosphorous by 20 to 40 percent (van der Zijpp 1991)
- Improving digestibility, by adding an enzyme (called phytase) that catalyzes the digestion of phosphates contained in feed and reduces waste loads on the environment
- Promoting feeding systems that reduce intake and stopping buffet-style ad libitum feeding, popular in the 1980s
- In mixed-farming systems of lower intensities, adding strategic supplementation for specific classes of animals such as lactating cows or growing animals
- Improving the efficiency of feed conversion with enhanced genetic potential, better health and environmental conditions, and improved management
- Improving efficiencies in feed conversion for the livestock sector as a whole by shifting to monogastrics, especially poultry and fish. This trend is likely to continue and will be particularly strong in the developing countries.

Biodiversity is best preserved by intensifying livestock production while intensifying crop production. Both will reduce the land requirements for given volumes of final product, alleviate pressures on habitats and biodiversity, and limit the use of land and water resources.

Waste Technologies

Historically, the raison d'être for keeping livestock was its use of resources for which there was no alternative use. This explains why efficiency per animal was not, and in many low-input systems still is not, a major concern.

The conversion of organic waste into livestock products, although associated with livestock waste, reduces the environmental hazards associated with crop and agroindustrial waste. Also, more and more food wastes are consumed by livestock, as urban agglomeration and changing eating habits offer a window of opportunity for the collection of food waste from catering units to be recycled as feed. Large amounts of straw, otherwise burned on the fields or left to decompose slowly with little nutritional benefit to the crops, may be turned into quality feed, for example through urea treatment (Li-Biagen, 1996, personal communication).

The cycles of matter can also be closed by using livestock waste as feed, energy, or fertilizer. Recycling manure—by feeding it to animals, livestock, and fish (Muller 1980)—is practiced only on a limited scale. Reluctance to use manure as feed originates mainly from the fear of health risks but is also due to the low nutritive value of manure, with the exception of poultry manure. Poultry manure is incorporated in the diets of intensive systems, particularly as ruminant feed. In Asia, especially, manure is fed to fish and pigs (China). The low quality of manure as feed and high processing costs limit its competitiveness vis-à-vis traditional commercial feeds. Sánchez (1995) provides a recent overview of the possibilities.

Technology also exists to use the energy content of manure. Biogas plants of all sizes and levels of sophistication not only recover the energy contained in manure but also eliminate most of the animal and human health problems associated with contamination of waste by micro-organisms (see box 21.8). Other methods of controlling the waste load are the purifying and drying of manure.

Promising approaches exist to reduce emissions from manure lagoons by recovering methane and using it for energy. Large confined animal operations allow such techniques to be profitable. This methane can be used for on-farm energy to generate electricity, and the slurry effluent can be used as animal feed, as aquaculture supplements, or as crop fertilizer. The controlled bacterial decomposition of the volatile solids in manure reduces the potential for contamination from runoff, significantly reduces pathogen levels, removes most noxious odors, and retains the organic nitrogen content of the manure.

Methane can be recovered in covered lagoons where manure solids are washed out of the livestock housing facilities with large quantities of water, and the resulting slurry flows into an anaerobic primary lagoon. Such anaerobic conditions result in signifi-

---

Box 21.8. Biogas in China

In China, more than 5.3 million rural biogas systems are in place, producing 1.25 million cubic meters annually. Biogas is used for household heating and cooking, poultry hatching, tea roasting, and grain and fruit storage. The slurry is used as fertilizer, for fish farming, and for pig feed, which shows good results in semi-intensive production systems (Henglian 1995).

---

cant methane emissions, particularly in warm climates, but the methane can be recovered by placing an impermeable floating cover over the lagoon and applying negative pressure. Alternatively, digesters can be used. Large-scale digesters are engineered vessels into which a mixture of manure and water is placed and retained for about 20 days. The digester is heated to about 60°C, after which the gas is drawn off and used for energy. Large dairy and pig farms with high energy requirements find these systems to be cost-effective. Small-scale digesters do not include heating and are appropriate for warm climates only.

The Need for Action

Novel concepts are being developed to integrate crop and livestock production in a farming area rather than on a mixed farm. This method of area-wide integration of crops and livestock allows individual enterprises to operate separately while linking the flow of energy and organic and mineral matter through markets and regulations. This allows for highest efficiencies at the enterprise level while maximizing social benefits. A policy framework is needed to organize crop and livestock production in this way.

Strategies for Livestock Production Systems

Putting the environment into the forefront does not mean that only environmental objectives count. On the contrary, environmental goals can only be tackled effectively if and when sound economic objectives are met.

Grazing systems will remain a source of extensively produced animal products. To some extent, these systems can intensify production by incorporating new technologies, especially in areas with higher potential (subhumid and highland areas). Often this can be facilitated by stronger organizations, local empowerment, and regulation of access to resources. Where there is potential for mixed farming, policies need to facilitate the transition of grazing systems into mixed-farming systems in the semiarid and subhumid tropics by integrating crops and livestock (manure management, animal draft, residue feeding, fodder crops, and so forth).

Grazing systems can also intensify by diversifying and opening up other uses. In grazing systems, livestock's role, in addition to providing a livelihood to pastoral people and market production, should be to protect the natural resource base, in particular land, water, and biodiversity.

Mixed-farming systems will see continued intensification and growth. Smallholder and family mixed farming will remain predominant for some time to come, with livestock based on crop by-products and surplus. Important productivity gains can be achieved by further enhancing the flow of nutrients and energy between the two components. Livestock's role, in addition to production, is to enhance and substitute natural resources. The environmental and economic stability of this system makes it the prime focus for the continuing transfer and development of technology. Where involution occurs, in areas of extreme population pressure, resource degradation, and poverty, this must be fought through accelerated uptake of technology (feed resources, animal traction, development of small ruminants, and so forth). Under more favorable agroecological and market conditions, new forms of industrial production will have to be established. These industrial systems will have to be based on the resource endowments of a region, if nutrient balances are to be maintained and the environment's ability to absorb waste respected. We are therefore projecting industrial systems integrated into a wider concept of land use, particularly for pig and poultry production. This trend is already under way in some industrial countries. This would blend resource-saving technologies with the absorptive capacities of the surrounding land. New organizational arrangements will have to be found to allow specialized units to capitalize on economies of scale. The strategy is thus to transform mixed-farming systems into specialized and commercial enterprises (through infrastructure and institutional development, animal production and health technologies, and processing) in rural areas where land pressure is on the increase and where the market allows.

Industrial systems in areas of high animal densities will have to cope with higher production costs as a result of more stringent regulations and pollution levies. In some cases, this will remove the competitive edge that industrial production has over land-based production. Potentially, this will also raise the prices for livestock products and reduce demand. Higher prices will provide incentives for the land-based systems to intensify. Scales of industrial production will grow further because of economies of scale for waste treatment. This system's purpose is mainly to satisfy the soaring demand in many parts of the developing world, which is often over and above the supply capacity of land-based production and at maximum efficiency of resource use.

As resources become more scarce, livestock producers must continue to search for technologies that increase the efficiency of resource use if the rapidly increasing demand is to be met without putting additional strain on natural resources. The challenge is to obtain more efficiency without concentrating animal production in a given area. Limiting the number of livestock while maintaining market mechanisms through, for example, tradable emission quotas seems to be an appropriate choice. Ideally, the advanced resource-saving technologies and the absorptive capacities of extended rural areas should be married. Thus the motto for most of the industrial world and the more densely populated parts of the developing world is to intensify, but not concentrate, animal production. Such an approach would promote the spread of processing into these areas and bolster economic development.

In a schematic way, we can thus identify pathways for livestock production systems. Intensification, specialization, and organization are the processes that characterize the different phases. As a result of the interaction between livestock production systems and natural resources, coupled with factors such as market access, there are development opportunities as well as threats to sustainability. They are sketched in figure 21.2 and lead to the identification of areas for strategic intervention.

To minimize environmental damage, governments should, in very general terms, intervene as suggested in figure 21.2. Strategic interventions need to focus on the phases of transition from one state to another, where entirely new sets of technology are introduced in an effort to intensify (where the agroecological and market potential allows), and on the fundamental pressures of poverty, population growth, and weak institutions.

It is evident that questions relating to livestock and the environment cannot be solved in environmental terms only. A comprehensive perspective is needed to ensure an enabling policy framework in which effective technologies can be introduced. Technology remains the key component because future development, including that of the livestock sector, will depend on technology to substitute for natural resources. This trend to knowledge-intensive systems

Figure 21.2 Development Pathways of Livestock Systems

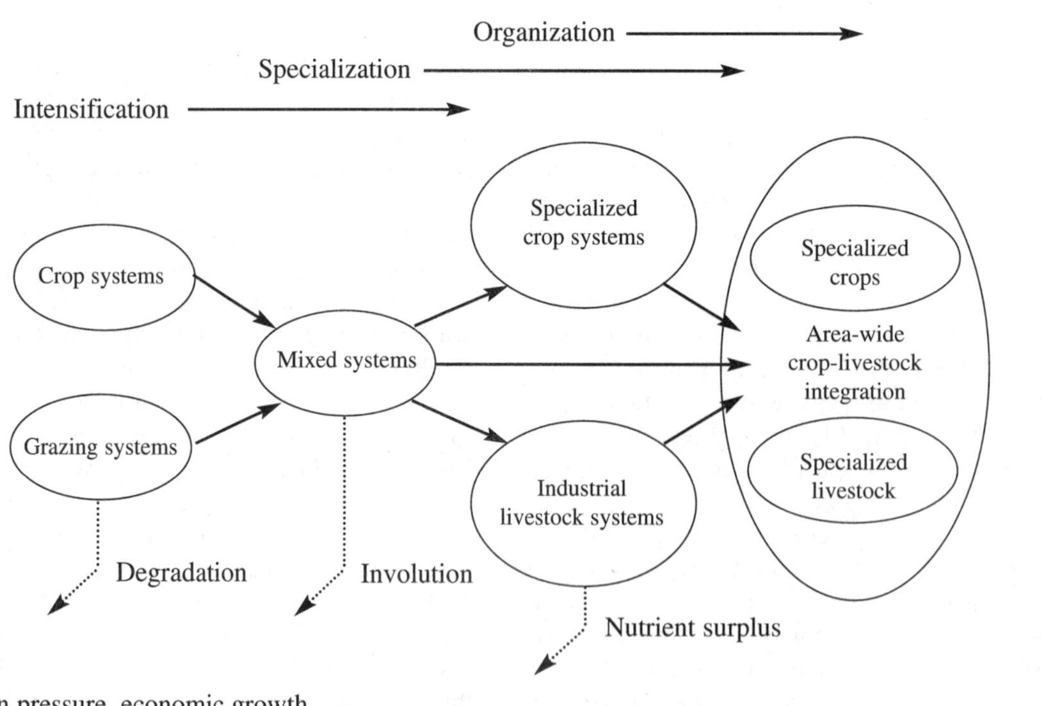

is widely observed: smart technologies, supported by astute policies, can help to meet future demands while maintaining the integrity of the natural resource base. Better information on which to base decisionmaking is urgently required.

Conclusions

Improved management is essential if the world's natural resources are to continue to provide the basis for life support and human well-being. Only with improved management can the dual objectives of sustainable agricultural production be fulfilled—to feed the world's growing population, while sustaining its natural resource base. Livestock production is the largest user of land and is about to become the most important agricultural activity in terms of economic output. Left to uncontrolled growth, the environment will suffer, compromising human welfare. However, this is unlikely to happen. The opportunities not only to mitigate environmental damage but also to tap the immense development potential offered by livestock are large: awareness, political will, and readiness to act are growing among all those involved and ensure that the problems will be tackled effectively.

References

The word "processed" describes informally reproduced works that may not be commonly available through libraries.

Alexandratos, N., ed. 1995. *World Agriculture: Towards 2010.* Rome: Food and Agriculture Organization of the United Nations; Chichester, U.K.: John Wiley and Sons.

Behnke, R. H., Ian Scoones, and C. Kerven. 1993. *Range Ecology in Disequilibrium.* London: Overseas Development Institute and the International Institute for Environment and Development Commonwealth Secretariat.

Boudet, G., M. Carriere, P. Christy, and H. Guerin. 1987. "Paturages et elevage au Sud de la Mauritanie. Etude integrée sur les paturages, leur conservation et leur restauration." Institut d'Elevage et de Médicine Vétérinaire des Pays Tropicaux (IEMVT), Maison Alfort, France. Processed.

Bowman, R. L., J. C. Croucher, and M. T. Picard. 1992. "Reducing Methane Emissions from Ruminant Livestock: India Prefeasibility Study." A.T. International for the U.S. Environmental Protection Agency, Global Change Division, Washington, D.C. Processed.

Brandjes, P. J., J. de Wit, H. G. van der Meer, and H. Van Keulen. 1996. "Environmental Impact of Animal Manure Management." Consultant's report for the multidonor study on livestock and the environment, Food and Agriculture Organization, World Bank, and U.S. Agency for International Development, Rome. Processed.

Byrne, P. V., C. Staubo, and J. G. Grootenhuis. 1996. "The Economics of Living with Wildlife in Kenya." In J. Bojo, ed., *The Economics of Wildlife.* AFTES Working Paper 19. World Bank, Washington, D.C. Processed.

Chiumenti, R., F. da Borso, and L. Donanton. 1994. "Dust and Ammonia Reduction in Exhaust Air from Piggeries Using a Mechanical Wet Separation System." In *Proceedings of the Seventh Technical Consultation of the European Cooperative Research Network on Animal Waste Utilization.* REUR Technical Series 34. Rome: Food and Agriculture Organization.

Christiaensen, L., E. Tollens, and C. Ezedinma. 1995. "Development Patterns under Population Pressure: Agricultural Development and the Cassava-Livestock Interaction in Smallholder Farming Systems in Sub-Saharan Africa." *Agricultural Systems* 48(1):51–72.

CIAT (Centro de Investigación en Agricultura Tropical). 1994. *Annual Report* 1993. Cali, Colombia.

de Wit, J., P. T. Westra, and A. J. Nell. 1996. "Environmental Assessment of Landless Ruminant Production Systems." Consultant's report for the multidonor study on livestock and the environment, Food and Agriculture Organization, World Bank, and U.S. Agency for International Development, Rome. Processed.

Dregne, H., M. Kassas, and B. Rozanov. 1992. "Status of Desertification and Implementation of the United Nations Plan to Combat Desertification." *Desertification Control Bulletin* 20:6–18.

English, J., M. Tiffen, and M. Mortimore. 1992. "Land Resource Management in the Machakos District, Kenya 1930–1990." Environment Paper 5. World Bank, Washington, D.C. Processed.

FAO (Food and Agriculture Organization of the United Nations). 1996. "Cereals: Feed for Animals or Food for People?" Fact sheet distributed at the World Food Summit, Rome, 13–17 November.

Processed.

———. 1997. "Current World Fertilizer Situation and Outlook 1994/95–2000/2001." Food and Agriculture Organization, United Nations Industrial Development Otganization, and World Bank Working Group on Fertilizers, Rome. Processed.

Fisher, M., I. M. Rao, M. A. Ayarza, C. E. Lascano, and J. I. Sanz. 1995. "Carbon Storage by Introduced Deep-Rooted Grasses in the South American Savannas." *Nature* 371(6494):236–38.

Grazingland Management Systems. 1996. "Environmental Impact Assessment of Livestock Production in Grassland and Mixed-Farming Systems in Temperate Zones and Grassland and Mixed-Farming Systems in Humid and Subhumid Tropic Zones." Consultant's report for the multidonor study on livestock and the environment, Food and Agriculture Organization, World Bank, and U.S. Agency for International Development, Rome. Processed.

Habib, G. S., S. Basit Ali Shah, S. Wahidullah, G. Jabbar, and Ghufranullah. 1991. "Importance of Urea Molasses Blocks on Animal Production (Pakistan)." International symposium on nuclear and related techniques in animal production and health, International Atomic Energy Agency, Vienna, Austria, 15–19 April. Processed.

Hendratno, C., J. V. Nolan, and R. A. Leng. 1991. "Importance of Urea Molasses Multinutrient Blocks for Ruminant Production in Indonesia." International symposium on nuclear and related techniques in animal production and health, International Atomic Energy Agency, Vienna, Austria, 15–19 April. Processed.

Henglian, X. 1995. "The Production of Rural Family Biogas in China." In *Proceedings of the Workshop of Peri-Urban Livestock Wastes in China, Beijing, China. 19–22 September 1994.* Rome: Food and Agriculture Organization of the United Nations.

Hiernaux, P. 1996. *The Crisis of Sahelian Pastoralism: Ecological or Economic.* Pastoral Development Network Paper 39a. London: Overseas Development Institute.

Houghton, J. T., B. A. Callander, and S. K. Verney, eds. 1995. *Intergovernmental Panel on Climate Change and an Evaluation of the IPCC IS 92 Emission Scenarios.* Cambridge, U.K.: Cambridge University Press.

ILCA (International Livestock Centre for Africa). 1994. *Annual Report 1993.* Addis Ababa.

Jahnke, H. E. 1982. *Livestock Production Systems and Livestock Development in Tropical Africa.* Kiel, Germany: Kieler Wissenschaftsverlag Vauk.

Jodha, N. S. 1992. *Common Property Resources: A Missing Dimension in Development Strategies.* Discussion Paper 169. Washington, D.C.: World Bank.

Kaimowitz, David. 1995. *Livestock and Deforestation in Central America.* EPTD Discussion Paper 9. Washington, D.C.: International Food Policy Research Institute; Coronado, Costa Rica: Instituto Interamericano de Cooperación Agrícola.

Leng, R. A. 1991. "Improving Ruminant Production and Reducing Methane Emissions from Ruminants by Strategic Supplementation." EPA/400/1-91/004. U.S. Environmental Protection Agency, Washington, D.C. Processed.

Mainguet, M., L. Guyot, M-C Chemin, and A. Braunstein. 1992. "Modifications to the Sahelian and Sudanian Ecosystems in Western Mali." In H. E. Dregne, ed., *Degradation and Restoration of Arid Lands.* Lubbock: Texas Tech University.

Milchunas, D. G., and W. K. Lauenrot. 1993. "Quantitative Effect of Grazing on Vegetations and Soils over a Global Range of Environments." *Ecological Monographs* 63(4):237–366.

Muller, Z. O. 1980. *Feed from Animal Wastes: State of Knowledge.* Animal Production and Health Paper 18. Rome: Food and Agriculture Organization of the United Nations.

Munasinghe, Mohan. 1994. "Economic and Policy Issues in Natural Habitats and Protected Areas." In Mohan Munasinghe and Jeffrey McNeely, eds., *Protected Area Economics and Policy: Linking Conservation and Sustainable Development.* Washington, D.C.: World Bank and the World Conservation Union.

Narjisse, H. 1996. "The Range Livestock Industry in Developing Countries: Current Assessments and Prospects." *In Proceedings of the Fifth International Rangeland Congress, Salt Lake City. 23–28 July 1995.* Vol. 2: *Invited Presentations.* Denver: Society for Range Management.

Nelson, R. 1990. *Dryland Management: The "Desertification Problem."* Technical Paper 116. Washington, D.C.: World Bank.

Oldeman, L. R., R. T. A. Hakkeling, and W. G. Sombroek. 1990. "World Map of the Status of Human-Induced Soil Degradation. An Explanatory Note." International Soil Reference and Information Centre/United Nations Environment Programme, Wageningen, The Netherlands. Processed.

Ottichilo, W. K. 1996. "Wildlife-Livestock Interactions in Kenya." Paper prepared for the World Bank/Food and Agriculture Organization workshop, balancing livestock and the environment, Washington, D.C., an event associated with the fourth conference on environmentally sustainable development, World Bank, Washington, D.C. Processed.

Preston, T. R., and R. A. Leng. 1994. "Agricultural Technology Transfer: Perspectives and Case Studies Involving Livestock." In J. Anderson, ed., *Agricultural Technology: Policy Issues for the International Community*. Oxon, U.K.: C.A.B. International.

Saadullah, M. 1991. "Importance of Urea Molasses Block Lick and Bypass Protein on Animal Production (Bangladesh)." International symposium on nuclear and related techniques in animal production and health, International Atomic Energy Agency, Vienna, Austria, 15–19 April. Processed.

Sánchez, M. 1995. "Use of Animal Excreta as Feed." In *Proceedings of the Workshop of Peri-Urban Livestock Wastes in China, Beijing, China, 19–22 September 1994*. Rome: Food and Agriculture Organization of the United Nations.

Scoones, Ian. 1994. *Living with Uncertainty: New Directions in Pastoral Development in Africa*. London: Intermediate Technology.

Sére, Carlos, and Henning Steinfeld. 1996. *World Livestock Production Systems: Current Status, Issues, and Trends*. Animal Production and Health Paper 127. Rome: Food and Agriculture Organization of the United Nations.

Sidahmed, A. E. 1996. "The Rangelands of the Arid/Semiarid Areas: Challenges and Hopes for the 2000's." Keynote address to the symposium on range management, international conference on desert development in the Arab gulf states, KISR, Kuwait, March. Processed.

Thomas, D., and D. Barton. 1995. "Interactions between Livestock Production Systems and the Environment: Impact Domain. Crop-Livestock Interactions." Consultant's report for the multidonor study on livestock and the environment, Food and Agriculture Organization, World Bank, and U.S. Agency for International Development, Rome. Processed.

Thomas, D. S. G., and N. J. Middleton. 1994. *Desertification: Exploding the Myth*. New York: John Wiley.

Tucker, C. J., H. E. Dregne, and W. W. Newcomb. 1991. "Expansion and Contraction of the Sahara Desert from 1980–1990." *Science* 253:299–301.

U.S. Environmental Protection Agency. 1995. "Study of Livestock-Environment Interactions: Global Impact Domain. Methane Emissions." Consultant's report for the multidonor study on livestock and the environment. Food and Agriculture Organization, World Bank, and U.S. Agency for International Development, Rome. Processed.

van der Zijpp, A. J. 1991. "Sustainable Animal Production: A Response to Environmental Problems." In J. Boyazoglu and J. Renaud, eds., *The Livestock Production Sector in Eastern Europe as Affected by Current Changes*, pp. 191–97. EAAP Publication 57. Wageningen: European Association for Animal Production.

Verheijen, L. A. H. M., D. Wiersema, L. W. Hulshoff Pol, and J. De Wit. 1996. "Management of Waste from Animal Product Processing." Consultant's report for the multidonor study on livestock and the environment, Food and Agriculture Organization, World Bank, and U.S. Agency for International Development, Rome. Processed.

Voorburg, J. H. 1994. "Farmers Options to Reduce Odor and Ammonia Emissions from Animal Buildings and Storage." In *Proceedings of Seventh Technical Consultation of the European Cooperative Research Network on Animal Waste Utilization*. REUR Technical Series 34. Rome: Food and Agriculture Organization.

World Resources Institute. 1992. *World Resources 1992/1993*. New York: Oxford University Press.

Young, M. D. 1996. "Maintaining Harmony: Equitable and Efficient Means to Minimize the Adverse Impact of Livestock on the Environment." Paper presented at the workshop on blancing livestock and the environment, an event associated with the fourth conference on environmentally sustainable development, World Bank, Washington, D.C. Processed.

22

Public Policies to Reduce Inappropriate Tropical Deforestation

David Kaimowitz, Neil Byron, and William Sunderlin

This chapter seeks to explain the ways in which, and the extent to which, public policy measures might be used to reduce inappropriate tropical deforestation. In this introductory section, we define the terms used (deforestation, agents of deforestation, and inappropriate deforestation) and present our central argument and the structure of our chapter.

In the last 8,000 years, world forest cover declined more than 40 percent, from 62 million to 33 million square kilometers (Bryant, Nielsen, and Tingley 1997). Contemporary concern about deforestation focuses on tropical countries because that is where the majority of forest cover is being removed. In the period 1980–90, an estimated 137.3 million hectares of tropical forests were cleared, about 7.2 percent of the total that existed in 1980 (FAO 1996).[1] Any claims regarding the magnitude of global deforestation must be made with caution, however, because there are serious problems with the data and definitions used.

Regardless of the exact magnitude and location of tropical deforestation, it is clear that deforestation often implies:

- The loss of livelihoods for forest-dependent people, many of whom may not wish to or be able to find other sources of employment
- Decreasing stocks of fuelwood and nontimber forest products as well as industrial timber
- Greater soil erosion and river siltation
- Substantial species and gene loss, in view of tropical forests' high level of endemic biodiversity
- Substantial emission of carbon dioxide, which contributes to global warming
- Other types of local and regional climate change.

We follow the Food and Agriculture Organization of the United Nations (FAO) in defining deforestation as "the sum of all … transitions from natural forest classes (continuous and fragmented) to all other classes" (FAO 1996: 22). This definition focuses attention on the loss of natural forests and is framed in the context of specific definitions of forest and nonforest classes of land use. Analysis of transitions among these classes can shed light on the cause-and-effect relations underlying deforestation and help to determine the optimal uses of land. Based on this definition, the amount of deforestation in the period 1980–90 was estimated at 8.2 percent in Asia, 6.1 percent in Latin America, and 4.8 percent in Africa (FAO 1996).

Forest degradation is a crucial adjunct to deforestation. Forest degradation is a "decrease of density or increase of disturbance in forest classes" (FAO 1996: 21). Degradation potentially implies a major loss of forest products and ecological services, even when there is little deforestation as such (outright conversion of forest to another land use). In what follows, we focus principally on deforestation, but concerns about degradation may be just as important in the long run.

The process of deforestation must be analyzed at two levels: agents and causes. The agents of deforestation are the people who physically (or through decisions over their labor forces) convert forests to nonforest uses: small farmers, plantation and estate owners, forest concessionaires, infrastructure construction agencies, and so forth. FAO (1996) attributes most deforestation in Latin America to medium- and large-scale operations (resettlement schemes, large-scale cattle ranching, hydroelectric dams) and notes that deforestation in that region is characterized by transitions from closed forest to nonforest land uses. In Africa, deforestation is related largely to the expansion of small-scale farming and rural population pressure (the growing number of smallholders) associated with the conversion of closed forest cover to short fallow farming. Deforestation in Asia is associated with both relatively large operations (as in Latin America) and rural population pressure (as in Africa) and involves conversion of closed forest to long and short fallow, plantations, and nonforest uses. In the larger countries, such as China, India, and Indonesia, different types of processes dominate depending on the specific context of the region involved. There are no simple, universal single-cause explanations.

The relative contribution that different agents make to deforestation is difficult to determine for two reasons. First, little reliable information exists on this subject. Second, even if it did, in many situations various agents operate in the same location and cannot be analytically separated. Agents are often interdependent (for example, cattle ranchers who supply smallholders with chain saws), and different activities can take place sequentially in the same location (small farmers frequently occupy abandoned timber concessions, and many farmers engage in salvage logging prior to clearing land for farming).

One reason for the intensity and confusion in the deforestation debate is that proponents of different views tend to emphasize the role of loggers, small farmers, or large ranchers and to portray these agents as being universally dominant, when in fact the different agents coexist and closely interact in many countries, and their relative importance varies over time and between regions. Given such complex relationships, there are no clear guilty or innocent parties, and no one should expect neat, simple solutions.

The "causes" of deforestation refer to the multiple factors that shape agents' actions and, in particular, their decision to deforest. These causes include market forces (international price fluctuations of agricultural export commodities), economic policies (currency devaluation), legal or regulatory measures (a change in land tenure laws), institutional factors (the decision to deploy more forest rangers to a particular area), and political decisions (a change in the way forest concessions are allocated), among others. This chapter discusses how certain causes can be manipulated to influence the behavior of agents, so as to lessen the rate of inappropriate deforestation.

Two main reasons lead us to believe that some deforestation is inappropriate and that inappropriate deforestation is a significant problem. First, deforestation generally causes negative externalities that generate costs to society not reflected in existing prices and has long-term consequences that individual producers rarely consider. Second, the relative importance of the negative externalities tends to grow over time if an increasing proportion of deforestation occurs in areas that have only marginal value for agriculture but that sequester large amounts of carbon, have fragile soils, or are high in biodiversity.

Any decision regarding which deforestation is appropriate or not is ultimately political in nature and cannot be justified on purely technical grounds. Nevertheless, many forestry policy experts agree that forest clearing is more likely to be inappropriate when it involves the following types of areas:

- Areas that currently have little value for agriculture by virtue of the quality of their soil or gradient of the land
- Areas that have large amounts of biodiversity, particularly endemic biodiversity, that is poorly represented in existing protected areas[2]
- Areas that have large numbers of forest-dependent people who show no inclination to abandon their existing livelihood strategies
- Areas that, by virtue of their rich potential and comparative advantage for timber production, make this the most profitable use of land, even after the area is logged for the first time
- Fragile areas where the ecological cost of conversion resulting from "downstream" effects outweighs any economic gain from nonforest land uses
- Periurban areas where forests play a key role in conserving aquifers, providing fuelwood, and supporting recreational and tourism activities.

It is important to distinguish inappropriate from appropriate deforestation for three reasons. First, it is often assumed (at least implicitly) that all tropical deforestation is inappropriate, and this is not necessarily the case. Second, to the extent that inappropriate deforestation can be defined, the geographical areas (and sometimes socioeconomic groups) can be specified that should be the targets of policy designed to reduce inappropriate deforestation. Third, by clearly defining the areas appropriate for conversion, pressure can conceivably be eased on forests where conversion is inappropriate.

Table 22.1 offers a rudimentary conceptual framework for distinguishing inappropriate from appropriate deforestation. It takes into account three cross-cutting categories of valuation: biophysical, economic, and political. No one category alone provides a sufficient basis for cogent decisionmaking on whether a forest should be protected or cleared. Decisionmakers must take all of them into account simultaneously.

The horizontal correspondences in this table are idealized and do not necessarily correspond to the "real world." For example, it can be assumed that over time, because an ever larger share of remaining tropical forests are in remote or hilly areas, these forests will be less appropriate for conversion because of their relative inadequacy for agriculture or their high biodiversity value.[3] These same forests, however, will likely be targeted by a growing number of stakeholders (precisely because they are scarce), and the chance of "win-lose" and "lose-lose" outcomes will increase as well. Identifying which forests should not be forested is a tall challenge, precisely because it requires giving attention to three (not always mutually consistent) criteria and because much remaining tropical forest cover falls into the "ambiguous" zone described in table 22.1. The three categories are not discrete but rather the end points and the midpoint of a gradient that ranges from clearly unsuitable to extremely suitable.

The chapter's basic thesis is as follows. Individuals and businesses deforest because it is their most profitable alternative. To get them not to deforest in situations where forest clearing is inappropriate, deforestation must be made less profitable, or other alternatives (either based on retaining forests or completely outside forest areas) must be made more profitable.

Deforestation can be made less profitable by
- Reducing the demand or prices for products produced from newly cleared land
- Increasing the unit costs and riskiness of activities associated with deforestation
- Eliminating speculative gains in land markets.

Alternatives to deforestation can be made more profitable by
- Increasing the income stream to be obtained from maintaining forests
- Reducing the costs of maintaining forests

Table 22.1. Conceptual Framework for Distinguishing "Appropriate" and "Inappropriate" Deforestation
(criteria influencing appropriateness)

Form of deforestation	*Biophysical*	*Economic*	*Political*
Inappropriate	High adverse local or downstream effects of conversion, high biodiversity, and high biomass	High utility to local people as forest and low or temporary agricultural (or other) potential	No potential stakeholders ultimately gain from the conversion ("lose-lose")
Ambiguous	Intermediate status between the characteristics listed above and below	Intermediate status between the characteristics listed above and below	Some stakeholders win, others lose from the conversion ("win-lose")
Appropriate	Low adverse local or downstream effects of conversion, low biodiversity, and low biomass	Low utility to local people as forest and high and lasting agricultural (or other) potential	All potential stakeholders gain from the conversion

Note: One might rightly ask, "If the situation is 'lose-lose,' why would someone in this situation deforest?" Here we mean "lose-lose" in two possible senses. First, the agent(s) deforesting expected a gain but had to move (and therefore "lost") because the soil could not support agriculture. Second, stakeholders appear to gain (because they obtain a source of livelihood through deforestation) yet lose in the relative sense (because they could have obtained a better livelihood through means other than deforestation).

- Increasing the opportunity costs of labor and capital that might otherwise be used in activities associated with deforestation.

The chapter's organization follows this typology of policy options for reducing inappropriate deforestation. The first section focuses on policies affecting the prices of and demand for agricultural and forestry products. The second section discusses policies that make production associated with deforestation more or less costly and risky. Section three examines policies that affect land speculation. Section four looks at different ways to increase the profitability of maintaining forests. Section five addresses the issue of the opportunity costs of capital and labor. All the policies analyzed are evaluated on the basis of the following six criteria: effectiveness, ability to be targeted, equity, political viability, direct cost, and indirect cost. In each section, a summary table evaluates the policy reforms discussed with respect to these criteria.

We conclude that there is no perfect or generalizable policy for reducing inappropriate deforestation. There are no "first-best" options. Each national situation is different, much uncertainty remains about key cause-and-effect relations, and tradeoffs usually exist between the criteria for evaluating the policies. Some policies are effective, but difficult to target, costly, or not politically viable. Other policies are easier to target, but more costly or not politically viable, and so forth.

Policies to Reduce Prices and Demand for Tropical Agricultural and Forestry Products

Demand for tropical products is a function of the size and incomes of the consuming population, relative prices, quantitative trade restrictions, and consumer preferences. Policies that affect any of these variables can also affect demand. This section first analyses the effect on forest clearing of policies that affect the demand for agricultural goods, then looks at policies that influence the demand for forest products.

High prices and demand for tropical agricultural commodities can stimulate the clearing of forests to produce these commodities, as illustrated by the "hamburger connection" in Central America or the cassava fodder export schemes in Thailand. Therefore, under certain circumstances, policies that reduce the demand for tropical agricultural products could limit the conversion of forests to croplands and pastures.

This type of approach has certain advantages. The policies involved are mostly national in scope, and governments can implement some of them easily and cheaply (at least with regard to their direct costs). Because the policies' direct effects are proportional in magnitude to the quantity of products sold, larger producers bear most of whatever negative effects the policies may have on producers' incomes.

Nevertheless, these policies are blunt instruments, which are difficult to target and thus likely to have perverse side effects. They rely entirely on the market to distinguish where and how supply should be reduced, and this is unlikely to meet environmental objectives efficiently. In some cases, lower prices may lead supply to fall more rapidly on lands that are economically marginal for agriculture, but this does not hold in other situations, and land that is economically marginal for agriculture is not necessarily the most important land to maintain under forest cover to meet the objectives of biodiversity, watershed protection, or timber production.[4] Often, it is preferable to find ways to meet the demand for agricultural products by using policies designed to promote more intensive agricultural production or concentrate production where it is most appropriate rather than by reducing aggregate demand. In fact, depressing the demand for tropical agricultural products lowers the incentives for technological changes that might lead such products to be produced more intensively, on less land. When deforestation is driven largely by land speculation or production of subsistence or nonmarketed goods, these policies will have little impact. Besides, it is difficult to win political support for policies that restrict the demand for or production of tropical products, particularly export products, and that restrict the incomes of primary producers in tropical developing countries.

Where high agricultural prices result from direct or indirect government subsidies, as is often the case, for example, with sugarcane production, the argument for reducing demand by eliminating those subsidies is somewhat stronger. Guaranteed price support schemes have, in the past, increased the expected profitability of deforestation both by increasing the expected sale price of outputs and by reducing the commercial risks associated with such ventures. Market promotion and marketing boards have similar effects: they both reduce market risk and volatility and increase long-run average returns to producers. Under these circumstances, eliminating subsidies or marketing assistance for products from newly cleared forestlands may be effective, low cost, and socially progressive, although in some countries it may be politically unacceptable because of the perceived impacts on farmers' incomes. In particular, the fact that most price subsidies accrue to large farmers, who produce most of the (subsidized) agricultural products, may make them politically difficult to eliminate.

Ceteris paribus, higher population implies more demand for agricultural products. Depending on the scope of the markets, the level of local, national, or international population may be relevant. However, income level and distribution, consumer preferences, and other factors that affect relative prices generally influence demand even more, at least in the short or medium term. Moreover, population growth can also influence forest clearing through numerous other causal relations, which often have conflicting effects (Bilsborrow and Geores 1994). After a considerable lag, policies that lead to lower population growth could potentially limit deforestation by lowering the demand for agricultural products, but the magnitude of that effect is unlikely to be large. Because population also affects many other important policy variables, governments are unlikely to design population policies principally with regard to their impact on deforestation.

Per capita income and economic growth greatly influence the demand for tropical products, although, again, this is only one way they affect deforestation. Several authors have speculated about the existence of an "environmental Kuznets curve," whereby at lower levels of per capita income economic growth leads to greater deforestation as a result of increased demand for tropical products, but after income levels reach a certain threshold, deforestation declines because the economies become less dependent on agricultural and forest products, primary production becomes more intensive, new employment opportunities arise, and demand for forest preservation grows (Capistrano 1990).

The empirical evidence on this issue is weak, contradictory, and based mostly on cross-sectional data and on the few "tiger economies of East Asia," which may not reflect how the majority of tropical countries are likely to evolve in the future. What evidence there is, however, suggests that given existing levels of per capita income in most tropical countries, without a change in the pattern of development, economic growth is likely to lead to greater forest clearing in the medium term (Stern, Common, and Barbier 1996). This implies that one way to avoid a tradeoff between economic growth and deforestation is for countries to promote patterns of economic development that are less dependent on primary agricultural and forestry commodities. The alternative is to develop a well-managed, ecologically and economically sustainable forest sector. In either case, it is essential to adopt policies that limit inappropriate forest conversion.

Governments influence the relative prices of, and hence demand for, tropical products through various measures, including exchange rate policies, tariffs, nontariff trade barriers, pricing of government utilities, and price controls, among others. The trend in tropical countries in recent years has been toward trade liberalization and more free market policies. The European and U.S. governments, in contrast, continue to use tariff and nontariff trade barriers to regulate imports of agricultural products to meet commercial, political, and environmental objectives.

Real currency devaluations, designed to increase the relative price of tradable goods and services compared with nontradables, generally increase real producer prices of tropical agricultural products, making it more profitable to convert forestland to agriculture (Capistrano 1990 and Wiebelt 1994). This problem is most likely to arise where export producers are important agents of deforestation, such as in certain countries of Central America (beef), South America (soybeans), and West Africa (cocoa; Kaimowitz, Thiele, and Pacheco 1996 and Kaimowitz 1997).

Because exchange rate policies are broad and far-reaching, under most circumstances concerns over inappropriate deforestation cannot be expected to be a major factor in their design, nor should they be. However, in situations where a large portion of anticipated export growth generated by devaluations comes from the expansion of agricultural production to areas currently under forests, the issue cannot be avoided. Policymakers should explicitly consider the tradeoffs between the possible benefits of short-term increases in exports and the costs associated with forest conversion. Alternatively, they might try to devise simultaneous parallel measures to minimize the predictable adverse impacts.

A similar argument applies to agricultural price policies focusing on production for domestic consumption, such as price controls, import restrictions, guaranteed minimum prices for food crops, or public marketing of foodstuffs. In most cases, the issue of deforestation will appropriately receive little attention when designing these policies. In some instances, however, policy changes can be expected to shift relative prices in a way that will promote rapid agricultural expansion in forested areas. In Ghana, for example, Munasinghe and Cruz (1994) have shown that policies that improve agricultural prices promote extensive, rather than intensive, agricultural growth and that the environmental costs may not justify the benefits. In such circumstances, policymakers should explicitly consider whether short-term agricultural growth is worth the long-term consequences and implement damage control measures alongside the necessary macroeconomic reforms.

The conversion of forests to pastures in Central America illustrates many of these points. Many authors have cited this as a case where strong demand for beef imports in the United States in the 1970s and high international beef prices led cattle ranching to expand rapidly at the expense of forests (Myers 1981). During the 1980s, however, Central American beef and dairy producer prices fell sharply as a result of increased protectionism and reduced per capita beef consumption in the United States, domestic price controls, overvalued exchange rates, and international donations of powdered milk. This, along with other factors, reduced the pace of forest conversion to pasture, but by no means stopped it. Ironically, when demand and prices declined, livestock production fell more rapidly in traditional cattle regions than along the agricultural frontier, because producers on the frontier had few alternatives to investing in cattle, could simultaneously speculate in land markets, and obtained high initial yields from pasture in recently deforested areas (Kaimowitz 1997). Thus, as with the cocoa example (Ruf and Siswoputranto 1995), it cannot be generally assumed that the "new" producers at the forest margin will be the first to become nonviable as demand contracts—although they are physically "at the margins," they are economically less marginal than the producers in "older" areas.

Similar arguments apply to tropical timber, but there are important differences. Because industrial logging in most tropical countries is selective and does not involve direct conversion of large forest areas to other uses, the effect on forest conversion of policies that affect the demand for forest products is much more indirect. Logging companies frequently facilitate forest conversion to other uses by building roads that assist farmers subsequently to enter and clear land or by providing investment capital to farmers or ranchers that can be used to clear land or purchase cattle. For example, the highest rate of population increase in the Philippine uplands was in the municipalities with logging concessions (Garrity, Kumar, and Guiang 1993). But in other situations, they do not affect forest conversion.[5] Moreover, policies that depress the demand for timber can be a disincentive for sustainable forest management, where that is a possibility, even though in the short term they limit some of the negative effects associated with current logging practices. The question, therefore, is not to reduce the aggregate demand for tropical timber, but rather to reduce the demand for the tropical timber that is closely and visibly associated with inappropriate tropical deforestation and degradation, while, if possible, stimulating the demand for tropical timber that can be proven not to be linked to deforestation. We return to this issue below.

As with agricultural products, larger populations, economic growth, exchange rate devaluation, and more free market pricing policies generally increase both the domestic and export demand for timber and hence logging. Large devaluations, in particular, have been associated with major increases in tropical timber exports from Bolivia, Central Africa, and Southeast Asia (Anderson, Constantino, and Kishor 1995 and Tchoungui and others 1995).

There has been much controversy over the use of log export taxes or bans to reduce the logging of natural forests, among other objectives (Barbier and others 1994).[6] These policies are relatively easy to enforce (although there always tends to be some slippage) and generally reduce the amount of logging in the short run, but at high economic cost. Their medium- and long-term impacts are less certain, but many experts believe that bans discourage efficient logging, wood processing, and investment in sustainable forest management and encourage the domestic consumption of timber. This sometimes leads to even greater logging than would have occurred without the ban (as occurred in Indonesia).

Bans and taxation of tropical timber imports, currently being discussed in certain industrial countries, are similar, with two notable differences. First, because not all countries that import timber are likely to adopt such a policy, its effect on the domestic prices of timber in exporting countries will probably be smaller than with an export ban (Buongiorno and Manurung 1992). Second, importing countries appear to be more prepared to distinguish timber taken from sustainably managed forests from timber taken from forests that are not sustainably managed.

This could provide an incentive to invest in sustainable forest management that does not exist with traditional log export bans. It could still prove difficult, however, to define sustainable forest management and monitor whether imported timber comes from forests that meet those standards.

Finally, brief mention should be made of the case of commercial firewood collection, which is sometimes directly responsible for clearing forests in drier periurban areas. Higher urban incomes generally lead to rapid substitution of fuelwood with other energy sources and hence can actually reduce the pressure on forests (and vice versa.) Attempts to promote the substitution of fuelwood by providing subsidies that lower the price of energy sources such as kerosene, butane, or electricity, however, have had little success in reducing the consumption of fuelwood and have been expensive for governments (Mercer and Soussan 1992).

In summary, as shown in table 22.2, any policies designed to restrict forest conversion by reducing the demand for primary products are blunt instruments, which make them alternatives of last resort. In some circumstances, however, they may be the only realistic alternative for achieving rapid reductions in inappropriate deforestation. Policymakers must become more conscious of the potential impact that their macroeconomic, agricultural, and trade policies have on forest clearing, actively promote patterns of economic growth that rely less on the liquidation of natural capital, and take a more long-term view in assessing policy tradeoffs.

Policies That Increase the Costs and Risks of Production Associated with Deforestation

In this section, we examine policy measures that can reduce inappropriate deforestation by increasing the costs and risks of production associated with deforestation. Often these policy measures eliminate government subsidies or other distortions generated by previous policies and thus actually save governments money, rather than involving additional expenses. First we review policies related to agricultural production, such as agricultural input subsidies, colonization schemes, and research and extension. Then we look at policies affecting logging in natural forests. Finally, we examine the implications of road construction and maintenance and of zoning and protected areas, which affect both agriculture and logging.

Basically, government actions that encourage agricultural intensification tend to slow deforestation at the forest-agriculture frontier, while actions that encourage extensification of agriculture (bringing

Table 22.2. Demand-Related Policies

Policies	*Effectiveness*	*Targetability*	*Direct costs*	*Indirect costs*	*Equity*	*Political viability*
Population control	Limited	None	Moderate	Low	Unknown	Moderate[a]
Limits on economic growth	Moderate	None	Low	High	Unknown	Low
Appreciated exchange rates	Moderate[b]	None	Variable[c]	High	Mixed	Variable[c]
Price controls on tropical products	Moderate	Limited	Moderate	High	Mixed	Low
Export bans and taxes on logging	Moderate[d]	Limited	Low	Variable[e]	Favorable	Variable
Import restrictions on primary products (coffee, cocoa, timber, or beef)	Moderate[d]	Moderate	Low[f]	Unknown	Favorable	Moderate

a. Population control is not politically viable in some highly religious countries.

b. Appreciated exchange rates will be more effective where forest conversion is for export production.

c. Appreciated exchange rates have higher direct costs for governments and are less politically viable when countries have balance of payments or debt problems.

d. Policies are more effective in the short run.

e. Indirect costs depend on the size of the sector and are higher in the medium and long run.

f. Targetability of import restrictions is higher if certification is involved, but direct costs are also.

new land into production) are likely to promote deforestation. We define intensification as increasing the amount of labor, capital, and information applied per hectare of land. Whether the expansion of market demands and higher product prices lead to more or less pressure on the forest-agriculture frontier will depend on the farmers' choice of technology (that is, how much to intensify and how much to expand an area) in response to perceived opportunities. Assuming that farmers respond rationally to the expected net returns from either means of increasing their output, any public policies that affect their input prices will affect their decisions. The effects of government subsidies for agricultural inputs such as fertilizer, water, seeds, pesticides, credit, fuel, and transport on agricultural intensification and extensification are far from uniform. Subsidies for supplying irrigation water generally facilitate intensification, because it is not feasible to supply irrigation water to remote areas at the forest frontier. Fuel and transport subsidies tend to favor extensification by making it more profitable to farm in remote areas. Livestock credit also tends to facilitate extensification, because tropical livestock production is typically an extensive activity. Fertilizers, seeds, and agricultural credit subsidies can facilitate either intensification or extensification. It is generally accepted that the agricultural green revolution, whose success was at least partially due to the subsidized availability of the inputs required for intensification (credit, fertilizer, pesticides, transport), helped to save considerable areas of forests in Southeast Asia by making the cultivation of unirrigated rice on forest hillsides much less profitable.

Governments of many countries have deliberately and explicitly encouraged forest conversion in the past through official resettlement and colonization programs, and some still do, for example the Trans Amazonia Highway and the Rondônia colonization scheme in Brazil and the transmigration program in Indonesia. Because so many governments have used these policies, it might seem at first glance that arguing for their removal would be hopeless. Moreover, many governments in Africa, Asia, and Latin America have used "national defense" arguments to justify their colonization schemes, and these arguments carry great political weight. Nevertheless, the current political climate favoring the elimination of public subsidies and less government intervention tends to favor these reforms.

Removing credit and input subsidies and eliminating land colonization and resettlement schemes to reduce inappropriate deforestation would probably be moderately effective, be readily targetable, have no cost to the national treasury, and result in substantial savings in budgetary outlays. Its political viability is unclear. We suggest that subsidies or programs that encourage extensive expansion of agriculture would promote deforestation, while subsidies that favor intensification would tend not to encourage deforestation. Thus it would be appropriate to distinguish among agricultural subsidies, before concluding that their abolition would help to slow deforestation. For example, removal of fertilizer subsidies in parts of Sub-Saharan Africa may well accelerate deforestation. It would be a mistake, therefore, to believe that eliminating subsidies alone is sufficient to solve the problem of inappropriate deforestation. Brazil and Central America, for example, greatly reduced the tax incentives and subsidized credit for cattle ranching in recent years, and in both cases inappropriate deforestation declined, but remained high (Ozório de Almeida and Campari 1995 and Kaimowitz 1997).

Although agricultural research and extension activities have rarely been aimed explicitly at promoting deforestation, nevertheless there are instances where farmers adopted the recommended new technologies not only on their existing fields but also on new fields that were formerly forest (even protected forests and national parks). As Angelsen (1995) notes, if a new agricultural technology is worth adopting on existing farmland, it may also be viable on nearby lands that were previously left in forests and may make it profitable to farm in places where previously it was not. Thus, for example, improvements in soybean technology in Bolivia and Brazil have been an important factor in promoting the clearing of forest areas to expand soybean production. New pasture varieties may also have made it profitable to clear forest for pasture in Latin America in places where otherwise it would not have been. One possible policy reform would be to ensure that the content of agricultural extension services is not biased to encourage deforestation. However, such measures are likely to be only moderately effective, they are difficult to target, and, although they have few cost or equity consequences, they are unlikely to find a strong political constituency.

Frequently recommended policy reforms that would reduce the profitability of industrial logging (and so presumably reduce the amount of inappropriate deforestation due to such logging) include increasing stumpage charges and other fees and reducing other direct and indirect subsidies to timber processors. Timber stumpage charges and other timber fees are often extremely low, and governments frequently justify this as a way to attract new investment or to capture social benefits like infrastructure development and employment creation. For example, the Bangladesh Forest Industries Development Corporation made only notional payments (a few cents per ton) for areas, logs, and bamboo for more than 20 years (FAO 1993). The Hindustan Paper Corporation in Kerala (India) paid $0.50 per ton of eucalyptus pulpwood compared to the Forest Department's direct production costs of almost $25 per ton (FAO 1993). Naturally, the industry operations expanded considerably, but when these stumpage charges were eventually increased to realistic levels, the company's activities contracted, leaving excess capacity, unemployment, demands for more plantations, and yet more logging of protected natural forests. Subsidizing log input prices is a very inefficient way of creating additional employment and discourages intensive recovery of finished products in mills.[7] The systems used for (noncompetitive) allocation of logging concessions and licenses have attracted almost as much criticism as the log-pricing issue, because of the effects on mismanagement of forests and accelerated deforestation as well as the revenues forgone by the state. The issuance of 500 timber concessions in Thailand by 1968, covering half of the country, opened up vast new areas for encroachment and official agricultural development (FAO 1993).

Substantial increases in the fees and charges set by governments for the exploitation of natural forests might reduce logging in inappropriate areas and eliminate some of the least efficient operators from the industry, who may also be among the most destructive. But their effectiveness for reducing inappropriate deforestation could be mixed if they also deterred sustainable forest management, so they would need to be carefully targeted. There would be no direct costs; rather they would increase government revenues if they did not lead to massive closures and reductions in volumes more than proportional to the price increases. Although the equity implications could be very favorable, the political viability is uncertain at best.

As one form of indirect subsidy to local industry, countries have operated a two-price scheme, where logs for local processing are considerably cheaper than logs for export. Accelerated industrialization has been accomplished in Indonesia and Malaysia, through a suite of policy measures linked to trade and tariff policies, tax incentives, log export restrictions, "cheap logs," worker training programs, and so forth (de Los Angeles and Idrus 1990). However, most independent studies suggest that the short-run economic cost to Indonesia has been extremely high (Lindsay 1989 and Douglas 1995). Even Malaysia's successful and widely acclaimed policies for accelerated industrialization and "downstream, value added processing" have recently been assessed by a World Bank team as economically inefficient and too ambitious. By making logging more profitable than it would otherwise be, the policy on subsidies contributed to the expansion of area logged annually. Further, if the prices of wood were set at artificially low levels, distorting wood processing decisions and leading to excessive physical wastage (Douglas 1995), then greater areas would have to have been logged to achieve the same final output of timber products.

Road construction plays a crucial role in deforestation and has been the topic of much analysis and debate. It is often argued that new road construction presents the greatest threat to remaining forests by providing access to forested areas (Bryant, Nielsen, and Tingley 1997 and Dudley, Jeanrenaud, and Sullivan 1995). Roads are often essential both to market farm outputs and to deliver inputs to the farm and can greatly increase the profitability of agricultural and wood production in frontier areas. In many countries one of the most common forms of assistance requested by farmers is the extension and upgrading of road transport systems.

Sometimes, however, roads do not create opportunities for deforestation; rather, the existence of those opportunities leads to the construction of roads. If prospective farmers perceive their best opportunity to be clearing some forest and converting the land to another use, they can and will ask governments to construct roads to facilitate that. In this case, roads facilitate deforestation, but they may not be its principal cause. Thus, in analyzing the impact of public road construction on forests, it is important to take into consideration the reasons why specific roads have been built. In some cases, the prime impetus is to provide access to timber for logging, in others it is for farming, and in still others (such as the Trans Amazonia or Trans Borneo highways) it is for military or "national development" reasons.

The road construction and transport policy reforms that might be required to reduce deforestation by increasing the costs of deforesting activities do not necessarily involve less road building, but rather a change in the nature, type, and location of roads built. In particular, existing road systems would have to be intensified, rather than expanding new roads into previously remote forested areas. Such reforms could be effective, targeted, and budget-neutral and could have low indirect costs. However, because increased road intensity could lead to capital appreciation of existing farmlands, the effects may be regressive. In addition, because of the strong political demand for roads by landholders in remote areas and many governments' interests in promoting colonization in those areas, such reforms may not be politically palatable.

Although the topics of land use zoning and the declaration and enforcement of protected areas are vast and warrant many volumes (Wells, Brandon, and Hannah 1992), we confine ourselves to summary comments. The rate of deforestation can be substantially reduced where protected areas are effectively managed and have the support of surrounding populations. However, where effective management and popular support do not exist, declaration of parks or national forests may actually accelerate deforestation. Many countries have claimed national forests, parks, and reserves but have been unwilling or unable to enforce such claims effectively. If governments were to enforce reserve boundaries more effectively and were consistent in not recognizing "squatters" within reserves, this could substantially increase the risks and reduce the expected profitability of deforestation. Although such measures could be targeted, they may be expensive, ineffective, and inequitable in developing countries with many poor landless people. Forest resource degradation has been made worse by restricting peoples' rights to use land and neglecting their traditional uses and their capacities to preserve forests. A new approach that actively engages local people in decisions and management, and ensures that they are not disadvantaged by forest conservation or production activities, is currently being applied and evaluated worldwide.

It is not difficult to conceive of a suite of policies that would retard or discourage deforestation of protected forest areas by increasing the risks and consequences to those who deforest. However, there seem to be few bases for optimism so long as the prevailing political tendency is to ignore deforestation and encourage the expansion of alternative land uses. As summarized in table 22.3, most such policy reforms

Table 22.3. Policies That Increase the Costs and Risks of Production Activities Associated with Deforestation

Policies	*Effectiveness*	*Targetability*	*Direct costs*	*Indirect costs*	*Equity*	*Political viability*
Reduced subsidies for certain agricultural inputs linked with expansion of farm areas	Moderate	Moderate to high	Saves money	Low	Favorable	Very low
Reduced support for colonization and settlement schemes	Moderate	High	Saves money	Low	Favorable to mixed	Very low
Reduced technical and advisory support for activities on newly cleared lands	Low	Low	None	None	Unclear	Low to moderate
Reduced road and transportation subsidies	Moderate	Moderate	Saves money	Low	Favorable to mixed	Variable
Removal of subsidies to logging and forest industries doing destructive exploitation	Moderate to high	High	Saves money	Some possible decline in exports	Favorable	Variable
Removal of tax and credit subsidies for agriculture on newly cleared lands	High	Moderate	Saves money	Low	Favorable	Low

appear to be moderately effective, equitable, and readily targeted and would reduce government expenditures. However, such subsidies currently exist because many governments apparently see the expansion of agriculture and (eventually) higher incomes for farmers as desirable goals, irrespective of the consequences for deforestation.

Policies That Discourage Forest Clearing to Establish Property Rights

Many people clear forests not only to reap the benefits of selling standing timber or producing crops and livestock on the cleared land but also to gain property rights over that land or to strengthen their existing rights (Ozório de Almeida and Campari 1995, Angelsen 1995, and Mendelsohn 1994). This allows them to earn an implicit rent from the land and perhaps later sell the acquired land at a substantial profit. In many Latin American countries, one condition for prospective landholders to obtain legal recognition of property rights over formerly public lands is to demonstrate that they are "productively" occupying those lands. They do so by clearing forest and planting pasture or crops. Deforested land is also better protected from expropriation by agrarian reform agencies or invasion by squatters (Utting 1992). In Southeast Asia and West Africa, farmers frequently clear forests to plant perennial crops such as cocoa or rubber, partly to establish or improve their property rights over that land (Angelsen 1995). Even when land is private, forests often belong to governments. This gives landholders an additional incentive to clear the forest to obtain broader property rights (Dorner and Thiesenhusen 1992) or greater legal recognition of de facto claims.

In much of Africa, communal land tenure regimes and ethnically segmented land markets traditionally have limited the relevance of land speculation as a factor in deforestation, although perennial crops are often planted to establish long-term usufruct rights (Dorner and Thiesenhusen 1992). This situation has slowly begun to change, however, as integration into the global market economy and demographic pressures have increased, and governments have sought to establish private property regimes.

To discourage the clearing of forests in order to gain or improve property rights, governments can use measures that break the link between forest clearing and tenure, can tax nonforested (cleared) land more heavily than forested land, or can reduce the opportunities for capital gains from land sales. This could potentially be achieved through a combination of land tenure, taxation, infrastructure, credit, and macroeconomic policies.

Some Latin American governments have changed their land titling and agrarian reform policies in an attempt to decouple forest clearing from increased tenure security. Formally, they no longer require public lands to be cleared for claimants to obtain or maintain property rights. They now deny the right to request land titles to farmers who have inappropriately deforested the land after the new policies have gone into effect (Sunderlin and Rodríguez 1996). Land titling programs for humid tropical areas have attempted to provide secure property rights to existing settlers, without obliging them to clear land for that purpose. Many Latin America countries require people who receive property rights over previously public lands in tropical forest areas to maintain a certain percentage of that land in forest. Brazil, for example, for many years required 50 percent of each rural property within Legal Amazonia to remain forest and now requires 60 percent (Burford 1991).

The impact of these policies has ranged from mildly effective to negative, with the main problems being weak implementation, forest clearing provoked by uncertainty surrounding the policy changes, and the infusion of new migrants attracted by the chance to obtain land titles. Governments find it somewhat expensive to implement these policies and hard to target them based on specific local conditions. Where governments are unwilling or unable to enforce legal property rights, land claimants keep clearing forests to ward off other potential claimants. Although there are typically few strong objections at the national or international level, local elites usually oppose any attempts to restrict their ability to gain property rights over deforested lands. As a result, the policies may be less equitable in practice than they are in theory.

Throughout the tropical world, there is increasing interest in establishing or supporting existing common property regimes for land and forests to protect indigenous and other forest-dependent people, to encourage sustainable forest management, and to limit the scope of land markets. The literature on the topic rarely stresses the third aspect, but it is potentially quite important.

Common property regimes can effectively limit deforestation related to land speculation in certain places. They are relatively cheap to establish legally where they already exist informally, and governments do not have to compensate other land claimants. However, they are difficult to create where informal common property regimes did not exist in the past (Richards 1997). Policies that promote common property regimes are equitable in the sense of supporting groups that have traditionally been poor and marginalized, but they often involve ensuring large amounts of resources to small numbers of individuals.

Theoretically, land taxes could be used to discourage forest clearing by taxing cleared lands at a higher rate than forested land. Land taxes and capital gains taxes could also potentially discourage land speculation by raising the cost of holding land as a hedge against inflation or as a source of capital gains, charging landowners for the costs of infrastructure improvement, and lowering the price of land (Strasma and others 1987). However, experience to date has not been encouraging. Because few countries have used punitive land and capital gains taxes to discourage land speculation, it is hard to assess their practical effectiveness. Most governments have found it difficult to administer such taxes effectively, because of the large amounts of information required and the high potential for evasion, or have avoided them, because of potential political opposition (Skinner 1991).

Because the purchase of land is a common hedge against inflation, macroeconomic policies that reduce inflation may also limit the incentives for land speculation. This seems to be particularly relevant for certain Latin American countries, such as Brazil, where high inflation rates were previously common but were not associated with general economic collapse. It is also true, however, that as long as infrastructure investments continue, regional markets grow, and public services gradually improve in agricultural frontier areas, real local land prices will rise and will encourage land speculation. There is ample evidence that similar processes are at work in Indonesia, the Philippines, and Thailand.

Certain government subsidies can increase land prices and thus stimulate deforestation for speculative purposes where property rights can be secured or improved by clearing land (Binswanger 1994). Road construction, in addition to facilitating access to forested areas, raises land values and offers opportunities for capital gains (Angelsen 1995). Credit subsidies or tax incentives for agriculture or livestock that are preferentially available to landholders also increase the value of holding land. Government purchases of land for protected areas, refugee resettlement, infrastructure construction, or other purposes often inflate local land markets. Removing these subsidies can greatly reduce inappropriate deforestation, especially where high subsidies existed in the past. Governments can actually save money by doing this, and there are few indirect costs. Because most such subsidies presumably go to wealthy families and corporations, except perhaps for road construction, eliminating them tends to favor equity. Important interest groups support these subsidies, but others often oppose them.

In summary, eliminating subsidies that promote land speculation and strengthening either private property or common property regimes can sometimes be effective tools against inappropriate deforestation (see table 22.4). Most other proposed policies that might limit forest clearing to gain secure tenure still have not been tested fully. Land taxation, in particular, merits greater consideration, given the notable gap between its theoretical potential and practical difficulties. Most

Table 22.4. Policies That Discourage Deforestation to Establish Property Rights

Policies	*Effectiveness*	*Targetability*	*Direct costs*	*Indirect costs*	*Equity*	*Political viability*
Land titling policies	Negative to low	Moderate	Moderate	Low	Unknown	Moderate
Support for common property regimes	Moderate	Moderate	Low to moderate	Low	Mixed	Variable
Land and capital gains taxes	Unknown	Low	High	Low	Unknown	Low
Fewer credit, tax, and road subsidies	Moderate	Moderate	Saves money	Low	Favorable to mixed	Variable
Anti-inflationary macro policies	Unknown	Low	Low	Variable	Mixed	Moderate to high

policies discussed in this section are easy to target to specific regions but are much more difficult to adapt to the conditions of individual properties. Generally, the more effective (and less purely symbolic) these policies are likely to be, the less politically viable they are, although eliminating certain credit subsidies and tax incentives is apparently an exception.

Policies That Increase the Profitability of Maintaining Forests

Having considered many policy instruments that might reduce the profitability of removing tropical forests, we turn now to possible interventions that could increase the profitability of retaining them and managing them on a long-term sustainable basis. We consider three aspects in this regard: increased revenues from timber, increased revenues from nontimber forest products, and the possibility of payments for global environmental services.

To the extent that excessive logging (both in area and in degree of damage caused) contributes to deforestation, some recent efforts to reduce deforestation have focused on policies to encourage forest management and logging practices on concessions and state forests that are less destructive and more sustainable (Bryant, Nielsen, and Tingley 1997 and Dudley, Jeanrenaud, and Sullivan 1995). Analysts (Douglas 1995) have suggested that market forces could provide a significant incentive for more sustainable forest management, as the expected future values of remaining commercial forests increase greatly and as accessible stocks of tropical forests are depleted. However, recent assessments (Byron and Ruiz Pérez 1996 and Sayer and Byron 1996) of how forestry practices may change in response to changing markets for forest products conclude that "frontier logging" of relatively remote areas in the tropics may become less important in the future. Once the forests in more accessible areas (close to roads, rivers, and ports) have been exploited, any cost advantage that these forests might have had will be rapidly eroded. Remaining natural forests will become less able to compete with the outputs of the rapidly expanding plantation sector in the tropics and subtropics, except in specialty markets. In contrast to rising costs and declining quality of logs from natural forests, the volume and quality of plantation material will continue to improve, and technological advances in plantation silviculture and wood processing will continue to lower unit production costs.

Thus although some market assessments have proposed that market forces will increase the potential timber revenues from retaining natural forests for high-value products, we have seen little basis for relying on that. The effect is likely to be too little, too late. Further, if timber concessionaires harbor doubts about their long-term security of tenure, they are likely to continue to behave in a short-term exploitative way; even if they agree that tropical timber values might soar in the future, they doubt that they will benefit from such price rises.

In the very short term, higher timber prices and more roads may encourage unsustainable practices but in the long term may be necessary for sustainable practices to be financially viable. Short-term licenses do not encourage long-term stewardship of forests, but even 25-year licenses are less than many cutting cycles, destroying any motivation to nurture regrowth. "Cleaning operations" every 5 to 10 years can be observed. Even with very long-term licenses, concessionaires in many tropical countries may behave as if their tenure were insecure, because of extreme uncertainty about political changes. In many countries, logging rights are not transferable, reducing the incentive to maintain a high residual value of the forest enterprise "as a going concern." Transferability of licenses might indirectly impose commercial penalties for owners who have not looked after the forest under their control and might provide financial rewards (as higher transfer prices) for those who have not high-graded, who have protected advanced growth, and who have installed good infrastructure. Systems like the "evergreen licenses" devised in British Columbia (Canada) with "clawback" provisions—providing strong commercial incentives for long-term sustainable forest management—may warrant consideration and application in tropical developing countries.

In 1990 the International Tropical Timber Organization (an agency of UNCTAD) reached agreement between producer and consumer countries that by the year 2000, all tropical timber entering international trade would come from sustainably managed forests. Other recent initiatives (such as those of the Forest Stewardship Council) advocate policies that might make sustainable timber management more profitable (through independent inspection and certification of secure and longer-term concessions, accompanied by ecolabeling of products) to achieve premium prices or preferential market access for timber from certified, well-managed forests. Since 1990 nongovernmental organizations (NGO) and, to a lesser extent, governments have urged consumers to purchase only certified or ecolabeled tropical timber. They wish to exclude or minimize the sale of timber from poorly managed forests, while simultaneously obtaining premium prices for certified timber.

The certification approach has generated much debate: Who certifies? On what basis? Who controls the chain of custody between the forest and the final consumer to guarantee validity? How many consumers would pay what sort of premium for certified timber? Would this system be voluntary or compulsory (and if compulsory, is it legal under the rules of the World Trade Organization)? In spite of the many unresolved issues, polices by both governments and NGOs to encourage and respect certification might give some positive commercial rewards for sustainable forest management. Although it is generally politically acceptable, it remains unclear how effective (or even relevant) certification will be in reducing inappropriate deforestation on a broad scale, where conversion to other land uses poses the primary threat to forests.

To summarize, the policy interventions that might reduce deforestation by increasing expected long-term timber values rely on the identification and development of niche markets for tropical timbers—small quantities of very high-value wood carefully removed from forests. Certification and ecolabeling are a special case of this approach that attempts to dissociate "sustainable forest management" from exploitative logging that causes deforestation. The feasibility, effectiveness, and costs of this approach remain to be determined. If one of the major disincentives to sustainable forest management is insecurity of tenure or fear of expropriation, removal of such disincentives would appear an attractive, even necessary, policy option. But there is no evidence that this is a sufficient condition for slowing commercial deforestation.

Numerous international NGOs (such as Cultural Survival Enterprises and Conservation International) have argued that finding new products in tropical forests, finding new markets for new or existing products, and marketing nontimber forest products more effectively will make tropical forests much too valuable to destroy. Both government institutions and NGOs have incorporated such efforts into their policies and programmatic agendas, ranging from the World Bank pilot project for Brazil (with a significant component aimed at supporting four extractive reserves in the Brazilian Amazon) to local NGO initiatives to promote new tree species of economic potential in the highlands of northern Sumatra. Numerous substantial policy and development initiatives are pursuing this proposition. Yet debate is raging about the extent to which efforts to promote the extraction, use, and marketing of nontimber forest products represent a real option for slowing deforestation and improving the livelihoods of forest people. Dove (1993) has argued that incorporating nontimber forest products into international markets will achieve neither socioeconomic nor environmental objectives and may actually be counterproductive in both senses—at best it diverts attention from what he sees as the real causes of deforestation and degradation and their political resolution.

The potential for these policy measures to slow deforestation depends on whether one accepts that forest people are destroying tropical forests because the stream of income from forests is too low (or because they simply do not understand the magnitude of these benefits). Dove believes that forest destruction is caused by loggers, migrants, and predatory outsiders—that forest people are the victims rather than the perpetrators of deforestation. Thus he argues that efforts to halt forest destruction should focus on the economic behavior of the former, not of the forest people. He asserts that increasing commercialization of nontimber forest products will foster predatory exploitation and destroy traditional management systems that might have helped to protect forests.

Recent research by the Center for International Forestry Research—CIFOR (Byron and Ruiz-Pérez 1996)—has shown how the distribution of property rights and the ability (or otherwise) of local people to claim and enforce such rights are crucial determinants of the outcome of commercialization of nontimber forest products. For a strong "community" of users and collectors of nontimber forest products with organization and political empowerment, commercialization of nontimber forest products can enhance welfare and forest conservation. Conversely, the existence of a strong political elite whose members may seek to expropriate the value of nontimber forest products as soon as they realize that significant potential rents exist may lead to serious social and environmental exploitation.

Policy proposals and development actions aimed at slowing deforestation by increasing the stream of revenues from nontimber forest products for forest peoples could be effective, targeted, equitable, low cost, and moderately politically acceptable to governments of countries whose tropical forests are inhabited, but only in specific institutional and legal contexts.

It has also been proposed that having beneficiaries pay for the environmental services provided by tropical forests could provide a market-based stimulus for retaining forests and slowing deforestation. The proposed policy reform is to pay those who retain forests for the environmental services their actions provide to the wider public. Analogous to the upstream-downstream analyses in which beneficiaries compensate the providers of environmental externalities in a watershed or on a national scale, there have been numerous proposals for capturing the "option and existence values" of tropical forests, as perceived by affluent residents of industrial countries, as a way of internalizing some of the global externalities associated with tropical deforestation. The U.N. Convention on Biological Diversity calls for further research on the *equitable distribution of benefits and costs of forest conservation*, which is seen as a prerequisite to devising suitable international or intranational compensatory schemes. To date, this has focused on two main areas: biodiversity prospecting and carbon sequestration.

The agreement between Merck Pharmaceuticals and the Biodiversity Institute (InBio) of Costa Rica for the commercial exploration and assessment of that country's biodiversity has attracted great publicity. It appears to be an example of using commercial, market-based transactions to make retaining tropical forests more valuable than destroying them. Despite the publicity about the potential of "biodiversity prospecting" to reward forest managers and indigenous forest people for retaining forest biodiversity, it has recently been argued that the realistic potential of such payments is very small, apart from a very few, very small, niche markets (Simpson 1997).

Similarly, the U.N. Framework Convention on Climate Change has authorized the so-called "activities jointly implemented" whereby carbon dioxide (CO_2) emitters (such as power utilities) in industrial countries pay for carbon sequestration in developing countries, as offsets for their CO_2 emissions. To date, these offsets have taken the form of payments for the establishment of new forests (such as the Dutch Forests Absorbing Carbon Emissions—FACE—Foundation) or the payment to logging companies for the adoption of reduced-impact logging techniques, which can reduce the amount of damaged and decaying material left in forests and increase the amount of future carbon sequestration by healthy, rapidly growing immature forests. As yet, there is no legal obligation for power utilities (in the United States) to expend shareholder funds in Central America or Malaysia to offset their U.S. emissions, nor are there tax credits at present.[8] Most studies have looked at the demand side based on the theoretical costs if companies have to pay carbon taxes or if they have to pay to reduce their emissions, which most do not, except in Denmark and Norway.

The Framework Convention on Climate Change has debated proposals for "international tradable carbon emissions permits" that appear to be an efficient and low-transaction costs approach to determining who would emit and sequester carbon dioxide and where. Yet countless practical difficulties remain. In the context of their efficacy in preventing tropical deforestation, it appears unlikely that payments (whether for carbon, biodiversity, or watershed protection) would have much impact unless they actually went to those making the deforestation decisions on the ground, rather than to national governments.

Another recent notion is that industrial countries, or the private sector, or NGOs should offer developing countries with tropical forests some "payment for carbon storage." Apart from the practical difficulties of paying local decisionmakers *not* to do something, this would have to be considered not only for forests but also for oil and coal. Basically, it would seem to violate the "polluter pays" principle, which has generally guided international environmental policy and financial transfers. But these proposals also await the results of research documenting and measuring the *physical* relationships, that is, measuring watershed outputs and soil erosion losses, quantifying biodiversity, or measuring the physical extent of carbon sequestration, which are still far from complete.

In summary, policies to encourage the retention of tropical forests by capturing payments for the global environmental services provided remain of questionable effectiveness, targetability, and cost. So many unknowns remain that the whole area is decidedly ambiguous (see table 22.5.)

Policies That Increase the Opportunity Costs of Labor and Capital

We have argued that any policies that reduce the relative profitability of clearing forests, as compared to retaining them, tend to slow deforestation rates.

Table 22.5. Policies That Increase the Profitability of Maintaining Forests

Policies	*Effectiveness*	*Targetability*	*Direct costs*	*Indirect costs*	*Equity*	*Political viability*
Niche marketing for tropical timbers, including certification and ecolabeling	Moderate	To be determined	Undetermined	Undetermined	Moderate	High in Europe, doubtful in exporting countries
Abolition of disincentives for long-term sustainable management	Moderate	High	Low?	Low; inequitable	Possibly	Variable
Greater security of tenure for forest people	Moderate to high	High	Low to moderate	Low	Favorable	Variable
Development of markets for nontimber forest products	Moderate to low	High	Low	Low	Favorable	High
Creation of a system for transfer payments for biodiversity conservation, carbon storage, and watershed management	Unknown	Unclear	Low	Low to moderate	Undetermined	Unclear to date

Because labor and capital are major factor inputs in forest conversion, but much less so in retention, increasing the opportunity costs of labor and capital could be expected to reduce deforestation rates. Policies that increase the opportunity costs of labor and capital include general policies that promote growth as well as specific policies that promote employment. In parts of East and South Asia (Korea and Malaysia), the opportunity costs of labor and capital have clearly increased, and deforestation seems to have slowed and even been reversed. In other places where alternative returns to labor and capital remain low (Cambodia and Laos), deforestation rates remain high. This suggests possible connections between these costs and the extent of inappropriate deforestation.

Even within Southeast and East Asian countries, there appears to be socioeconomic stratification and differentiation: remote, poor, rural forest villages are being left out of the economic miracle. Most have experienced very low income growth, if any; people who have little formal education, limited access to or control over valuable resources, and poor transport and communications infrastructure struggle to maintain traditional living standards. One of the options for those remaining in the villages and forests is to clear forests for marginal agriculture, which is often temporary and unsustainable. The forest populations have been willing to clear land, to log illegally, and to hunt and collect nontimber forest products to earn cash incomes. If the prices they receive are low, they harvest more in order to earn the required incomes!

In cities and towns, high-technology industries, tourism, and the service sector are booming as prosperous enclaves. This rapid economic growth in urban, nonland-based sectors is already having profound impacts on forests and deforestation rates. Although rapid growth in agriculture, timber industries, and plantation estates may contribute to deforestation, employment growth in sectors like tourism, textiles, and manufacturing has increased labor opportunity costs to such an extent that it has reduced deforestation. However, where the deforesters have been left out of that growth, and no more attractive livelihoods have emerged, deforestation has continued. The evidence to date is circumstantial and anecdotal, but the policy implications could be that the promotion of sustainable, nonrural employment options may reduce the types of deforestation caused by poor people trying to earn a meager living from deforesting. Similarly, where deforestation is driven by the interests of large capital, the implication could be that with the emergence of more lucrative options (through the development of capital markets), these parties might lose interest in deforestation (see table 22.6).

This diagnosis of who is clearing forests, where, and why, and the observation that where more attractive alternative employment options have emerged, people (both large-scale and peasants) have voluntarily moved out of forest-degrading activities suggest a possible policy measure. Economic growth, with reasonable equity of access to new industries and resources, may make certain types of deforestation in particular contexts sufficiently unremunerative as to slow it voluntarily. The pursuit of equitable and sustainable economic development is a stated goal of most developing countries, so it would be fortuitous indeed if that process inherently encouraged forest conservation. Unfortunately, there is little evidence that the solution is so simple. In terms of the criteria used here, policies to promote industrialization and employment generation in nonland-based sectors are already on the political agenda; although they might be effective in reducing deforestation, and equitable if they could be achieved, relatively few countries have achieved this. Indeed, promoting sustainable economic growth and employment creation has been more elusive than policies to slow deforestation directly but, in the long run, might be the eventual solution.

Conclusions

There is no perfect or generalizable policy for reducing inappropriate deforestation. There are no "first-best" options. Each national situation is different, much uncertainty remains about key cause-and-effect relations, and there are usually tradeoffs among effectiveness, ability to be targeted, political viability, and direct and indirect costs of policies. Most of the policy instruments discussed here are very blunt instruments with regard to stopping deforestation—governments will be forced to choose from a mix of measures specifically crafted to local conditions. There is no magic policy reform that will slow all types of tropical deforestation. Most panaceas that have been advocated have been revealed as merely partial solutions applicable under limited conditions.

It is crucial to resolve the interrelations between deforestation, logging, road construction, and conversion to alternative land uses. Logging will lead to increased forest conversion by an influx of migrants, *if* the following conditions apply simultaneously:

- Roads are constructed that open up new areas of forests.
- The nonforest land use is much more profitable than the retention of forests (in part due to the policy distortions described).
- Forest boundaries are poorly enforced by government agencies (the forest service or national parks service) *and* an institutional or legal context exists in which people can expect that the land they occupy, claim, or "stake out" will eventually be recognized or even legalized by the government, for example, by creating an open-access frontier.
- A large pool of people who are unemployed or landless or who have very low incomes and prospects constitute potential migrants (whether moving spontaneously or sponsored by the state or private agoindustries). We hypothesize that the pace of colonization might be related to the difference between current incomes of potential migrants and the amount they expect to earn by colonizing forest areas (hence the importance of the opportunity costs of labor).

All of the instances, of which we are aware, of rapid forest clearance by "squatters" appear to be consistent with this hypothesis. Conversely, where these conditions do not apply (which admittedly is still rare in tropical developing countries), forests can

Table 22.6. Policies That Increase the Opportunity Costs of Labor and Capital Used in Deforestation

Policies	*Effectiveness*	*Targetability*	*Direct costs*	*Indirect costs*	*Equity*	*Political viability*
Increased urban employment and wages	Moderate	Low	Ambiguous	Low; high potential indirect gains	Ambiguous	High
Improved capital markets	Low	Low	Low	Low	Unclear	High

be logged but subsequently remain under management and not be cleared or seriously degraded (in peninsular Malaysia).

This assessment, if correct, suggests that the answer to the forest conversion issue is not necessarily to stop logging per se, or to stop logging in all new areas, or to ban all new road construction in forest areas, but rather to reform those policies and institutions that at present make forest colonization seem more attractive than the potential migrants' current activities. This might include the pull factors (to reduce the profitability of illegally clearing forests or of speculating in land that was supposed to be kept as forest) or the push factors (to increase the limited livelihood options outside of forests). The evidence from the rapid economic growth of Asian tiger economies is that as employment and income prospects outside the agricultural sector improve, fewer people want to undertake the dangerous, illegal, difficult, and often unprofitable activities of temporary agriculture in forestlands. However, if the new land use is very profitable (growing cocoa, coffee, cinnamon, rubber or fruit trees, or even timber trees like eucalyptus, teak, or gmelina) and if the potential capital gains from "capturing" some real estate from the government forests are high, it might be very difficult to slow the rate of forest conversion by such people. The case for intensification of agriculture as a primary instrument for reducing deforestation has been cogently summarized by Waggoner. (1995).[9]

Thus the principal government policy reforms that could reduce inappropriate tropical deforestation are likely to be:

- Eliminate subsidies to agricultural and pastoral industries that encourage deforestation and, if necessary, offset these by schemes targeted at intensification of existing agricultural areas. This also applies to road construction; that is, intensification of existing rural road networks rather than expansion into hinterlands to open up new areas (unless they are specifically zoned for conversion from forest to agriculture).
- Eliminate existing legal and institutional incentives or requirements to clear forests as a basis for gaining recognized land tenure and to reward "squatting" or colonization of protected forest reserves.
- Reform forest industry concessions and licenses to provide incentives for long-term sustainable management, through increasing the duration and security of licenses and making them tradable, while collecting more of the potential economic rents to remove the "gold-rush" mentality that has characterized the industry for much of the past 40 years.
- Develop innovative institutional arrangements for devolving more decisionmaking authority and responsibility to those whose livelihoods and quality of life are linked directly to the extent and quality of tropical forests. In association with this, both production and conservation activities (for the full range of consumptive and nonconsumptive goods and services from forests) can be fostered, or at least not discouraged, from both government forests and private property outside of government forest reserves.
- Encourage voluntary market differentiation by consumers that discriminates positively toward those products that have been sustainably produced from forests and penalizes those products that have contributed to deforestation. This could be applied not only to tropical timber and to nontimber forest products like Brazil nuts, but also to other products such as chocolate, coffee, and leather.
- Facilitate the recognition and compensation for environmental services provided by forests—at global, national, and local scales—through mechanisms linked directly to the amount of environmental services generated (whether carbon sequestration or biodiversity conservation or watershed protection) and ensure that transfer payments are received by those making the decisions at the forest frontiers.

Most of the major environmental problems in developing countries are due not to the pursuit of economic development but rather to incorrect economic policies, such as poorly defined property rights, underpricing of resources, state allocations and subsidies, and neglect of nonmarketed social benefits. Instead of trying to devise new policies to stop further resource and environmental deterioration while promoting real development, we should first eliminate those (legal, social, political, and institutional) factors that cause or exacerbate the problems.

Notes

1. These figures are derived from the data in FAO (1996: table 4.2.13). They are based on the F3 definition of "forest," which includes closed, open, long fallow, and fragmented forests.

2. The official target of the International Union for the Conservation of Nature (IUCN) is for at least 10 percent of each major forest ecological zone to be given protected area status. However, Iremonger and others (1997) conclude that of 138 regional and ecoregional zones worldwide, only in insular Southeast Asia has this target been achieved.

3. This is not to imply that the areas of highest biodiversity value are necessarily remote and hilly sites undesirable for agriculture (or vice versa). Indeed, conflicts arise because the high biodiversity areas frequently *are* the remaining forest areas of high agricultural potential.

4. Ruf and Siswoputranto (1995) analyze the economics of growing cocoa on newly cleared forestlands, relative to existing farmlands. They find that newly cleared lands have a consistent price advantage of $300–$400 per hectare (the "natural fertility subsidy"). This means that if world prices fall, the supply from growers on existing farmlands contracts first, so clearing new forest becomes the only viable option as prices fall.

5. There are numerous examples, covering vast areas around the world, where forests have been logged but are still reasonably productive and ecologically stable. Thus it is overly simplistic, and incorrect, to equate commercial logging with deforestation. Industrial logging is most likely to lead to deforestation when associated with spontaneous small-scale settlement and conversion or with large-scale agroindustry conversion.

6. In fact, log export bans are rarely intended as conservation measures, but rather as part of a package to encourage accelerated domestic processing, with employment generation, value added, and, it is hoped, higher net foreign exchange earnings.

7. Manasan (1989) has observed that although Southeast Asian governments justified their direct and indirect subsidies by the expectation that the expansion of forest industries would generate jobs, only the Philippines offered a direct financial incentive to employment, via tax rebates. Yet even this was ineffective because it was neutralized by other tax concessions, which promoted capital-for-labor substitution. Overall, relatively few jobs have been created despite the substantial subsidies, and the impact on forests has been severe (Douglas 1995 and Lindsay 1989).

8. Although spending $1 million on offsets and tree planting in tropical developing countries is cheaper than spending $10 million on controlling emissions at home, there are no legal obligations to do either.

9. Waggoner (1995) concludes that if during the next 60–70 years the world farmer could reach the average yield of today's average U.S. corn grower (8 tons of grain equivalent per hectare), then 10 billion people—twice the world's current population—would need only half the current area of cropland, while eating at today's average calorie levels in the United States.

References

The word "processed" describes informally reproduced works that may not be commonly available through libraries.

Anderson, R. J., L. F. Constantino, and N. M. Kishor. 1995. "Stabilization, Structural Adjustment, and Bolivia's Forestry Exports." Environment Division Dissemination Note 13. Latin America Technical Department, World Bank, Washington, D.C. Processed.

Angelsen, Arild. 1995. "The Emergence of Private Property Rights in Traditional Agriculture: Theories and a Study from Sumatra." Paper presented at the fifth common property conference: reinventing the commons, International Association for the Study of Common Property, Bodo, Norway, 24–28 May. Processed.

Barbier, Edward B., J. C. Burgess, J. Bishop, and B. Aylward. 1994. *The Economics of the Tropical Timber Trade*. London: Earthscan Publishers.

Bilsborrow, Richard, and M. Geores. 1994. "Population, Land-Use, and the Environment in Developing Countries: What Can We Learn from Cross-National Data?" In D. Pearce and K. Brown, eds., *The Causes of Tropical Deforestation: The Economic and Statistical Analysis of Factors Giving Rise to the Loss of Tropical Forests*, pp. 106–33. London: University College London Press.

Binswanger, Hans. 1994. "Brazilian Policies That Encourage Deforestation in the Amazon." *World Development* 19(7):821–29.

Bryant, D., D. Nielsen, and L. Tingley. 1997. *The Last Frontier Forests: Ecosystems and Economies on the Edge*. Washington, D.C.: World Resources Institute.

Buongiorno, J., and Togu Manurung. 1992. "Predicted Effects of an Import Tax in the European Community on International Trade in Tropical Timbers." *Journal of World Forest Resource Management* 6:117–37.

Burford, Nicolette. 1991. "Land Scarcity in Amazonia:

Migrant Farmers' Access to Land and Forest Conservation in the State of Rondônia." M.Sc. thesis, Oxford Forestry Institute, Oxford.

Byron, R. N., and M. Ruiz Pérez. 1996. "What Future for the Tropical Moist Forests 25 Years Hence." *Commonwealth Forestry Review* 75(2):124–29.

Capistrano, Ann D. 1990. "Macro-Economic Influences on Tropical Forest Depletion: A Cross-Country Analysis." Ph.D. diss. University of Florida, Gainesville.

de los Angeles, M., and R. Md Idrus. 1990. "Investment Incentives of Southeast Asian Countries in the Wood-based Sector." In *Proceedings of the Malaysian Timber Industry Board Asia-Pacific Conference, Kuala Lumpur, 24–26 September 1990.* Kuala Lumpur: Malaysian Timber Industry Board.

Dorner, P., and W. C. Thiesenhusen. 1992. *Land Tenure and Deforestation: Interactions and Environmental Implications.* Discussion Paper 34. Geneva: United Nations Research Institute for Social Development.

Douglas, J. J. 1995. "The Economics of Long-Term Management of Indonesia's Natural Forest." World Bank, Jakarta. Processed.

Dove, M. R. 1993. "A Revisionist View of Tropical Deforestation and Development." *Environmental Conservation* 20:1–17.

Dudley, N., J. P. Jeanrenaud, and F. Sullivan. 1995. *Bad Harvest: The Timber Trade and the Degradation of the World's Forests.* London: Earthscan.

FAO (Food and Agriculture Organization of the United Nations). 1993. *Forestry Policies of Selected Countries in Asia and the Pacific.* FAO Forestry Paper 115. Rome.

———. 1996. *Forest Resources Assessment 1990: Survey of Tropical Forest Cover and Study of Change Processes.* FAO Forestry Paper 130. Rome.

Garrity, D. P., D. M. Kumar, and E. S. Guiang. 1993. "The Philippines." In *Sustainable Agriculture and the Environment in the Humid Tropics,* pp. 549–624. Washington, D.C.: National Academy Press.

Iremonger, S., V. Kapos, J. Rhind, and R. Luxmore. 1997. "A Global Overview of Forest Conservation." Paper prepared for the world forestry conference, Turkey, October. Processed.

Kaimowitz, David. 1997. *Livestock and Deforestation in Central America in the 1980s and 1990s: A Policy Perspective.* Bogor, Indonesia: Center for International Forestry Research.

Kaimowitz, David, Graham Thiele, and Pablo Pacheco. 1996. "The Effects of Structural Adjustment on Deforestation and Forest Degradation in Lowland Bolivia." Center for International Forestry Research, Bogor, Indonesia. Processed.

Lindsay, H. R. 1989. "Indonesian Forestry Policy: An Economic Analysis." *Bulletin of Indonesian Economic Studies* 25(1):99–115.

Manasan, R. G. 1989. "A Review of Investment Incentives in ASEAN Countries." Working Paper 88-27. Phillipines Institute of Development Studies, Manila. Processed.

Mendelsohn, R. 1994. "Property Rights and Tropical Deforestation." *Oxford Economic Papers* 46 (October):750–56.

Mercer, D. E., and J. Soussan. 1992. "Fuelwood Problems and Solutions." In N. P. Sharma, ed., *Managing the World's Forests, Looking for Balance between Conservation and Development,* pp. 177–214. Dubuque, Iowa: Kendall/Hunt Publishing Company.

Munasinghe, Mohan, and Wilfrido Cruz. 1994. "Economy-wide Policies and the Environment." Environment Paper 10. Environment Department, World Bank, Washington, D.C. Processed.

Myers, Norman. 1981. "The Hamburger Connection: How Central America's Forests Became North America's Hamburgers." *Ambio* 10(1):3–8.

Ozório de Almeida, A. L., and J. S. Campari. 1995. *Sustainable Settlement in the Brazilian Amazon.* New York: Oxford University Press.

Richards, M. 1997. "What Future for Forest Management Based on Common Property Resource Institutions in Latin America?" *Development and Change* 28(1):95–118.

Ruf, F., and P. S. Siswoputranto, eds. 1995. *Cocoa Cycles: The Economics of Cocoa Supply.* Cambridge, U.K.: Woodhead Publishing.

Sayer, J. A., and R. N. Byron. 1996. "Technological Advance and the Conservation of Forest Resources." *Journal of World Ecology and Sustainable Development* 3(3):43–53.

Simpson, R. D. 1997. "Biodiversity Prospecting: Shopping the Wild Is Not the Key to Conservation." *Resources* 126:12–15.

Skinner, J. 1991. "If Agricultural Land Taxation Is So Efficient, Why Is It So Rarely Used?" *The World Bank Economic Review* 5(1):113–33.

Stern, D. I., M. S. Common, and Edward B. Barbier. 1996. "Economic Growth and Environmental Degradation: The Environmental Kuznets Curve and Sustainable Development." *World Devel-*

opment 24(7):1141–160.

Strasma, J. D., J. Alm, E. Shearer, and A. Waldstein. 1987. "Impact of Agricultural Land Revenue Systems on Agricultural Land Usage." Associates in Rural Development, Burlington, Vt. Processed.

Sunderlin, W., and J. Rodríguez. 1996. *Cattle, Broadleaf Forests, and the Agricultural Modernization Law of Honduras, The Case of Olancho.* CIFOR Occasional Paper 7. Bogor: Center for International Forestry Research.

Tchoungui, R., and others. 1995. "Structural Adjustment and Sustainable Development in Cameroon." Working Paper 83. Overseas Development Institute, London. Processed.

Utting, P. 1992. *Trees, People, and Power: Social Dimensions of Deforestation and Forest Protection in Central America.* Geneva: United Nations Research Institute for Social Development.

Waggoner, Paul E. 1995. "How Much Land Can Ten Billion People Spare for Nature? Does Technology Make a Difference?" *Technology in Society* 17(1):17–34.

Wells, Michael, Katrina Brandon, and Lee Hannah. 1992. *People and Parks: Linking Protected Area Management with Local Communities.* Washington, D.C.: World Bank, World Wide Fund for Nature, and U.S. Agency for International Development.

Wiebelt, M. 1994. "Protecting Brazil's Tropical Forest, A CGE Analysis of Macroeconomic, Sectoral, and Regional Policies." Kiel Working Paper 638. Kiel: Kiel Institute of World Economics.

23

Costs, Benefits, and Farmer Adoption of Agroforestry

Dean Current, Ernst Lutz, and Sara Scherr

In Central America and the Caribbean, where the empirical part of our work was done, the forest cover in each of the eight countries studied declined between 10 and 24 percent during the 1980s (table 23.1). In the two countries with the highest population density (El Salvador and Haiti), only 6.2 and 1.3 percent of forest cover, respectively, remained by 1990.

Deforestation means increasing scarcity of wood—a principal source of fuel and construction materials in rural areas—and contributes to environmental degradation of fragile agricultural lands. One response of governments, backed by nongovernmental organizations (NGOS) and international donors, has been to encourage agroforestry: a land use system in which trees, shrubs, palms, and bamboos are cultivated on the same land as agricultural crops or livestock for economic and environmental reasons: to replenish wood stocks while upgrading land through diminishing erosion, water loss, and other natural phenomena. The system includes managing natural regrowth; seeding, planting, and maintaining trees as border plantings; and interplanting them in agricultural crops, in woodlots, in home gardens, or other systems. Agroforestry reinforces traditional reforestation efforts by reducing the pressure on remaining natural forests.

Traditional agroforestry practices already play an important role in many farming systems—fruit trees are planted in home gardens, pastures, and fields and among permanent tree crops; live shrubs and bushes are used for fencing; coffee and cocoa plants are cultivated for shade; and fallow (shifting cultivation) systems are maintained that sequentially rotate trees with crops.

Agroforestry systems introduce the deliberate use of woody perennials (trees, shrubs, palms, bamboos) on the same land management unit as agricultural crops, pastures, and animals. This may consist of spatial arrangement in the same place at the same time or a sequence over time; it may include woodlots, taungya, and managed woody fallows for forest tree production. For farm areas devoted to perennial crops, systems include deliberate perennial intercrops and home gardens. The main systems introduced for annual cropland are trees planted in lines on boundaries, alley cropping of annual crops between tree rows, windbreaks, barrier hedges along the contours of fields, and trees planted on terrace risers.

Although agroforestry activities has attracted solid support and are likely to receive further substantial investment from public agencies, NGOS, and international donors, they have generally been poorly documented. Monitoring of projects, where it exists, has been weak or focused narrowly on administrative concerns (Scherr and Muller 1991 and Current, Rivas, and Gómez 1994).

Most research and analysis has dealt with the physical and biological aspects of agroforestry systems, neglecting systematic analysis of agroforestry's economic contributions at the farm level (Swinkels

Table 23.1. Key Indicators in Case Study Countries

Country	Population density (population per square kilometer)	Rate of population growth (percent)	GNP per capita (1992 U.S. dollars)	Forest cover in 1990 (percent)	Forest cover per capita (hectares)	Percentage change in forest cover, 1981–90	Rate of deforestation, 1981–90 (percent)[a]	Reforestation as a percentage of area deforested per year[a]
Costa Rica	62.7	2.8	1,960	28.8	0.5	–23.8	2.9	7.5
Dominican Republic	149.0	2.1	1,050	22.5	0.2	–24.2	2.8	1.1
El Salvador	257.1	1.4	1,170	6.2	0.0[b]	–16.8	2.2	16.1
Guatemala	89.0	2.9	980	39.3	0.5	–15.6	1.7	3.1
Haiti	—	—	—	1.3	0.0[b]	–10.3	4.8	73.3
Honduras	48.2	3.3	580	41.2	0.9	–19.4	2.1	0.4
Nicaragua	30.0	2.7	340	50.8	1.6	–16.8	1.9	1.5
Panama	32.5	2.1	2,420	41.1	1.3	–17.0	1.9	0.8
Total	77.3	—	—	36.9	—	–18.1	—	2.3
Latin America and the Caribbean	24.2	1.9	2,180	56.2	2.3	– 7.1	0.8	5.0

— Not available.

a. Deforestation is defined as the change of land use from forest to other land use or depletion of forest crown cover to less than 10 percent.

b. Extremely small values.

Source: World Bank 1994, BID 1993, and FAO 1993.

and Scherr 1991 and Scherr and Muller 1991), even though the absence of a profit motive is a root cause of many project failures (Tschinkel 1987 and Murray 1991). The appropriateness of particular agroforestry systems to subregional economic conditions is rarely assessed (Scherr 1992). Nor has there been much comparison of different extension approaches and their influence on farmer adoption of agroforestry (Vergara and MacDicken 1990).

Important questions remain about which strategies to pursue for agroforestry development. Which agroforestry practices, under different environmental and economic conditions, are profitable for farmers? Under what conditions are farmers willing to invest in agroforestry systems oriented to environmental rehabilitation? Do certain conditions justify subsidies and nonmarket incentives? What level and type of field extension assistance are needed to disseminate new agroforestry species and information and to encourage adoption? What are the principal constraints at the farm level to further adoption of agroforestry? What kinds of policy action would encourage adoption?

To answer these questions, this study analyzed the experiences of 21 projects in eight countries of Central America and the Caribbean (see figure 23.1) and drew together the lessons learned for the design and execution of future efforts. Consultants from the region and staff at the Tropical Agricultural Research and Higher Education Center (CATIE) in Costa Rica carried out eight country studies on the basis of an earlier pilot study (Current and Lutz 1992) under the overall direction of the World Bank, with methodological support from the International Food Policy Research Institute. Each study included a country overview and case studies of two or three agroforestry projects, analyzing in detail the technical interventions and their adoption, their economic performance at the farm level, the project's approach to extension, and policies that influenced implementation of the project and its adoption by farmers. National workshops and a regional workshop were held for farmers, extensionists, project managers, and decisionmakers, with representatives from farmer organizations, NGOs, national agencies, and international donors to review preliminary findings. To our knowledge, this was the first large-scale study of farm-level profitability of agroforestry in the tropics.

The specific objectives of the study were to:

- Document earlier agroforestry efforts and their

Figure 23.1. Approximate Location of Case Studies

lessons for tree planting on farms

- Determine costs and benefits of agroforestry systems from the farmers' perspective
- Determine the advantages and disadvantages for farmers of different types of agroforestry systems under different site and resource conditions
- Identify the effects of institutional factors and government policies on the adoption of agroforestry systems.

This chapter summarizes the principal findings of the country studies, comparing them where possible with the experience of other regions. Beginning with a brief background history of agroforestry in the region, the article describes the investigative methods used, outlines the characteristics of the projects reviewed, and discusses the findings on household benefits, the pros and cons of different extension approaches, and the effects of different policies.

Evolution of Agroforestry in Central America and the Caribbean

Numerous projects have been established in the past two decades throughout Central America and the Caribbean to promote communal and individual tree plantings and agroforestry systems. Agroforestry development in the region generally reflected—and in some cases led—international trends in thinking and action about agroforestry. In the late 1970s and early 1980s when the emphasis in forestry was on fuelwood production, agroforestry systems were used on farms for planting trees for fuelwood (Eckholm 1975 and King 1979). Fuelwood projects were implemented worldwide (U.S. Comptroller General 1982 and USAID/ROCAP 1979), and research was undertaken on fuelwood species (National Academy of Sciences 1980). But the response to fuelwood projects in rural communities was discouraging: farmers were not enthusiastic about contributing the sizable investment in labor—for planting, felling, sectioning, and splitting—needed to produce a product with a relatively low market value. Furthermore, the fact that fuelwood has traditionally been collected as a free good (even on private lands) lessened its attraction as a product for farmers. In El Salvador, with its high population density and low forest cover, a recent national study showed that only 33 percent of the rural population considered the acquisition of fuelwood a problem (Current and Juárez 1992).

Subsequently, the emphasis shifted from fuelwood production to the use of multipurpose tree species in agroforestry systems. The establishment of multipurpose tree species on farmland can provide a wide range of benefits. For farm families, they may help to meet household consumption needs, provide a source of income, or improve land husbandry to ensure future production. As well as fuelwood, multipurpose tree species supply green manure, wood for construction, fence posts, fuel, raw materials for local or industrial processing, and cash income and savings from any of these. Conflicts do arise between the cultivation of trees and agricultural crops, but many agroforestry systems allow farmers to integrate trees into their farming systems, with only small drawbacks for crop production or even an increase in farm productivity overall (Gregersen, Draper, and Elz 1989 and Raintree 1991).

Social benefits from establishing trees on farms can be significant and are often cited to justify tree planting projects supported by national governments and donors. Trees can provide protection from soil erosion generated by water and wind, protecting future productive potential and preserving important watersheds that feed hydroelectric projects and supply water for population centers; they provide wood products for the farm, thus relieving the pressure on natural forest areas (Current and Lutz 1992); they provide raw materials for rural industries that generate employment for rural communities (Arriola and Herrera 1991 and Campos, Rodríguez, and Ugalde 1991); and they provide environmental and social benefits, such as bird habitats, water retention capacity, or shade for dwellings.

Regionally, the most important applied research initiative has been the Tree Crop Production (Madeleña) Project financed by the U.S. Agency for International Development (USAID) Regional Office for Central America and Panama and executed by CATIE (Belaunde and Rivas 1993). Other internationally important research programs include those of the Forestry Fuelwood Research and Development Project in Asia (French and van den Beldt 1994), the International Center for Research in Agroforestry (Steppler and Nair 1987), and the Nitrogen Fixing Tree Association (Nitrogen Fixing Tree Association 1993). Numerous national and international NGOS have also undertaken applied research and extension activities (Cook and Grut 1989 and Nair 1990).

The shift in emphasis toward multipurpose tree species was accompanied by a corresponding realization that a project is more likely to succeed if project planners consult local communities about their perceived needs and design projects to meet those needs instead of imposing schemes that communities may not consider a priority.[1] In line with this thinking, research focused on traditional agroforestry systems (Budowski 1987), and many forestry agencies have been changing their emphasis from "guarding" the forest to helping communities produce tree plantings that, in turn, may help to protect the remaining forest. This stronger community role in the design of projects and technology has parallels in Africa and Asia (Kerkhof 1990, Budd and others 1990, French and van den Beldt 1994, and Raintree 1991).

Methods of the Study

The study compared the economic and technical performance of the most prevalent agroforestry systems used in the 21 projects.

Analysis of Household Benefits

The analysis identified the role—both intended and actual—of trees in the household's livelihood strategy: to satisfy household needs, to generate cash income or savings, or to supply insurance. It also assessed the effects of the agroforestry practice on labor, land, and other household resources (including household risks) and used cost-benefit and related analyses to assess the profitability of the agroforestry practice relative to farmers' alternative systems. This was done using sensitivity analysis to determine the range of conditions (such as input or output prices, productivity levels) under which an agroforestry technology was likely to be profitable (Scherr 1995).

The data on inputs and outputs and costs and prices required to develop the multiyear agroforestry enterprise budgets were generated in various ways, among them recall data and plot measures from single-visit farm surveys, multivisit studies of case study farms, and results of on-farm research trials in the project area. Costs and benefits were valued at the prices that these agents actually face, with no attempt to adjust for economic distortions. Farmers were interviewed about the means and the opportunity costs of using land, labor, and capital resources in agroforestry. They also provided qualitative assessments and anecdotal information on social and environmental costs and benefits; quantitative informa-

Table 23.2. Farming and Agroforestry Systems in the Case Study Projects

Project and country	*Agroforestry systems promoted*	*Principal tree species used*
Costa Rica		
Community Agricultural Center: Hojancha	A,E,H	5,21,31
PRODAF Forestry /Agroforestry Development Project	A,E,H,L	5,11,17
COOPELDOS—windbreaks for coffee and dairy products	F	9,14,15,19
Traditional Agroforestry Systems of the Humid Tropics	A,C,L	13,17,20
El Salvador		
Agroforestry Support to Low-Income Rural Communities	D,E,G,H,I	18,20,24,31
Community and Family Nursery Program	D,E,G,H	8,18,31
Guatemala		
Agroforestry Program of DIGEBOS, CARE, Peace Corps, and DIGEBOS/CATIE Madeleña	3A,C,D,G,L	3,9,17,18,20,22,25
La Máquina Settlement—Guatemala	E,G,H	6,18
Honduras		
Development of the Broadleaved Forest (BLFDP)	A,C,E,G	20,23,29
COHASA Rural Development	B,E,H	20,24
Sierra de Omoa Management Unit	A,I	13,20
Land Use Planning (LUPE)	C,D,E,H	13,20
Nicaragua		
Control of Erosion in the West and Reforestation	F	18,24,30
Support to Farm Forestry and Rehabilitation of Windbreaks	E,F,G,H	18,20,24,30
Demonstration Farm "La Esperanza"	C,E,L	7,8,16,20,27
Panama		
Agroforestry for Community Development (INRENARE/CARE)	B,E,F,G,H	2,18,24,26
Food Production and Community Development—Aquaculture (MIDA/WFP)	H	24,26,31
Dominican Republic		
ENDA—Caribe: Desarrollo Rural Integrado de Zambrana	C,G	2,7
Fondo Rotativo Agroforestal y Comercial (FORESTA)	C,E,H	2,17,18,20
Haiti		
Agroforestry Outreach Project—Pan American Development Foundation	D,E,H	4,8,10,18,24
Maissade Watershed Management Project—Save the Children	E	1,8,12,28

Code	*Agroforestry systems*
A	Perennial crop with trees
B	Tree intercropped with agriculture
C	Alley cropping
D	Contour plantings
E	Border plantings and live fences
F	Windbreak
G	Taungya
H	Woodlot
I	Fallow
J	Home garden
K	Fodder trees
L	Trees in pasture (silvopastoral)

Code	*Principal tree species (common name)*
1	*Acacia auriculiformis* (Japanese acacia)
2	*Acacia mangium* (Mangium)
3	*Alnus acuminata* (Alder)
4	*Azadirachta indica* (Neem)
5	*Bombacopsis quinatum* (Pochote–C. Amer.)
6	*Caesalpinia velutina* (Aripin—C. Amer.)
7	*Calliandra calothyrsus* (Calliandra)
8	*Cassia siamea* (Yellow cassia)
9	*Casuarina cuninghamiana* (Casuarina)
10	*Catalpa longissima* (Catalpa)
11	*Cedrela odorata* (Spanish cedar)
12	*Colubrina arborescens* (Coffee colubrina)
13	*Cordia alliodora* (Onion cordia)
14	*Croton niveus* (Colpachi —C. Amer.)
15	*Cupressus lusitanica* (Mexican cypress)
16	*Enterolobium cyclocarpum* (Earpod tree)
17	*Erythrina sp.* (Mountain immortelle)
18	*Eucalyptus camaldulensis* (Eucalyptus)
19	*Eucalyptus saligna* (Eucalyptus)
20	*Gliricidia sepium* (Mother of cocoa)
21	*Gmelina arborea* (Gmelina)
22	*Grevillea robusta* (Silk oak)
23	*Inga vera* (Guaba)
24	*Leucaena leucocephala* (Leucaena)
25	*Melia azederach* (Chinaberry)
26	*Pinus caribaea* (Caribbean pine)
27	*Pitehcolobium saman* (Raintree)
28	*Simaruba glauca* (Aceituno—C. Amer.)
29	*Swietenia macrophyla* (Honduras mahogany)
30	*Tecoma stans* (Yellow trumpet flower)
31	*Tectona grandis* (Teak)

Table 23.3. Project Extension Approaches

Country and project	Number of communities	Number of families participating	Number of technical staff	Extension tools: Information and training: Extension visits	Food for work	Marketing, processing, or infrastructure	Demonstration plots	Radio	Local promoters	Field trips, training courses	Incentives: Material inputs	Cash incentive or credit
Costa Rica												
Traditional	3	280	None	○	○	○	○	○	○	○	○	○
Community Agricultural Center	53	300	2	●	○	●	●	○	●	●	○	●
PRODAF	69	144	5	●	○	○	●	○	●	●	○	●
COOPELDOS	19	119	1.5	●	○	○	○	○	●	●	○	●
Dominican Republic												
ENDA—Caribe	66	800	5	●	○	●	○	○	●	●	○	○
FORESTA	—	89	5	●	○	●	●	○	○	○	●	●
Guatemala												
La Máquina	—	—	—	●	○	○	○	●	○	○	●	○
DIGEBOS/CARE	350	10,000	86	●	●	●	●	○	○	●	●	○
Madeleña 3	22	1,309	22	●	○	○	●	○	○	●	○	○
Haiti												
Agroforestry Outreach	6[a]	—	200 NGOS	●	○	○	○	○	●	○	○	○
Maissade Watershed	Associations	2,400	12–15	●	○	○	○	○	●	●	●	○
Honduras												
Broadleaf forest	10	656	32	●	○	●	●	○	○	●	●	○
COHASA	6	1,500	5 (—)	●	●	●	○	○	●	●	●	○
LUPE	44	5,427	193 (45)	●	○	○	●	○	●	○	●	●
Omoa	7	109	3	●	●	○	○	○	○	○	●	●

Nicaragua												
Erosion	—	3,000	30 + 600[b]	○	○	○	○	○	○	○	○	○
Windbreaks	16	500	20	●	○	○	○	○	○	○	●	○
La Esperanza	11	765	3	●	○	○	○	○	●	○	○	●
Panama												
INRENARE/CARE	15 (10)	157 (80)	(10)	●	●	○	●	○	○	●	●	○
MIDA/WFP	11 districts	4,380 (600)	(57)	●	●	○	○	○	○	○	●	○
El Salvador												
Agroforestry support	100	3,302	20	●	○	○	○	○	○	●	●	○
Nursery program	—	22,470	—	●	●	○	○	○	○	○	●	○

— Not available.

Note: Numbers in parentheses represent the first phases of the projects. Numbers outside the parentheses represent the totals to 1992.

a. Associations.

b. Workers were hired only during the establishment phase.

Source: World Bank data.

tion was generally not available.

The study deliberately emphasized household benefits because, since farmers make their decisions on land use in light of their own objectives, production possibilities, and constraints, understanding their incentives is essential to understanding patterns of resource use and formulating appropriate responses to problems. Ultimately, the success of any agroforestry system depends on how farmers, not project planners, view its costs and benefits. Furthermore, the farm-level approach places the focus firmly on the effects on farm productivity, which are particularly important in this region. (This is not to minimize the importance, in some situations, of off-farm effects.)

Analysis of Technology Adoption

Researchers collected information on indigenous agroforestry practices, farming systems, and land use problems as well as the farmers' criteria for selecting agroforestry interventions. Characteristics of technology adoption (the systems selected and the extent, durability, and determinants of adoption) were assessed through informal farmer surveys. Formal and informal surveys of farmers, focus groups, and investigations by project staff identified specific problems, and researchers assessed the experience of the project with the selection of tree species, tree-crop interactions, and methods of establishing trees.

Policy Assessment

The study identified policy influences on agroforestry adoption, project implementation, and the profitability of agroforestry from several sources. Country overviews drew on public documents, key informants, and unpublished literature. Project case studies used interviews with farmers and project staff as well as project reports to gain policy insights and to document constraints caused by specific legal, institutional, and policy problems. National and regional workshops provided a forum for project executives and policymakers to develop specific recommendations for policy action.

Characteristics of the Projects Reviewed

The 21 projects analyzed were selected on the basis of the age of the project (older projects being preferred), the availability of economic data, the willingness of project staff to cooperate in the study, and the focus on smallholders (average farm size was less than 20 hectares in most projects and less than 5 hectares in five projects). Together, the projects cost approximately $150 million and represented more than 50,000 farm households; they covered a variety of agroforestry systems and site and socioeconomic conditions (see table 23.2).

Table 23.3 summarizes key elements of the extension strategies used by the projects. Such strategies can generally be divided into four broad categories: using existing forestry or agricultural extension networks; training and providing logistical support to extension agents hired specifically for the project; integrating project extension agents with national extension agencies; and supporting extension staff organized and managed by NGOs. The projects studied represented all four of these categories. The most common methods used to provide information and training to the farmers were visits by extension agents or paratechnicians, field trips and training courses, and demonstration plots. Six projects provided marketing assistance. Most projects used material incentives (free seedlings or other inputs), and many others used cash incentives, credit, or food-for-work incentives. (Food for work has been used for communal and privately owned nurseries, as well as for tree planting on both communal and private lands. Plant survival rates have been much higher in community nurseries where trees are planted on private land and for private tree plantings.)

Findings: Incentives for Adopting Agroforestry, Adoption Patterns, and Implications for Extension

The success or failure of the effort to promote agroforestry depends ultimately on how many farmers take up the system and persevere with it. Perceived profitability for the household is a critical influence on a farmer's initial decision to include agroforestry in the farm's activities; the systems adopted and the scale and pace of adoption will then be determined by practical matters, such as the resources available to farmers and the specific constraints they face. All of these have important implications for the design of extension efforts.

Incentives for Adopting Agroforestry

The study assessed the profitability of agroforestry at the household level in light of financial benefits and

Table 23.4. Averages of Indicators by Agroforestry System

Agroforestry system	*Number of systems studied*	*Net present value* 10 percent	*Net present value* 20 percent	*Cost-benefit ratio*	*Return to labor*	*Payback period*	*Average person-days per year*
Trees with agricultural crops	5	2,863	1,300	1.8	5.03	3.4	165
Alley cropping	9	1,335	847	2.1	6.98	1.9	56
Contour planting	4	1,426	761	1.6	3.75	2.0	116
Perennial with trees	4	2,867	1,405	1.8	4.19	4.0	139
Home garden	4	—	372	2.2			
Taungya	8	6,797	2,868	2.5	15.98	4.9	53
Woodlot	10	764	–33	1.0	3.41	9.2	12

— Not available.

Note: This table should be used for general comparison only. The systems analyzed represent only a small sample of each system and cannot be considered a random representative sample. Net present values and return to labor are influenced by localized market conditions and do not fairly represent differences between planting systems. The payback periods and average man-days required per year can more reliably be compared. (Current, Lutz, and Scherr 1995 present this information by each individual system analyzed.)

Source: Current, Lutz, and Scherr 1995.

costs, risk management, and the nature and accessibility of markets for agroforestry products.

Financial results. Financial analyses were undertaken for the major agroforestry systems used in the region—trees intercropped with agricultural crops, alley cropping, contour plantings, perennial crops with interplanted trees, home gardens, taungya, and woodlots (see Current, Lutz, and Scherr 1995, for detailed results).[2] Table 23.4 shows averages of the financial indicators calculated for these systems and net present values for each technology—that is, all costs and benefits at the plot level estimated at 10 and 20 percent real discount rates and compared to the net present values for one or more alternative practices. This approach was adopted because the systems often differed considerably. The net present values were found to be positive at a 20 percent discount rate in all but one (woodlots) of the seven systems analyzed.

Local scarcities of timber, fuelwood, and poles, which differ not only from country to country but from region to region within countries, are critical to the economics and adoption of agroforestry. The higher prices and correspondingly higher profits that often accompany scarcity can be an important incentive for farmers to turn to agroforestry.[3] This finding is consistent with those from qualitative assessments and empirical case studies from other parts of the world. Summarizing results for South Asia and East Africa, Arnold (1995: 272–73) concludes that

> With most situations experiencing reductions in access to off-farm supplies, growing demand, declining site productivity, and increased exposure to risk, it should not be surprising that tree-planting activity does increase as agriculture and land use become more intensive. Though it would be incorrect to assume that this always happens, there is a general progression towards more planted trees as agriculture and pressures on land intensify, and existing tree stocks diminish, within most systems.

A system's financial value may derive from the value of the tree products it provides (woodlots, taungya, timber species in coffee and cacao); the protection and subsequent increase in agricultural production it affords (windbreaks, shade trees in perennial crops); its contribution to soil improvement through nitrogen fixation, organic matter, and improved soil structures; or a combination of these. The outputs most commonly produced in the eight countries and seven systems analyzed were fuelwood, charcoal, and roundwood products for local markets and household use. In Costa Rica and the Dominican Republic, sawlogs were major products.

Prices for tree products were highly variable across the region and within each country. Fuelwood prices provide an example of this variability and the effect it may have on a system's profitability: according to our estimates, fuelwood prices ranged from $2 per cubic meter in Panama to almost $6 in Guatemala. In Guatemala City, fuelwood prices were more than twice the price of the same fuelwood in other parts of the country. In contrast, despite a relatively low forest cover around León, Nicaragua, a cubic meter of fuelwood cost only $2.14, one of the lowest prices in the region. This apparent contradiction can probably best be explained by differences in the employment situation, the amount of fuelwood marketing, and relative access to the tree resource. One of the social benefits of windbreaks in Nicaragua is that they serve as a source of fuelwood, providing gainful (though illegal) employment for the surrounding communities and creating a competitive market that maintains prices at a low level.

Financial analyses for all of the seven practices studied showed average net present values of more than $300 per hectare (the exception was woodlots). Findings on profitable returns from taungya and from intercropping of multipurpose tree species in perennial crops confirmed findings of earlier regional case studies of the same region. The on-farm profitability of agroforestry in Central America appears to be consistently high, under a wide range of conditions and for different systems. Perhaps the reason for this high profitability is that farmers modify systems to ensure profitability under their own farm and local conditions.

Returns for perennial intercrops of coffee and black pepper were almost all high; low prices for plantain and cacao reduced the returns to those systems in Costa Rica. On cropland, net present values for alley cropping and trees on contours were high except in Nicaragua. Analysis of the alley cropping systems used conservative assumptions about the medium- and long-term effects on yield of green manuring and nitrogen fixation; high returns were often due to the added value of the wood product. Except in Costa Rica, windbreaks brought low returns because of the high costs of establishment and the suppression of yields in adjacent crops. In the case of government-sponsored windbreaks in Nicaragua, they were valued only for their fuelwood, although more valuable products had been harvested from them, and this valuation underestimated the value of the windbreak. In Costa Rica, increased yields of coffee and milk production alone made the establishment of windbreaks profitable in three years. (Windbreaks protect pasture grass from desiccation and reduce the stress on cows.)

Cost-benefit ratios were above 1.6 for six of the seven agroforestry systems, with woodlots once more the exception. Payback periods were quite long for the woodlot systems, averaging nine years compared with two to five years for the other systems. Thus the long gestation period commonly assumed for tree-based systems does not seem to be a problem for most of the agroforestry practices used in this region.

Of course, absolute profitability is less important to a farmer's decisionmaking than the profitability relative to the available alternatives. Comparing the financial returns for agroforestry with those for alternative farming practices, the case studies found that agroforestry systems produced significantly higher returns per hectare—at least 25 percent—in approximately 40 percent of the cases and at least 10 percent greater returns in more than half the cases. In more than 75 percent of the cases, the returns to labor were higher for systems including agroforestry than the local agricultural wage.

Financial results were generally most sensitive to yields and product prices, followed in importance by labor; the results were least sensitive to the prices of purchased inputs such as seedlings, agricultural seed, and agrochemicals. Including trees in agricultural systems makes tree, crop, and livestock systems less sensitive to changes in yields and prices and improves overall profitability.

Risk management. Adopting agroforestry systems can both reduce and increase the risks to the security of a household's livelihood. Income diversification is a strongly positive benefit that reduces the risks for cropping systems, whether annual or perennial, and for livestock farming. But risks are associated with the cultivation of trees, as with any other crop. Unusual drought conditions or poor planting materials can mean failure to establish the trees; uncertain market prospects—most projects were weak in developing or improving farmers' access to tree product markets—increase risk; badly managed trees, or overly competitive tree species, can destroy associated crops; and where permit procedures and government regulations are restrictive, there is a risk of not being able to harvest trees—a major obstacle to initial adoption. These risks, and perceived risks, reduce the rates of adoption and expansion, even where average financial returns are good.

Typically the benefits received from the trees offset reported losses in agricultural production, but data on the impacts of trees on agricultural crops are scarce. Positive effects of trees on crops were reported for agroforestry systems that produced organic mulch and green manure or wind protection. This was true of trees (*Alnus*) planted on contours in

Guatemala, *Leucaena* intercropped with corn in Honduras, and windbreaks in Costa Rica. In Honduras, the financial analyses reflected a decline in agricultural production in the alternative systems without trees.

Markets. Most farmers in the study initially adopted agroforestry systems to meet household subsistence needs, a result consistent with the findings of household surveys in six countries in Asia in 1991, reported by the Forestry/Fuelwood Research and Development Project (French and van den Beldt 1994). In 12 of the projects, a significant part of total output continued to be reserved for use on the farm. Once subsistence needs were satisfied, farmers became interested in market opportunities; in seven projects, local markets were important, and for 12 projects, considerable output was directed to regional or urban markets (usually through middlemen who purchased locally).

The market for farm-grown tree products varies within the region. In El Salvador and on the Pacific coast of Guatemala, markets have developed for rural construction; poles and roundwood have also found a market in Guatemala for tobacco-drying racks. In La Máquina, Guatemala, when the roundwood market for construction became saturated, the market for tobacco racks picked up the slack. But farmers in Costa Rica who had been led to believe that they would be able to sell products from thinnings (wood three to five years old) encountered a limited market that quickly became saturated, leaving many without an outlet. The market for sawlogs is strong in Costa Rica but requires wood 11 to 15 years old. Similar problems with saturation in certain markets have also arisen in Asia.

Local organizations are working to develop new local, national, and international markets—notably in Guanacaste, Costa Rica, where tree farmers have formed a regional association of local organizations and cooperatives that is studying markets for a regional forest industry based on their new and existing plantations. On the Pacific coast of Guatemala, tree farmers recently formed a similar organization to explore markets for their tree products.

In Hojancha, Costa Rica, early work to improve nursery practices for some of the most commonly planted species engendered a nursery industry that now supplies plants for individual and industrial tree planting throughout the country and provides a source of employment and income for Hojancha. The market has had to adapt to the normal vicissitudes of supply and demand; as new nurseries have developed outside Hojancha, the demand and prices for certain species have declined, so that some nurseries have had to sell plants at a loss, while others have specialized their product to suit industrial growers who have special requirements for their plants.

Studies by Cook and Grut (1989) and Warner (1993) for Africa and the collection edited by Raintree and Francisco (1994) for Asia and by Arnold (1995) for Africa and Asia show the importance of local markets in providing incentives for agroforestry. Nevertheless, they still conclude that household use is the driving force for most adoption. Arnold (1995: 278) further argues that distinctions between the two are blurred:

> In most of the situations studied self-sufficiency, in particular tree products, proved to be the primary objective. However, with the growing dependence of farm households on income to meet at least part of their needs, as forest products such as fuelwood, fodder, and fruits become progressively commoditized, and the marketplace provides opportunities to substitute purchased inputs such as fertilizer for inputs previously supplied by growing trees, the distinction between production for subsistence or sale has progressively less meaning. Not only will producers sell what is surplus to their subsistence needs, but they will sell a commodity needed in the household if the opportunity cost of doing so is advantageous.

The role of local and regional woodfuel markets in encouraging farmers to adopt agroforestry appears to be more important in much of Central America and Asia than in Africa, possibly because farming systems are more highly commercialized in Central America and the Caribbean than in much of Africa, and the region is more integrated economically. Thus farmers may be more inclined to specialize in particular tree species and tree products; the multiproduct feature of agroforestry, which appears to be so highly valued in Africa and Asia (Raintree and Francisco 1994), is generally less valued in Central America.

Adoption Patterns

Potential profitability may be a key incentive in farmers' decisions to adopt agroforestry, but the viability of a system for a particular household or area will be determined by the resources available to them and the constraints they face.

Management factors. Land, labor, and capital resources are as influential as profit considerations in farmers' decisions about agroforestry (Dewees 1993 and Scherr 1992). The resources available to the farm family limit the type of agroforestry systems that can be adopted, because different systems require different resources. Home gardens and alley cropping, for instance, may be very labor-intensive, whereas timber trees or woodlots require little labor but much land.

An advantage of agroforestry is that tree management operations can often be scheduled for periods of slack labor demand; indeed, the case studies suggested that this is a significant incentive for adoption. In El Salvador, fuelwood prices follow the agricultural calendar: during slack agricultural periods, farmers dedicate more time to extracting and processing fuelwood, and the increase in supply is reflected in a seasonal decline in fuelwood prices (Current and Juárez 1992). One constraint on the adoption of alley cropping is that it needs labor for pruning, which frequently conflicts with major agricultural operations. Intensive agroforestry systems, such as alley cropping, are less attractive to farmers with available land for fallowing.

Management of agroforestry ranges in scope and sophistication from simple gathering to active management of naturally growing resources, domestication and establishment of trees in selected sites, expansion in the scale of tree planting, and intensive management (Scherr 1992). The level of intensity that is economic for farmers depends on factors such as scarcity of forest resources, level and composition of demand, returns to alternative uses of land, labor, and capital resources, and the costs of production relative to the costs of available substitutes for tree products and services (bricks for building poles, kerosene for fuelwood). The Panama project demonstrates that a fallow system for recuperating soils and the extraction of forest products are a very profitable traditional alternative where land is available for that type of management. The study found the highest intensity of agroforestry management—for example, trimming branches to limit shading of agricultural crops—in areas of greatest land shortage, such as El Salvador and the Guatemalan highlands.

Agroforestry practices and tree species. The most prevalent agroforestry systems found on farms were taungya or block plantings, followed by trees in some kind of linear arrangement (border plantings, contour plantings, live fences). Evidence of recent adoption was mixed for alley cropping, home gardens, windbreaks, green manuring, and tree-pasture systems. The choice of particular tree species for use in agroforestry technologies was dictated by considerations such as familiarity to farmers, growth performance, market value, ease of propagation and management, multiple-use options, interaction with crops (see table 23.5), availability, project recommendations, and environmental effects. The study identified 32 different forest tree species promoted by at least one project, along with a large number of well-known or exotic fruit trees. The most frequently used species were *Eucalyptus camaldulensis, Gliricida spp*., and *Leucaena leucocephala. Tectonis grandis* (teak), *Acacia mangium*, and *Cassia sp*. were also mentioned by at least four projects (see table 23.2 for common names). Projects had tended in the past to limit the options they provided to introduced species. More recently, there was a shift in the opposite direction, with some projects offering only native species. A mix of species appeared to be desirable, from the farmers' point of view.

Scale and pace of adoption. Most empirical studies have found that farmers tend to adopt agroforestry gradually, and in the projects studied here, significant adoption commonly required 5 to 10 years. This would seem an especially wise course of action for smallholders on resource-poor farms, given the need to ensure food security and reduce risks. Continuous, small-scale adoption seems a more appropriate objective than crash planting programs for smallholders. Studies documenting cases of rapid adoption (for example, of eucalyptus in India) have found poorer economic returns. Several theoretical cost-benefit modeling exercises also show that enterprises are more profitable when costs and risks of tree establishment are spread over several years, obtaining early benefits from multiproduct systems and a more regular flow of final products over time (Swinkels and Scherr 1991; selected cases).

Farm size clearly influences the pace and scale of adoption: even within the limited range of resource-poor farms targeted by the project, researchers found that adoption was greater on larger farms. Farmers selected and adapted agroforestry practices and management to suit their particular labor, land, and capital resources. Female participation was significant in

Table 23.5. Interaction of Agroforestry Systems with Other Crops

Agroforestry system	*Positive tree-crop interactions*	*Negative tree-crop interactions*
Perennial crop with trees	Complementary	Competes for water and nutrients; shade
Trees intercropped with agricultural crops	Complementary	Competes for water and nutrients; shade
Alley cropping	Complementary	Competes for water and nutrients; shade
Contour planting	Increases production by providing protection from soil erosion and contributing organic material	Competes for space, water, and nutrients
Windbreak	Increases production by providing protection from winds	Competes for space, water, and nutrients
Taungya	Crop production helps to cover the cost of plantation establishment	Shade and competition eliminate crop production in three to four years
Woodlot	Isolates trees from crops, avoiding potential negative tree-crop interactions	Competes for crop production area; poor matching of site-species and management has led to soil erosion
Home garden	Generally does not compete for crop production area; provides for tree product and fruit production on the same area	No tree-crop interface, so little or no interaction

Note: Choice of species and management may determine whether interaction is positive or negative and may limit or enhance the magnitude of the impact (positive or negative). In a complementary interaction, trees shade crops so that inputs improve tree growth, and trees provide organic material and nutrients to crops. In a competetive interaction, agricultural production decreases as a result of competition for sunlight, water, and nutrients, and tree harvesting may damage perennials.

Source: Authors' data.

several projects. Information from four projects in El Salvador and Guatemala indicates that the average number of trees planted ranged between 440 on farms with an average size of 2.2 hectares (or 200 trees per hectare) and 1,160 on farms with an average of 10.4 hectares (or 112 trees per hectare). Although the farmers with more land planted more trees in total, farmers with less land tended to plant more trees per hectare and dedicated a greater percentage of their total land area to trees. Although farm size may affect adoption, it was not a limiting factor. For example, extension monitoring data from El Salvador show that 40 percent of the farmers participating in a community nursery program owned less than 1 hectare of land.

Effects of land tenure and land use regulation. Harvesting laws and regulations, created to protect forests, are a significant barrier to the adoption of agroforestry in many places. In several countries, farmers are reluctant to plant trees because they fear possible expropriation of the land or future inability to harvest trees without permits. Most projects have developed mechanisms to overcome this, such as contracts and written agreements giving the farmer explicit rights to harvest trees, but such agreements are often not applicable nationally. Farmers without clear land rights—renters, squatters, or residents of land that is de jure under forest reserve or protection—face substantial obstacles to agroforestry.

Lack of formal land tenure generally is not a binding constraint to adoption (Cook and Grut 1989 and Godoy 1992), except in certain situations in Africa, where constraints to women on tree planting or difficulties in protecting trees established in fields used by pastoralists during the fallow period may be a deterrent. By contrast, tree ownership and disposal rights, particularly the restrictions and insecurities created by state claims of control over harvesting, have been particularly problematic in dry areas in West Africa

(Shepherd 1992) as well as in Central America. Terms of sharecropping arrangements also appear to be a constraint in parts of Asia (Arnold 1995).

Implications for Extension

The forces that influence the spread of agroforestry, and the patterns of adoption that have emerged, should be a central concern in designing or modifying projects, particularly the amount of extension assistance offered and the form it should take. Sometimes tree planting on farms has been promoted irrespective of the profitability of the agroforestry system or how it fits into management of the household and farm resource. Of course, there is room for extension and supply-driven innovation to act as a catalyst by promulgating new species or agroforestry systems, for instance through demonstration plots or farms. But extension and technical assistance need to be highly responsive to farmers' concerns and be able to provide a range of options and to outline their pros and cons. They should not push tree planting as something that is intrinsically good.

Information on available options. Projects should provide farmers with information on resource requirements and performance of a variety of agroforestry designs and species rather than with standard designs. This would require promotion programs to broaden the basket of choice of species and systems to provide detailed information and performance under different types of management. Farmers could then select those that best meet household needs and resources and modify the basic design to suit their own preferences.

Demonstration, technical assistance, and training. Agroforestry development is clearly occurring in parts of the region as a natural response to economic incentives and subsistence needs. Qualitative information from farmers, project staff, and informed observers suggests that the demonstrated benefits of fast-growing tree species and systems often have been a strong impetus to the expansion of agroforestry activities (see box 23.1). Nonetheless, even where agroforestry systems are demonstrably profitable, some technical assistance is needed to facili-

Box 23.1. The La Máquina Project in Guatemala: A Case of Participatory Field Research with Demand-Driven Diffusion of Technology

La Máquina is a farming area of approximately 345 square kilometers on the Pacific coastal plain of Guatemala. Activities were initiated there in 1983 under the Lena Project, a regional project with core professional staff located at CATIE in Costa Rica, operating in participating countries through a national coordinator working with a counterpart from the forestry agency. Under their direction, two or three professionals assisted with the activities, along with support and field personnel provided by the forestry agency.

The principal activity was to establish silvicultural experiments to evaluate the performance of species in different systems. Rather than being restricted to experimental fields, research plots were established on farmers' lands by agreement with the owners, and the local population was involved in the research effort. In 1987 the work was expanded to include plots on demonstration farms managed by the farmers under various agroforestry systems. Detailed information was collected on the inputs and returns for both agricultural and forestry/agroforestry plots.

A nursery was also established in the area, managed by two resident technicians who also provided technical assistance to farmers interested in planting or managing multipurpose tree species. Between 1986 and 1991, about 238,000 seedlings of multipurpose tree species were provided to 550 farmers.

La Máquina's success is principally attributable to three aspects of the project: the demonstration effect, the quality of the product, and the evolution of incentives.

- The demonstration effect meant that dissemination was driven initially by demand, followed later by directed extension activities to support the effort.
- As the product of a research project, planting stock was reliable and technical assistance in planting and management helped to guarantee quality.

Incentives were modified as the project developed. To launch tree planting, free seedlings were given to farmers and were often also transported to the farm. Once the community became interested, farmers arrived at the nursery to request seedlings and received technical assistance from the staff. Later, seedlings were provided in exchange for an equivalent number of plastic bags. Most recently, the project and a new effort in an adjacent community began promoting family nurseries. Thus the project gradually shifted the costs and responsibilities for tree production to the farmers.

tate adoption and to provide information about tree management for unfamiliar species and configurations. The process of diffusion is slow, in part because farmers take a long time to evaluate perennial systems and in part because they are frequently unfamiliar with product markets at the farm or group level. Technical assistance can accelerate diffusion. This can be phased out after the technology has been transferred, although, as the La Máquina example indicates, continued applied research to produce the best tree species and to discover which conditions suit them best remains very important even after public extension on agroforestry has been phased out.

Training and employing local people as paratechnicians was judged to be a successful, low-cost means of promoting technology in nine of the projects. Paratechnicians could generally turn to professional staff for technical expertise. Use of paratechnicians develops local human resources and the capacity to continue diffusing agroforestry beyond the project period.

Provision of planting materials. Of the variety of strategies for providing germ plasm—direct seeding of native and introduced species, family nurseries, community nurseries, and specialized centralized nurseries producing containerized seedlings in "root trainers" for easier transport—community and family nurseries proved superior to centralized nurseries, when backed by good technical assistance and logistical support. Farmers in many projects indicated a willingness to purchase seedlings at market prices.

Financial and material incentives. The case studies suggested that financial incentives and subsidies may not, in fact, be needed for adoption, at least for the size and duration found. The important objective for extension is to put forward agroforestry technologies that will be profitable for the farmer without subsidies. There may be an argument for providing time-limited financial incentives to early adopters of unfamiliar technologies, to be reduced and eventually eliminated once farmers are aware of the benefits.

Small, in-kind material inputs (seedlings, watering cans, planting bags, and fertilizer), which were generally provided with extension services, proved widely effective in persuading farmers to try agroforestry. The advantage of these incentives was not so much that they had an effect, which was often quite modest, on profitability but that they reduced the perceived risks of new planting and the need for out-of-pocket cash expenditures. Food-for-work incentives, by contrast, were very problematic, commonly leading to dependency, failure to maintain systems once they were established and incentives were no longer given, and reluctance to undertake subsequent agroforestry efforts without payment.

Credit for agroforestry. Evidence from the study suggests that formal lines of credit may not be suitable for promoting smallholder agroforestry. Poor farmers seldom use formal production credit, because of the risks involved, and when they do, they may reserve it for other priorities. Credit is mainly needed to cover the costs of hired labor for planting relatively large acreages—an activity not relevant for smallholders. Where credit is made available, integrated or nontargeted systems (rather than separate lines of credit for specific crops or trees) would benefit farmers most.

Social and Environmental Benefits

Quantitative data on the social (as distinguished from private) benefits provided by agroforestry systems are sparse. But there is some information and anecdotal evidence that these projects are making social and environmental contributions that could justify further promotion for that purpose. Agroforestry systems are replacing less stable systems in watersheds, protecting the remaining forest resource by providing an alternative supply of tree products, protecting cities from the effects of airborne dust, and providing sources of employment and income generation for rural communities, thus helping to slow outmigration. These benefits, of course, generally do not accrue to the farmer and therefore tend not to figure prominently in adoption decisions. As in Central America, Cook and Grut (1989) found for Africa that environmental benefits, although appreciated by farmers, are not sufficient to spur adoption, except in areas of intensive production where yields are declining.

Some specific examples of environmental and social benefits were noted in Central America:

- Cacao plantings with tree shade introduced in fragile watersheds in Choloma, Honduras, are replacing shifting agricultural systems with perennial tree crops that protect and build soil. In Choluteca, Honduras, communal and individual plantings combined with soil conservation measures are providing a similar benefit. In addition, in most areas where projects have been operating, the burning of agricultural fields and pastures has declined.

- In Honduras, El Salvador, and Guatemala, tree planting is providing an alternative source of tree products and thus reducing the higher social costs, such as loss of biodiversity and increased downstream flooding, entailed in "mining" a natural forest. Data from three projects demonstrated that in Honduras, farmers who depended on the natural forest for up to 100 percent of their tree products are now meeting the bulk of that demand from their own tree plantings. In El Salvador, a community close to the capital city, which used to produce charcoal for sale from natural vegetation, is now using tree plantings for that purpose. In Guatemala, tree plantings on the Pacific coast are being used as posts for an expanding tobacco industry instead of the remaining mangrove forest, which would have been the most probable source of that material.
- In Costa Rica, the Monteverde Conservation League has for several years promoted the establishment of windbreaks, which have improved the productivity of dairy and coffee for participating farmers. A recent study by researchers from Duke University has demonstrated that those windbreaks are acting as biological corridors connecting remnant forest patches. In addition, because their productivity has increased as a result of their participation with the Monteverde Conservation League, participating farmers are more receptive to protecting and improving their natural forest and expanding the width of the windbreaks to serve environmental purposes.
- In Nicaragua, extensive windbreaks were established to protect León from dust carried into the city by winds. Reportedly, the windbreaks have reduced the dust and, with it, the health costs for the population.
- Finally, as this study demonstrates, these projects improved rural welfare by providing needed tree products and generating income and employment. A documented example is the effort of the Centro Agrícola Cantonal in Hojancha, Costa Rica. According to a report prepared in 1990, the commercial nurseries in Hojancha that developed from the reforestation efforts generated 30,000 person-days of employment in 1989, and seed collection from plantations provided an additional 4,900 person-days (Campos, Rodríguez, and Ugalde 1991). The increased employment opportunity is an advantage for Hojancha, which is a region whose outmigration adds people to already overburdened urban slums. Another social benefit derives from the fact that agroforestry activities are often undertaken in periods when the demand for labor is low. Recent efforts to process small diameters and explore markets for tree resources will provide further possibilities for employment and income generation.

In order to get a better empirical handle on social and environmental impacts, much more data will have to be collected than are currently available. More monitoring of projects is needed in general, and that monitoring should include data collection to allow social and environmental impact hypotheses to be tested.

Conclusions

The attitude of the individual farmer to agroforestry is crucial to the success of agroforestry projects. Farmers' perception of the role that the system will play in their farm's production system and of its costs, benefits, and profitability will determine the extent and durability of adoption and should guide project strategies on extension and institutional and policy issues.

The Profitability of Agroforestry and Its Role in the Farming System

Agroforestry, in many ways, is comparable to other parts of the farming system. Its special characteristics are that it includes a large number of species, configurations, and management intensities, has longer gestation than most agricultural crops, and produces outputs with multiple uses.

In addition to the estimated financial return, farmers attach considerable importance to how an agroforestry system fits into the overall farm production system and the existing land, labor, and capital constraints. Even more important for a farmer's decision than the absolute profitability of agroforestry systems may be the returns relative to alternative options. The most profitable agroforestry systems sometimes have the most market risks. Some marginally profitable systems are widely used to meet specific household subsistence needs. Once those are met, there is an interest in market opportunities. The existence or development of markets is crucial for expansion (see similar findings in Alvarez-Brylla, Lazos-Chavero, and García-Barrios 1989 for Mexico; Mussak and Laarman 1989 for Ecuador; French and van den Beldt 1994 for Asia).

Many agroforestry systems are profitable to farmers under a considerable range of economic conditions, and various types of (low-intensity) traditional agroforestry are indeed practiced in many areas. Many agroforestry systems are profitable at real discount rates of 20 percent or higher. Intercropping, if managed properly, tends to be superior to woodlots. Soil-improving interventions could be profitable even with conservative assumptions of environmental effects. The performance of windbreaks tends to be poor, perhaps because design and local participation are unsatisfactory. Perennial intercrops diversify income, add value per unit of land, improve cash flow, and cause only limited loss of the main product.

Agroforestry Adoption

Local scarcity of wood products is, as might be expected, a key motivator in adopting nontraditional agroforestry. Projects must begin by assessing the scarcity of wood, as well as the existence of local markets for products, when locating agroforestry efforts.

Taungya, perennial intercrops, trees on contours, and tree lines have proved to be the easiest systems to introduce. Results are mixed for alley cropping, home gardens, windbreaks, green manuring, dispersed trees in cropland, and tree-pasture systems. Farmers are willing to invest in rehabilitating their land where systems also produce products or income, and they prefer less-intensive systems.

Even when agroforestry is profitable, smallholders will and should adopt agroforestry incrementally and gradually because of management and resource constraints. Poorer farmers, in particular, are often hampered by limited land, labor, and capital resources and their need to ensure food security and reduce risks. Programs to promote rapid, large-scale adoption of agroforestry may put smallholders at risk or bias adoption and benefits heavily toward higher-income farmers; continuous, small-scale adoption is a more appropriate objective for smallholder programs. Preferred agroforestry systems offer short-term and intermittent benefits that permit farmers to self-finance investments.

Adoption patterns differ between small and large farms. For medium and large farms, fallow and extensive grazing are still important, and intensive agroforestry systems may not yet be economic. For smallholders, in contrast, intensive systems may be more interesting, but food security and risk issues are more critical.

Extension Strategies

The demonstration effect of fast-growing tree species on farms, and of benefits on demonstration plots, has helped to expand agroforestry activities, reducing the costs of extension and increasing its effectiveness.

Rather than offering standard designs, programs would serve farmers best if they would offer a broad selection of species and systems from which to choose those most suitable to the household's needs and resources.

Technical assistance is needed to facilitate adoption and to provide information about managing unfamiliar species and configurations. An endogenous process of agroforestry is clearly developing in parts of the region in response to economic incentives and subsistence needs. But this process is slow, and its acceleration would require access to limited, but continuous, farm- or group-level technical assistance. Where new markets for products are being accessed or developed, extension services may also provide critical information and assistance in marketing.

Involving local people as paratechnicians is often a successful, low-cost approach to promoting technology. Their participation makes projects more sustainable by developing local human resources and thus the capacity to diffuse agroforestry beyond the project period.

Financial incentives and subsidies should be kept to a minimum. Agroforestry technologies promoted by extension should be financially profitable, and thus adoptable, for the farmer without subsidies, with the possible exception of time-limited financial incentives for early adopters of unfamiliar technologies. In-kind, material inputs encourage farmers to experiment with and adopt agroforestry, but the experiences with food-for-work incentives are generally not promising.

Institutional and Policy Implications

There has been considerable regional (Alfaro and others 1994) and international (FAO 1993 and Oram and Scherr 1993) discussion recently about the types of policy reforms necessary to promote agroforestry as well as forestry. In general, the issues raised concerning macro- and intersectoral policy linkages, trade policies, institutional development, and conservation policy are relevant across-the-board. For Central America and the Caribbean, particular attention should be paid to improving the institutional structure for tree product markets (information, monitoring, grading, and standards); modifying regulations that restrict markets for farm-produced products; providing public support services for decentralized NGO extension and paraextension efforts; and incorporating agroforestry into planning efforts.

Institutional responsibilities for agroforestry extension and support need to be defined. Because agroforestry falls between the ministries of forestry and agriculture, the institutional "home" for agroforestry activities has been uncertain. Nongovernmental organizations have taken a leading role in providing information and support but have sometimes undertaken isolated and uncoordinated efforts. Semi-autonomous projects coordinated closely with host-government agencies have been effective.

Lack of land titles does not in and of itself appear to be a significant constraint to agroforestry adoption in most areas. The important point is how secure farmers feel about their property rights with or without an official title. De facto property rights generally provide farmers with enough security. However, in general, no tree planting is taking place on rented land.

Future Research

More precise assessment is needed of the effects of specific policies and extension approaches. For that purpose, empirical studies are needed on a large number of farms, with specific agroforestry practices sampled across a range of environments, farm size, and market access. Monitoring of on-farm environmental effects would be particularly interesting, given the dearth of information in that area.

Notes

Funding from United Nations Development Programme and institutional support from Hans Binswanger and Michael Baxter are gratefully acknowledged. The authors are indebted to the consultants who prepared the country case studies: Dean Current and Ricardo Hernández (Costa Rica), Carlos Reiche (El Salvador), Otto Samoyoa (Guatemala), Carlos Rodríguez (Honduras), Rodolfo Vieto (Nicaragua), Manuel Gómez (Panama), Abel Hernández (Dominican Republic), and John Jickling and T. Andrew White (Haiti). The case studies are presented in Current, Lutz, and Scherr (1995). Thanks also go to the many individuals interested in agroforestry issues who provided information or who participated in the national workshops or in the regional meeting in El Salvador. The comments by Chona Cruz, Peter Dewees, John McIntire, Augusta Molnar, and Stefano Pagiola on earlier drafts are much appreciated.

1. Although many trees have multiple potential uses, management for one use may preclude other uses. One cannot produce high-quality wood in alley cropping systems. Fodder banks and living fences generally do not produce timber.

2. These results include information on three windbreak systems for which averages could not be calculated because the systems were not fully comparable, as well as a number of other, less frequently used systems. They also show that certain systems are more prevalent and important in some countries than in others owing to climatic, ecosystem, or cultural factors.

3. Financial profitability was determined through a cost-benefit analysis of the agroforestry systems reviewed in their entirety, not the incremental net benefit of the agroforestry system over an alternative system of land use. The profitability of the agroforestry systems was compared with that of an alternative.

Glossary

This glossary is based on Rocheleau, Weber, and Field-Juma (1988).

Agroforestry. The deliberate use of woody perennials (trees, shrubs, palms, bamboos) on the same land management unit as agricultural crops, pastures, and animals. This may consist of a mixed spatial arrangement in the same place at the same time or a sequence over time.

Alley cropping. Cultivation of annual crops between rows of trees or hedgerows. It is sometimes called hedgerow intercropping.

Boundary planting. Lines of multipurpose trees or shrubs planted along borderlines and boundaries dividing properties or land uses.

Contour planting. Rows of trees or shrubs planted along the contour of the field.

Home garden. A complex collection of woody and

herbaceous plants deliberately grown in small plots in or near home compounds, often associated with the production of small domestic animals.

Live fenceposts. Use of living trees, rather than dead posts, as fenceposts. In Central America, these are usually used to support barbed wire fencing. The fenceposts may also be managed for fuelwood or poles.

Multipurpose trees or shrubs. A woody perennial grown to provide more than one product or service.

Taungya. A system in which new forest plantations are established together with food and cash crops, which continue to be intercropped until shaded out by the maturing plantation.

Trees intercropped with annual crops. Closely spaced trees planted or maintained in cropland for their products or their positive effects on associated crop production (for example, nitrogen fixation or microclimate improvement).

Trees intercropped with perennial crops. Closely spaced trees planted or maintained in perennial crops for their products or their positive effect on associated crop production (for example, shade or nitrogen fixation).

Windbreak. Strips of trees or shrubs planted to protect fields, homes, canals, or other areas from wind and blowing soil or sand.

Woodlot. Stands of trees planted and managed to produce various tree products, associated plants, or services.

Woody fallow. Fallow is land resting from cropping, which may be grazed or left unused, often colonized by natural vegetation. For woody fallows, the fallow is left uncultivated for sufficient time that woody plants come to predominate; farmers may enrich, manage, or harvest from this woody fallow.

References

The word "processed" describes informally reproduced works that may not be commonly available through libraries.

Alfaro, Marielos, Ronnie de Camino, María Ileana Mora, and Peter Oram, eds. 1994. *Taller regional: Necesidades y prioridades de investigación en políticas forestales y agroforestales para Latinoamérica.* San José, Costa Rica: Instituto Interamericano de Cooperación Agrícola.

Alvarez-Brylla, M. E., E. Lazos-Chavero, and J. R. García-Barrios. 1989. "Home Gardens of a Humid Tropical Region in Southeast Mexico: An Example of an Agroforestry Cropping System in a Recently Established Community." *Agroforestry Systems* 8(2):133–56.

Arnold, J. E. M. 1995. "Retrospect and Prospect." In J. E. M. Arnold and Peter A. Dewees, eds., *Tree Management in Farmer Strategies: Responses to Agricultural Intensification*, pp. 272–87. Oxford: Oxford University Press.

Arriola, Francisco, and Rudy Herrera. 1991. *Casa Cabrican, Quetzaltenango, Guatemala. Agricultura sostenible en las laderas Centroamericanas: Oportunidades de colaboración interinstitutional.* Coronado, Costa Rica: Instituto Interamericano de Cooperación Agrícola.

Belaunde, Elvira, and A. C. Rivas. 1993. *Responding to Practice and Affecting Policy: The National and Regional Impact of Local Experience in the Madeleña Project in Central America.* Rural Development Forestry Network Paper 16a. London: Overseas Development Institute.

BID (Banco Interamericana de Desarrollo). 1993. "Progreso económico y social en América Latina." *Informe 1993.* Special theme: *Inversión en recursos humanos.* Washington, D.C.

Budd, W. William, Irene Duchhart, L. H. Hardesty, and Frederick Steiner, eds. 1990. *Planning for Agroforestry.* Amsterdam: Elsevier.

Budowski, Gerardo. 1987. "The Development of Agroforestry in Central America." In Howard A. Steppler and P. K. Ramachandran Nair, eds., *Agroforestry: A Decade of Development*, pp. 69–88. Nairobi: International Center for Research in Agroforestry.

Campos, Oscar, Emel Rodríguez, and Luis Ugalde. 1991. "Desarrollo agropecuario sostenible en la región de Hojancha, Guanacaste, Costa Rica." In *Agricultura sostenible en las laderas centroamericanas: Oportunidades de colaboración interinstitucional. Memorias del taller bajo patrocinio de Cooperación Suiza para el Desarrollo, organizado por el Centro Internacional de Agricultura Tropical, Instituto Interamericano de Cooperación para la Agricultura, Centro Agronómico Tropical de Investigación y Enseñanza, Centro Internacional de Mejoramiento de Maíz y Trigo*, pp. 245–77. Costa Rica: Instituto Interamericano de Cooperación Agrícola.

Cook Cynthia, and Michael Grut. 1989. *Agroforestry in Sub-Saharan Africa: A Farmer's Perspective.*

Technical Paper 112. Washington, D.C.: World Bank.

Current, Dean, and Modesto Juárez. 1992. *The Present and Future Status of Production and Consumption of Fuelwood in El Salvador*. Tropical Agricultural Research and Higher Education Center–U.S. Agency for International Development–El Salvador. Turralba, Costa Rica: Tropical Agricultural Research and Higher Education Center.

Current, Dean, and Ernst Lutz. 1992. "A Preliminary Economic and Institutional Evaluation of Selected Agroforestry Projects in Central America." Divisional Working Paper 1992-38. Environment Department, World Bank, Washington, D.C. Processed.

Current, Dean, Ernst Lutz, and Sara Scherr. 1995. *Costs, Benefits, and Farmer Adoption of Agroforestry: Project Experience in Central America and the Caribbean*. Washington, D.C.: World Bank.

Current, Dean, Carlos Rivas, and Manuel Gómez. 1994. *El sistema estandarizado de registros para actividades de extensión forestal MIRAEXT: Una herramienta para seguimiento y evaluación*. Turralba, Costa Rica: Tropical Agricultural Research and Higher Education Center.

Dewees, Peter A. 1993. *Trees, Land, and Labor*. Environment Paper 4. Washington, D.C.: World Bank.

Eckholm, Eric. 1975. *The Other Energy Crisis: Firewood*. Worldwatch Paper 1. Washington, D.C.: Worldwatch Institute.

FAO (Food and Agriculture Organization of the United Nations). 1993. *Forestry Policies of Selected Countries in Asia and the Pacific*. Forestry Paper 115. Rome.

French, James H., and Rick J. van Den Beldt, eds. 1994. *Forestry/Fuelwood Research and Development Project. Final Report*. Bangkok, Thailand: Winrock International.

Godoy, Ricardo A. 1992. "Determinants of Smallholder Commercial Tree Cultivation." *World Development* 20(5):713–25.

Gómez, Manuel, and Thomas McKenzie. 1991. "Resúmen anual de operaciones económicas en fincas demostrativas, del Proyecto Madeleña, en América Central." Internal technical report for the Madeleña Project. Turrialba, Costa Rica: Tropical Agricultural Research and Higher Education Center. Processed.

Gregersen, Hans, Sydney Draper, and Dieter Elz. 1989. *People and Trees: The Role of Social Forestry in Sustainable Development*. Washington, D.C.: Economic Development Institute and the World Bank.

Kerkhof, P. 1990. *Agroforestry in Africa: A Survey of Project Experience*. London. U.K.: The Panos Institute.

King, K. F. S. 1979. "Concepts of Agroforestry." In M. T. Chandler and D. Spurgeon, eds., *International Cooperation in Agroforestry. Proceedings of an International Conference*, pp. 1–14. Nairobi, Kenya: International Center for Research in Agroforestry.

Murray, Gerald F. 1991. "The Tree Gardens of Haiti: Agroforestry among Caribbean Peasants." In David Challinor and Margaret Hardy Frondorf, eds., *Social Forestry: Communal and Private Management Strategies Compared*, pp. 35–44. Washington, D.C.: U.S. Agency for International Development; Baltimore, Md.: Johns Hopkins University Press.

Mussak, M. F., and J. G. Laarman. 1989. "Farmer's Production of Timber Trees in the Cacao-Coffee Region of Coastal Ecuador." *Agroforestry Systems* 9(2):155–70.

Nair, P. K. R. 1990. *The Prospects of Agroforestry in the Tropics*. Technical Paper 131. Washington, D.C.: World Bank.

National Academy of Sciences. 1980. *Firewood Crops: Shrub and Tree Species for Energy Production*. Vol. 1. Washington, D.C.: National Academy Press.

Nitrogen Fixing Tree Association. 1993. *Nitrogen Fixing Tree Research Report*. Vol. 11. Hawaii.

Oram, Peter, and Sara J. Scherr, eds. 1993. *Report of the African Regional Workshop on Forestry and Agroforestry Policy Research*. Nairobi, Kenya, 20 November–4 December. Washington, D.C.: International Food Policy Research Institute.

Raintree, John B. 1991. *Socioeconomic Attributes of Trees and Tree Planting Practices*. Rome: Food and Agriculture Organization of the United Nations.

Raintree, John B., and Hermina A. Francisco, eds. 1994. *Marketing of Multipurpose Tree Products in Asia. Proceedings of an International Workshop Held in Baguio City, Philippines, 6–9 December, 1993*. Bangkok, Thailand: Winrock International.

Rocheleau, Dianne, Fred Weber, and Alison Field-Juma. 1988. *Agroforestry in Dryland Africa*. Nairobi, Kenya: International Council for

Research in Agroforestry.

Scherr, Sara J. 1992. "Not Out of the Woods Yet: Challenges for Economics Research on Agroforestry." *American Journal of Agricultural Economics* 74(3):802–08.

———. 1995. "Economic Analysis of Agroforestry Systems from the Farmers' Perspective." In Dean Current, Ernst Lutz, and Sara Scherr, eds., *Farmer Adoption and Economic Benefits of Agroforestry: Project Experience in Central America and the Caribbean*. Washington, D.C.: World Bank.

Scherr, Sara J., and Eva Muller. 1991. "Technology Impact Evaluation in Agroforestry Projects." *Agroforestry Systems* 13(3):235–58.

Shepherd, Gill, ed. 1992. *Forest Policies*. London: Overseas Development Institute.

Steppler, Howard A., and P. K. Ramachandran Nair, eds., *Agroforestry: A Decade of Development.* Nairobi: International Center for Research in Agroforestry.

Swinkels, Rob A., and Sara J. Scherr, comps. 1991. *Economic Analysis of Agroforestry Technologies: An Annotated Bibliography*. Nairobi, Kenya: International Center for Research in Agroforestry.

Tschinkel, Henry. 1987. "Tree Planting by Small Farmers in Upland Watersheds: Experience in Central America." *International Tree Crops Journal* 4(4):249–68.

U.S. Comptroller General. 1982. *Changes Needed in U.S. Assistance to Deter Deforestation in Developing Countries. Report to the Congress of the United States*. Washington, D.C.: General Accounting Office.

USAID/ROCAP (U.S. Agency for International Development/Regional Office for Central American Programs). 1979. "Fuelwood and Alternative Energy Sources." Washington, D.C. Processed.

Vergara, Napoleon T., and Kenneth G. MacDicken. 1990. "Extension and Agroforestry Technology Delivery to Farmers." In Kenneth G. MacDicken and Napoleon T. Vergara, eds., *Agroforestry: Classification and Management*, pp. 354–75. New York: John Wiley and Sons.

Warner, Katherine. 1993. *Patterns of Farmer Tree Growing in Eastern Africa: A Socioeconomic Analysis*. Nairobi, Kenya: Oxford Forestry Institute and International Center for Research in Agroforestry.

World Bank. 1994. *World Development Report 1994*. New York: Oxford University Press.

Part IV
Conclusions

Meeting the food and fiber needs of growing populations in developing countries while conserving the resource base is and will continue to be a great challenge for everyone involved in rural development. This volume brings together recent findings and current thinking intended to help meet the challenge.

Past agricultural growth has been associated with a variety of environmental and resource problems including excessive deforestation and forest degradation, water depletion, waterlogging and salinization, fish stock depletion, soil degradation, health effects, and biodiversity losses. Comprehensive, empirically strong estimates for the costs of resource degradation associated with agriculture in developing countries do not seem to exist; in their absence, Hazell and Lutz (chapter 2) present indicative estimates from various sources.

A key message in this volume is that environmental problems do not have to be the inevitable outcome of agricultural growth. Problems are usually associated with inappropriate incentive systems (broadly defined), inappropriate institutional settings, insufficient investment in resource-poor areas, inadequate poverty and social concerns, and political economy systems in which the rich and powerful extract rents.

Although no precise definition is offered, sustainable agricultural development is conceptualized as agricultural income traditionally measured in national income accounts corrected for changes in the value of the stock of capital, which includes natural, manmade, and human capital. We also adopt a definition of sustainability under which capital can be changed from one category into another, while keeping the total constant. Yet we also recognize that there may be critical levels of each type of capital (critical ecosystems) beyond which capital should not be substituted.

Sustainable rural development must appropriately integrate environmental concerns (Hazell and Lutz, chapter 2). Such integration includes correcting price distortions that encourage excessive use of modern inputs in intensive agriculture, ensuring that farmers have secure property rights over their resources, strengthening community management systems and empowering local organizations, improving the performance of relevant public institutions that manage and regulate natural resources by devolving management decisions to resource users, or groups of users, wherever possible, and giving greater attention to sustainability features of recommended technologies (Hazell, chapter 8).

It is evident that environmentally and socially sustainable rural development depends on appropriate policy frameworks, appropriate institutional approaches, as well as appropriate technologies. We have structured our conclusions into these three parts, which mirror the three main parts of the volume.

Appropriate Policy Frameworks

It essential that an appropriate policy framework be in place that is conducive to sustainable rural development. In too many countries, there are still too many obvious distortions in the incentive frameworks. Most of these should be removed simply on the basis of sound economics and efficiency.[1] In many cases, these reforms would have direct or indirect positive impacts on the environment; that is, taking environmental concerns into account would often *add* to the benefits of reforms that should have been undertaken on the basis of economics. But many reforms are not carried out for political economy reasons, where perhaps small but important groups or individuals derive benefits from the existing distortions that they are not willing to give up voluntarily. Government leaders either may themselves benefit from the distortions or may feel that they cannot afford to lose the support of a particular stakeholder group that is benefiting from the current policies.

Resource and environmental problems are quite different for the array of agroecological conditions that one encounters. For example, in irrigated agriculture the main problem is the inappropriate use of inputs (irrigation water, seeds, fertilizers, pesticides, and machinery), often in response to a faulty incentive framework, while in rainfed agricultural areas, the main problem may be resource degradation due to insufficient yield growth in relation to population growth (Hazell and Lutz, chapter 2).

Sustainable agriculture seeks the integrated use of a wide range of pest, nutrient, soil, and water management technologies. It also aims for increased diversity of enterprises within farms so that by-products or wastes from one component or enterprise become inputs to another (Pretty, chapter 4), thereby reducing negative impacts on the environment.

The diversity of situations implies that sustainable agriculture cannot be a fixed model that is to be imposed; rather, it should be a process of learning (Pretty, chapter 4; Bunch, chapter 12). What needs to be made sustainable, therefore, is the social process of innovation itself, which implies an enhanced capacity to adapt to unexpected changes and emerging uncertainties. For such innovation to occur, it is necessary to have an enabling policy environment. Chapter 3 (Heath and Binswanger) and chapter 5 (Holden and Binswanger) elaborate on this point, supported by empirical evidence that conducive policy and institutional environments are a key condition for inducing innovations.

The Boserup hypothesis states that, under the right conditions, higher population and market access can lead to improvements in natural resources rather than their deterioration. But Boserup effects are far from automatic; they are an outcome of investment decisions made by farmers, who require a positive incentive regime and access to markets and to soil and water resources. Adverse policy regimes force impoverished peasants to mine, rather than augment, the land resources. This mining is especially damaging if they only have access to marginal land or to humid tropical forest frontiers. Therefore, the removal of adverse policies should be given top priority.

Adverse policies often reduce economic efficiency, increase poverty, and degrade natural resources (Heath and Binswanger, chapter 3). Faulty policies have tended to favor the modernization of large-scale farming at the expense of more efficient and employment-intensive family farms. Reforming adverse policies would be a *win-win-win* situation because the same policy changes could lead to more growth, less poverty, and more sustainable natural resource management.

Although we emphasize here the importance of appropriate agricultural policies including overall incentive frameworks (including macro and trade policies), we also recognize the importance of proper infrastructure (related to market access), education (in particular, the education of girls), health, and macroeconomic stability. In some contexts—for example, where farmers are poorly integrated into markets or where there are significant market imperfections and farmers are facing significant risks—it may also be necessary to assist in additional ways, such as by interlinked contracts (Holden and Binswanger, chapter 5).

New policies to stimulate sustainable rural development are much needed, and, to succeed, these policies have to build on an understanding of the decisionmaking environment and behavioral responses of small farmers. Small farmers largely are rational and respond to changes in the set of constraints and opportunities they face. But inefficiencies may still accrue due to economies of scale, poverty and subsistence constraints, unequal access to markets, and various policies disfavoring small farmers. These inefficiencies may also lead to underinvestments in conservation. There may thus be grounds for interventions from the viewpoints of efficiency and equity as well as sustainability. Given the complexity of small-farm problems, it is necessary to start with pilot projects and to adopt a flexible project design that is accompanied by systematic monitoring and evaluation to allow for maximum learning and consequent adjustment and upscaling.

Part of improving the overall incentive framework usually means reducing trade distortions. Many countries began to reduce such distortions during the past decade. In general, trade liberalization leads to overall net gains, but some negative environmental effects are possible (Anderson, chapter 6). Where they occur, environmental policy or regulatory actions may be needed. To support these, one should make more efforts to quantify the impacts that are likely to be associated with trade and other economic reforms.

Unilaterally removing the policies that keep domestic food prices artificially low in developing countries would boost incentives to invest in agricultural research. But at the higher price levels, the cost—in terms of degrading the farm sector's natural resource base—to those countries of failing to reduce distortions in the markets for farm chemicals and irrigation water would be greater than when product prices, and hence input intensities, were lower. This suggests the need to introduce simultaneous reform in the pricing of outputs and inputs.

The outcomes of policy reforms and development activities depend on institutional arrangements in general and property relations in particular (Bromley, chapter 7). Property regimes are central in development because they connect people to one another with respect to land and related natural resources. Development projects are sometimes less successful than they would otherwise be precisely because the existing property relations have been ignored or misunderstood.

Bromley offers nine main lessons for development practitioners: (1) property regimes are part of the larger institutional structure, (2) all property regimes require external legitimacy, (3) ambiguous rights regimes have ambiguous efficiency and distributional consequences, (4) property relations must be specified prior to the implementation of development projects, (5) resource degradation is the result of problems that precede property regimes, (6) ecological variability demands flexible institutions and actions, (7) resource degradation is contextual, (8) land titles are not necessary for efficient investments in productivity (security of tenure is much more important), and (9) mobilizing local interests can improve a program's chances of success.

An elaboration of lesson number 4 may be important for development practitioners. When development projects increase the flow of income from natural resources over which rights previously may have been ambiguous, powerful individuals at the local level will usually devise a way to expropriate at least part—if not the majority—of these new income streams. Hence, although ambiguity can often work to the advantage of the dispossessed, it is unlikely to do so when development projects are introduced into this kind of institutional environment. Therefore, development projects introduced into settings in which there is great ambiguity about property relations will most probably have economic and social impacts quite different from those postulated in the proposals and feasibility analyses that led to the project's acceptance and implementation in the first place. Therefore, every development intervention must be preceded by a concerted effort to ensure that the institutional arrangements have been modified so as to ensure that the benefits go to the intended beneficiaries. This modification of the "working rules" of the local economy must occur before the new benefit stream begins to materialize (Bromley, chapter 7).

Eliminating the privileges of rural elites that are embedded in adverse policies is an extremely difficult task. These policies did not come about by accident, but as the consequence of a historical evolution that involved bargaining among politically strong groups looking out for their interests at each point in time (Heath and Binswanger, chapter 3). In practice, policy is the net result of the actions of different interest groups pulling in complementary and opposing directions, making the outcome uncertain (Pretty, chapter 4). Analyzing the joint consequences of the policies, using sector studies, for example, may aid the reform effort. But it is unlikely to be sufficient. Steps must be taken to provide peasants with greater input into policymaking (Heath and Binswanger, chapter 3). Unless more political actors with enlightened hearts and minds see it as in their interest to bring poor peasants more strongly into the policy process, reforms of the policy environment and of land may continue to progress slowly.

How can development agencies encourage reforms when political economy constraints are blocking them? First, they should seek unity among themselves as to what reforms are to be undertaken. Second, they should take a firm approach, supported by credible analyses of current policies that show who are the gainers from the current policies and who are the losers (producers, consumers, treasury, environment). New lending or grants should be provided after progress has been made in specific areas. Unless a government demonstrates its commitment to improving economic development, outsiders should not be "enablers" of inefficiency.

In addition to building a basis for seeking unity among donors and holding discussions with governments, applied, empirical analyses may inform and activate public opinion, strengthen reform-oriented constituencies, and create more local ownership through workshops, seminars, and meetings with civil society and nongovernmental organizations (NGOS).

Appropriate Institutional Approaches

The approach to rural development must be participatory and decentralized. A sound policy environment is essential for sustainable agricultural development, among other things, because it creates the incentive framework that induces farmers to innovate. Appropriate institutional frameworks and consideration of the social dimensions are equally important because they provide the structure or fabric within which policies are applied and form an important part of the context within which farmers work.

Projects must be participatory and decentralized because building local ownership is necessary for sustainable efforts (Narayan, chapter 9). It has sometimes been argued that participation is time-consuming and that decentralized approaches are difficult because of the generally limited institutional capacity at local or regional levels. But experience shows that these objections are not well founded if the intention is to further locally driven assistance that produces sustainable impacts over the long term.

Limited development success of central, top-down approaches has led to recognition of the need for participatory, community-driven development (Narayan, chapter 9). Community-driven development is a process in which community groups initiate, organize, and take action to achieve common interests and goals. No single model is appropriate for all places and times. But successful community-driven development generally has five main characteristics: local organizational capacity or the existence of viable community groups, the appropriate fit of technology to community capacity, effective agency outreach strategies, client-responsive agencies, and enabling policies.

Local organizational capacity is embedded in the norms that enable groups of people to trust one another, work together, organize, solve problems, mobilize resources, resolve conflicts, and network with others to achieve agreed goals. Viable community groups are key to the success of community-driven development. Getting local groups and organizations to become self-managing organizations can extend over several years and does not happen without investment in capacity building. Local elites often take leadership roles, and, although not necessarily bad, this can result in the hijacking of resources unless transparency and accountability are somehow enforced.

The appropriateness of technology is determined by the local management capacity and how the introduction of a technology influences social cohesion and benefit flows. Technology needs to be viewed in a holistic way to determine its fit with the existing social system and the direction of change that will be set into motion by its introduction.

Outreach mechanisms can be classified broadly into two basic approaches: the *extension* approach and the *empowerment* approach. The two often overlap, and an effective outreach strategy contains elements of both. The extension approach is characterized by dissemination of information, promotion of demand, and delivery of particular services or inputs. The empowerment approach is characterized by the empowerment of local groups or the creation of local organizational capacity for self-management through involvement in decisionmaking.

Client-centered agencies are needed to ensure that demand is met and community self-management is adequately supported by community fieldworkers. To become client-centered, agencies must be convinced to change their rules and incentive structures. Interest and commitment to change can be generated by pilot projects that demonstrate alternative strategies and use a variety of participatory techniques, including field visits to other regions and countries.

Community-driven development on a large scale requires enabling policies as well as political support to protect agencies so that they can initiate the reform process and give it time to take root. Although individuals play key roles in starting or protecting the reform process in the early stages, change can be sustained only if the rules and regulations at different levels provide appropriate incentives to work and organize.

To increase the probability of success with participatory community-based development projects, the following 10 lessons from worldwide experience should be reflected in the project design, subject to testing with pilot activities: (1) start with clearly stated objectives logically linked to strategies, outputs, indicators of success, and physical or capacity-building outcomes; (2) identify the key actors and their capacity and interests (using social assessment, for example); (3) assess demand (as measured by commitment before construction—for example, with a financial investment up-front in capital cost); (4) craft a self-selection process (such as by requiring a financial or organizational contribution up-front); (5) structure subsidies that do not distort demand; (6) restructure fund release to support demand (such as by implementing fiscal decentralization); (7) plan for learning and a plurality of models (the project must be seen as likely to adjust and change over time as local priorities change and as local organizations mature); (8) invest in outreach mechanisms and social organization (investments are needed, and there are no shortcuts to strengthening local social organization for collective action); (9) use participatory monitoring and evaluation (listening to people's suggestions is empowering and encourages innovation and responsibility); and (10) redefine procurement rules, where necessary (so that they do not hinder community initiative).

Conserving or improving natural resources often requires collective action by groups of users, but there is evidence that self-organizing groups have been weakened and undermined by the growth of government intervention, with negative consequences for resource management. Involving local organizations in rural development contains principles of both participation (Narayan, chapter 9) and decentralization (Caldecott and Lutz, chapter 14). Ashby and others (chapter 10) explain these principles at the watershed level and illustrate them with practical experiences and a case study.

There are three main advantages in involving local organizations in watershed management (Ashby and others, chapter 10). First, local organizations can often be very effective in generating and securing compliance with rules for the use of common property such as water, common grazing land, or forest and the management of buffer zones around conservation areas, all of which may be important features of watershed management. This is in contrast to the difficulties experienced by nonlocal agencies, especially in low-income countries, in imposing sanctions on undesirable management practices or in providing incentives that are lasting. Second, organizations that involve local stakeholders in the development of management practices and the selection of technologies aimed at conserving watershed resources often promote innovation by identifying locally appropriate technologies and securing their adoption more effectively than external agencies. Third, devolving responsibility to local organizations can both externalize some costs of enforcing conservation from the state to local communities and also reduce costs overall, by creating conditions in which nonlocal agencies become more efficient and effective through their collaboration with local organizations.

Although many lessons can be learned and principles derived from experience, the last word on what works, and what does not work, is by no means decided (Ashby and others, chapter 10). Indeed, the dynamic nature of the process of involving local organizations in watershed management is one of its most important characteristics and requires focusing on managing the process rather than on defining a blueprint for how to go about it. An interesting case study in chapter 10 illustrates this.

Cooperation in watershed management is not necessarily easy to initiate or sustain, even when local organizations are involved. Multiple uses for any given watershed resource create multiple stakeholders, often with competing priorities. For this reason, conflict among stakeholders over the rights and conditions for use of a given resource is a common feature of watershed management. Thus, one of the key functions of local organizations in watershed management is to build the social capital required for people to build consensus about and enforce the agreed use of watershed resources for diverse, multiple, and often conflicting purposes.

Few local organizations at the watershed scale have arisen without some intervention from outside agencies to catalyze their formation and action. At the local level, outsiders usually play a significant role in enabling stakeholders to arrive at a joint plan of action that takes into account transboundary effects not readily perceived or measured by those who are not directly affected. Another important point of interface between local organizations and external agencies in watershed management is the need for external intervention in identifying the relevant stakeholders and bringing them to the table, whether this is literally a negotiating table or participation in collective labor, monitoring, or enforcement of sanctions.

The involvement of local organizations in watershed management does not necessarily ensure equitable participation of all relevant stakeholders in a resource management initiative (chapters 9 and 10). Case studies show that women, the landless, marginal ethnic groups, and the laboring poor are unlikely to be represented in participatory watershed management. Leadership in local organizations is generally captured by the higher status, more well-to-do members of dominant elites, and the priorities of local organizations tend to reflect their interests. Therefore, watershed development projects should deliberately seek to overcome this bias by setting up catchment committees with representation of underprivileged groups.

Participation is also one of the keys to successful generation of technology among low-income farmers (Farrington, chapter 11). Participatory approaches are easy enough to implement on a small scale, but specific institutional arrangements have to be made to permit wide-scale implementation. These can best be based on partnerships between public agencies, private organizations (especially NGOS), and farmers themselves, both to bring additional resources to bear and to reorient the public sector. Chapter 11 argues that, especially under unreliable rainfed conditions where the private commercial sector is likely to remain weak, public sector research and dissemination continue to have an important role. However, in such areas, it has been difficult to maintain an effective public sector capacity on the ground, and research and extension have generally been oriented only weakly toward clients' needs.

Multiagency approaches—not replacement by the private sector or complete abolition—are one way of overcoming the limitations of the public sector. Innovative multiagency approaches are now being tested in various ways. They seek synergistic interactions in which different types of organizations work together to analyze problems, contribute jointly to their solution, review progress in an iterative fashion, and make course corrections by mutual agreement as necessary.

Although some success has been achieved in introducing participatory approaches at the diagnostic stage of the research cycle (for instance, participatory rural appraisal is now widely used, although often badly), much less attention has been paid to ensuring continued interaction between researchers and their clients during subsequent identification, testing, and dissemination of technologies (Farrington and Thiele, chapter 11).

Participatory principles are also key to successful extension with long-run impact (Bunch, chapter 12). One such approach called participatory people-centered agricultural development (PCAD) has been developed and tested extensively in Central America by World Neighbors and has spread to mostly nongovernmental organizations in many nations around the world.

Government extension services have tended to neglect farmers in resource-poor areas. One way to overcome this is to enlist the farmers' own efforts in solving their problems of low productivity. A participatory process teaches villagers how to experiment and teach each other through the most efficient teaching process: learning by doing.

PCAD consists of a series of principles with much room for adaptation to local circumstances, farmer needs, and the institutional context. They include motivating and teaching farmers to experiment with new technologies on a small scale, thereby reducing the risk of adoption and providing them with a means to continue developing, adopting, and adapting new technologies in a permanent process of innovation. This principle is frequently referred to as participatory technology development. Another key principle is the importance of training village leaders as extensionists and supporting them as they teach additional farmers, thereby creating and nurturing a community-based multiplier effect. This principle is called farmer-to-farmer extension.

In many instances, where PCAD has been used successfully, yields of major crops have tripled, and new ones have been introduced. But success has not been uniform. One survey showed that, in the 25 to 30 percent of villages where the response was best, yields continued to increase after program termination, the number of organizations increased, land prices increased dramatically, incomes increased, and outmigration either diminished drastically or was reversed. In 40 to 50 percent of the average villages, yields more or less maintained or increased only marginally after project completion, whereas in the 20 to 35 percent of the villages where impact was poorest, yields decreased, although they continued to be better than the yields at program initiation.

One of the major conclusions of PCAD is that sustainability does not reside in technologies, which have a half-life of perhaps five years. Markets change, input prices increase, new technological opportunities appear, varieties degenerate, pests spread, and competition becomes stiffer. The hope for sustainability of agricultural development is in the nature of an ongoing social process of widespread experimentation and sharing of information, of innovation, and of group problem-solving. In PCAD, it is more important how things are taught than what particular technologies are taught. Thereby the PCAD process creates a sustainable process of agricultural development. By its very nature, it also empowers people in ways useful for rural development efforts in general.

One central condition for moving toward more sustainable rural development is for farmers to have access to land and to have security of tenure. Therefore, even though land reform is politically sensitive and difficult to implement, it is very important that these issues be tackled. Theoretical reasons and empirical evidence suggest that land reform may provide benefits both for equity and for efficiency. First, a large body of research has demonstrated the existence of a robustly negative relationship between farm size and productivity due to the supervision cost associated with employing hired labor. This implies that redistributing land from large to small farms can increase productivity. Second, in many situations, landownership is associated with improved access to credit markets, providing benefits as an insurance substitute to smooth consumption intertemporally. By enabling the poor to undertake indivisible productive investments (or preventing them from depleting their base of assets), this could lead to higher aggregate growth. Third, even in aggregate cross-country regression, the distribution of productive assets—more than the distribution of income—seems to have an impact on aggregate growth. Finally, land reform is expected to increase the environmental sustainability of agricultural production and to reduce rural violence (Deininger, chapter 13).

Notwithstanding this apparent potential, actual experience with land reform has been disappointing in all but a few exceptional cases such as Japan, Korea, Taiwan (China), and to some extent Kenya. Despite—or because of—this, land reform remains a hotly debated issue in a number of countries (Brazil, Colombia, El Salvador, Guatemala, South Africa, and Zimbabwe).

A new type of negotiated land reform is now being tried in a number of countries (Brazil, Colombia, South Africa) where land transfers are based on voluntary negotiation and agreement between buyers and sellers and the government's role is restricted to the provision of a land purchase grant to eligible beneficiaries (Deininger, chapter 13).

As Heath and Binswanger (chapter 3) indicate, small farmers in Colombia were often driven off their traditional lands to eke out a living in marginal and environmentally fragile areas, and much of the best agricultural land continued to be devoted to extensive livestock grazing or was not farmed at all due to violence. Only 25 percent of the land suitable for crop production was actually devoted to this use, while the rest was left to pasture. This suggests that there are indeed large tracts of unused or underused land that could be subjected to land reform in order to increase agricultural productivity—a notion in line with available empirical evidence.

Centralized land reform was attempted in Colombia but did not have much success. Now negotiated, market-based land reform is being tried. It is characterized by three main principles. Potential beneficiaries (with assets below a certain minimum level) can negotiate independently with landowners and, once a deal has been struck, are eligible for a grant of up to 70 percent of the land purchase price (subject to an upper limit). The government's role is limited to that of regulatory oversight and grant disbursement. A decentralized institutional structure ensures that land reform is driven by demand and is coordinated with other government programs. Despite favorable preconditions, and the government's expressed determination to distribute 1 million hectares within four years, this new type of land reform has had a disappointingly slow start.

Brazil and South Africa have, under different conditions, recently initiated programs of negotiated land reform. With an institutional background very similar to that of Colombia (presence of land reform legislation and a central land reform institute dating from the early 1960s), negotiated land reform in Brazil has been initiated by individual states. The purpose of the Brazilian interventions is to establish cheaper, more agile policy alternatives to centralized land reform in an environment where the issue of land reform is high on the political agenda and potential beneficiaries have at least some idea of what to do with the land. By contrast, negotiated land reform in South Africa has been adopted in the context of a national reconstruction program in an environment in which productive small-scale agriculture was eradicated almost a century ago. Here, even greater effort is required to establish the decentralized infrastructure necessary to implement land reform, to provide complementary services such as marketing and technical assistance, and to increase beneficiaries' agricultural and entrepreneurial capacity.

Several of the chapters in part II (particularly chapters 9–12) stress the importance of participation to successful, sustainable rural development. Some also touch on the need to pursue decentralized approaches, which are important for many reasons, including their role in facilitating participation.

Decentralization is one aspect of the greater issue of good governance, and one question is whether it promotes or hinders biodiversity conservation and under what conditions (Caldecott and Lutz, chapter 14).

Many past attempts to conserve biodiversity failed. One reason was an overly centralized approach, often involving top-down planning by technicians and bureaucrats without concern for the opinions or well-being of the people affected by their decisions. But because of the externality issues involved, complete decentralization can also be counterproductive to conservation. Under complete decentralization, local people, perhaps in cooperation with outside entrepreneurs, may simply degrade and deplete the resources faster and more efficiently. So central governments clearly continue to have a role. The challenge is to find the degree of decentralization that is most appropriate.

Decentralization and conservation are complex processes that interact with one another in many ways. Decentralizing by giving new responsibilities to local government units and NGOs creates both opportunities and potential problems. To take advantage of the former while avoiding the latter, a cluster of arrangements must be made as a whole if conservation is to work well in a (moderately) decentralized setting. Of these, seven institutional and incentive elements merit special attention: (1) local participation, especially in a way that allows local people to understand and endorse the boundaries and management plans of nature reserves and that promotes clear tenure over land and other resources in and around the reserves, (2) capacity building, especially to increase skills and accountability, among local government units and nongovernmental organizations to enable them to work together to promote conservation and rural development, (3) incentive structures, especially those that allow local communities to keep income from the sustainable use of nature reserves and other biodiversity assets, (4) conditional subsidies, especially where divergent costs and benefits of conservation are experienced by local and nonlocal groups, making it necessary for global and national society to bridge the gap with livelihood investments or grants, (5) appropriate enforcement, especially against powerful local or central interests and always in the context of education and public relations activities, (6) stakeholder forums and ecoregional executives, which need decisionmaking and fiscal authority to fulfill their three main roles of avoiding conflict through dialogue, authorizing conservation action, and requesting help from nonlocal society to meet local development priorities, and (7) enabling policies, laws, and institutions to provide a clear and supportive framework for conservation on behalf of national government, thus creating incentives at the local level to harmonize development with conservation and reducing the need for enforcement action (Caldecott and Lutz, chapter 14).

In addition to participation and decentralization, appropriate attention to gender issues is also needed to achieve sustainable rural development. For several decades, food, agricultural, and natural resource management policies have been designed without acknowledging that rural men and women may have different preferences, face different constraints, and respond differently to incentives. Indeed, women tend to be "invisible" within agriculture, and the predominant assumption is that the male head of household makes most, if not all, farm allocation and production decisions.

Quisumbing and others (chapter 15) show how neglecting the gender dimension in development policy design has led to failures in project implementation, failure to adopt new agricultural technologies, or adoption of the new technology but with negative unanticipated impacts. Therefore, policymakers would do well to take gender issues into account when formulating food, agricultural, and natural resource management policies. A review of empirical evidence shows that reducing inequalities between men and women in human and physical capital can lead to major efficiency and productivity increases in agriculture.[2] Moreover, addressing inequalities in the underlying distribution of property rights between men and women may be essential not only for increased agricultural productivity but also for environmental and social sustainability. The second-round effects are also important. When controlled by women, income increases have a greater impact both on household food security and on investments in child health and schooling.

Although participatory approaches to project

design (such as discussed in chapter 9) have been recognized as a way of ensuring community involvement, commitment, and ownership of development projects, they are not synonymous with incorporating gender analysis. Participatory approaches to project design by themselves do not guarantee that women will participate in either assessment of their needs or design of the proposed project. This means that a special effort needs to be made to have gender properly addressed, such as by gender analysis. This goes beyond a women-in-development approach, which identifies the differences in access to, ownership of, and control of resources between men and women. A gender-and-development approach recognizes the determinants of these asymmetries and their consequences for individuals, households, communities, and economic development. In essence, the women-in-development approach recognizes an outcome, whereas a gender-and-development approach recognizes both the outcome and its consequences as well as the process leading to those outcomes. The understanding gained from the gender-and-development approach facilitates understanding not only how a planned project or policy will affect men and women but also how it will affect the underlying processes that condition the allocation of rights, resources, and responsibilities within communities and households. Projects can integrate women's concerns by mainstreaming them or by given them an identifiable budget and reporting structure within a project. The appropriate choice depends on the community's characteristics and the project's goals.

Appropriate Technologies

Sustainable rural development must be technically sound and innovative and must consider environmental and social impacts. In order to move closer toward environmentally and socially sustainable rural development, much technical innovation is needed. Farmers have been innovating under difficult conditions and often under poor incentive frameworks. A question is how they can be better supported in their efforts. As discussed in parts I and II of this volume, the main support must come through conducive policy and institutional frameworks, including proper infrastructure. It must also come through appropriate research and empowering extension.

Our technical discussion covered the following issues: soil conservation, mainstreaming of biodiversity, integrated pest management and other strategies to reduce pesticide use, economic and environmental aspects of irrigation and drainage, livestock-environment interactions, inappropriate deforestation, and agroforestry.

Soil is one of the basic inputs into the agricultural production system. Soil degradation, in turn, affects productivity. As soil is degraded, crop yields decline or input levels (and hence costs) rise in an effort to keep or restore productivity. Despite long-standing concern about these problems, surprisingly little hard evidence exists on their magnitude (Lutz and others, chapter 16).

Degradation can be slowed or arrested by a large range of options, including cultural practices such as contour plowing and minimum tillage, vegetative practices such as grass strips, strip cropping, and vegetative barriers, and mechanical measures such as terraces and cutoff drains. Adoption of any of these techniques can be costly, either directly in investment requirements or indirectly in production forgone, and some measures are better suited to some conditions than to others. The critical question facing farmers, and society as a whole, is whether the benefits of a given conservation measure or set of measures are sufficient to make the costs worth bearing.

Decisions about land use are ultimately made by the farmers themselves in light of their own objectives, production possibilities, and constraints. Understanding the incentives facing individual farmers is necessary, therefore, if patterns of resource use are to be understood and appropriate responses to problems formulated. Farmers seek to maximize the present value of the stream of expected net returns to agricultural production. With regard to adoption of conservation measures, the issue is whether returns under the optimal path of the new, more conserving system are sufficiently greater than returns under the path of the current, more degrading system to justify the cost of switching.

Estimated productivity losses of soil erosion and degradation vary considerably, as do the estimated effects on yields of conservation practices. Returns to conservation depend on the specific agroecological conditions faced, on the technologies used, and on the prices of inputs used and outputs produced. Generally, the farmers' decision to invest in conservation is based on normal considerations of benefit and cost: farmers tend to adopt conservation measures when it is in their interest to do so, unless some constraint is present. Cases in which returns to conservation are estimated to be low or negative correlate well with low rates of adoption. Profitability of conservation practices is a necessary but not always a sufficient condition for their adoption. Factors other than strict cost-benefit considerations also play a role. Institutional and motivational issues also must be considered together with the results of the cost-benefit analysis.

Advocates of soil conservation often argue that subsidies are indispensable to induce farmers to adopt conservation measures. But such statements assume that conservation is inherently desirable whether or not there is concrete evidence that the benefits outweigh the costs. Lutz and others (chapter 16) show that this may be far from the case; frequently, the benefits of specific conservation techniques (such as mechanical structures) do not justify their costs. Unless there are important off-site effects or the price signals received by farmers are significantly distorted, subsidies to induce adoption will not bring increased economic efficiency.

One important way for governments to help is to make sure that constraints such as insecure tenure do not prevent farmers from adopting conservation measures. Also, governments already conduct some research on soil conservation and provide, through extension services, some assistance to farmers who undertake conservation work. However, research in experiment stations has tended to favor technical efficiency (including structural measures such as terraces) over cost-effectiveness. Further, government extension work is often ineffective. In many cases, nongovernmental organizations, such as Vecinos Mundiales in Central America, have proven to be more effective than government at stimulating technological development (see chapter 12).

Some of the beneficial effects of conservation may come through better moisture conservation and better conservation of organic substances, both of which may also benefit soil biodiversity. Chapter 14 discusses biodiversity conservation with a focus on protected areas. But much of the world's biodiversity exists in human-managed or -modified systems (Srivastava and others, chapter 17). Biodiversity has two important dimensions: the genetic variation within species and populations and the preservation of habitat. The significance of variation within a species is not widely appreciated but is critical, particularly for agriculture. The continued productivity of existing crops and livestock hinges in large part on harnessing the genetic variation found within each species. Habitat conservation seeks to safeguard natural habitats for wild species and populations and to manage habitats that have been modified for human use, such as farmland. Agrobiodiversity is found in habitats that have been modified for crop and livestock production. Agrobiodiversity includes all plants and animals that contribute directly or indirectly to the raising of crops and livestock.

Although it is conceptually useful to differentiate agrobiodiversity from the larger array of species and habitats, the boundaries between biodiversity and agrobiodiversity are not clear-cut. The interface between wild and domesticated plants and animals is constantly shifting. This fact underscores the importance of conserving as much biodiversity as feasible for agricultural development in the future.

How agriculture can be intensified without damaging biodiversity is a critical question for rural development. Environmentally inappropriate intensification of agriculture has led to eutrophication of lakes and estuaries, loss of soil micro-organisms, accelerated soil erosion, contamination of groundwater, and draining of wetlands. All of these activities trigger a potentially dangerous loss of biodiversity.

A balanced conservation strategy includes in situ conservation (maintaining animal and plant genetic resources in places where they occur naturally) as well as ex situ conservation (maintaining genetic resources in seed or field gene banks). In situ conservation of crops and livestock can be supported by (a) emphasizing the safeguarding of plants and animals for the future improvement of agriculture as a new dimension of existing wilderness parks and biological reserves, (b) creating world heritage sites for genes for agricultural development, (c) integrating agrobiodiversity in ecotourism where appropriate, (d) finding markets for lesser-known crops and local varieties under the motto "use it or lose it," and (e) finding ways for livestock owners to generate more revenues from threatened breeds. Better in situ conservation can also be supported by rapid agrobiodiversity assessment teams (Srivastàva and others, chapter 17).

The new vision for agricultural research adopts a holistic approach that is more sensitive to environmental concerns, while still addressing the need to boost yields and incomes of rural producers and caretakers of the land. It includes, but is not restricted to, (a) integrated pest management (IPM), (b) a participatory approach with farmers, (c) better use of farmer knowledge, (d) greater support for research, development, and dissemination of lesser-known crops and animals, (e) support for research on new crops and livestock, (f) greater sensitivity to the value of a mosaic of land uses, (g) greater diversity of habitats within land use systems, (h) greater reliance on recycling of organic matter, (i) shift of the research focus from individual traits to lifetime and herd productivity characteristics, (j) determination of the critical number of breeds for conservation purposes, and (k) an effort to learn about genetic components of adaptation in livestock.

The notion of a new research paradigm has implications for institutional development and the exploration of new ways of doing business. Innovative institutional arrangements would include more effective partnerships among agricultural research centers, NGOS, growers association, private companies involved in the manufacture and sale of agricultural technologies, universities and agricultural extension agencies, and development lending institutions. To some degree, all of these partnerships are being explored and tested.

Generating the gains in agricultural productivity necessary to secure food availability and livelihoods in the developing world over the coming decades requires an approach in which the intensification of agricultural systems is consistent with the conservation of the natural resource base. This approach requires less reliance on the intensive use of external inputs and greater dependence on management skills and location-specific knowledge of agroecosystems. Integrated pest management constitutes one such approach and is critical to sustainable rural development (Schillhorn and others, chapter 18; Pingali and Gerpacio, chapter 19).

Agricultural intensification—the movement from an extensive to an intensive production system or from a subsistence production system to a commercial one—generally has increased the use of agrochemicals. Indiscriminate use of pesticides in the production of various crops has often impaired health due to direct or indirect exposure to hazardous chemicals; contaminated ground and surface waters through runoff and seepage; transmitted pesticide residues through the food chain; increased resistance of pest populations to pesticides, thereby reducing their efficacy and causing pest outbreaks; and reduced the population of beneficial insects ("good bugs" like parasites and predators), thereby reducing the effectiveness of pest control strategies that attempt to minimize pesticide use.

A reappraisal of the role of pesticides in agriculture has been under way for some time. This has been associated with a wider shift in focus from agricultural production per se to environmentally sustainable management of production systems in which IPM is a critical element. Integrated pest management is a key component of integrated farming practices that are based on an understanding of ecology and the interaction between crops or animals and their pests, as well as an understanding of the environments in which pests operate.

Significant advances have been made in the last three decades in the development and dissemination of crop varieties with resistance to the major cereal pests. Much of the advance has come through the use of conventional breeding approaches, although substantial gains in the development of resistance could come through the use of modern biotechnology tools. The introduction of varieties with host-plant resistance has dramatically reduced the need for insecticides in rice and maize and the need for fungicides in wheat.

Given the extent of breeding efforts in producing pest-resistant crop cultivars and evidence of insignificant productivity benefits of pesticide use, why are high levels of agrochemicals still used in crop production? First, the dissemination of crop varieties resistant to pest pressures has not been accompanied by extension messages on the reduced need for pesticides. Second, farmers' pest control decisions, scientists' research priorities, and policymakers' prescriptions, which are all based largely on perceived pest-related losses in yield (in turn often not related to actual losses) have led to the promotion of pesticide use. Consequently, high and injudicious applications of broad-spectrum pesticides have continued as before, causing the breakdown in varietal resistance. New varieties generated to replace cultivars with resistance breakdown have subsequently been overcome through further biotype changes in pest populations. The breeding treadmill could only be overcome through dramatic changes in crop management practices, especially in the use of pesticides.

Under low levels of pest infestation, natural control is the economically dominant strategy for pest management. (This is true for insects and diseases for rice, wheat, and maize.) Natural control relies on predator populations to control pest infestations under normal circumstances, when pest-resistant varieties are used. Pesticides may have to be used as a last resort in the rare instance of high infestations.

Natural control does not imply doing nothing;

rather, it requires in-depth farmer knowledge of the pest-predator ecology and frequent monitoring of field conditions by the farmer. In this regard, natural control can be considered the ultimate goal of an IPM program, and farmers who are well versed in IPM techniques would converge toward it (Pingali and Gerpacio, chapter 19). This implies a paradigm shift in the traditional IPM strategies from when best to apply pesticides to when not to apply.

IPM is knowledge-based; it presents by far the most difficult challenge to traditional, small-scale farmers in the developing countries who are making the transition to scientific farming. IPM requires farmers to grasp complex sets of data that are often anything but self-evident, unitary, standardized, or amenable to trial-and-error learning. The institutional and economic structure in the rural sector of developing economies also requires some policy intervention to reconcile long-term societal goals with short-term individual objectives in pest control. Promoting sustainable pest management within an IPM framework requires improved research and extension linkages, effective farmer training methods, community action, and an undistorted price structure.

The IPM concept is holistic; it requires farmers to take a systems view of the farm enterprise and to understand the interlinkages among various components of the system. Therefore, farmer training in sustainable pest management is an essential component of a strategy to achieve minimum use of insecticides. The eventual goal is to build farmer capacity to identify and solve problems based on a thorough understanding of field ecology. The experience of the Food and Agriculture Organization's farmer field schools has shown that trained farmers use significantly lower levels of pesticides than untrained farmers. They are also more likely to experiment with other components of sustainable production systems, such as improved fertilizer management and more efficient water management.

Training millions of farmers will be costly. In that regard, there are essentially two options, which are not mutually exclusive, for reducing training costs. The first is to train a core group of farmers within a geopolitical unit, such as a municipality, and then to rely on farmer-to-farmer training for disseminating the IPM message to a wider group of farmers. The second is to condense the complex message into several simple rules that are easy for the farmer to implement. An example, from rice, of such a rule is "Do not spray insecticides against leaf-feeding insects for the first 40 days of crop growth." The rule is based on detailed pest ecology studies showing that the predominant insect pests during the first 40 days of crop growth are leaf-feeding insects and that even very high levels of infestations rarely lead to any loss in yield (Pingali and Gerpacio, chapter 19).

Even with a well-established IPM program, pesticides may have to be kept as a technology of last resort. Essentially, the idea of pesticide use in IPM is to spray only when imperative, using the smallest amount possible to do the job. The one area where agrochemicals continue to have a significant impact on productivity is in the management of weeds; herbicides will continue to be the preferred alternative in the foreseeable future, even when the health costs of herbicides are explicitly taken into account.

A number of policy and regulatory instruments are available to governments to encourage environmentally sound and economically rational pest management practices. The more important of these are (a) development of a system that increases the awareness of policymakers, consumers, and producers of the hazards of pesticide use, (b) development of a regulatory framework to ensure appropriate and safe production, distribution, and use of pesticides, (c) introduction of appropriate economic incentives, including taxes and special levies on pesticide use, to account for negative externalities, and short-term subsidies, to account for positive externalities in the use of IPM, (d) orientation of research and technology policies to generate a steady supply of relevant pest management information and technologies, including adequate budget allocations for research, extension, and training, and (e) signing of and adherence to international agreements and conventions (Schillhorn and others, chapter 18).

Over the past three decades, the dramatic spread of irrigated agriculture was mainly responsible for increases in food production that kept pace with growth in population. This was achieved in part through the large-scale development of new water resources and construction of new irrigation capacity (Chakravorty, chapter 19). The International Irrigation Management Institute has estimated that to meet future demand, irrigated agriculture may need to deliver output increases of more than 3.5 percent annually—a daunting task by any measure of performance. The major challenge is that this growth must come primarily from increases in irrigation efficiency and not from additions to new irrigation capacity. Moreover, as is becoming increasingly common in the industrial countries and even in many developing countries, rising demand for residential use of water and rising environmental concerns are beginning to put serious limitations on new water development projects. Producing more food with less water—placing a smaller financial burden on the taxpayer and reducing environmental costs—will be the major challenge for irrigation in the twenty-first century.

Irrigation projects around the world are in trouble for a number of reasons: ex post project benefits have often been far lower than projected ex ante returns; full-cost pricing of water is rare, with most projects recouping only a fraction of their operation and maintenance (O&M) costs; low tax collections and poor O&M of project structures have led to high rates of water loss through seepage and percolation; and inadequate investments in drainage have rendered large tracts of prime agricultural land unusable because of salinity and waterlogging. The low price of irrigation water, which is often unrelated to use, has led farmers to withdraw too much, giving rise to salinity and waterlogging. Investments in the irrigation sector in developing countries have relied too much on bureaucratic federal and state agencies that have paid little attention to the economic pricing of irrigation services, the creation of reliable delivery systems, and the participation of users in the operation and maintenance of projects. Adverse health and environmental effects of water use have received little attention, partly because of a lack of coordination and interaction between water and public health agencies.

These problems call for a new approach that is more comprehensive and integrates the elements of irrigation policy, which include the economic and environmental effects and the associated intersectoral linkages (Chakravorty, chapter 20). This approach must be built on pricing of resources that better reflects opportunity costs whenever possible and creation of delivery infrastructure that promotes the efficient allocation and, if feasible, trading of the resource among users.

The introduction of markets that allow for trading of water among users would lead to a more efficient allocation of the resource. But water markets only work under very selective conditions: (a) the presence of effective water users associations or water authorities, (b) the removal of legal and institutional regulations that prohibit or limit water trading, and (c) the creation of infrastructure (canals) and control systems for the transfer and measurement of water.

Partly because of budgetary pressures, some developing countries are more willing to undertake institutional reforms in an effort to improve the pricing and trading of water. New "holistic" thinking is also reflected in an emerging emphasis on "basin-wide" management promoted by agencies such as the International Irrigation Management Institute (Chakravorty, chapter 20). These agencies argue that water efficiency must be measured not within a project but over an entire basin so as to include the reuse of drainage water from seepage and percolation. This suggests that policies such as the adoption of sprinkler irrigation within an irrigation project may improve the efficiency of water use within the project but reduce the availability of water elsewhere in the basin. Similarly, salt and chemical pollution of water through application of inputs may adversely affect agriculture downstream of the project. These issues can only be handled if the basin is adopted as the relevant unit of analysis and the relevant externalities are internalized.

Livestock have often been seen as being a significant cause of environmental degradation. Indeed, they can damage the global natural resources in a number of ways, but they can also contribute to environmental balance (Steinfeld and others, chapter 21). About 34 million square kilometers, or 26 percent of the world's land area, are used for grazing livestock. In addition, 3 million square kilometers, or about 21 percent of the world's arable area, are used for producing cereal for livestock feed. Livestock, in turn, generate 13 billion tons of waste per year. A large part of this is recycled, but where animal concentrations are high, waste poses an enormous environmental hazard.

Livestock grazing can affect the water balance in certain areas. Water is needed to produce fodder and feed concentrate, to supply drinking water for animals, and to drain surplus waste and chemicals. Livestock interact directly and indirectly with biodiversity, and livestock and livestock waste cause gaseous emissions with important local and global impact on the environment.

Policies need to be designed to correct the negative environmental effects of livestock production. These policies should address the underlying causes of environmental degradation and must be flexible, site-specific, and well targeted. Instruments to enhance positive and mitigate negative environmental impacts include pricing, regulations, and institutional development. The collective purpose of these instruments is to establish feedback mechanisms to ensure that the use of livestock is consistent with overall social objectives.

Within an enabling policy framework, a wide range of available technologies can be employed, while others still need to be developed. Technologies can be grouped into four sets (Steinfeld and others, chapter 21). Although there is some overlap among categories, they help to provide a good picture of livestock-resource interactions:

- Technologies that reduce the environmental damage by alleviating the direct pressure on natural resources or by reducing the pollution load through modifying the chemical or physical characteristics of products
- Technologies that enhance natural resources by making them more productive or richer
- Technologies that save natural resources by allowing farmers to generate more revenue from the same resource or to generate the same from less
- Technologies that turn waste into products by closing cycles (a point also made by Pretty, chapter 4).

Novel concepts are being developed to integrate crop and livestock production in a farming area rather than on a mixed farm (Steinfeld and others, chapter 21). This method of area-wide crop-livestock integration allows individual enterprises to operate separately while linking the flows of energy and organic and mineral matter through markets and regulations. This allows for highest efficiencies at the enterprise level, while maximizing social benefits. Thus the motto for the more densely populated parts of the developing world is to intensify, but not concentrate, animal production.

As a result of the interaction between livestock production systems and natural resources, coupled with factors such as market access, there are development opportunities as well as threats to sustainability. A comprehensive perspective is needed to ensure an enabling policy framework in which to introduce effective technologies (Steinfeld and others, chapter 21). Technology remains a key component because future development, including that of the livestock sector, will depend on technology to substitute for natural resources. This trend to knowledge-intensive systems has been widely observed.

Expansion of livestock production has often been named as one of the main causes of tropical deforestation. Contemporary concern about deforestation focuses on tropical countries because that is where the majority of forest cover is being removed (Kaimowitz and others, chapter 22). In the period 1980–90, 137.3 million hectares of tropical forests were cleared, about 7.2 percent of the total that existed in 1980. Any claims regarding the magnitude of global deforestation must be taken with caution, however, because there are serious problems with the data and definitions used.

Deforestation often implies:

- The loss of livelihoods for forest-dependent people, many of whom may not wish to or be able to find other sources of employment
- Decreasing stocks of fuelwood and nontimber forest products as well as of industrial timber
- Greater soil erosion and river siltation
- Substantial loss of species and genes, in view of the high level of endemic biodiversity in tropical forests
- Substantial emission of carbon dioxide, which contributes to global warming
- Other types of local and regional climate change.

The concern is not only with deforestation but also with forest degradation, which can be defined as a decrease of density or increase of disturbance in forest classes. In the long run, this may be just as important as deforestation itself.

The process of deforestation must be analyzed at two levels: agents and causes. The agents of deforestation are the people who physically (or through decisions over their labor forces) convert forests to nonforest uses: small farmers, plantation and estate owners, forest concessionaires, infrastructure construction agencies, and so forth. In Latin America, most deforestation is attributed to medium- and large-scale operations (resettlement schemes, large-scale cattle ranching, hydroelectric dams), and deforestation is characterized by transitions from closed forest to nonforest land uses. In Africa, deforestation is largely related to the expansion of small-scale farming and rural population pressure (growing number of smallholders) associated with the conversion from closed forest cover to short fallow farming. In Asia, deforestation is associated with both relatively large operations (as in Latin America) and rural population pressure (as in Africa). There are no simple, universal single-cause explanations. Different agents coexist and closely interact in many countries, and their relative importance varies over time and between regions. Given such complex relationships, there are no clear guilty or innocent parties, and no one should expect neat, simple solutions.

Individuals and businesses deforest because it is their most profitable alternative. To get them not to deforest in situations where forest clearing is inappropriate, deforestation must be made less profitable, or other alternatives (either based on retaining forests or completely outside forest areas) must be made more profitable. This is the main thesis of Kaimowitz and others (chapter 22). A key question is how certain causes can be manipulated to influence the behavior of agents, so as to lessen the rate of inappropriate deforestation.

There is no perfect or generalizable policy for reducing inappropriate deforestation (Kaimowitz and others, chapter 22). Each national situation is different, much uncertainty remains about key cause-and-effect relations, and there are usually tradeoffs among effectiveness, ability to be targeted, political viability, and direct and indirect costs of policies. Most of the policy instruments are very blunt instruments for stopping deforestation, and governments will be forced to choose from a mix of measures, specifically crafted to local conditions.

The answer to the issue of forest conversion is not necessarily to stop logging per se, or to stop logging in all new areas, or to ban all new road construction in forest areas, but rather to reform those policies and institutions that make forest colonization seem more attractive than the potential migrants' current activities (Kaimowitz and others, chapter 22). These might include the pull factors (to reduce the profitability of illegally clearing forests or of speculating in land that was supposed to be kept as forest) or the push factors (to increase the livelihood options outside of forests).

Thus the principal reforms that could reduce inappropriate tropical deforestation are likely to be a combination of the following government policies:

- Eliminate subsidies to agricultural and pastoral industries that encourage deforestation
- Eliminate legal and institutional incentives or requirements to clear forests as a basis for gaining recognized land tenure
- Reform forest industry concessions and licenses to provide incentives for long-term sustainable management
- Develop innovative institutional arrangements for devolving more decisionmaking authority and responsibility to those whose livelihoods and quality of life are linked directly to the extent and quality of tropical forests
- Encourage voluntary market differentiation by consumers that discriminates positively toward products that have been sustainably produced from forests
- Facilitate recognition of and compensation for environmental services provided by forests and ensure that transfer payments are received by the persons making decisions at the forest frontiers.

One response to deforestation has been to encourage agroforestry: a land use system in which trees, shrubs, palms, and bamboos are cultivated on the same land as agricultural crops or livestock for economic and environmental reasons. The system includes managing natural regrowth, seeding, planting, and maintaining trees as border plantings, and interplanting them in agricultural crops, in woodlots, in home gardens, or in other systems. Agroforestry's special characteristics are that it includes a large number of species, configurations, and management intensities, has longer gestation than most agricultural crops, and produces outputs with multiple uses.

In addition to the estimated financial return, farmers attach considerable importance to how an agroforestry system fits into the overall farm production system and the existing land, labor, and capital constraints (Current and others, chapter 23). Many agroforestry systems are profitable to farmers under a considerable range of economic conditions, and various types of (low-intensity) traditional agroforestry are practiced in many areas. Even when agroforestry is profitable, smallholders will and should adopt agroforestry incrementally and gradually because of management and resource constraints; poorer farmers, in particular, are often hampered by limited land, labor, and capital resources and by their need to ensure food security and reduce risks.

Local scarcity of wood products is, as might be expected, a key motivator in adopting nontraditional agroforestry. Projects must begin by assessing the scarcity of wood and nontimber products, as well as the existence of local markets for products. Taungya, perennial intercrops, trees on contours, and tree lines have proved to be the easiest systems to introduce. Results are mixed for alley cropping, home gardens, windbreaks, green manuring, dispersed trees in cropland, and tree-pasture systems. Farmers are willing to invest in rehabilitating their land where systems also produce products or income, and they prefer less-intensive systems.

The demonstration effect of fast-growing tree species on farms, and of benefits on demonstration plots, has helped to expand agroforestry activities, reducing the costs of extension and increasing its effectiveness. Rather than offering standard designs, programs serve farmers best if they offer a broad selection of species and systems from which to choose those most suitable to their household's needs and resources. Involving local people as paratechnicians is often a successful, low-cost approach to promoting technology.

Financial incentives and subsidies should be kept to a minimum. Agroforestry technologies promoted by extension should be financially profitable, and thus adoptable, for the farmer without subsidies. The possible exceptions are time-limited financial incentives for early adopters of unfamiliar technologies. In-kind, material inputs encourage farmers to experiment with and adopt agroforestry, but the experiences with food-for-work incentives are mixed.

Implementing many of the needed changes to move toward more sustainable rural development will not be easy. As discussed in this volume, it will require significant efforts by farmers and many others on many fronts. It will also require a greater commitment to the goal of sustainable agricultural development by many national governments (which often have an urban bias) as well as by development assistance agencies.

Notes

1. Examples of necessary reforms can be given for many subject areas and for many countries. With regard to *pesticide subsidies*, many countries have reduced subsidization. Yet despite these improvements, many countries still implement a variety of policies that directly or indirectly favor the use of pesticides and impede the adoption of ecologically sound pest control techniques. Negative side effects of pesticides with medium and high toxicity include groundwater contamination, chronic or acute health problems for farmers and farmworkers, losses of nontarget crops and other species, pesticide resistance, and pesticide residues in food. *Subsidization of water*, mainly in the form of a lack of cost recovery, also remains a problem with serious long-term consequences. Only a handful of countries have made strides in increasing cost recovery and moving toward full-cost pricing, which would help to improve management, reduce waste, reduce overuse, and thereby reduce the problems of misallocation, waterlogging, and salinity. Finally, over the past decade or two, many *parastatals and marketing monopolies* have been abolished or reformed. Yet many still remain. Such intervention often results in welfare losses for producers or consumers or both, marketing operations may not be efficient, and there is potential for misuse of funds.

2. In many developing countries, women have lower rates of school enrollment, literacy, and schooling attainment. This disparity continues to be larger in rural areas, where low educational attainment persists despite high private rates of return to women's schooling and high social returns to women's education.

Index

DISTRIBUTORS OF WORLD BANK PUBLICATIONS

Prices and credit terms vary from country to country. Consult your local distributor before placing an order.

ARGENTINA
Oficina del Libro Internacional
Av. Cordoba 1877
1120 Buenos Aires
Tel: (54 1) 815-8354
Fax: (54 1) 815-8156
E-mail: olilibro@satlink.com

AUSTRALIA, FIJI, PAPUA NEW GUINEA, SOLOMON ISLANDS, VANUATU, AND SAMOA
D.A. Information Services
648 Whitehorse Road
Mitcham 3132
Victoria
Tel: (61) 3 9210 7777
Fax: (61) 3 9210 7788
E-mail: service@dadirect.com.au
URL: http://www.dadirect.com.au

AUSTRIA
Gerold and Co.
Weihburggasse 26
A-1011 Wien
Tel: (43 1) 512-47-31-0
Fax: (43 1) 512-47-31-29
URL: http://www.gerold.co/at.online

BANGLADESH
Micro Industries Development Assistance Society (MIDAS)
House 5, Road 16
Dhanmondi R/Area
Dhaka 1209
Tel: (880 2) 326427
Fax: (880 2) 811188

BELGIUM
Jean De Lannoy
Av. du Roi 202
1060 Brussels
Tel: (32 2) 538-5169
Fax: (32 2) 538-0841

BRAZIL
Publicacões Tecnicas Internacionais Ltda.
Rua Peixoto Gomide, 209
01409 Sao Paulo, SP.
Tel: (55 11) 259-6644
Fax: (55 11) 258-6990
E-mail: postmaster@pti.uol.br
URL: http://www.uol.br

CANADA
Renouf Publishing Co. Ltd.
5369 Canotek Road
Ottawa, Ontario K1J 9J3
Tel: (613) 745-2665
Fax: (613) 745-7660
E-mail: order.dept@renoufbooks.com
URL: http:// www.renoufbooks.com

CHINA
China Financial & Economic Publishing House
8, Da Fo Si Dong Jie
Beijing
Tel: (86 10) 6333-8257
Fax: (86 10) 6401-7365

China Book Import Centre
P.O. Box 2825
Beijing

Chinese Corporation for Promotion of Humanities
52 You Fang Hu Tong,
Auan Nei Da Jie
Beijing
Tel: (86 10) 660 72 494
Fax: (86 10) 660 72 494

COLOMBIA
Infoenlace Ltda.
Carrera 6 No. 51-21
Apartado Aereo 34270
Santafé de Bogotá, D.C.
Tel: (57 1) 285-2798
Fax: (57 1) 285-2798

COTE D'IVOIRE
Center d'Edition et de Diffusion Africaines (CEDA)
04 B.P. 541
Abidjan 04
Tel: (225) 24 6510; 24 6511
Fax: (225) 25 0567

CYPRUS
Center for Applied Research
Cyprus College
6, Diogenes Street, Engomi
P.O. Box 2006
Nicosia
TTel: (357 2) 59-0730
Fax: (357 2) 66-2051

CZECH REPUBLIC
USIS, NIS Prodejna
Havelkova 22
130 00 Prague 3
Tel: (420 2) 2423 1486
Fax: (420 2) 2423 1114
URL: http://www.nis.cz/

DENMARK
SamfundsLitteratur
Rosenoerns Allé 11
DK-1970 Frederiksberg C
Tel: (45 35) 351942
Fax: (45 35) 357822
URL: http://www.sl.cbs.dk

ECUADOR
Libri Mundi
Libreria Internacional
P.O. Box 17-01-3029
Juan Leon Mera 851
Quito
Tel: (593 2) 521-606; (593 2) 544-185
Fax: (593 2) 504-209
E-mail: librimu1@librimundi.com.ec
E-mail: librimu2@librimundi.com.ec

CODEU
Ruiz de Castilla 763, Edif. Expocolor
Primer piso, Of. #2
Quito
Tel/Fax: (593 2) 507-383; 253-091
E-mail: codeu@impsat.net.ec

EGYPT, ARAB REPUBLIC OF
Al Ahram Distribution Agency
Al Galaa Street
Cairo
Tel: (20 2) 578-6083
Fax: (20 2) 578-6833

The Middle East Observer
41, Sherif Street
Cairo
Tel: (20 2) 393-9732
Fax: (20 2) 393-9732

FINLAND
Akateeminen Kirjakauppa
P.O. Box 128
FIN-00101 Helsinki
Tel: (358 0) 121 4418
Fax: (358 0) 121-4435
E-mail: akatilaus@stockmann.fi
URL: http://www.akateeminen.com/

FRANCE
Editions Eska
5, avenue de l'Opéra
75001 Paris
Tel: (33 1) 42-86-56-00
Fax: (33 1) 42-60-45-35

GERMANY
UNO-Verlag
Poppelsdorfer Allee 55
53115 Bonn
Tel: (49 228) 949020
Fax: (49 228) 217492
URL: http://www.uno-verlag.de
E-mail: unoverlag@aol.com

GHANA
Epp Books Services
P.O. Box 44
TUC
Accra
Tel: 223 21 778843
Fax: 223 21 779099

GREECE
Papasotiriou S.A.
35, Stournara Str.
106 82 Athens
Tel: (30 1) 364-1826
Fax: (30 1) 364-8254

HAITI
Culture Diffusion
5, Rue Capois
C.P. 257
Port-au-Prince
Tel: (509) 23 9260
Fax: (509) 23 4858

HONG KONG, CHINA; MACAO
Asia 2000 Ltd.
Sales & Circulation Department
302 Seabird House
22-28 Wyndham Street, Central
Hong Kong, China
Tel: (852) 2530-1409
Fax: (852) 2526-1107
E-mail: sales@asia2000.com.hk
URL: http://www.asia2000.com.hk

HUNGARY
Euro Info Service
Margitszgeti Europa Haz
H-1138 Budapest
Tel: (36 1) 350 80 24, 350 80 25
Fax: (36 1) 350 90 32
E-mail: euroinfo@mail.matav.hu

INDIA
Allied Publishers Ltd.
751 Mount Road
Madras - 600 002
Tel: (91 44) 852-3938
Fax: (91 44) 852-0649

INDONESIA
Pt. Indira Limited
Jalan Borobudur 20
P.O. Box 181
Jakarta 10320
Tel: (62 21) 390-4290
Fax: (62 21) 390-4289

IRAN
Ketab Sara Co. Publishers
Khaled Eslamboli Ave., 6th Street
Delafrooz Alley No. 8
P.O. Box 15745-733
Tehran 15117
Tel: (98 21) 8717819; 8716104
Fax: (98 21) 8712479
E-mail: ketab-sara@neda.net.ir

Kowkab Publishers
P.O. Box 19575-511
Tehran
Tel: (98 21) 258-3723
Fax: (98 21) 258-3723

IRELAND
Government Supplies Agency
Oifig an tSoláthair
4-5 Harcourt Road
Dublin 2
Tel: (353 1) 661-3111
Fax: (353 1) 475-2670

ISRAEL
Yozmot Literature Ltd.
P.O. Box 56055
3 Yohanan Hasandlar Street
Tel Aviv 61560
Tel: (972 3) 5285-397
Fax: (972 3) 5285-397

R.O.Y. International
PO Box 13056
Tel Aviv 61130
Tel: (972 3) 649 9469
Fax: (972 3) 648 6039
E-mail: royil@netvision.net.il

Palestinian Authority/Middle East
Index Information Services
P.O.B. 19502 Jerusalem
Tel: (972 2) 6271219
Fax: (972 2) 6271634

ITALY, LIBERIA
Licosa Commissionaria Sansoni SPA
Via Duca Di Calabria, 1/1
Casella Postale 552
50125 Firenze
Tel: (39 55) 645-415
Fax: (39 55) 645-415
E-mail: licosa@ftbcc.it
URL: http://www.ftbcc.it/licosa

JAMAICA
Ian Randle Publishers Ltd.
206 Old Hope Road, Kingston 6
Tel: 876-927-2085
Fax: 876-977-0243
E-mail: irpl@colis.com

JAPAN
Eastern Book Service
3-13 Hongo 3-chome, Bunkyo-ku
Tokyo 113
Tel: (81 3) 3818-0861
Fax: (81 3) 3818-0864
E-mail: orders@svt-ebs.co.jp
URL: http://www.bekkoame.or.jp/~svt-ebs

KENYA
Africa Book Service (E.A.) Ltd.
Quaran House, Mfangano Street
P.O. Box 45245
Nairobi
Tel: (254 2) 223 641
Fax: (254 2) 330 272

KOREA, REPUBLIC OF
Dayang Books Trading Co.
International Division
783-20, Pangba Bon-Dong, Socho-ku
Seoul
Tel: (82 2) 536-9555
Fax: (82 2) 536-0025
E-mail: seamap@chollian.net

Eulyoo Publishing Co., Ltd.
46-1, Susong-Dong
Jongro-Gu
Seoul
Tel: (82 2) 734-3515
Fax: (82 2) 732-9154

LEBANON
Librairie du Liban
P.O. Box 11-9232
Beirut
Tel: (961 9) 217 944
Fax: (961 9) 217 434

MALAYSIA
University of Malaya Cooperative
Bookshop, Limited
P.O. Box 1127
Jalan Pantai Baru
59700 Kuala Lumpur
Tel: (60 3) 756-5000
Fax: (60 3) 755-4424
E-mail: umkoop@tm.net.my

MEXICO
INFOTEC
Av. San Fernando No. 37
Col. Toriello Guerra
14050 Mexico, D.F.
Tel: (52 5) 624-2800
Fax: (52 5) 624-2822
E-mail: infotec@rtn.net.mx
URL: http://rtn.net.mx

Mundi-Prensa Mexico S.A. de C.V.
c/Rio Panuco, 141-Colonia Cuauhtemoc
06500 Mexico, D.F.
Tel: (52 5) 533-5658
Fax: (52 5) 514-6799

NEPAL
Everest Media International Services (P.) Ltd.
GPO Box 5443
Kathmandu
Tel: (977 1) 472 152
Fax: (977 1) 224 431

NETHERLANDS
De Lindeboom/Internationale Publicaties b.v.–
P.O. Box 202, 7480 AE Haaksbergen
Tel: (31 53) 574-0004
Fax: (31 53) 572-9296
E-mail: lindeboo@worldonline.nl
URL: http://www.worldonline.nl/~lindeboo

NEW ZEALAND
EBSCO NZ Ltd.
Private Mail Bag 99914
New Market
Auckland
Tel: (64 9) 524-8119
Fax: (64 9) 524-8067

Oasis Official
P.O. Box 3627
Wellington
Tel: (64 4) 499 1551
Fax: (64 4) 499 1972
E-mail: oasis@actrix.gen.nz
URL: http://www.oasisbooks.co.nz/

NIGERIA
University Press Limited
Three Crowns Building Jericho
Private Mail Bag 5095
Ibadan
Tel: (234 22) 41-1356
Fax: (234 22) 41-2056

NORWAY
SWETS Norge AS
Book Department, Postboks 6512 Etterstad
N-0606 Oslo
Tel: (47 22) 97-4500
Fax: (47 22) 97-4545

PAKISTAN
Mirza Book Agency
65, Shahrah-e-Quaid-e-Azam
Lahore 54000
Tel: (92 42) 735 3601
Fax: (92 42) 576 3714

Oxford University Press
5 Bangalore Town
Sharae Faisal
PO Box 13033
Karachi-75350
Tel: (92 21) 446307
Fax: (92 21) 4547640
E-mail: ouppak@TheOffice.net

Pak Book Corporation
Aziz Chambers 21, Queen's Road
Lahore
Tel: (92 42) 636 3222; 636 0885
Fax: (92 42) 636 2328
E-mail: pbc@brain.net.pk

PERU
Editorial Desarrollo SA
Apartado 3824, Ica 242 OF. 106
Lima 1
Tel: (51 14) 285380
Fax: (51 14) 286628

PHILIPPINES
International Booksource Center Inc.
1127-A Antipolo St, Barangay, Venezuela
Makati City
Tel: (63 2) 896 6501; 6505; 6507
Fax: (63 2) 896 1741

POLAND
International Publishing Service
Ul. Piekna 31/37
00-677 Warzawa
Tel: (48 2) 628-6089
Fax: (48 2) 621-7255
E-mail: books%ips@ikp.atm.com.pl
URL: http://www.ipscg.waw.pl/ips/export/

PORTUGAL
Livraria Portugal
Apartado 2681, Rua Do Carmo 70-74
1200 Lisbon
Tel: (1) 347-4982
Fax: (1) 347-0264

ROMANIA
Compani De Librarii Bucuresti S.A.
Str. Lipscani no. 26, sector 3
Bucharest
Tel: (40 1) 613 9645
Fax: (40 1) 312 4000

RUSSIAN FEDERATION
Isdatelstvo <Ves Mir>
9a, Kolpachniy Pereulok
Moscow 101831
Tel: (7 095) 917 87 49
Fax: (7 095) 917 92 59

SINGAPORE; TAIWAN, CHINA
MYANMAR; BRUNEI
Hemisphere Publication Services
41 Kallang Pudding Road #04-03
Golden Wheel Building
Singapore 349316
Tel: (65) 741-5166
Fax: (65) 742-9356
E-mail: ashgate@asianconnect.com

SLOVENIA
Gospodarski Vestnik Publishing Group
Dunajska cesta 5
1000 Ljubljana
Tel: (386 61) 133 83 47; 132 12 30
Fax: (386 61) 133 80 30
E-mail: repansekj@gvestnik.si

SOUTH AFRICA, BOTSWANA
For single titles:
Oxford University Press Southern Africa
Vasco Boulevard, Goodwood
P.O. Box 12119, N1 City 7463
Cape Town
Tel: (27 21) 595 4400
Fax: (27 21) 595 4430
E-mail: oxford@oup.co.za

For subscription orders:
International Subscription Service
P.O. Box 41095
Craighall
Johannesburg 2024
Tel: (27 11) 880-1448
Fax: (27 11) 880-6248
E-mail: iss@is.co.za

SPAIN
Mundi-Prensa Libros, S.A.
Castello 37
28001 Madrid
Tel: (34 91) 4 363700
Fax: (34 91) 5 753998
E-mail: libreria@mundiprensa.es
URL: http://www.mundiprensa.com/

Mundi-Prensa Barcelona
Consell de Cent, 391
08009 Barcelona
Tel: (34 3) 488-3492
Fax: (34 3) 487-7659
E-mail: barcelona@mundiprensa.es

SRI LANKA, THE MALDIVES
Lake House Bookshop
100, Sir Chittampalam Gardiner Mawatha
Colombo 2
Tel: (94 1) 32105
Fax: (94 1) 432104
E-mail: LHL@sri.lanka.net

SWEDEN
Wennergren-Williams AB
P. O. Box 1305
S-171 25 Solna
Tel: (46 8) 705-97-50
Fax: (46 8) 27-00-71
E-mail: mail@wwi.se

SWITZERLAND
Librairie Payot Service Institutionnel
Côtes-de-Montbenon 30
1002 Lausanne
Tel: (41 21) 341-3229
Fax: (41 21) 341-3235

ADECO Van Diermen EditionsTechniques
Ch. de Lacuez 41
CH1807 Blonay
Tel: (41 21) 943 2673
Fax: (41 21) 943 3605

THAILAND
Central Books Distribution
306 Silom Road
Bangkok 10500
Tel: (66 2) 235-5400
Fax: (66 2) 237-8321

TRINIDAD & TOBAGO
AND THE CARRIBBEAN
Systematics Studies Ltd.
St. Augustine Shopping Center
Eastern Main Road, St. Augustine
Trinidad & Tobago, West Indies
Tel: (868) 645-8466
Fax: (868) 645-8467
E-mail: tobe@trinidad.net

UGANDA
Gustro Ltd.
PO Box 9997, Madhvani Building
Plot 16/4 Jinja Rd.
Kampala
Tel: (256 41) 251 467
Fax: (256 41) 251 468
E-mail: gus@swiftuganda.com

UNITED KINGDOM
Microinfo Ltd.
P.O. Box 3, Omega Park, Alton,
Hampshire GU34 2PG
England
Tel: (44 1420) 86848
Fax: (44 1420) 89889
E-mail: wbank@microinfo.co.uk
URL: http://www.microinfo.co.uk

The Stationery Office
51 Nine Elms Lane
London SW8 5DR
Tel: (44 171) 873-8400
Fax: (44 171) 873-8242
URL: http://www.theso.co.uk/

VENEZUELA
Tecni-Ciencia Libros, S.A.
Centro Cuidad Comercial Tamanco
Nivel C2, Caracas
Tel: (58 2) 959 5547; 5035; 0016
Fax: (58 2) 959 5636

ZAMBIA
University Bookshop, University of Zambia
Great East Road Campus
P.O. Box 32379
Lusaka
Tel: (260 1) 252 576
Fax: (260 1) 253 952

ZIMBABWE
Academic and Baobab Books (Pvt.) Ltd.
4 Conald Road, Graniteside
P.O. Box 567
Harare
Tel: 263 4 755035
Fax: 263 4 781913